# Introduction to
# Adaptive Arrays

# Introduction to
# Adaptive Arrays

**Robert A. Monzingo**
**Thomas W. Miller**

*Hughes Aircraft Company*
*Fullerton, California*

**A WILEY-INTERSCIENCE PUBLICATION**

**John Wiley & Sons**
**New York · Chichester · Brisbane · Toronto**

**Library of Congress Cataloging in Publication Data:**

Monzingo, Robert A. 1938–
Introduction to adaptive arrays.

"A Wiley-Interscience publication."
Includes index.
1. Antenna arrays. I. Miller, Thomas W. 1946–, joint author. II. Title.
TK7871.6.M66 621.38′028′3 80-10220
ISBN 0-471-05744-4

Printed in the United States of America

10 9 8 7 6 5 4 3 2 1

# Preface

This book is intended to serve as an introduction to the subject of adaptive array sensor systems whose principal purpose is to enhance the detection and reception of certain desired signals. Array sensor systems have well-known advantages for providing flexible, rapidly configurable, beamforming and null-steering patterns. The advantages of array sensor systems are becoming more important, and this technology has found applications in the fields of communications, radar, sonar, radio astronomy, seismology, and ultrasonics. The growing importance of adaptive array systems is directly related to the widespread availability of compact, inexpensive digital computers that make it possible to exploit certain well-known theoretical results from signal processing and control theory to provide the critical self-adjusting capability that forms the heart of the adaptive structure.

The field of adaptive array sensor systems is now a maturing technology, and with applications of these systems growing more and more numerous there is a wealth of widely scattered literature available on various aspects of such systems. There are even a few textbooks that briefly treat certain aspects of adaptive array systems, but until now there have been no books devoted entirely to presenting an integrated treatment of such systems that provide the reader with the perspective to organize the available literature into easily understood parts. The decision to write this book was made as a result of this lack. The primary emphasis of the book is to cover those principles and techniques that are of fundamental importance in modern adaptive array systems. Most of the contents are derived from readily available sources in the literature, although a certain amount of original material has been included.

This book is intended for use both as a textbook at the graduate level and as a reference work for engineers, scientists, and systems analysts. The material presented will be most readily understood by readers having an adequate background in array theory, signal processing (communication theory and estimation theory), optimization techniques, control theory, and probability and statistics. It is not necessary, however, that a reader have such a complete background since the book presents a step-by-step discussion of the basic theory and important techniques required in the above topics, and appropriate references are also given for readers interested in pursuing these topics further. Fundamental concepts are introduced and illustrated with examples before more current developments in adaptive array techniques are introduced. Problems at the end of each chapter have been chosen to illustrate and extend the material presented in the text. These extensions introduce the reader to actual adaptive array engineering problems and provide motivation for further reading of the background reference material. In this manner both students and practicing engineers may easily gain familiarity with the modern contributions that adaptive arrays have to offer practical signal reception systems.

The book is organized into three parts. Chapters 1 to 3 introduce the advantages that obtain with the use of adaptive array sensor systems, define the principal system components, and develop the optimum steady-state performance limits that any array system can theoretically achieve. A considerable variety of adaptation algorithms have been reported in the literature, but with few exceptions only meager efforts have been made to directly compare the various approaches. Part Two (Chapters 4 through 10) provides the designer with a survey of adaptive algorithms and a performance summary for each algorithm type. With this information available, the designer may then quickly identify those approaches most likely to lead to a successful design for the signal environment and system constraints that are of concern. Part Three (Chapters 11 and 12) considers the problem of compensation for adaptive array system errors that inevitably occur in any practical system and introduces current trends in adaptive array research. It is hoped that this book succeeds in presenting the adaptive array material using mathematical tools that make the subject interesting, accessible, and appealing to a wide audience.

ROBERT A. MONZINGO

THOMAS W. MILLER

*Fullerton, California*
*July 1980*

# Acknowledgments

The authors are deeply indebted to a number of individuals who assisted in the preparation of the manuscript. During the period when the book was being written inspiration and encouragement were received from Prof. Martin S. Roden, the Associate Dean of Engineering at the California State University, Los Angeles. Appreciation is owed to William Pedler and James Sawyers who helped review portions of the manuscript, and also to Nancy Ledford who generated and prepared some of the figures used in the text. Robert Nelson provided the opportunity to complete the computer simulation work required to evaluate the adaptive algorithms. Thanks must also go to Lynda Schenet who efficiently transformed handwritten notes into a finished manuscript. Finally, special recognition is hereby expressed to our families for their understanding in giving up so many weekends and evenings of our time that the effort in completing this work involved.

# Contents

**Introduction to**

# Adaptive Arrays

# Part I. Adaptive Array Fundamental Principles: System Uses, System Elements, Basic Concepts, and Optimum Array Processing

# Chapter 1. Introduction

Signal reception using an array of sensor elements has long been an attractive solution to severe problems of signal detection and estimation because an array offers a means of overcoming the directivity and beamwidth limitations of a single sensor element. The advent of highly compact, inexpensive digital computers has now made it possible to exploit well-known results from statistical detection and estimation theory and from control theory to develop array systems that automatically respond to a changing signal environment. This self-adjusting or adaptive capability renders the operation of such systems more flexible and reliable and (more importantly) offers improved reception performance that would be difficult to achieve in any other way.

## 1.1 MOTIVATION FOR USING ADAPTIVE ARRAYS

Conventional signal reception systems are susceptible to degradation in signal-to-noise ratio (SNR) performance because of the inevitable presence in the signal environment of undesired "noise" signals that may enter the system either by the beam pattern sidelobes or by the mainlobe. These noise signals may consist of deliberate electronic countermeasures (ECM), nonhostile RF interference (RFI), clutter scatterer returns, and natural noise sources. Such SNR degradation may be further aggravated by antenna motion, poor siting conditions, multipath ray effects, and a constantly changing interference environment. As radar and communication traffic increases, the suppression of interference becomes more important in all applications.

Adaptive arrays are currently the subject of extensive investigation as a means for reducing the vulnerability of the reception of desired signals to the presence of these interference signals in radar, sonar, seismic, and communications systems. The principal reason behind this widespread interest in adaptive array systems derives from their ability to sense automatically the presence of interference noise sources and to suppress these noise sources while simultaneously enhancing desired signal reception without prior knowledge of the signal/interference environment. Adaptive arrays can be designed to complement other interference suppression techniques so the actual suppression achieved is greater than that which can be obtained solely through more conventional means (e.g., by use of spread-spectrum techniques or the use of a highly directive sensor device).

An adaptive array is a system consisting of an array of sensor elements and a real-time adaptive signal receiver-processor that automatically adjusts the array beam sensitivity pattern so that a measure of the quality of the array performance is improved. An adaptive array offers enhanced reliability compared to that of a conventional array. When a single sensor element in a conventional array fails, the sidelobe structure of the array sensitivity pattern may be significantly degraded because of sidelobe increase. With an adaptive array, however, the response of the remaining operational sensors in the array can be automatically adjusted until the array sidelobes are reduced to an acceptable level. Adaptive arrays therefore fail gracefully compared to conventional arrays, and increased reliability results. Frequently an array beam pattern is determined more by near-field scattering effects than by its own inherent pattern in free space. For example, if a very low sidelobe antenna is placed on an aircraft, the tail and wings of the aircraft will completely change the antenna directional sensitivity pattern. An adaptive array will often yield successful operation even when antenna patterns are severely distorted by near-field effects.

The operation of an adaptive array can be most easily visualized by considering the response in terms of the array beam sensitivity pattern. Interference signal suppression is obtained by appropriately steering beam pattern nulls and reducing sidelobe levels in the directions of interference sources, while desired signal reception is maintained by preserving desirable mainlobe (or signal beam) features. An adaptive array system therefore relies heavily on spatial characteristics to improve the output SNR. Since it is possible to form very deep sensitivity pattern nulls over a narrow bandwidth region, very strong interference suppression can be realized. This exceptional interference suppression capability is a principal advantage of adaptive arrays compared to waveform processing techniques, which generally require a large spectrum-spreading factor to obtain comparable levels of interference suppression. Sensor arrays possessing this key automatic response capability are sometimes referred to as "smart" arrays, since they respond to far more of the signal information available at the sensor outputs than do more conventional array systems.

The capabilities provided by the adaptive array techniques to be discussed in this book offer practical solutions to the realistic interference problems mentioned above by virtue of their ability to sort out and distinguish the various signals in the spatial domain, in the frequency domain, and in polarization. At the present time adaptive nulling is considered to be the principal benefit of the adaptive techniques employed by adaptive array systems, and automatic cancellation of sidelobe jamming provides a valuable electronic counter-countermeasure (ECCM) capability for radar systems.

Adaptive arrays can be designed to incorporate more traditional capabilities such as self-focusing on receive and retrodirective transmit. In addition to automatic interference nulling and beam-steering, adaptive imaging arrays may also be designed to obtain microwave images having high angular resolution. It is useful to call self-phasing or retrodirective arrays *adaptive transmitting arrays* in order to distinguish the principal function of such systems from an *adaptive receiving array*, the latter being the focus of this book.

## 1.2 HISTORICAL PERSPECTIVE

The term "adaptive antenna" was first used by Van Atta [1] and others [2] to describe a self-phasing antenna system that automatically reradiates a signal in the direction from which it was received thereby acting as a "retrodirective" system, without any prior knowledge of the direction in which it is to transmit. Retrodirective arrays have found application for point-to-point satellite communications by use of a strong pilot signal; the retrodirective system can overcome the usual beamwidth (and consequent directivity) limitations since the array automatically returns the pilot signal to its station of origin thereby permitting many elements and high directivity to be used.

The development of the phase-lock loop was another major step that made possible the self-steering (or self-phasing) type of adaptive array [3]. A self-phased array may be defined as one in which each of the array elements is independently phased, based on information obtained from the received signals. For example, several large-aperture receiving antennas with slaved-steering can be self-phased on received signals from satellites or space vehicles so the effective receiving aperture is the sum of the individual apertures of all participating antennas.

In the early 1960s the key capability of adaptive interference nulling was recognized and developed by Howells [4], [5]. Subsequently, Applebaum established the control law associated with the Howells adaptive nulling scheme by analyzing an algorithm that maximizes a generalized SNR [6]. Concurrently, the capability of self-training or self-optimizing control was applied to adaptive arrays by Widrow and others [7]–[9]. The self-optimizing control work established the least mean square error (LMS) algorithm that was based on the method of steepest descent. The Applebaum and the Widrow algorithms are very similar, and both converge toward the optimum Wiener solution.

The use of sensor arrays for sonar and radar signal reception had long been common practice by the time the early adaptive algorithm work of Applebaum and Widrow was completed [10], [11]. Early work concerning the processing of array outputs usually concerned achieving a "desirable" directional beam pattern, and attention later shifted to the problem of obtaining an improvement in the SNR [12]–[14]. Seismic array development commenced about the same period, so papers describing applications of seismic arrays to detect remote seismic events appeared during the late 1960s [15]–[17].

The major area of current interest in adaptive arrays is their application to problems arising in radar and communications systems, where the designer almost invariably faces the problem of interference suppression [18]. A second example of the use of adaptive arrays is that of direction finding in severe interference environments [19], [20]. Another area in which adaptive arrays are proving useful is for systems that require adaptive beamforming and scanning in situations where the array sensor elements must be organized without accurate knowledge of element location [21]. Furthermore, large, unstructured antenna array systems may employ adaptive array techniques for high angular resolution imaging [22], [23].

It behooves the designer to be familiar with the strengths and weaknesses of the various types of adjustment algorithms and to tailor the algorithm selection according to any special circumstances the adaptive array system will be expected to respond to successfully. One of the principal thrusts of research over the last decade has been directed toward enabling adaptive arrays to achieve satisfactory SNR performance in environments that may be only partially characterized. The other principal concern of research has been toward achieving more rapid transient response thereby making it possible to adapt swiftly to nonstationary signal environments—especially those in which an intelligent jammer is seeking, by adroit manipulation of several jamming sources acting in coordination, to defeat the adaptive array adjustment algorithm. Work has also begun on adaptive filtering techniques that, when complemented with adaptive sidelobe rejection, have the potential to yield highly refined adaptive systems that can overcome most forms of RFI, clutter, and jamming.

## 1.3 PRINCIPAL SYSTEM ELEMENTS

The adaptive array functional diagram of Figure 1.1 shows the principal system elements that an adaptive array must possess if it is to achieve successfully the twin objectives of enhancing desired signal reception and rejecting undesired interference signals. The principal adaptive array system elements consist of the sensor array, the pattern-forming network, and the adaptive pattern control unit or adaptive processor that adjusts the variable weights in the pattern-forming network. The adaptive pattern control unit may furthermore be conveniently subdivided into a signal processor unit and an adaptive control algorithm. The manner in which these elements are actually implemented depends on the propagation medium in which the array is to operate, the frequency spectrum of interest, and the user's knowledge of the operational signal environment.

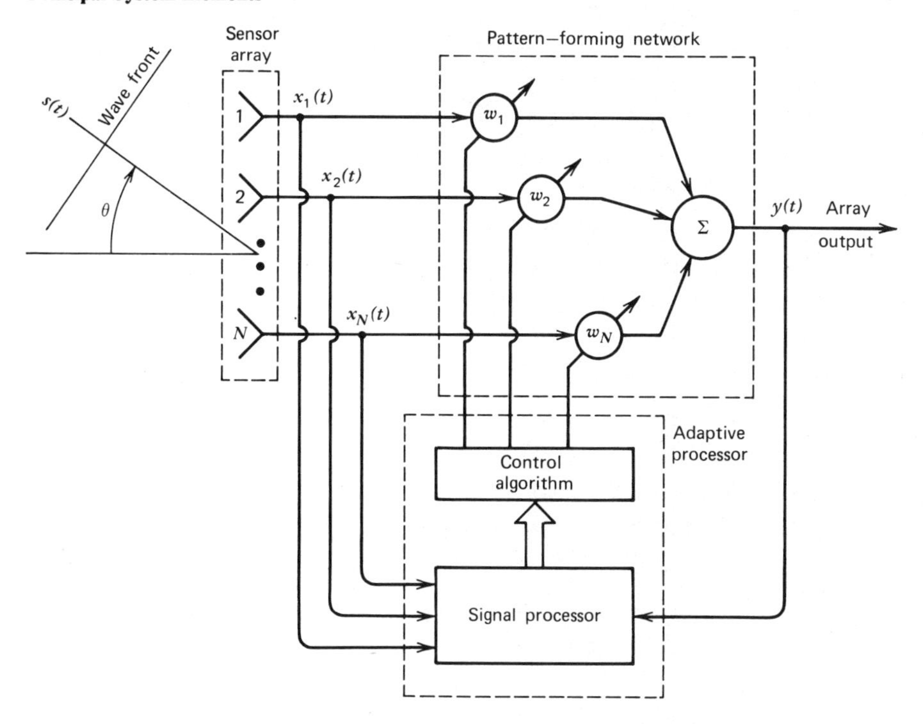


**Figure 1.1** Functional diagram of an $N$-element adaptive array.

The array itself consists of $N$ sensors designed to receive (and transmit) signals in the propagation medium of interest. The sensors are arranged to give adequate coverage (pattern gain) over a certain desired spatial region. The selection of the sensor elements and their physical arrangement place fundamental limitations on the ultimate capability of the adaptive array system. The output of each of the $N$ elements is directed to the pattern-forming network, where the output of each sensor element is first multiplied by a complex weight (having both amplitude and phase) and then summed with all other weighted sensor element outputs to form the overall adaptive array output signal. The weight values within the pattern-forming network (in conjunction with the sensor elements and their physical arrangement) then determine the overall array beam sensitivity pattern; it is the ability to shape this overall array pattern that in turn determines how well the specified system requirements can be met for a given signal environment.

The exact structure of the adaptive processor unit is critically dependent on the degree of detailed information about the operational signal environment that is available to the array. As the amount of *a priori* knowledge (e.g., desired signal location, jammer power levels, etc) concerning the signal environment decreases, the adaptive control algorithm selected for the adaptive processor unit becomes critical to a successful design. In the event that the precise nature of all signals present is known, and the signal directions with respect to the sensor array as well as the characteristics of the sensor elements themselves are known, then the adaptive processor unit may be dispensed with altogether and a straightforward calculation will determine how the pattern-forming network should be adjusted in order to enhance or suppress the signal environment so defined. Since such detailed a priori information is not available in practice, the adaptive processor must be capable of responding automatically to whatever signal environment (within broad limits) is presented. If any signal environment limits are known or can reasonably be construed, such bounds may prove helpful in determining the form that the adaptive processor unit should take.

## 1.4 ADAPTIVE ARRAY PROBLEM STATEMENT

The fundamental problem facing the adaptive array designer is to improve the reception of a certain desired signal in the presence of undesired interfering signals. The terms "desired signal" and "interfering signals" imply that the characteristics of these two signal classes are different in some respect, and that this difference may be exploited in order to realize the desired signal reception improvement. For example, if the direction of arrival of the desired signal is known (or can be deduced), then any signals arriving from different directions can be suppressed by the introduction of array pattern nulls in those directions. Likewise, if the desired signal occupies a certain known frequency spectrum, then interference signals outside the desired frequency spectrum can be rejected by bandpass filtering. Because the desired signal must be known well enough so that it can be distinguished from interference signals, it is reasonable to assume that the nature of the desired signal is known even though certain signal parameters (such as direction of arrival, amplitude, and phase) may have to be estimated. If the designer were solely concerned with suppressing interfering signals, then desired signal reception might suffer. Likewise, if desired signal enhancement were the sole focus of attention, then interference signal reception might also be enhanced. Therefore the twin (and sometimes conflicting) objectives of desired signal enhancement and interference signal suppression must be sought so that the overall desired signal reception performance is improved. In many cases the overall reception performance is best measured by the output SNR. For passive sensor systems, however, the basic problem is that of determining whether or not a desired signal is present in a background of ambient noise and interfering signals. Determining signal presence or absence then requires a decision that is not provided simply by maximizing the output SNR, and statistical decision theory provides solutions to problems of this kind that minimize the risk associated with incorrect decisions. The optimum processors prescribed by statistical decision theory are closely related to those obtained by maximizing the output SNR, so there is an underlying unity to problems that initially appear to be quite different.

To realize the maximum practical improvement in the overall desired signal reception performance, the adaptive array designer must select a sensor array configuration, pattern-forming network implementation, signal processor, and adaptive weight control algorithm that enable the system to meet several different requirements on its resulting performance in as simple and inexpensive a manner as possible. The system performance requirements may be conveniently divided into two types: transient response and steady-state response. Transient response time refers to the length of time required for the adaptive array to adjust successfully from the time it is turned on until steady-state conditions have been reached or to adjust successfully to a change in the signal environment. Steady-state response refers to the state of the adaptive array after it has been allowed to reach the steady-state condition. Various measures of this state are commonly employed, including the shape of the overall directivity pattern and the output signal to interference plus noise ratio. Several popular performance measures are considered in detail in Chapter 3. The speed of response of an adaptive array system depends in a complex way on the type of control and the control algorithm selected and the nature of the operational signal environment. The steady-state array response, however, can easily be formulated in terms of the complex weight settings, the signal environment, and the sensor array structure.

A fundamental trade-off exists between the rapidity of change in a nonstationary noise field and the steady-state performance of an adaptive system: generally speaking, the slower the variations in the noise environment the better the steady-state performance of the adaptive system. Any adaptive system design should therefore seek to optimize the trade-off between the speed of adaptation and the accuracy of adaptation.

System requirements place limits on the allowable range of transient response speed. In an aircraft communication system, for example, the fastest response speed is limited by the signal modulation rate (since if the response is too fast, the adaptive weights interact with the desired signal modulation). The slowest speed is determined by the need for the array to respond fast enough to compensate for aircraft motion.

The weights in an adaptive array may be controlled by any one of a variety of different algorithms, some of which can be implemented in either analog or digital form, whereas others can be implemented only in digital form and therefore require at least a small digital computer. The "best" algorithm for a given application is chosen on the basis of a host of factors including the signal structures, the a priori information available to the adaptive processor, the performance characteristics to be optimized, the required speed of response of the processor, the allowable circuit complexity, any device or other technological limitations, and cost effectiveness. For many practical communication systems, analog implementation methods appear to be preferable to digital methods because they are much faster and less complex. Furthermore, the analog-to-digital conversion of wideband signals, which is required for some digital algorithms, does not appear to be feasible at sufficiently high rates for many communication systems. However, if required data rates are sufficiently low—on the order of one megabit per second or less as with sonar arrays, for example—digital implementation may be preferred.

Referring to Figure 1.1, the received signal impinges on the sensor array and arrives at each sensor at different times as determined by the direction of arrival of the signal and the spacing of the sensor elements. The actual received signal for many applications consists of a modulated carrier whose information carrying component consists only of the complex envelope. If $s(t)$ denotes the modulated carrier signal, then $\tilde{s}(t)$ is commonly used to denote the complex envelope of $s(t)$ (as explained in Appendix B) and is the only quantity that conveys information. Rather than adopt complex envelope notation, however, it is simpler to assume that all signals are represented by their complex envelopes so the common carrier reference never appears explicitly. It is therefore seen that each of the $N$ channel signals $x_k(t)$ represents the complex envelope of the output of the element of a sensor array that is comprised of a signal component and a noise component, that is,

$$x_k(t) = s_k(t) + n_k(t), \qquad k = 1, 2, \ldots, N \tag{1.1}$$

In a linear sensor array having equally spaced elements and assuming ideal propagation conditions, the $s_k(t)$ are determined by the direction of the desired signal. For example, if the desired signal direction is located at an angle $\theta$ from mechanical boresight, then (for a narrowband signal)

$$s_k(t) = s(t)\exp\left\{ j\frac{2\pi kd}{\lambda}\sin\theta \right\} \tag{1.2}$$

where $d$ is the element spacing, $\lambda$ is the wavelength of the incident planar wavefront, and it is presumed that each of the sensor elements is identical.

For the pattern-forming network of Figure 1.1, the adaptive array output signal can be written as

$$y(t) = \sum_{k=1}^{N} w_k x_k(t) \tag{1.3}$$

Equation (1.3) can be conveniently expressed in matrix notation as

$$y(t) = \mathbf{w}^T\mathbf{x} = \mathbf{x}^T\mathbf{w} \tag{1.4}$$

where the superscript $T$ denotes transpose, and the vectors $\mathbf{w}$ and $\mathbf{x}$ are given by

$$\mathbf{w}^T = [w_1 w_2 \ldots w_N] \tag{1.5}$$

$$\mathbf{x}^T = [x_1 x_2 \ldots x_N] \tag{1.6}$$

Throughout this book the boldface lower case symbol (e.g., **a**) denotes a vector while a boldfaced upper case symbol (e.g., **A**) denotes a matrix.

The problem facing the adaptive processor is to select the various complex weights $w_k$ in the pattern-forming network so that a certain performance criterion is optimized. The performance criterion that governs the operation of the adaptive processor must be chosen to reflect the steady-state performance characteristics that are of concern. The most popular performance measures that have been employed include the mean square error [9], [24]–[27]; SNR ratio [6], [14], [28]–[30]; output noise power [31]; maximum array gain [32], [33]; minimum signal distortion [34], [35]; and variations of these criteria that introduce various constraints into the performance index [16], [36]–[39]. In Chapter 3 selected performance measures are formulated in terms of the signal characterizations of (1.1)–(1.4). Solutions are found that determine the optimum choice for the complex weight vector and the corresponding optimum value of the performance measure. It will be found that the operational signal environment plays a crucial role in determining how well it is possible for the adaptive array to operate, and the description of that environment implicitly includes the effects of the sensor array configuration. Since the array configuration can have pronounced effects on the resulting system performance, it is useful to directly consider sensor spacing effects before proceeding with an analysis utilizing an implicit description of such effects. The consideration of array configuration is undertaken in Chapter 2.

## 1.5 EXISTING TECHNOLOGY

Although the overwhelming majority of adaptive array systems in use today is designed for application in communications, radar, and sonar systems, the techniques employed involve features that are also common to the fields of radio astronomy, seismology, and ultrasonics [40]. To obtain an overview of existing technology, it is reasonable to consider briefly the elements of radar and sonar systems. With both active radar and sonar systems a signal is transmitted, and a target is then detected by the echo that it reflects. It is therefore hardly surprising that the fundamental principles applied in designing radars also apply equally to the design of sonar systems [41].

In any echo-ranging system, the maximum detection range $R_{max}$ determines the minimum period between consecutive pulses $T_{min}$ [and hence the pulse repetition frequency (PRF)], according to

$$T_{min} = \frac{2R_{max}}{\mathfrak{v}} \tag{1.7}$$

where $\mathfrak{v}$ is the velocity of propagation of the transmitted signal.

For underwater applications the velocity of sound in water varies widely with temperature, although a nominal value of 1500 m/sec can be used for rough calculations [42]. The velocity of electromagnetic wave propagation in the atmosphere can be taken approximately to be the speed of light or $3\times10^8$ m/sec.

If the range discrimination capability between targets is to be $r_d$, then the maximum pulse length $t_{\max}$ (in the absence of pulse compression) is given by

$$t_{\max}=\frac{2r_d}{\mathfrak{v}} \tag{1.8}$$

It will be noted that $r_d$ also corresponds to the "blind range"— that is, the range within which target detection is not possible. Since the signal bandwidth $\cong 1/$pulse length, it is seen that range discrimination capability also determines the bandwidth that transducers and their associated electrical channels must have if successful operation is to result.

The generated pulses form a pulse train in which each pulse modulates a carrier frequency as shown in Figure 1.2. The carrier frequency $f_0$ in turn determines the wavelength of the propagated wavefront since

$$\lambda_0=\frac{\mathfrak{v}}{f_0} \tag{1.9}$$

where $\lambda_0$ is the wavelength. For sonar systems, frequencies in the range 100–100,000 Hz are commonly employed [42], whereas for radar systems the range can extend from a few megahertz up into the optical and ultraviolet regions, although most equipment is designed for microwave bands between 1 and 40 GHz. The wavelength of the propagated wave front is important because the array element spacing (in units of $\lambda$) is an important parameter in determining the array directivity pattern.

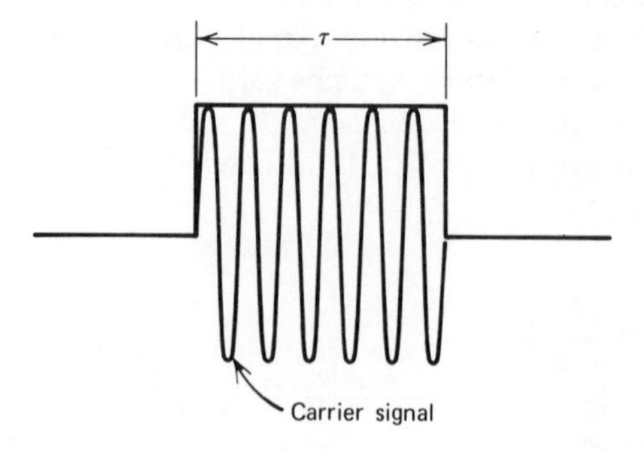


**Figure 1.2** Pulse modulated carrier signal.

### 1.5.1 Radar Technology

There has been a steady increase in the demand for increased radar system performance and additional capability for both military and civilian purposes, and the vast number of applications of modern radar technology precludes anything more than the briefest mention of the major areas in which radar systems are found. Military applications may very well involve a number of requirements that in the past would have involved a separate radar system for each different requirement. For example, a fire control system radar may be required to search large volumes of space, detect and track both high and low speed targets ranging from very low to extremely high altitudes, provide fire control for both missiles and guns against both airborne and ground (or sea) targets, and in addition provide navigational aid and perform reconnaissance. Current civil aeronautical needs include air traffic control, collision avoidance, instrument approach systems, weather sensing, and navigational aids. Additional applications in the fields of law enforcement, transportation, and earth resources are just beginning to grow to sizable proportions [43].

The demand for increased performance and flexible capability has dictated a high level of complexity in modern radar systems. This complexity in turn requires that radar systems be designed so that functional options can easily and effectively be selected. The advent of reliable inexpensive digital microelectronic circuitry has made possible the introduction of a digital subsystem to provide signal processing, decision, and control capabilities that were considered impractical and too costly only a few years ago. Despite the sophistication and complexity that so often characterizes modern radar systems, it is still convenient to think of a radar system as being composed of a group of blocks, with each block performing a specific function in the system operation. A simplified block diagram of a digitally controlled radar system consisting of five major blocks is shown in Figure 1.3. These major

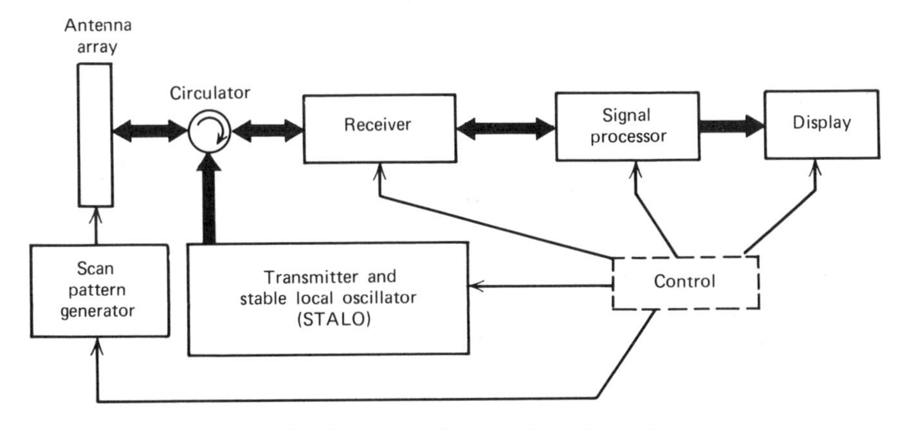


**Figure 1.3** Simplified block diagram of a typical radar system.

blocks and their corresponding functions are described as follows [43]:

| Block | Function |
|---|---|
| Transmitter | generates high power R.F. waveform |
| Antenna | determines direction and shape of transmit and receive beam |
| Receiver | provides frequency conversion and low noise amplification |
| Signal processing | provides target detections, target and clutter tracking, and target trajectory estimates. |
| Display | converts processed signals into meaningful tactical information |

The antenna, receiver, and signal processing blocks are of primary interest for our purposes, and these are now each briefly discussed, in turn.

***Radiating Elements and Antenna Arrays.*** The vast frequency range and power over which modern radar systems operate have led to an astonishing variety of radiator elements ranging from huge parabolic reflectors to relatively tiny horns, simple dipoles, bow ties, loops, multiturn loops [44], spirals [45], and log periodic radiators [46]. Large antenna apertures result in narrow beamwidths that are required for long-range detection and high resolution if targets close to one another are to be distinguished. Microwave frequencies are by far the most popular for radar applications since antenna apertures of relatively small physical size (but large in terms of wavelengths) can quite conveniently be obtained.

The antenna types selected for use in radar applications usually differ from those selected for communications systems. Shaped directive beam patterns that can be scanned are most popular for radar uses, whereas most communication applications require beams designed for omnidirectional coverage or for fixed point-to-point transmission. The earliest radars (developed during World War II) operated in the VHF and UHF frequency bands and used array antennas. Microwave frequency radar systems on the other hand make extensive use of the parabolic reflector, although microwave lenses have also found applications. For airborne-radar applications, conformal arrays and surface-wave antennas are useful since the antenna then does not protrude beyond the airframe skin.

An array antenna differs in fundamental concept from lens and reflector type antennas. In both lens and reflector type antennas, the shape of a transmitted wavefront on leaving the feed is spherical and is converted to a plane wave by the action of the reflector or lens. In an array antenna, however, proper phase relationships are applied to the signal before it is radiated so the radiated wavefront is at all times planar. The two most common forms of antenna arrays for radar applications are the linear array and the planar array.

A linear array simply consists of antenna elements arranged in a straight line. A planar array, on the other hand, is a two-dimensional configuration in which the antenna elements are arranged to lie in a plane. The linear array is most often used to generate a fan beam and is useful where broad coverage in one plane and narrow beamwidth in the orthogonal plane are desired. The planar array is most frequently used in radar applications since it is the most versatile of all antenna types. A fan shaped beam is easily produced by a rectangular shaped aperture. A pencil beam may easily be generated by a square or circular shaped aperture. With proper weighting, an array can be made to simultaneously generate multiple search and/or tracking beams with the same aperture. The flexibility offered by many individually controlled elements in an array antenna make it attractive (although complex and expensive) for radar applications. The emergence of relatively low-cost digital control and the possibility of employing partially adaptive control concepts [47]–[49] may soon drastically reduce the expense associated with array control and signal processing.

Array beam scanning requires that proper phase shifts at each of the array elements be provided. Phase shifting devices may be classified as (1) fixed phase shifters, (2) mechanically variable phase shifters, and (3) electronically variable phase shifters [50]. Electronically controlled phase shifters are the most important for modern systems, and phase shifting devices at microwave frequencies can be obtained with ferrite materials, PIN diodes, gaseous discharges, or traveling-wave tubes [51].

Antenna arrays can be made extremely compact and are therefore very attractive for shipboard applications [52]. An example of a rotating linear-array antenna consisting of 80 waveguide horns is described in Reference [53]. Beam steering through the use of phase shifters with a planar array was applied during World War II in a microwave radar called FH MUSA [54], [55]. An example of an electronically scanned array radar is ESAR, which is a 50 ft planar array containing over 8000 elements that employs a frequency-conversion phasing scheme to accomplish the scanning [50].

***Receivers.*** Noise and its elimination is the most important consideration in radar receiver design. Consequently receiver design is most often based on maximizing the SNR in the linear portion of the receiver. This philosophy leads to a receiver design based on regarding the receiver as a matched filter or as a cross correlator. Different types of receivers that have been employed in radar applications include the superheterodyne, superregenerative, crystal video, and tuned radio frequency (TRF) [56]. The most popular and widely applied receiver type is the superheterodyne, which is useful in applications where simplicity and compactness are especially important [50]. A received signal enters the system through the antenna, then passes through the circulator and is amplified by a low-noise RF amplifier. Following RF amplification, a mixer stage is entered to translate the RF to a lower intermediate frequency (IF) where the necessary gain is easier to obtain and filtering is easier to synthesize. The gain and filtering are then accomplished in an IF amplifier section.

***Signal Processing.*** Having maximized the SNR in the receiver section, the next step is to perform two basic operations by signal processing as follows: (1) detection of the presence of any targets, and (2) extraction of information from the received waveform to obtain target trajectory data such as position and velocity. The problem of the detection of the presence of a signal imbedded in a noise field is treated by means of statistical decision theory. Similarly, the problem of the extraction of information from radar signals can be regarded as a problem concerning the statistical estimation of parameters. While many of the methods of mathematical statistics have been known for some time, their direct application to problems of radar signal processing was considered impractical and too costly only a few years ago. The development of reliable, inexpensive digital microelectronic circuitry has dramatically changed the design of modern radar signal processing circuitry, with the result that rather sophisticated processors including highly advanced moving target indicator (MTI) receivers and pipeline fast Fourier transform (FFT) processors are now in use [57]. Basic operations that are now routinely exploited as a result of the advances in digital technology include in-phase/quadrature processing using bipolar video channels, the use of coherent processors for MTI, pulse doppler, and pulse compression waveforms, and the digital generation of programmable filters.

While current MTI successfully suppress clutter echoes from fixed ground targets, and adaptive MTI can suppress moving clouds of rain and chaff at ranges beyond ground clutter, nevertheless many desired targets are also suppressed, and some undesired moving objects (e.g. birds and automobiles) succeed in passing through the MTI filter [58]. Technology to correct the current deficiencies of MTI has been introduced, and a moving target detector (MTD) has demonstrated the practical feasibility of realizing enhanced detection performance of small targets in clutter [59].

### 1.5.2 Sonar Technology

Currently operational active sonar systems may be classified as (1) search light sonar, (2) scanning sonar, or (3) rotational directional transmission (RDT) sonar [60]. Searchlight sonar is characterized by having very narrow transmitting and receiving beam patterns and provides an azimuth search capability by mechanically rotating the directional hydrophone array. Since the array is mechanically scanned and aimed, the data rate is correspondingly low, and the system does not provide a multiple target detection and tracking capability. The requirement for mechanical directional training also limits the array size, so that operational frequencies are usually greater than 15 kHz, thereby increasing attenuation loss.

Scanning sonar systems overcome the data rate limitation of searchlight sonars by transmitting an omnidirectional, short-duration pulse and using electronic means to rapidly rotate a narrow receive beam continuously over a 360° azimuthal sector. The receiving beam output is presented to a panoramic display called a plan position indicator (PPI) that is used extensively in both radar and sonar systems. Scanning type sonar systems thereby provide a multiple target detection and tracking capability, and lower operating frequencies can be used thereby decreasing attenuation losses. The scan speed of the receive beam is a compromise between the desired target resolution and the maximum receiver bandwidth (or minimum input SNR) that is permissible.

An RDT sonar system is characterized by RDT and a scanned preformed beam (PFB) receiver. Consequently high transmitting directivity and a high data rate are achieved together along with a low operational frequency. An RDT system therefore combines the best features of searchlight and scanning sonars. A PFB receiver can have a smaller bandwidth than a scanning receiver thereby improving the SNR. Furthermore a PFB receiver can be corrected for doppler due to own ship's motion employing a method called "own doppler nullifying" (ODN), whereas a scanning receiver cannot be so corrected.

The principal elements of a sonar receiver are shown in the block diagram of Figure 1.4. These elements are the hydrophone array, the beamformer, the signal processor, and the information processor. Each of these principal elements (except for the information processor, which involves display formatting and other command and control functions) are briefly discussed in turn.

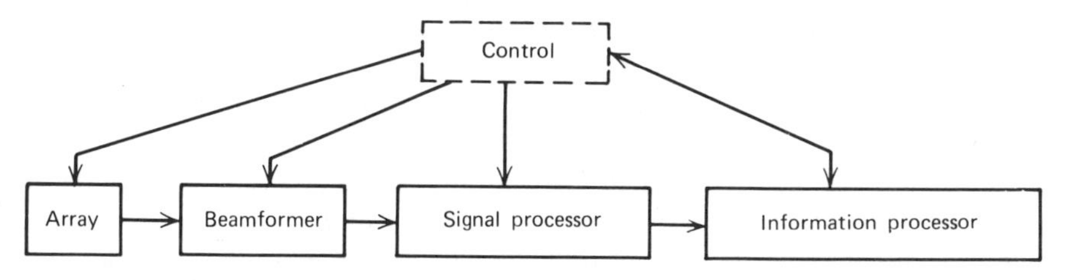


**Figure 1.4** Sonar receiver block diagram.

***Sonar Transducers and Hydrophones.*** If the underwater sensor device only listens for sounds in the ocean, it is called a *hydrophone*; whereas if it both generates and receives sound, it is referred to as a *transducer*. For underwater applications, a very wide frequency range is involved—from about 10 Hz to more than 1 MHz [61]. A transmitting power ranging from a few acoustic watts up to several thousand acoustic watts at ocean depths up to 20,000 ft can be achieved [62].

The basic physical mechanisms most widely used in transducer technology include the following:

1. Moving coil. This is long familiar from use as a loud speaker in music reproduction systems and used extensively in water for applications requiring very low frequencies.
2. Magnetorestrictive. Magnetic materials vibrate in response to a changing magnetic field. Magnetorestrictive materials are rugged and easily handled, and magnetorestrictive transducers were highly developed and widely used during World War II.
3. Piezoelectric. The crystalline structure of certain materials results in mechanical vibration when subjected to an alternating current or an oscillating magnetic field. Certain ceramic materials also exhibit a similar effect and have outstanding electromechanical properties. Consequently over the last decade the great majority of underwater sound transducers have been piezoceramic devices that can operate over a wide frequency band and have both high sensitivity and high efficiency.

Variable reluctance and hydroacoustic transducers have also been used for certain experimental and sonar development work, but these devices have not challenged the dominance of piezoceramic transducers for underwater sound applications [61].

***Sonar Arrays.*** Greatly improved arrays of sonar transducers have come into use having lower sidelobes and provisions for beam steering over wide angular sectors. A wide variety of underwater acoustic array configurations now exist which include: linear, planar, cylindrical, spherical, conformal, volumetric, reflector, and acoustic lens [63]. These different array types lend themselves to towing, conformal mounting on hulls (where the array surface conforms to the shape of the underwater hull, so that no appendage is required), beam steering, and to sidelooking sonar and synthetic-aperture applications.

A simple 2 ft×4 ft planar array having more than 500 sensor elements for deep submergence applications is shown in the diagram of Figure 1.5. In the quest for larger power and lower frequency (with attendant lower attenuation losses) some array assemblies have been built having very large physical dimensions (amounting to several wavelengths). In one case a 35 ft×50 ft low-frequency planar array was built weighing 150 tons and having a transmit power capability of close to $10^6$ watts [63]. No adequate theory exists for radiators having dimensions comparable with one wavelength, and it has been found that large arrays with many closely spaced elements develop "hot spots" that are due to mutual impedance terms that were ignored by most designers. Consequently the concept of "velocity control" was developed [64],

**Figure 1.5** Rectangular planar array for searchlight type sonar system having more than 500 elements.

[65] to protect individual transducer elements against extreme local impedance variations.

Current surface-ship sonar technology employs a bubble-shaped bow dome in which is placed a cylindrical array like that shown in Figure 1.6. This array uses longitudinal vibrator-type elements composed of a radiating front end of light weight and a heavy back mass, with a spring having active ceramic rings or disks in the middle [63]. The axial symmetry of a cylindrical array renders beam steering fairly simple for, independent of the azimuth direction in which the beam is formed, the symmetry allows identical electronic equipment for the phasing and time delays required to form the beam. Planar arrays do not have this advantage, since each new direction in space (whether in azimuth or

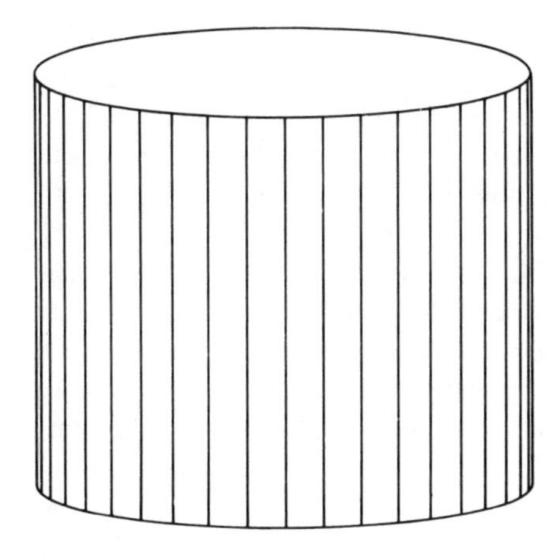

**Figure 1.6** Cylindrical sonar array used in bow-mounted dome.

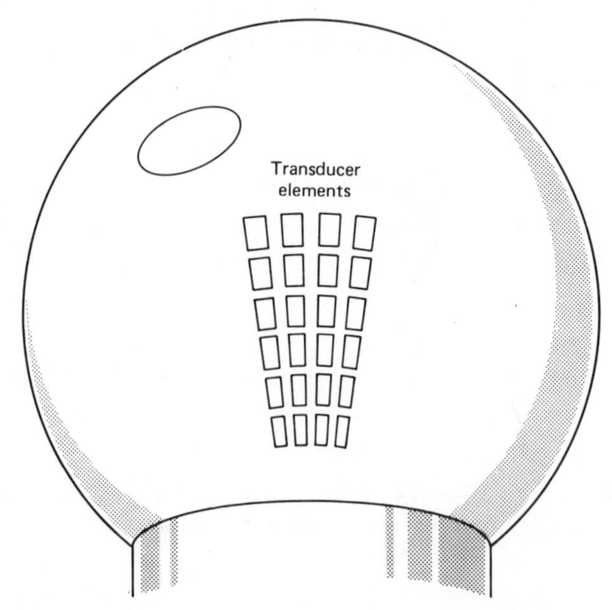


**Figure 1.7** Spherical array having 15-ft diameter and more than 1000 transducer elements.

in elevation) requires a new combination of electronic equipment to achieve the desired pointing.

Symmetry considerations lead to the conclusion that the ideal shape for the broadest array coverage in all directions is spherical. Spherical arrays like that shown in the diagram of Figure 1.7 have been built with a diameter of 15 ft and more than 1000 transducer elements. This spherical arrangement can be integrated into the bow of a submarine by means of an acoustically transparent dome that provides minimum beam distortion.

***Beamformer.*** Rapid advances in solid-state physics have revolutionized electronics engineering. These advances have made it possible for digital computers to become sufficiently compact to find their way through the hatches of submarines and into the control rooms of destroyers. The impact of this development has been to make available the considerable resources of digital technology for the processing and interpretation of acoustic signals, with tremendous implications for both beamforming and signal processing.

Beamforming ordinarily involves forming multiple beams from multielement arrays through the use of appropriate delay and weighting matrices. Such beams may be directionally fixed or steerable. The classical sonar systems of the 1950s and 1960s consisted largely of independent sonar sets for each transducer array. More recently, the sophisticated use of multiple sensors and advances in computer technology have led to integrated sonar systems that allow the interaction of data from different sensor arrays [66], [67]. Such integrated sonar systems required the introduction of a large general purpose computer that made it possible to realize programmable delay and weighting matrices, thereby generalizing the structure of digital time domain beamformers. Consequently several units of a single (programmable) beamformer design may be used for all the arrays in an integrated system. Furthermore, programmable beamformer matrices make it possible to adapt the receiver beam pattern to the changing structure of the masking noise background.

***Signal Processor.*** Signal processing involves filtering, spectral analysis, correlation, and the related operations of dynamic range compression and normalization (DRCN) that "match" the received array signals to the display/decision functions contained in the information processing block of Figure 1.4 [60], [68]. The DRCN removes some of the spatial and temporal characteristics of the acoustic channel that are impressed on the propagating signal. Whether or not a sufficient degree of DRCN is achieved usually determines whether a system design will succeed or fail since it specifies the integration of a theoretically designed optimum signal processor to its system interfaces.

Current technology trends in signal processing involve large-scale programmable signal processors that emphasize reliability [67]. One of the principal benefits resulting from the use of programmable signal processing elements is to allow a number of identical hardware modules to efficiently share a processing load.

## 1.6 ORGANIZATION OF THE BOOK

*Part One:* Chapter 2 introduces the adaptive array concept by first considering the nature of the signal environment in which an adaptive array is expected to improve the overall reception of a desired signal. The question of array structure is considered, and the impact of the structure selection on the ultimate system performance that can be realized is discussed. The potential of an array to enhance the output SNR performance by adjusting the pattern-forming network is demonstrated.

To provide the array with an adaptive automatic response capability requires that a system performance measure be defined. This performance measure should be chosen to reflect a quality it is desired the array should maximize (or minimize). Several widely used performance measures for both narrowband and broadband applications are presented in Chapter 3 that are formulated in terms of the elements found within the adaptive array functional model.

Any adaptive array system must contend with two types of performance limits as follows:

1 Performance limits imposed by the array physical structure.
2 Performance limits resulting from the nature of the signal environment.

The consideration of performance limits leads to a discussion of the array performance that results after the automatic adaptation process has been permitted to operate long enough to reach a steady-state solution. A steady-state analytic solution to the adaptive array control problem can be found for each performance measure, which enables the designer to determine ultimate system performance limits. The mathematical foundation required to develop the steady-state analytic solution for various performance measures is laid, and the relationships among the solutions obtained to the optimal "Wiener solution" are presented.

*Part Two:* The heart of the adaptive capability within an adaptive array system is the adaptive control algorithm used to adjust the array beam pattern in response to the signal information found at the sensor element outputs. Part Two, consisting of Chapters 4 through 10, introduces different classes of adaptation algorithms. In some cases adaptation algorithms can be selected according to the kind of signal information available to the receiver as in the following when:

1 The desired signal to be received by the array is known.
2 The desired signal to be received is unknown, but its direction of arrival with respect to the array is known.
3 The desired signal to be received is unknown and its direction of arrival with respect to the array is unknown, but the desired signal power level is known.
4 No signal information is available at the outset, but as array operation progresses such information must be "learned" by the adaptive processor.

The selection of an adaptation or adjustment algorithm has important consequences for the system transient performance.

The convergence properties of the various adaptation algorithms are analyzed and performance comparisons are made. Furthermore, the shortcomings of the algorithms under certain conditions are discussed. These results are summarized for convenience and provide the designer with a means for assessing which candidate algorithms are most appropriate for the signal conditions and system requirements with which he is concerned.

*Part Three:* The adaptive array operating conditions considered so far were nonideal only in that interference signals were present with which the array had to contend. In actual practice, however, the effects of several other nonideal operating conditions often result in unacceptable degradation of array performance unless compensation of such effects is undertaken. Such nonideal operating conditions include: processing of broadband signals, multipath effects, channel mismatching, and array propagation delay effects. Compensation for these factors by means of tapped delay-line processing is considered, and the question of how to design a tapped delay line in order to achieve a desired degree of compensation is addressed. Finally, current trends in adaptive array research that provide an indication of the direction that future developments are likely to take are discussed.

## 1.7 SUMMARY AND CONCLUSIONS

The motivation for and actual use of adaptive array systems are presented. The principal elements of an adaptive array system are defined, and the fundamental problems facing an adaptive array designer are given. Adaptive array design is a compromise among such factors as [69]:

1. Hardware complexity and cost.
2. Data rate.
3. Maximum range of detection (for radar and sonar).
4. Resolution in angle (and range and doppler for radar and sonar).
5. Precision in the measurement of range, bearing, and doppler (for radar and sonar).
6. Ability of the adaptive array to meet both transient and steady-state system performance requirements.

Since adaptive nulling of unwanted interference signals is presently considered to be the principal benefit of adaptive array use, it is worthwhile to mention some actual cancellation performances that have been realized by means of current technology.

A suboptimal acoustical array processor known as the DICANNE processor operated in sea tests against ship-generated interferences and consistently formed cancellation nulls 10–15 dB deep [60]. Use of an optimal wideband processor based on the minimum signal distortion performance measure in a computer simulated sonar experiment resulted in effectively suppressing a strong coherent interfering signal by forming cancellation nulls 50 dB deep [60]. Such deep cancellation nulls were found, however, to be quite sensitive to (1) small changes in interference signal bearing, (2) small errors in the adaptive weight values, and (3) statistical fluctuations of measured correlations due to finite integration time.

A lightweight four-element adaptive array using hybrid microwave integrated circuitry and weighing only one pound, intended for communication applications, was built and tested [70]. This unit employed a null-steering algorithm appropriate for a coherent sidelobe canceller and succeeded in forming broadband nulls over a 60–100 MHz bandwidth having a cancellation depth of 25–30 dB under weak desired signal and strong interference signal conditions. To attain this degree of interference signal cancellation, it was essential that the element channel circuitry be very well matched over a 20% bandwidth.

Another experimental four-element adaptive array system for eliminating interference in a communication system was also tested [44]. Pattern nulls of 10–20 db for suppressing interference signals over a 200–400 MHz band were easily achieved so long as the desired signal and interference signal had sufficient spatial separation (greater than the resolution capability of the antenna array), assuming the array has no way to distinguish between signals on the basis of polarization. Exploiting polarization differences between desired and interference signals by allowing full polarization flexibility in the array, an interference signal located at the same angle as the desired signal can be suppressed without degrading the reception of the desired signal. Yet another system employing digital control was developed for UHF communications channels and found capable of suppressing jammers by 20–32 dB [71].

In summary, interference suppression levels of 10–20 dB are consistently achieved in practice. It is more difficult but nevertheless practicable to achieve suppression levels of 20–35 dB, and usually very difficult to form cancellation nulls greater than 35 dB in a practical operating system.

The rapid development of digital technology is presently having the greatest impact on signal reception systems. The full adaptation of digital techniques into the processing and interpretation of received signals is making possible the realization of practical signal reception systems whose performance approaches that predicted by theoretical limits. Digital processors and their associated memories have made possible the rapid digestion, correlation, and classification of data from larger search volumes, and new concepts in the spatial manipulation of signals have been developed. Adaptive array techniques started out with limited numbers of elements in the arrays, and the gradual increase in the numbers of elements and in the sophistication of the signal processing will likely result in an encounter with techniques employed in optical and acoustical holography [68], [72]. Holography techniques are approaching such an encounter from the other direction, since they start out with a nearly continuous set of spatial samples (as in optical holography) and move down to a finite number of samples (in the case of acoustic holography).

## PROBLEMS

**1** Suppose it is desired to design a radar pulse waveform that would permit two Ping-Pong balls to be distinguished when placed only 6.3 cm apart in range up to a maximum range from the radar antenna of 10 m.

(a) What is the maximum PRF of the resulting pulse train?

(b) What bandwidth is required for the radar receiver channel?

(c) If it is desired to maintain an array element spacing of $d=2$ cm where $d=\lambda_0/2$, what pulse carrier frequency should the system be designed for?

**2** In the design of an actual sonar system many factors must be considered —all the sonar parameters (source level, target strength, etc.) and the environment parameters. The effect of environmental parameters depends largely on frequency. Suppose in a highly oversimplified example that only the factors of transmission loss (due to attenuation) and ambient noise are of concern. Let the attenuation coefficient $\alpha$ be given by

$$\log_{10}(\alpha)=\tfrac{1}{4}\left[-21+5\log_{10}(f)\right]$$

Furthermore, let the ambient noise spectrum level $N_0$ be given by

$$10\log_{10}(N_0)=\tfrac{1}{3}\left[20-50\log_{10}(f)\right]$$

If the cost to system performance is given by $J=C_1\alpha+C_2N_0$ where $C_1$ and $C_2$ denote the relative costs of attenuation and noise to the system, what value of pulse carrier frequency $f$ should be selected to optimize the system performance?

## REFERENCES

[1] L. C. Van Atta, "Electromagnetic Reflection," U.S. Patent 2908002, October 6, 1959.

[2] *IEEE Trans. Antennas Propag.* (Special Issue on Active and Adaptive Antennas), Vol. AP-12, March 1964.

[3] D. L. Margerum, "Self-Phased Arrays," in *Microwave Scanning Antennas*, Vol. 3, *Array Systems*, edited by R. C. Hansen, Academic Press, New York and London, 1966, Ch. 5.

[4] P. W. Howells, "Intermediate Frequency Sidelobe Canceller," U.S. Patent 3202990, August 24, 1965.

[5] P. W. Howells, "Explorations in Fixed and Adaptive Resolution at GE and SURC," *IEEE Trans. Antennas Propag.*, Special Issue on Adaptive Antennas, Vol. AP-24, No. 5, pp. 575–584, September 1976.

[6] S. P. Applebaum, "Adaptive Arrays," Syracuse University Research Corporation, Rep. SPL TR66-1, August 1966.

[7] B. Widrow, "Adaptive Filters I: Fundamentals," Stanford University Electronics Laboratories, System Theory Laboratory, Center for Systems Research, Rep. SU-SEL-66-12, Tech. Rep. 6764-6, December 1966.

[8] B. Widrow, "Adaptive Filters," in *Aspects of Network and System Theory*, edited by R. E. Kalman and N. DeClaris, Holt, Rinehard and Winston, New York, 1971, Ch. 5.

[9] B. Widrow, P. E. Mantey, L. J. Griffiths, and B. B. Goode, "Adaptive Antenna Systems," *Proc. IEEE*, Vol. 55, December 1967.

[10] L. Spitzer, Jr., "Basic Problems of Underwater Acoustic Research," National Research Council, Committee on Undersea Warfare, Report of Panel on Underwater Acoustics, September 1, 1948, NRC CUW 0027.

[11] R. H. Bolt and T. F. Burke, "A Survey Report on Basic Problems of Acoustics Research," Panel on Undersea Warfare, National Research Council, 1950.

[12] H. Mermoz, "Filtrage adapté et utilisation optimale d'une antenne," Proceedings NATO Advanced Study Institute, September 14–26, 1964. "Signal Processing with Emphasis on Underwater Acoustics," (Centre d'Etude des Phénomenès Aléatoires de Grenoble, Grenoble, France), 1964, pp. 163–294.

[13] F. Bryn, "Optimum Signal Processing of Three-Dimensional Arrays Operating on Gaussian Signals and Noise," *J. Acoust. Soc. Am.*, Vol. 34, March 1962, pp. 289–297.

[14] S. W. W. Shor, "Adaptive Technique to Discriminate Against Coherent Noise in a Narrowband System," *J. Acoust. Soc. Am.*, Vol. 39, January 1966, pp. 74–78.

[15] P. E. Green, Jr., R. A. Frosch, and C. F. Romney, "Principles of an Experimental Large Aperture Seismic Array (LASA)," *Proc. IEEE*, Vol. 53, December 1965, pp. 1821–1833.

[16] R. T. Lacoss, "Adaptive Combining of Wideband Array Data for Optimal Reception," *IEEE Trans. Geosci. Electron.*, Vol. GE-6, May 1968, pp. 78–86.

[17] B. D. Steinberg, "On Teleseismic Beam Formation by Very Large Arrays," *Bull. Seismol. Soc. Am.*, August 1971.

[18] *IEEE Trans. Antennas Propag.* (Special Issue on Adaptive Antennas), Vol. AP-24, September 1965.

[19] A. J. Berni, "Angle of Arrival Estimation Using an Adaptive Antenna Array," *IEEE Trans. Aerosp. Electron. Syst.*, Vol. AES-11, No. 2, March 1975, pp. 278–284; see errata *IEEE Trans. Aerosp. Electron. Syst.*, Vol. AES-11, May and July 1975.

[20] R. C. Davis, L. E. Brennan, and I. S. Reed, "Angle Estimation with Adaptive Arrays in External Noise Fields," *IEEE Trans. Aerosp. Electron. Syst.*, Vol. AES-12, No. 2, March 1976, pp. 179–186.

[21] B. D. Steinberg, *Principles of Aperture and Array System Design*, Wiley, New York, 1976, Ch. 11.

[22] B. D. Steinberg, "Design Approach for a High-Resolution Microwave Imaging Radio Camera," *J. Franklin Inst.*, Vol. 296, No. 6, December 1973.

[23] E. N. Powers and B. D. Steinberg, "Valley Forge Research Center Adaptive Array," USNC/URSI Annual Meeting, Boulder, CO, October 1975.

[24] L. J. Griffiths, "A Simple Adaptive Algorithm for Real-Time Processing in Antenna Arrays," *Proc. IEEE*, Vol. 57, No. 10, October 1969.

[25] J. Chang and F. Tuteur, "Optimum Adaptive Array Processor," Proceedings of the Symposium on Computer Processing in Communications, April 8–10, 1969, Polytechnic Institute of Brooklyn, pp. 695–710.

[26] O. L. Frost, "An Algorithm for Linearly Constrained Adaptive Array Processing," *Proc. IEEE*, Vol. 60, No. 8, August 1972.

[27] C. A. Baird, "Recursive processing for Adaptive Arrays," Proceedings of Adaptive Antenna Systems Workshop, Vol. 1, Naval Research Laboratory, Washington, D.C., September 27, 1974, pp. 163–180.

[28] R. T. Adams, "An Adaptive Antenna System for Maximizing Signal-to-Noise Ratio," Wescon Conference Proceedings, Session 24, 1966, pp. 1–4.

[29] L. J. Griffiths, "Signal Extraction Using Real-Time Adaptation of a Linear Multichannel Filter," SEL-68-017, Tech. Rep. No. 6788-1, System Theory Laboratory, Stanford University, February 1968.

[30] L. E. Brennan, E. L. Pugh, and I. S. Reed, "Control Loop Noise in Adaptive Array Antennas," *IEEE Trans. Aerosp. Electron. Syst.*, Vol. AES-7, March 1971, pp. 254–262.

[31] C. A. Baird, Jr. and J. T. Rickard, "Recursive Estimation in Array Processing," Proceed-

ings of the Fifth Asilomar Conference on Circuits and Systems, 1971, Pacific Grove, CA., pp. 509–513.

[32] J. J. Faran and R. Hills, Jr., "Wide-Band Directivity of Receiving Arrays," Harvard University Acoustics Research Laboratory, Tech. Memo. 31, May 1, 1953.

[33] H. H. Az-Khatib and R. T. Compton, Jr., "A Gain Optimizing Algorithm for Adaptive Arrays," *IEEE Trans. Antennas Propag.*, Vol. AP-26, No. 2, March 1978, pp. 228–235.

[34] N. Wiener, *Extrapolation, Interpretation, and Smoothing of Stationary Time Series*, MIT Press and Wiley, New York, 1949.

[35] H. L. Van Trees, "Optimum Processing for Passive Sonar Arrays," IEEE 1966 Ocean Electronics Symposium, August 29–31, Honolulu, HI, pp. 41–65.

[36] K. Takao, M. Fujita, and T. Nishi, "An Adaptive Antenna Array Under Directional Constraint," *IEEE Trans. Antennas Propag.*, Vol. AP-24, No. 5, September 1976, pp. 662–669.

[37] J. Capon, R. J. Greenfield, and R. J. Kolker, "Multi-Dimensional Maximum-Likelihood Processing of a Large Aperture Seismic Array," *Proc. IEEE*, Vol. 55, February 1967, pp. 192–217.

[38] J. F. Clearbout, "A Summary, by Illustration, of Least-Square Filter With Constraints," *IEEE Trans. Inf. Theory*, Vol. IT-14, March 1968, pp. 269–272.

[39] A. Booker and C. Y. Ong, "Multiple-Constraint Adaptive Filtering," *Geophysics*, Vol. 36, June 1971, pp. 498–509.

[40] Proceedings of the Symposium on Signal Processing in Radar and Sonar Directional Systems (with Special Reference to Systems Common to Radar, Sonar, Radio Astronomy, Ultrasonics, and Seismology), July 6–9, 1964, published by The Institution of Electronics and Radio Engineers, 8–9 Bedford Square, London, W.C. 1.

[41] B. Wyndham, "Demonstrating Radar Using Sonar, Parts 1 and 2," *Wireless World*, August 1968, pp. 248–251, and September 1968, pp. 325–328.

[42] R. L. Deavenport, "The Influence of the Ocean Environment on Sonar System Design," EASCON 1975 Record, IEEE Electronics and Aerospace Systems Convention, September 29–October 1, Washington, DC, pp. 66.A–66.E.

[43] W. G. Bruner, W. N. Case, and R. C. Leedom, "Digital Control in Modern Radar," EASCON 1971 Record, IEEE Electronics and Aerospace Systems Convention, October 6–8, 1971, Washington, DC, pp. 200–207.

[44] R. T. Compton, Jr., "An Experimental Four-Element Adaptive Array," *IEEE Trans. Antennas Propag.*, Vol. AP-24, No. 5, September 1965, pp. 697–706.

[45] W. L. Curtis, "Spiral Antennas," *IRE Trans. Antennas Propag.*, Vol. AP-8, May, 1960, pp. 298–306.

[46] R. H. DuHamel and D. G. Berry, "Logarithmically Periodic Antenna Arrays," IRE WESCON Convention Record, 1958, Vol. 2, Pt. 1, pp. 161–174.

[47] R. Nitzberg, "OTH Radar Aurora Clutter Rejection when Adapting a Fraction of the Array Elements," EASCON 1976 Record, IEEE Electronics and Aerospace Systems Convention, Washington, DC, September 26–29, 1976, paper 62-A.

[48] A. Vural, "A Comparative Performance Study of Adaptive Array Processors," Proceedings of the IEEE International Conference on Acoustics, Speech, and Signal Processing, May 9–11, 1977, Hartford, CT, paper 20.G.

[49] D. R. Morgan, "Partially Adaptive Array Techniques," *IEEE Trans. Antennas Propaga.*, Vol. AP-26, No. 6, November 1978, pp. 823–833.

[50] M. I. Skolnik, *Introduction to Radar Systems*, McGraw-Hill, New York, 1962, Ch. 7.

[51] H. R. Senf, "Electronic Antenna Scanning," Proceedings of the National Conference on Aeronautical Electronics, 1958, Dayton, OH, pp. 407–411.

[52] H. G. Byers and M. Katchky, "Slotted-Waveguide Array for Marine Radar," *Electronics*, Vol. 31, December 5, 1958, pp. 94–96.

[53] A. M. McCoy, J. E. Walsh, and C. F. Winter, "A Broadband, Low Sidelobe, Radar Antenna," IRE WESCON Convention Record, 1958, Vol. 2, Pt. 1, pp. 234–250.

[54] D. H. Ring, "The FH MUSA Antenna," Report of Conference on Rapid Scanning," MIT Radiation Laboratory Report 54–27, June 15, 1943, pp. 3–15.

[55] H. T. Friis and W. D. Lewis, "Radar Antennas," *Bell Syst. Tech. J.*, Vol. 26, April, 1947, pp. 219–317.

[56] S. N. Van Voorhies, (ed.), *Microwave Receivers*, MIT Radiation Laboratory Series, Vol. 23, McGraw-Hill, New York, 1948.

[57] G. A. Gray and F. E. Nathanson, "Digital Processing of Radar Signals," EASCON 1971 Record, IEEE Electronics and Aerospace Systems Convention, October 6–8, 1971, Washington, DC, pp. 208–215.

[58] D. K. Barton, "Radar Technology for the 1980's," *Microwave J.*, November 1978, pp. 81–86.

[59] R. M. O'Donnell et. al., "Advanced Signal Processing for Airport Surveillance Radars," EASCON 1974 Record, IEEE Electronics and Aerospace Systems Convention, 1974, Washington, DC, pp. 71A–71F.

[60] A. A. Winder, "Underwater Sound—A Review: II. Sonar System Technology," *IEEE Trans. Sonics and Ultrason.*, Vol. SU-22, No. 5, September 1975, pp. 291–332.

[61] C. H. Sherman, "Underwater Sound—A Review: I. Underwater Sound Transducers," *IEEE Trans. Sonics and Ultrason.*, Vol. SU-22, No. 5, September 1975, pp. 281–290.

[62] E. A. Massa, "Deep Water Transducers for Ocean Engineering Applications," Proceedings, IEEE 1966 Ocean Electronics Symposium, August 29–31, Honolulu, HI, pp. 31–39.

[63] T. F. Hueter, "Twenty Years in Underwater Acoustics: Generation and Reception," *J. Acoust. Soc. Am.*, Vol. 51, No. 3 (Part 2), March 1972, pp. 1025–1040.

[64] J. S. Hickman, "Trends in Modern Sonar Transducer Design," proceedings of the Twenty-Second National Electronics Conference, October 1966, Chicago, IL.

[65] E. L. Carson, G. E. Martin, G. W. Benthien, and J. S. Hickman, "Control of Element Velocity Distributions in Sonar Projection Arrays," Proceedings of the Seventh Navy Science Symposium, May 1963, Pensacola, FL.

[66] J. F. Bartram, R. R. Ramseyer, and J. M. Heines, "Fifth Generation Digital Sonar Signal Processing," IEEE EASCON 1976 Record, IEEE Electronics and Aerospace Systems Convention, September 26–29, Washington, DC, pp. 91.A–91.G.

[67] T. F. Hueter, J. C. Preble, and G. D. Marshall, "Distributed Architectures for Sonar Systems and Their Impact on Operability," IEEE EASCON 1976 Record, IEEE Electronics and Aerospace Convention, September 26–29, Washington, DC, pp. 96.A–96.F.

[68] V. C. Anderson, "The First Twenty Years of Acoustic Signal Processing," *J. Acoust. Soc. Am.*, Vol. 51, No. 3 (Part 2), March 1972, pp. 1062–1063.

[69] M. Federici, "On the Improvement of Detection and Precision Capabilities of Sonar Systems," Proceedings of the Symposium on Sonar Systems, July 9–12, 1962, University of Birmingham, pp. 535–540. Reprinted in *J. Br. I.R.E.*, Vol. 25, No. 6, June 1963.

[70] G. G. Rassweiler, M. R. Williams, L. M. Payne, and G. P. Martin, "A Miniaturized Lightweight Wideband Null Steerer," *IEEE Trans. Antennas Propag.*, Vol. AP-24, September 1976, pp. 749–754.

[71] W. R. Jones, W. K. Masenten, and T. W. Miller, "UHF Adaptive Array Processor Development for Naval Communications," Digest AP-S Int. Symp., *IEEE Ant. and Prop. Soc.*, June 20–22, 1977, Stanford, CA, IEEE Cat. No. 77CH1217-9 AP, pp. 394–397.

[72] G. Wade, "Acoustic Imaging with Holography and Lenses," *IEEE Trans. Sonics Ultrason.*, Vol. SU-22, No. 6, November 1975, pp. 385–394.

# Chapter 2. Adaptive Array Concept

To understand why an array of sensor elements has the potential to improve the overall reception performance of some desired signal in an environment having several sources of interference and how this potential can be realized, it is necessary to understand the nature of the signals it is desired to receive and the properties of an array of sensor elements. Furthermore, the characteristics of the array elements and the element arrangement used to form the array have a direct impact on adaptive array performance. To gain this understanding the desired signal characteristics, interference characteristics, and signal propagation effects are first discussed. The properties of sensor arrays are then introduced, and the possibility of adjusting the array response to enhance the desired signal reception is demonstrated. Trade-offs for linear and planar arrays are presented to aid the designer in finding an economical array configuration.

In arriving at an adaptive array design, it is necessary to consider the system constraints imposed by the nature of the array, the associated system elements the designer has to work with, and the system requirements that the design is expected to satisfy. Adaptive array system requirements may be classified as either (1) steady-state or (2) transient depending on whether it is assumed that the array weights have reached their steady-state values (assuming a stationary signal environment) or are being adjusted in response to a change in the signal environment. If the system requirements are to be realistic, they must not exceed the ultimate theoretical performance limits that can be predicted for the adaptive array system being considered. Formulation of the constraints imposed by the sensor array itself is quite straightforward and is addressed in this chapter. Steady-state performance limits are considered in Chapter 3. The formulation of transient performance limits, which is considerably more involved, is addressed in Part 2. In order for the performance limits of adaptive array systems to be analyzed it is necessary to develop a generic analytic model for the system. The development of such an analytic model will be concerned with the signal characteristics and the subsequent processing necessary to obtain the desired system response.

## 2.1 SIGNAL ENVIRONMENT

We are concerned with the general problem of processing data from arrays so the overall reception performance of some desired signal is improved. The adaptive array designer must exploit significant differences between the desired signal and any interference signals in order to realize such an improvement. The signal parameters that may be exploited include direction-of-arrival, amplitude, phase, spectral characteristics (including frequency and power), modulation characteristics, and polarization. It is therefore worthwhile to consider the desired signal characteristics and the nature of spurious interfering signals in different contexts.

### *2.1.1 Signals in Active and Passive Sensors*

Active sensing devices such as radar and sonar systems generate a known pulse (or pulse train) that propagates through a transmission medium and is reflected by some target back to the original sender. During most of the listening time interval the desired signal is absent (in contrast with communication systems where the desired signal is usually present), although there is a good idea of the signal structure and the direction-of-arrival so the desired signal can be quite easily recognized when it is present. In addition to the desired signal echo, there may also be spurious returns due to clutter and multipath that contribute to the overall noise field [1]. For radar systems diffuse scattering of multipath can give rise to spurious signals, and a jammer may deliberately generate an interference signal. For active sonar systems the two main types of interference signals are due to ambient noise and reverberation return [2].

Reverberation is analogous to clutter in radar systems and can be defined as signal generated noise that results when a sonar signal transmitted into the ocean encounters many kinds of diffuse scattering objects that reradiate part of the signal back to the receiver. Reverberation returns are classified as surface, bottom, or volume reverberation depending on whether the unwanted reflections originate from the ocean surface, the ocean bottom, or some point in between. Furthermore, multipath signals may arise from signals that reflect from nondiffuse reflectors located at different reflecting angles and impinge on the receiving array. Other propagation effects that cause sonar signal distortion are geometric spreading, attenuation, multiple propagation paths, doppler frequency shift, finite amplitude, medium coherency, and time dispersion [3].

In the case of passive sensing systems, the desired signal is generated by the target (or event of interest) itself and in many cases may be present for most of the listening interval. The basic problem for a passive system is that of distinguishing the desired target signal from the background noise [4]. In contrast to active sensing devices, however, the direction-of-arrival of the desired signal may not be known beforehand, and the desired signal itself must be unknown in some respect if it is to convey useful information to the receiver. The most common means for distinguishing a desired signal from an interference signal is the existence of a known frequency band within which the desired signal must fall. In some cases the power level of the desired signal may be known and used as a distinguishing characteristic. Spread spectrum communication systems commonly employ a known PN (pseudo-noise) code to modulate the transmitted waveform, and this code then provides a convenient means for distinguishing the desired signal.

### 2.1.2 Signal Models

In the passive sonar context, the signal generated by the target may arise from a variety of sources such as engine noise or propeller noise [5]. Consequently the signal source is inherently random. Likewise, if reception of an unknown communications signal is desired, the signal source may once again be regarded as random. Thermal sensor noise, ambient noise, and interference signal sources are also random in nature. These noises typically arise from the combined effect of many small independent sources, and application of the central limit theorem of statistics [6] permits the designer to model the resulting noise signal as a Gaussian (and usually stationary) random process. Quite frequently the physical phenomena responsible for the randomness in the signals of concern are such that it is plausible to assume a Gaussian random process. The statistical properties of Gaussian signals are particularly convenient because the first and second-moment characteristics of the process provide a complete characterization of the random signal.

In other cases of interest, it may be difficult to assign statistical properties to the signal. It may then be desirable to design an array processor that is optimum for a selected nonrandom signal. It turns out in many cases of practical interest that by designing the array processor for a particular reference signal, quite good results can be achieved even though the reference signal is not a perfect replica of the desired signal; any reference signal that is reasonably well correlated with the desired signal and uncorrelated with the interference signals will suffice. In these situations, a nonrandom model is appropriate to use.

The desired signal may sometimes be known exactly, as in the case of a coherent radar with a target of known range and character. In other cases the desired signal may be known except for the presence of uncertain parameters such as phase and signal energy. The case of a signal known except for phase occurs with an ordinary pulse radar having no integration and with a target of known range and character. Likewise, the case of a signal known except for phase and signal energy occurs with a pulse radar operating without integration and with a target of unknown range and known character. The frequency and bandwidth of communication signals are typically known, and such signals may be given a signature by introducing a pilot signal.

For a receiving array comprised of $N$ sensors, the received waveforms correspond to $N$ outputs; $x_1(t), x_2(t), \ldots, x_N(t)$. These $N$ outputs are conveniently described by the received signal vector $\mathbf{x}(t)$ where

$$\mathbf{x}(t) \triangleq \begin{bmatrix} x_1(t) \\ x_2(t) \\ \vdots \\ x_N(t) \end{bmatrix} \quad \text{for} \quad 0 \leqslant t \leqslant T \tag{2.1}$$

and where the range of $t$ describes the observation time interval. The desired signal component in the received signal vector may be designated by $\mathbf{s}(t)$ and the noise component by $\mathbf{n}(t)$. Therefore when the desired signal is present the received signal vector can be written as

$$\mathbf{x}(t) = \mathbf{s}(t) + \mathbf{n}(t) \quad \text{for} \quad 0 \leqslant t \leqslant T \tag{2.2}$$

For either random or nonrandom signals, the desired signal vector can be represented by

$$\mathbf{s}(t) = \begin{bmatrix} s_1(t) \\ s_2(t) \\ \vdots \\ s_N(t) \end{bmatrix} \quad \text{for} \quad 0 \leqslant t \leqslant T \tag{2.3}$$

where the various signal components may be either known exactly, known only to a rough approximation, or known only in a statistical sense.

Interference noise fields at best are stationary and unknown, but they usually exhibit variations with time. Adaptive array processors that automatically respond to the interference environment must cope with such time variations, and current adaptive processing techniques are based on considering slowly varying Gaussian ambient noise fields. Adaptive processors designed for Gaussian noise are distinguished by the pleasant fact that they depend only upon second-order noise moments. Consequently when non-Gaussian noise fields must be dealt with, the most convenient approach is to design a Gaussian-equivalent suboptimum adaptive system based upon the second-order moments of the non-Gaussian noise field.

In general, the slower the variations that occur in the noise environment the better the resulting performance of the adaptive system will be. At the present time the reverberation field associated with a moving sonar platform is considered to exhibit variations that are too rapid to be amenable to current adaptive techniques, although the reverberation field corresponding to a stationary sonar platform may exhibit variations that are slow enough to permit adaptive techniques to be applied [7].

### 2.1.3 *Ideal Propagation Model*

The signal vector $\mathbf{s}(t)$ is usually assumed to be related to a scalar signal $s(t)$ generated at some point source in space by an equation of the form [8]

$$\mathbf{s}(t)=\int \mathbf{m}(t-\tau)s(\tau)\,d\tau \tag{2.4}$$

where the $i$th component of $\mathbf{m}(t)$ is $m_i(t)$ and represents the propagation effects from the source to the $i$th sensor as well as the response of the $i$th sensor. For the ideal case of nondispersive propagation and distortion-free sensors, then $m_i(t)$ is a simple time delay $\delta(t-\tau_i)$, and the desired signal component at each sensor element is identical except for a time delay so that (2.4) can be written as

$$\mathbf{s}(t)=\begin{bmatrix} s(t-\tau_1) \\ s(t-\tau_2) \\ \vdots \\ s(t-\tau_N) \end{bmatrix} \tag{2.5}$$

A special case of great practical significance occurs when the desired signal propagation may be regarded as a plane wave from the direction $\boldsymbol{\alpha}$ as shown in Figure 2.1 (where $\boldsymbol{\alpha}$ is taken to be a unit vector).

In this case the various time delays are simply given by [7]

$$\tau_i=\frac{\boldsymbol{\alpha}\cdot\mathbf{r}_i}{\mathfrak{v}} \tag{2.6}$$

where $\mathfrak{v}$ is the propagation velocity, each sensor coordinate is given by the vector $\mathbf{r}_i$, and $\boldsymbol{\alpha}\cdot\mathbf{r}_i$ denotes the dot product

$$\boldsymbol{\alpha}\cdot\mathbf{r}_i \triangleq \boldsymbol{\alpha}^T\mathbf{r}_i \tag{2.7}$$

and $T$ denotes transpose.

Measuring the relative time delays experienced at each sensor element therefore provides a means of determining the unknown direction-of-arrival of the desired signal $s(t)$. In the following sections on array properties, it is found that the plane wave propagation effects described above play a

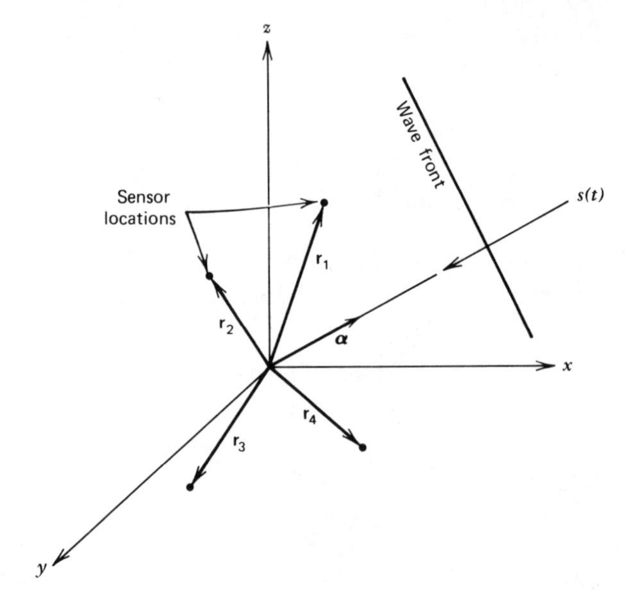


**Figure 2.1** Three-dimensional array with plane wave signal propagation.

fundamental role in determining the nature of the array response. Furthermore, it should be noted that the relative time-delay properties suggested by (2.5) and (2.6) are not restricted to nonrandom signals and hold for any signal experiencing plane-wave propagation.

## 2.2 ARRAY ELEMENT SPACING CONSIDERATIONS

It was noted earlier that an array of sensors offers the possibility of overcoming the sensitivity and beamwidth limitations of a single sensor as well as offering the possibility of modifying the array beam shape. To see how these possibilities arise, a single pair of identical sensor elements are first considered. The discussion then proceeds to consider the one-dimensional linear array of equally spaced elements and finally extends the array beam concept to a two-dimensional array.

After considering the effects of sensor element arrangement within the array, the impact on array performance limits that are determined by a selected array configuration are examined. The position of the elements within the array determines array resolution and interferometer (grating lobe) [9] effects. In general, resolution increases as the array dimension (or separation between elements) increases. High array resolution improves the maximum output signal-to-noise ratio (SNR) when the angular separation between the desired and the undesired signals is small. High resolution capability also implies sharp array pattern nulls, however, thereby reducing the array ability to place a broad null on clustered interference sources. An $N$-element linear array has $N-1$ degrees of freedom so that up to $N-1$ array beam pattern nulls can be independently adjusted for array operation.

### 2.2.1 Pair of Identical Sensors

Consider the pair of identical omnidirectional sensors shown in Figure 2.2 spaced apart by a distance $d$. Let a signal $x(t)$ impinge on the two sensor elements in a plane containing the two elements and the signal source from a direction $\theta$ with respect to the array normal. It is seen in Figure 2.2 that element 2 experiences a time delay with respect to element 1 of

$$\tau = \frac{d \sin\theta}{\mathfrak{v}} \tag{2.8}$$

Let the array output signal $y(t)$ be given by the sum of the two sensor element signals so that

$$y(t) = x(t) + x(t-\tau) \tag{2.9}$$

If $x(t)$ is a narrowband signal having center frequency $f_0$, then the time delay $\tau$ corresponds to a phase shift of $2\pi(d/\lambda_0)\sin\theta$ radians where $\lambda_0$ is the wavelength corresponding to the center frequency,

$$\lambda_0 = \frac{\mathfrak{v}}{f_0} \tag{2.10}$$

The overall array response may then be found merely by considering the

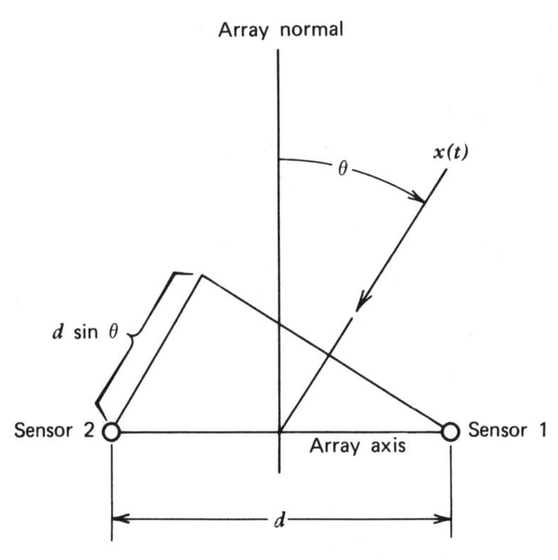


**Figure 2.2** Pair of identical omnidirectional sensor elements.

phasor sum of the signal contributions from the two array elements. That is,

$$y(t)=\sum_{i=1}^{2} x(t)e^{j(i-1)\psi} \tag{2.11}$$

where

$$\psi=2\pi(d/\lambda_0)\sin\theta \tag{2.12}$$

The directional pattern of the array (that is, the relative sensitivity of response to signals for a specified frequency from various directions) may be found by considering only the term

$$A(\theta)=\sum_{i=1}^{2} e^{j(i-1)\psi} \tag{2.13}$$

The normalized directional pattern in decibels for the two elements is then given by

$$G(\theta)(\text{decibels})=10\log_{10}\left\{\frac{|A(\theta)|^2}{4}\right\} \tag{2.14}$$

A plot of $G(\theta)$ for this two-element example is given in Figure 2.3 for

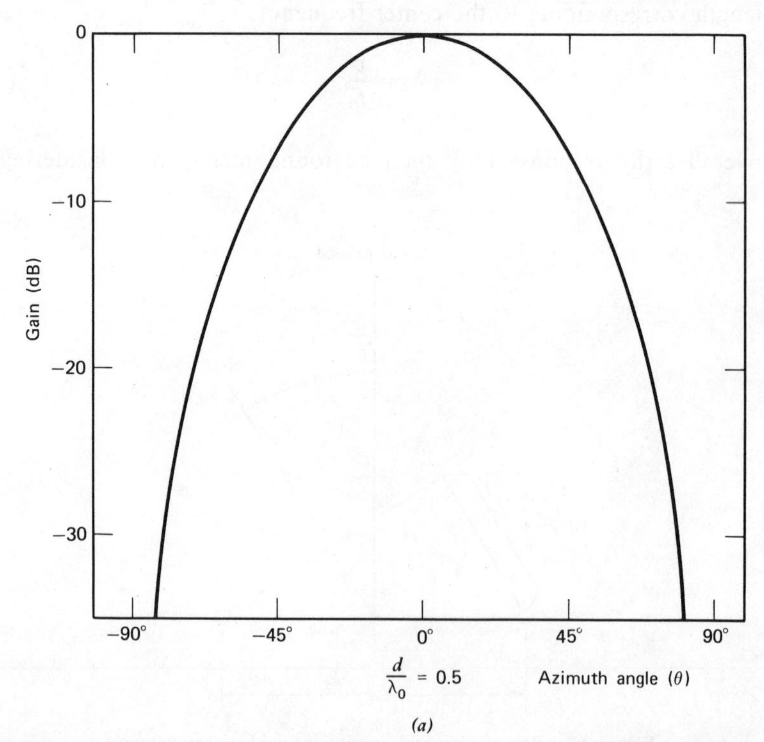


(*a*)

**Figure 2.3** Array beam patterns for two-element example: (*a*) $d/\lambda_0=0.5$. (*b*) $d/\lambda_0=1$. (*c*) $d/\lambda_0=1.5$.

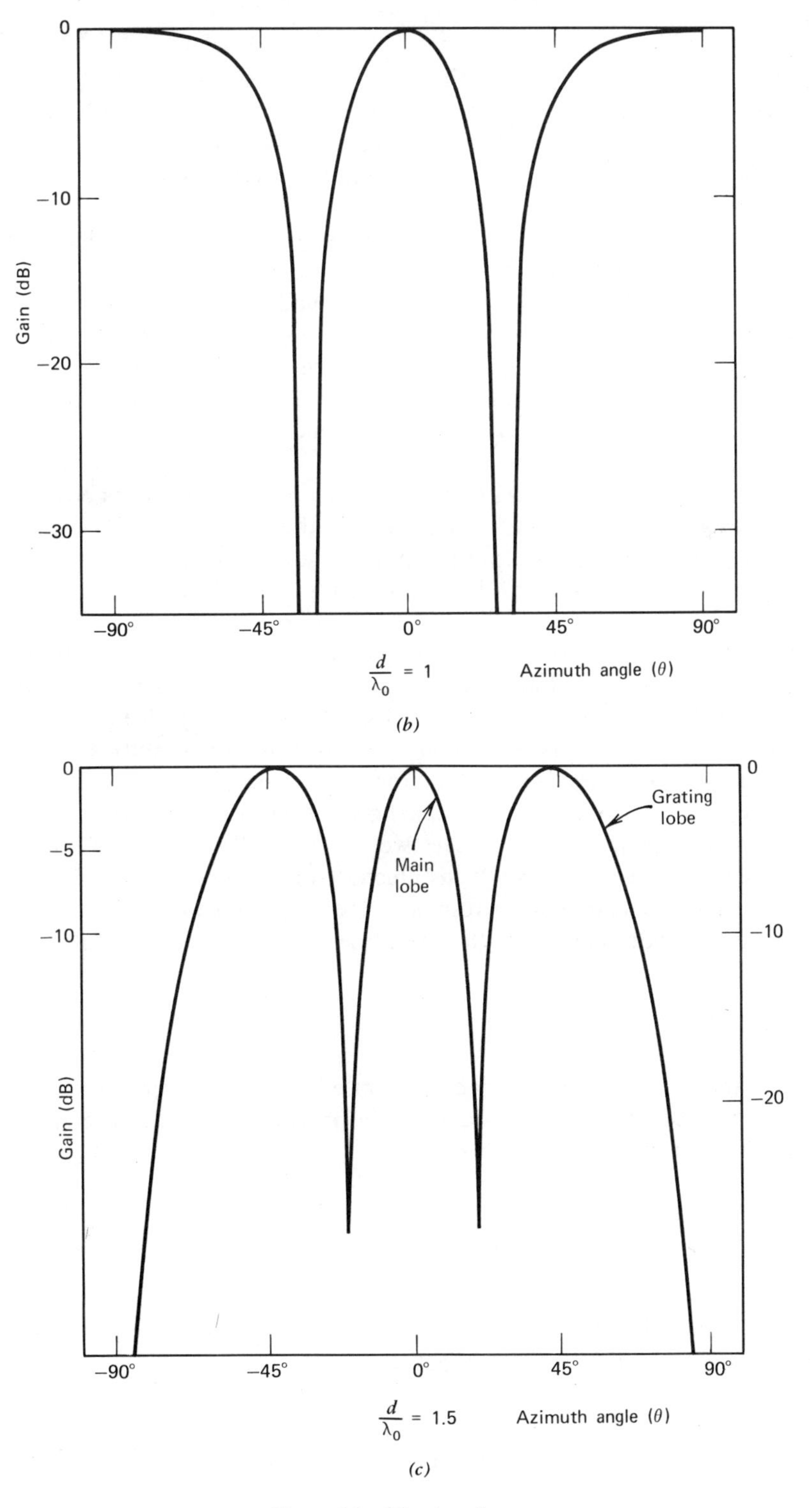


**Figure 2.3** (*Continued*).

$d/\lambda_0=0.5$, 1.0, and 1.5. From Figure 2.3*a* it is seen that for $d/\lambda_0=0.5$, there is one principal lobe (or main beam) having a 3 dB beamwidth of 60° and nulls at $\theta=\pm 90°$ off broadside. The nulls at $\theta=\pm 90°$ occur because at that direction-of-arrival the signal wavefront must travel exactly $\lambda_0/2$ between the two sensors, which corresponds to a phase shift of 180° between the signals appearing at the two sensors and therefore yields exact cancellation of the resulting phasor sum. If the element spacing is less than $0.5\lambda_0$, then exact cancellation at $\theta=\pm 90°$ does not result, and in the limit as the element spacing approaches zero (ignoring mutual coupling effects) the directional pattern becomes the omnidirectional pattern of a single element. It is well known that there is very little difference in the directional pattern between a single element and two closely spaced elements (less than $\lambda/4$ apart), and consequently arrays employing many elements very closely spaced are considered "inefficient" if it is desired to use as few array elements as possible for a specified sidelobe level and beamwidth. If the element spacing increases to greater than $0.5\lambda_0$, the two pattern nulls migrate in from $\theta=\pm 90°$, occuring at $\theta=\pm 30°$ when $d=\lambda_0$, as illustrated in Figure 2.3*b*. The nulls at $\theta=\pm 30°$ in Figure 2.3*b* occur because at that angle-of arrival the phase path difference between the two sensors is once again 180°, and exact cancellation results from the phasor sum. Two sidelobes at $\theta=\pm 90°$ having an amplitude equal to the principal lobe at $\theta=0°$ also appear because the phase path difference between the two sensors is then 360°, the two phasors exactly align, and the array response is the same as for broadside angle-of-arrival. As the element spacing increases to $1.5\lambda_0$, the main lobe beamwidth decreases still further, thereby improving resolution, the two pattern nulls migrate further in, and two new nulls appear at $\pm 90°$, as illustrated in Figure 2.3*c*. Further increasing the interelement spacing results in the appearance of even more pattern nulls and sidelobes and a further decrease in the main lobe beamwidth.

### 2.2.1 *Linear Arrays*

For a linear array of $N$ equispaced sensor elements, the overall array response may again be found by considering the phasor sum of signal contributions from each array element as in (2.11):

$$y(t)=\sum_{i=1}^{N} x(t)e^{j(i-1)\psi} \tag{2.15}$$

The directional pattern in a plane containing the array may therefore be found by considering the array factor

$$A(\theta)=\sum_{i=1}^{N} e^{j(i-1)\psi} \tag{2.16}$$

and the normalized directional pattern is given by

$$G(\theta) = 10\log_{10}\left\{ \frac{|A(\theta)|^2}{N^2} \right\} \tag{2.17}$$

For nonisotropic sensor elements, it is necessary to introduce an additional factor $F(f_0, \theta)$ in (2.16) to include the directional response pattern introduced by each sensor element. Alternatively, the response of an arbitrary configuration of identical elements having the same orientation can be derived by using the principle of pattern multiplication [9]. The directional pattern of an array of identical spatial elements may be found by (1) replacing each of the elements by an omnidirectional element at the same point, (2) determining the directional array pattern of the resulting array of omnidirectional elements, and (3) multiplying the array pattern resulting from step 2 by the beam pattern of the individual elements of the original array. The phasor sum corresponding to (2.16) is a maximum when $\sin\theta = 0$ or $\theta = k2\pi$, for then all the phasors line up. If the ratio $(d/\lambda_0)$ is large enough, then for $N$ elements the phasors will be equispaced in the complex plane yielding a sum of zero whenever

$$2\pi(d/\lambda_0)\sin\theta = 2\pi/N \tag{2.18}$$

Such zero phasor sums occur in a direction $\theta_1$ given by

$$\sin\theta_1 = \frac{1}{N}\left(\frac{\lambda_0}{d}\right) \tag{2.19}$$

Letting $L = (N-1)d$ be the length of the array, (2.19) can be rewritten as

$$\theta_1 = \arcsin\left(\frac{\lambda_0}{L+d}\right) \tag{2.20}$$

Maintaining the interelement spacing at $d/\lambda_0 = 0.5$ and increasing the number of identical omnidirectional elements, the normalized array directional (or beam) pattern may be found from (2.17), and the results are shown in Figure 2.4 for three and four elements. It is seen that as the number of elements increases the main lobe beamwidth decreases, and the number of sidelobes and pattern nulls increases.

To illustrate how element spacing affects the directional pattern for a seven-element linear array, Figures 2.5*a* through 2.5*d* show the directional pattern in the azimuth plane for values of $d/\lambda_0$ ranging from 0.1 to 1.0. So long as $d/\lambda_0$ is less than $\frac{1}{7}$, the beam pattern has no exact nulls as the $-8.5$ dB null occurring at $\theta = \pm 90°$ for $d/\lambda_0 = 0.1$ illustrates. As the interelement spacing is permitted to increase beyond $d/\lambda_0 = \frac{1}{7}$, pattern nulls and sidelobes

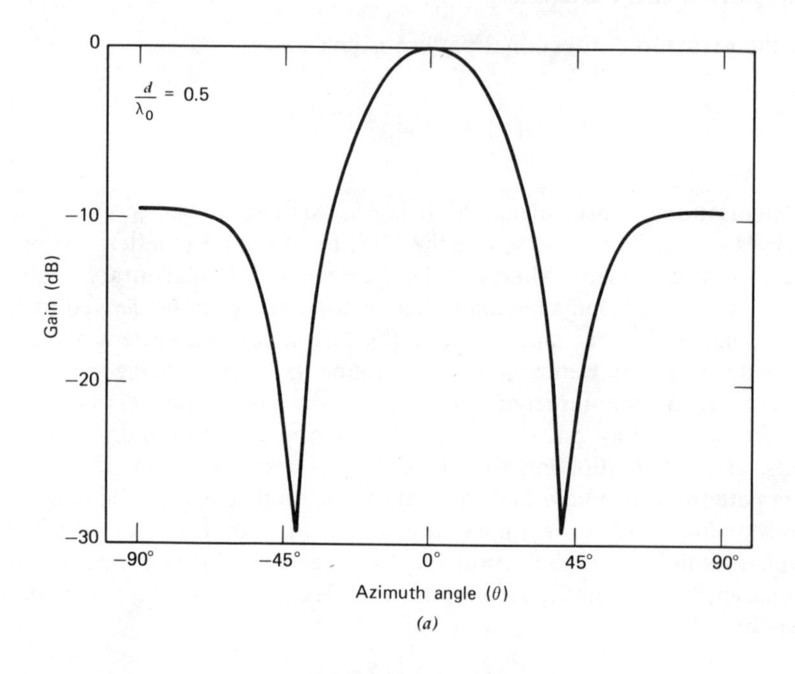


*(a)*

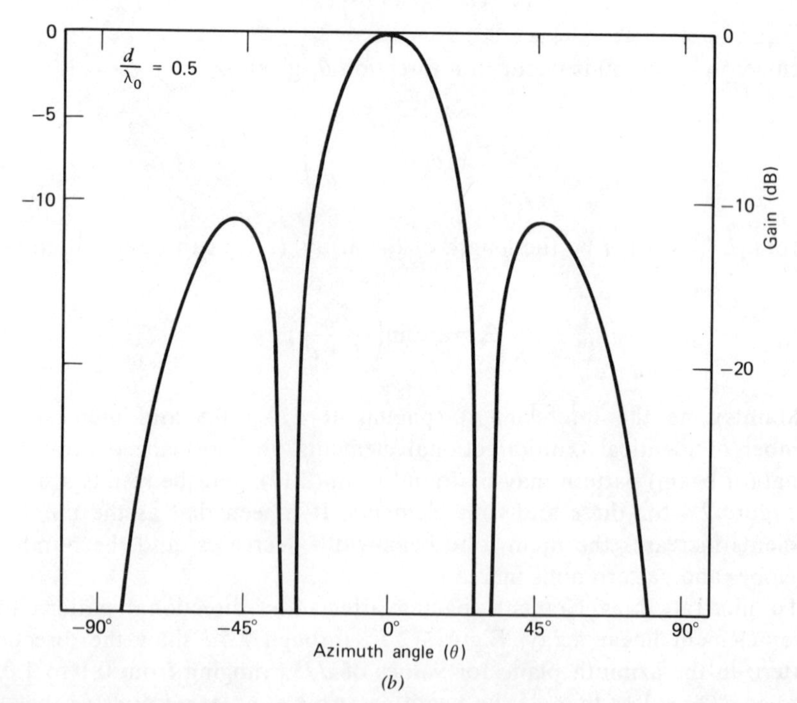


*(b)*

**Figure 2.4** Linear array beam patterns for $d/\lambda_0 = 0.5$. *(a)* Three-element array. *(b)* Four-element array.

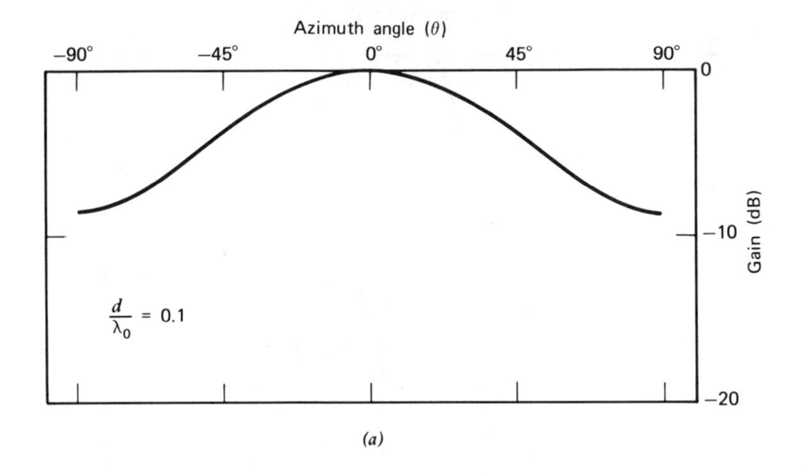


(a)

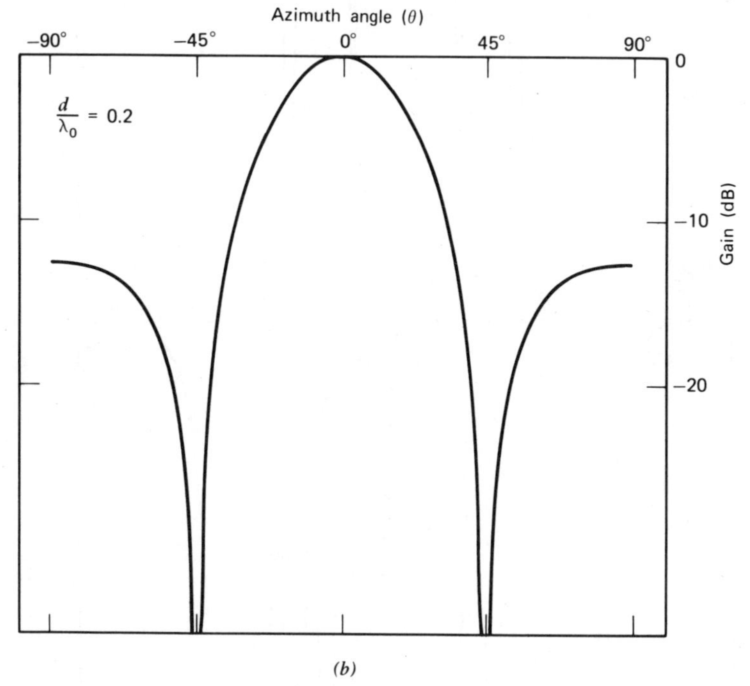


(b)

**Figure 2.5** Seven-element linear array beam patterns. (*a*) $d/\lambda_0 = 0.1$. (*b*) $d/\lambda_0 = 0.2$. (*c*) $d/\lambda_0 = 0.5$. (*d*) $d/\lambda_0 = 1.0$

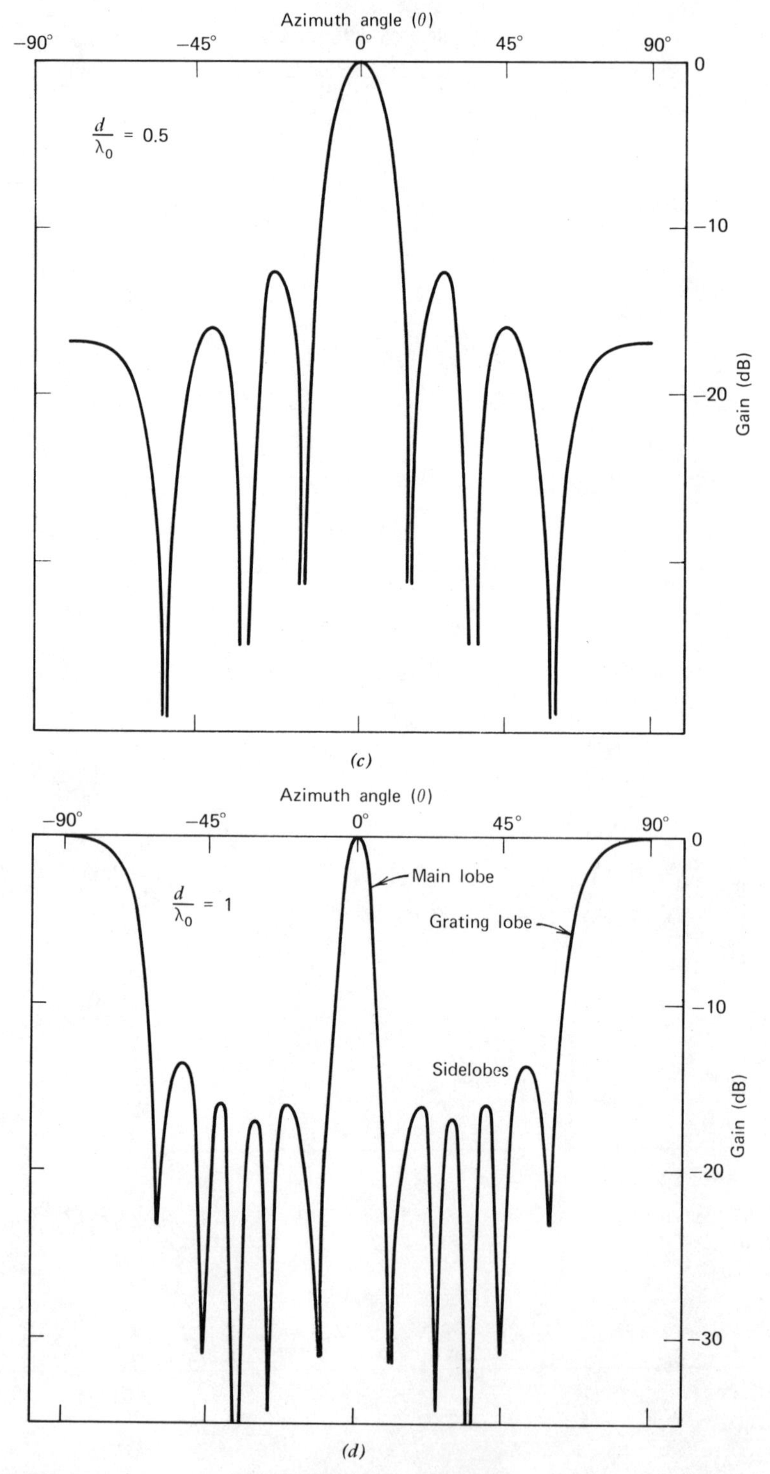


**Figure 2.5** (*Continued*).

(and grating lobes) begin to appear, with more lobes and nulls appearing as $d/\lambda_0$ increases and producing an interferometer pattern. When $d/\lambda_0 = 1$ the endfire sidelobes at $\theta = \pm 90°$ have a gain equal to the mainlobe since the seven signal phasors now align exactly and add coherently.

Suppose for the linear array of Figure 2.6 that a phase shift (or an equivalent time delay) of $\delta$ is inserted in the second element of the array, a phase shift of $2\delta$ in the third element, and a phase shift of $(n-1)\delta$ in each succeeding $n$th element. The insertion of this sequence of phase shifts has the effect of shifting the principal lobe (or main beam) by

$$\theta = \arcsin\left[\frac{1}{2}\left(\frac{\lambda_0}{d}\right)\delta\right] \tag{2.21}$$

so the overall directional pattern has in effect been "steered" by insertion of the phase shift sequence. This effect is illustrated in Figure 2.7 for $\delta = 30°$ and $d/\lambda_0 = 0.5$, which can be directly compared with Figure 2.5$c$ to see the degree of shift resulting in the array pattern.

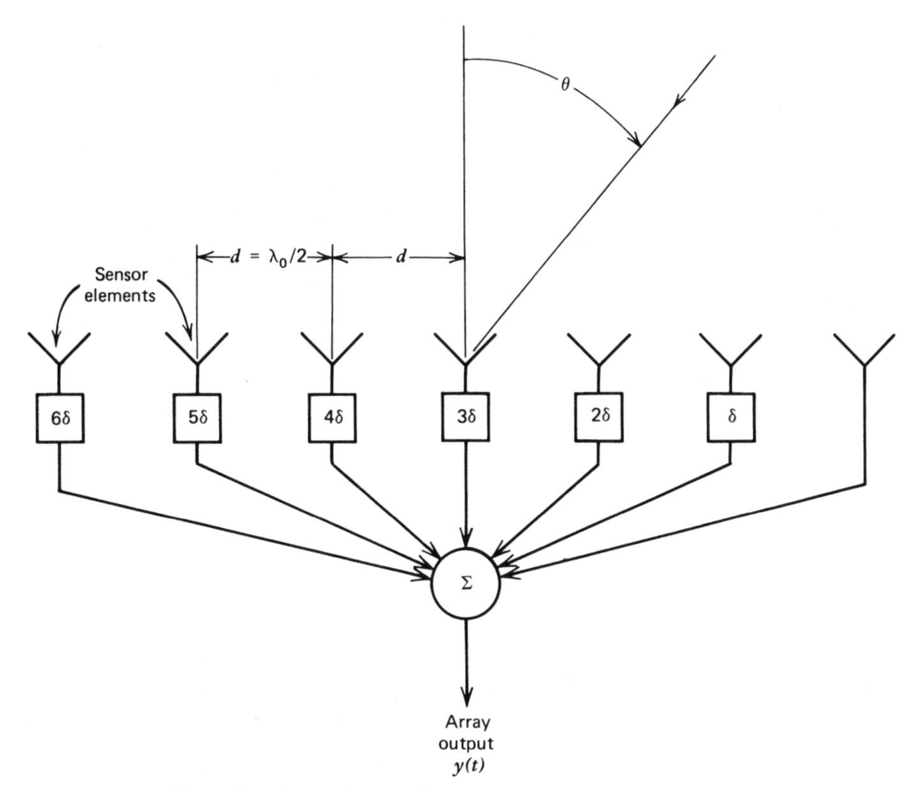


**Figure 2.6** Seven-element linear array steered with phase shift elements.

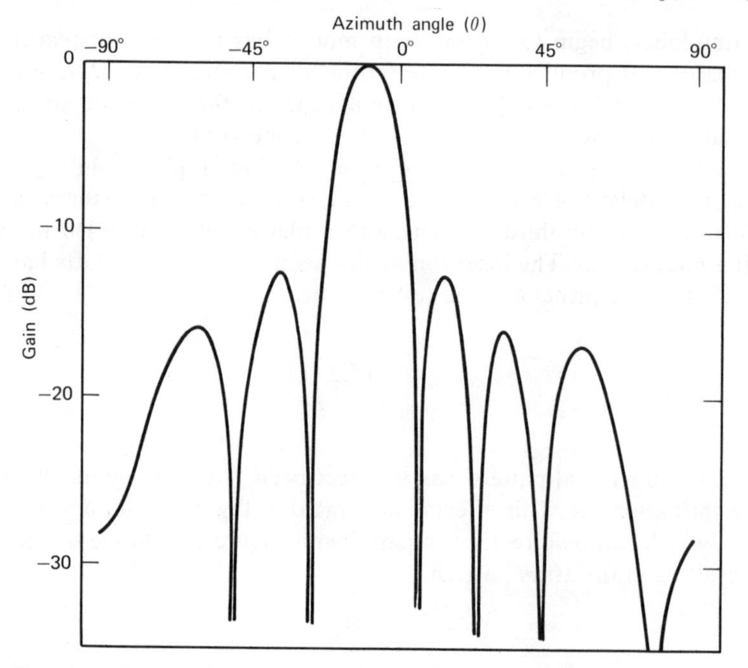


**Figure 2.7** Seven-element linear array directivity pattern with $d/\lambda_0=0.5$ and $\delta=30°$.

### 2.2.2 *Planar Arrays*

Consider the rectangular-shaped planar array of discrete sensor elements arranged in the $x-y$ plane as shown in Figure 2.8, where the coordinate origin is chosen at the central element. With $N_x$ elements in each $x$ axis parallel column and uniform spacing $d_x$, and $N_y$ elements in each $y$ axis parallel row with uniform spacing $d_y$, the entire array has $N_x \times N_y$ elements.

The phasor sum of signal contributions from each array element in a single column is the same as that for a linear array and is therefore given by

$$y(t)=\sum_{i=1}^{N_x} x(t)e^{j(i-1)\psi_x} \tag{2.22}$$

where now

$$\psi_x=2\pi\left(\frac{d_x}{\lambda_0}\right)\sin\theta\cos\phi \quad \text{and} \quad \psi_y=2\pi\left(\frac{d_y}{\lambda_0}\right)\sin\theta\sin\phi \tag{2.23}$$

It is seen that the output signal now depends on both the projected azimuth angle $\phi$ and the elevation angle $\theta$. The total phasor sum of signal contributions from all array elements is given by

$$y(t)=\sum_{i=1}^{N_x}\sum_{k=1}^{N_y} x(t)e^{j(i-1)\psi_x}e^{j(k-1)\psi_y} \tag{2.24}$$

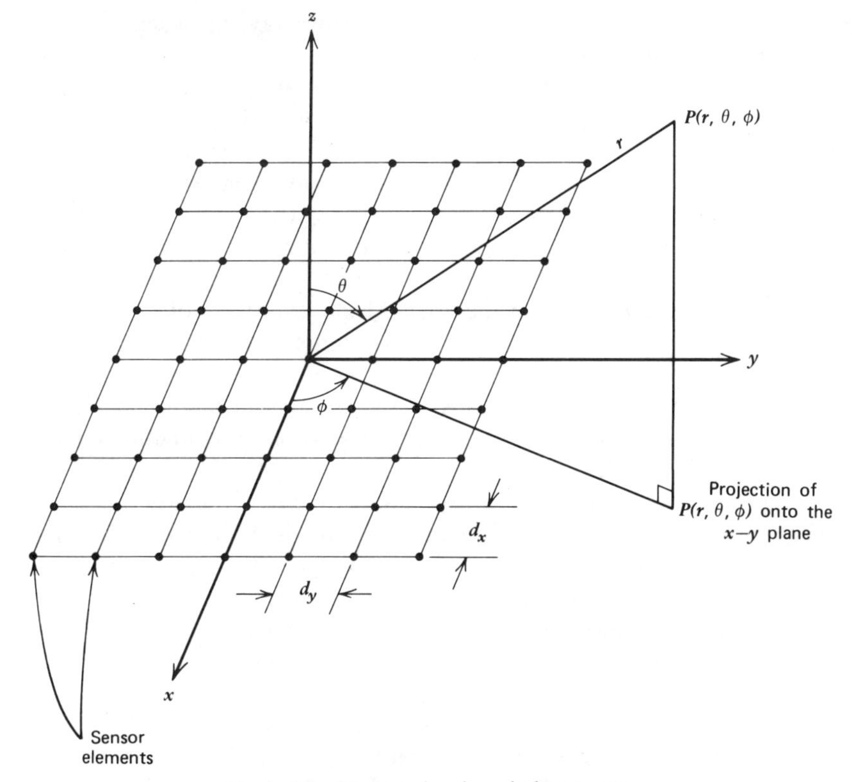


**Figure 2.8** Rectangular-shaped planar array.

Consequently the directional pattern of the array may be found by considering the term

$$A(\theta,\phi)=\sum_{i=1}^{N_x}\sum_{k=1}^{N_y} e^{j(i-1)\psi_x}e^{j(k-1)\psi_y}=A_x(\theta,\phi)A_y(\theta,\phi) \tag{2.25}$$

where

$$A_x(\theta,\phi)=\sum_{i=1}^{N_x} e^{j(i-1)\psi_x}$$

and

$$A_y(\theta,\phi)=\sum_{k=1}^{N_y} e^{j(k-1)\psi_y} \tag{2.26}$$

From (2.25) it follows that the planar array beam pattern can be considered to be the product of array factors of two linear arrays, one along the $x$ axis and the other along the $y$ axis. Consequently the requirement on the element spacings $d_x$ and $d_y$ is the same as that for linear arrays in order to avoid the formation of grating lobes.

As with linear arrays, the row and column array factors $A_x(\theta,\phi)$ and $A_y(\theta,\phi)$ of the rectangular array can each be arbitrarily steered by introduction of the appropriate phase shift sequences for the row and column elements. Practical application of a planar array demands that the main beams of $A_x(\theta,\phi)$ and $A_y(\theta,\phi)$ intersect, although it is certainly possible to introduce phase shift sequences so they do not intersect. Steering the row and column array factors so that the respective conical beams point at the same position $(\theta_0,\phi_0)$ is the principal idea behind the utilization of a planar array. With the row and column beams aligned at the same position two pencil beams result, one directed above the array plane and the other below. The pencil beam directed below the array plane can be eliminated either by appropriate choice of a directive sensor element pattern function or by the use of a ground plane.

In addition to the rectangular arrays, there are circular arrays and elliptical arrays, which are useful when angular symmetry is desired in a two-dimensional operation. The detailed analysis and synthesis of planar arrays is beyond the scope of concern here, and the interested reader may refer to reference [10] for a thorough treatment of this important subject.

From the foregoing discussion it is clear that the beamwidth, sidelobe levels, and main beam directions for a planar array can be analyzed in the same manner employed for linear arrays. Since the resulting two-dimensional beam is generated in an azimuth-elevation $(\phi,\theta)$ space, it is natural to plot the directional pattern in a polar projection diagram as shown in Figure 2.9*a* where the range of $\theta$ is $0°\rightarrow 90°$ and the range of $\phi$ is $0°\rightarrow 360°$. A polar projection diagram, however, has two shortcomings: it is difficult to interpolate accurately between discrete points recorded on such a diagram be-

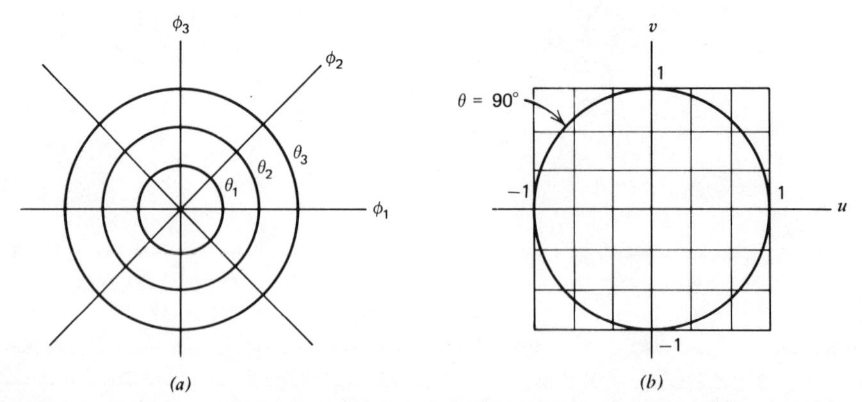


**Figure 2.9** Alternate diagrams for presenting two-dimensional array directivity patterns. (*a*) Polar projection diagram. (*b*) Rectilinear diagram.

cause of the nonlinear character of the coordinate system, and it is not possible to exploit computer controlled plotting routines that are invariably designed for rectilinear coordinate plotting.

As a consequence of the difficulties encountered with a polar coordinate representation of a planar array beam pattern, it is convenient to introduce the transformation

$$\left.\begin{aligned} u &= \sin\theta\cos\phi \\ v &= \sin\theta\sin\phi \end{aligned}\right\} \tag{2.27}$$

which reduces the polar coordinate $\theta-\phi$ representation to a rectilinear coordinate $u-v$ diagram as shown in Figure 2.9*b*, for which the ranges of $u, v$ are the interval $(-1 \rightarrow 1)$. The rectilinear coordinate diagram is very convenient for plotting and interpolation, and it is consequently the most frequently selected diagram for presenting two-dimensional beam patterns for planar arrays.

## 2.3 ARRAY PERFORMANCE

Since we have found that an array of sensors can yield increased resolution and sensitivity compared to a single sensor, it is now appropriate to consider how an array also offers the possibility of enhancing reception performance by simultaneously suppressing undesirable interference and preserving a desired signal. Furthermore, in selecting an array configuration, it is necessary to keep in mind the following critical factors:

1. Resolution capability.
2. Angular coverage.
3. Number of array elements.
4. Sidelobe level.

If the number of array elements is small, then forming an array null to suppress an interference signal can significantly degrade the array sensitivity over the desired angular coverage region. One way of overcoming such coverage degradation while retaining the adaptive nulling capability is to ensure high resolution by using a large number of elements. As the number of elements increases, however, the cost and complexity of the array also increases. Therefore there is a fundamental trade-off between the array resolution capability, sidelobe level, and the number of array elements for a specified desired angular coverage region. This trade-off is considered for linear and planar arrays.

### 2.3.1 Enhanced Signal Reception by Adjustment of the Array Response

The possibility of steering and modifying the array directional pattern to enhance desired signal reception and simultaneously suppress interference signals by complex weight selection is illustrated by the following example: Consider the two omnidirectional-element array of Figure 2.10 in which a desired signal arrives from the normal direction $\theta=0°$, and the interference signal arrives from the angle $\theta=\pi/6$ radians. For simplicity, both the interference signal and the desired pilot signal are assumed to have the same frequency $f_0$. Furthermore, assume that at the point exactly midway between the array elements the desired signal and the interference are in phase (this assumption is not required but simplifies the development). The output signal from each element is input to a variable complex weight, and the complex weight outputs are then summed to form the array output.

Now consider how the complex weights can be adjusted to enhance the reception of $p(t)$ while rejecting $I(t)$. The array output due to the desired signal is

$$Pe^{j\omega_0 t}\{[w_1+w_3]+j[w_2+w_4]\} \tag{2.28}$$

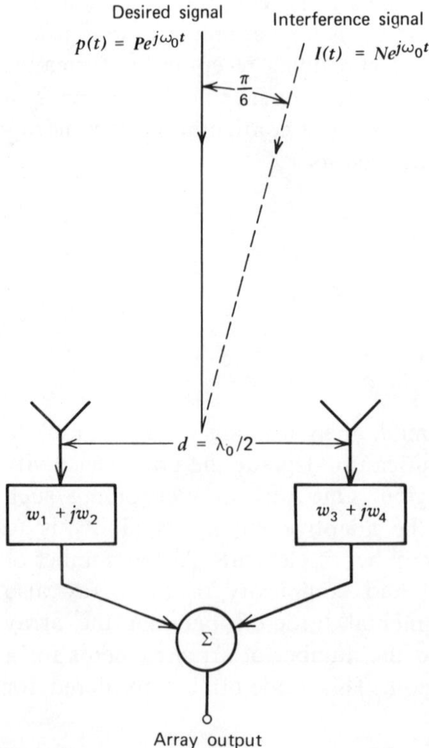


**Figure 2.10** Two-element array for interference suppression example.

For the output signal of (2.28) to be equal to $p(t) = Pe^{j\omega_0 t}$, it is necessary that

$$\left.\begin{array}{l} w_1 + w_3 = 1 \\ w_2 + w_4 = 0 \end{array}\right\} \tag{2.29}$$

The incident interfering noise signal exhibits a phase lead with respect to the array midpoint when impinging on the element with complex weight $w_3 + jw_4$ of value $2\pi(\frac{1}{4})\sin(\pi/6) = \pi/4$ and a phase lag when striking the other element of value $-\pi/4$. Consequently the array output due to the incident noise is given by

$$Ne^{j(\omega_0 t - \pi/4)}[w_1 + jw_2] + Ne^{j(\omega_0 t + \pi/4)}[w_3 + jw_4] \tag{2.30}$$

Now

$$e^{j(\omega_0 t - \pi/4)} = \frac{1}{\sqrt{2}}\left[e^{j\omega_0 t}(1 - j)\right]$$

and

$$e^{j(\omega_0 t + \pi/4)} = \frac{1}{\sqrt{2}}\left[e^{j\omega_0 t}(1 + j)\right]$$

so for the array noise response to be zero, it is necessary that

$$\left.\begin{array}{r} w_1 + w_2 + w_3 - w_4 = 0 \\ -w_1 + w_2 + w_3 + w_4 = 0 \end{array}\right\} \tag{2.31}$$

Solving (2.29) and (2.31) simultaneously then yields:

$$w_1 = \tfrac{1}{2}, w_2 = -\tfrac{1}{2}, w_3 = \tfrac{1}{2}, w_4 = \tfrac{1}{2} \tag{2.32}$$

With the above weights, the array will accept the desired signal while simultaneously rejecting the interference.

While the above computation for complex weight selection will yield an array directional pattern that achieves the desired system objectives, it is not a very practical way of approaching adaptive array problems. The method used in the above example exploits the fact that there is only one directional interference source, that the signal and interference sources are monochromatic, and uses the *a priori* information concerning the frequency and the direction of arrival of each signal source. A practical processor should not require such detailed information about the location, number, and nature of signal sources. Nevertheless, this example has demonstrated that the control afforded by complex weight adjustment does offer the possibility of realizing array system objectives once a practical adaptive array processor is employed to accomplish the complex weight adjustment. The development of such a practical adaptive array processor is undertaken in Part Two.

Instead of using one complex weight per array element for pattern control as in Figure 2.10, another approach sometimes employed to reject interference signals through the formation of array pattern nulls is the use of a nulling tree [11]. If $K$ distinct interference sources are to be rejected, and if the number of elements to be used in forming the main beam is $M$, then the total number of elements required in the entire array is $N = M + K$. A nulling tree employs identical phase shifters in a multilayer arrangement such that each layer will produce one pattern null and the number of phase shifters required in the nulling tree is

$$N_{\phi} = KN - \frac{K(K+1)}{2} \tag{2.33}$$

Each null produced by a nulling tree can be controlled independently of the other nulls. The array beamforming and scanning for the main beam can then be accomplished by a layer of $M$ complex weights following the layers of phase shifters. This arrangement has the drawback that the number of controls required to adjust the phase shifters for null formation exceeds the number of degrees of freedom and is therefore in this sense inefficient. For this reason the pattern control approach represented by the use of complex weights is usually preferred over the nulling tree approach, and complex weighting is the pattern forming approach that is assumed throughout this book.

The foregoing discussion on modifying and adapting the overall array directional pattern has pointed out that interferometer effects occur whenever the element spacing exceeds $\lambda_0/2$, where $\lambda_0$ is the center frequency wavelength. In the case of a linear array with equispaced elements, if it is desired to produce a pattern null at an angle $\theta_0$ (corresponding to an interference signal location), then the process of modifying the array beam pattern to produce the desired null will also yield other pattern nulls. For example, if it is desired to "steer" an array directional pattern to an angle $\theta_0$ (corresponding to an interference signal location), then $\theta_0$ satisfies (2.19). Recognizing that the condition for a zero phasor sum can be rewritten as

$$2\pi\left(\frac{d}{\lambda_0}\right)\sin\theta_k = \frac{2\pi}{N} \pm 2\pi k \tag{2.34}$$

where $k$ is any integer, it then follows that pattern nulls will also occur at directions $\theta_k$ satisfying the relationship

$$\sin\theta_k = \underbrace{\frac{1}{N}\left(\frac{\lambda_0}{d}\right)}_{\sin\theta_0} \pm k\left(\frac{\lambda_0}{d}\right) \tag{2.35}$$

The above equation shows that no real values of $\theta_k$ occur other than the desired direction $\theta_0$ so long as the element spacing $d$ is less than $\lambda_0/2$. It may likewise be seen that the number of additional pattern nulls beside that in the direction $\theta_0$ is proportional to the element spacing. If an array has additional nulls beside that in the desired direction $\theta_0$, it may develop that a desired signal happens to be located at (or near) one of the additional nulls. In this case, a degradation in the SNR array performance will ensue. It is for this reason that grating null effects are considered undesirable in adaptive array applications, and a maximum element spacing of $\lambda_0/2$ is usually maintained (resulting in what is termed a "filled array"). Occasionally, for special applications, a high degree of array resolution is required without recourse to a large number of array elements, in which case the element spacing of $\lambda_0/2$ may be exceeded (resulting in a "thinned array"). When element spacing is permitted to exceed $\lambda_0/2$, then the effects of the resulting grating nulls can be reduced by employing nonuniformly spaced array elements to disrupt the periodic array structure [12]–[17].

### 2.3.2 *Universal Trade-off Curves for Array Design*

Recent demands for high-performance signal reception systems have made apparent a great need for antennas possessing high resolution, high gain, and low sidelobe level. To achieve these goals, steerable reflector antennas as large as 200–300 ft in diameter have been built, and considerable effort has also been devoted to the design of large phased arrays, which are not so prohibitively expensive as large reflector antennas. For conventionally designed arrays where all elements are uniformly spaced, the interelement spacing must not exceed an upper limit if grating lobes are not to be present in the visible region. Consequently the required number of elements for such an array becomes astronomically large if a beamwidth on the order of one minute of arc is desired.

For any array design there is an inevitable trade-off to be determined between the number of elements required and the resulting resolution and sidelobe level that can be achieved. It is well known that by using an array design employing randomly spaced elements in which the average element spacing is not too small (more than a few wavelengths so that mutual coupling effects can be ignored), then arrays can be designed to have a narrow beam, moderately low sidelobe level, and very wide bandwidth without the need for a large number of array elements [11],[18],[19]. The theory for such arrays with randomly spaced elements was developed by Lo and provides a means for predicting (in a probabilistic sense) the results for various element spacings. Such a prediction can be made before carrying out any detailed computations, so the odds of achieving a successful design are high. Actual evaluation of the final array design is then carried out after arriving at the preliminary design. An array with randomly spaced elements has the following properties:

1. The number of elements required depends primarily on the desired sidelobe level—this number is in general much less than the corresponding number required for uniform element spacings.
2. The array resolution depends principally on the aperture dimension.
3. The array directive gain is proportional to the number of elements used.
4. When the number of elements is fixed, the resolution and bandwidth corresponding to an aperture can be improved by a factor of 10, 100, or more using random element spacing, without substantial risk in having a higher sidelobe level.

If we recognize that the number of elements required in an array having randomly spaced elements is less than that required for uniformly spaced array elements, the theory that describes the trade-off between the number of elements required and the resulting resolution and sidelobe level that can be achieved provides a convenient measure for determining the relative efficiency of any large array design. The two most critical assumptions in applying the random element spacing array theory are that most elements are widely spaced and that the number of elements involved is moderately large (greater than 50).

For a linear array the distribution of the sidelobe level is given by [18]

$$\Pr(\text{sidelobe level below } r) = (1 - e^{-Nr^2})^{[4a]} \tag{2.36}$$

where $\Pr(\cdot)$ denotes "probability of"
- $[4a]$ denotes the larger of the two integers nearest to $4a$
- $N$ = total number of array elements
- $a$ = aperture dimension in wavelengths
- $r$ = the sidelobe level with the main beam maximum normalized to unity.

Equation (2.36) can be plotted as in Figure 2.11 to form a family of universal curves for the probability distribution that the sidelobe level is below $r$ versus $\sqrt{N}\ r$ for various aperture dimension indexes, $q = \log_{10}(a)$. A plot of $\sqrt{N}\ r$ versus aperture dimension index for a constant probability of 90% as in Figure 2.12 then gives the trade-off between sidelobe level and the required number of array elements for a given array size.

Figure 2.11 can also be applied to two-dimensional arrays having array dimensions (in wavelengths) of $a = 10^q$ and $b = 10^p$ by replacing $q$ with $(p+q)$ and modifying the cumulative probability scale in the following manner: For a given probability level $\Pr_1$ determine a corresponding probability level $\Pr_2$ for the one dimensional results of Figure 2.11 by finding a value $x$ that satisfies

$$\Pr_1 = 1 - e^{-x^2} \tag{2.37}$$

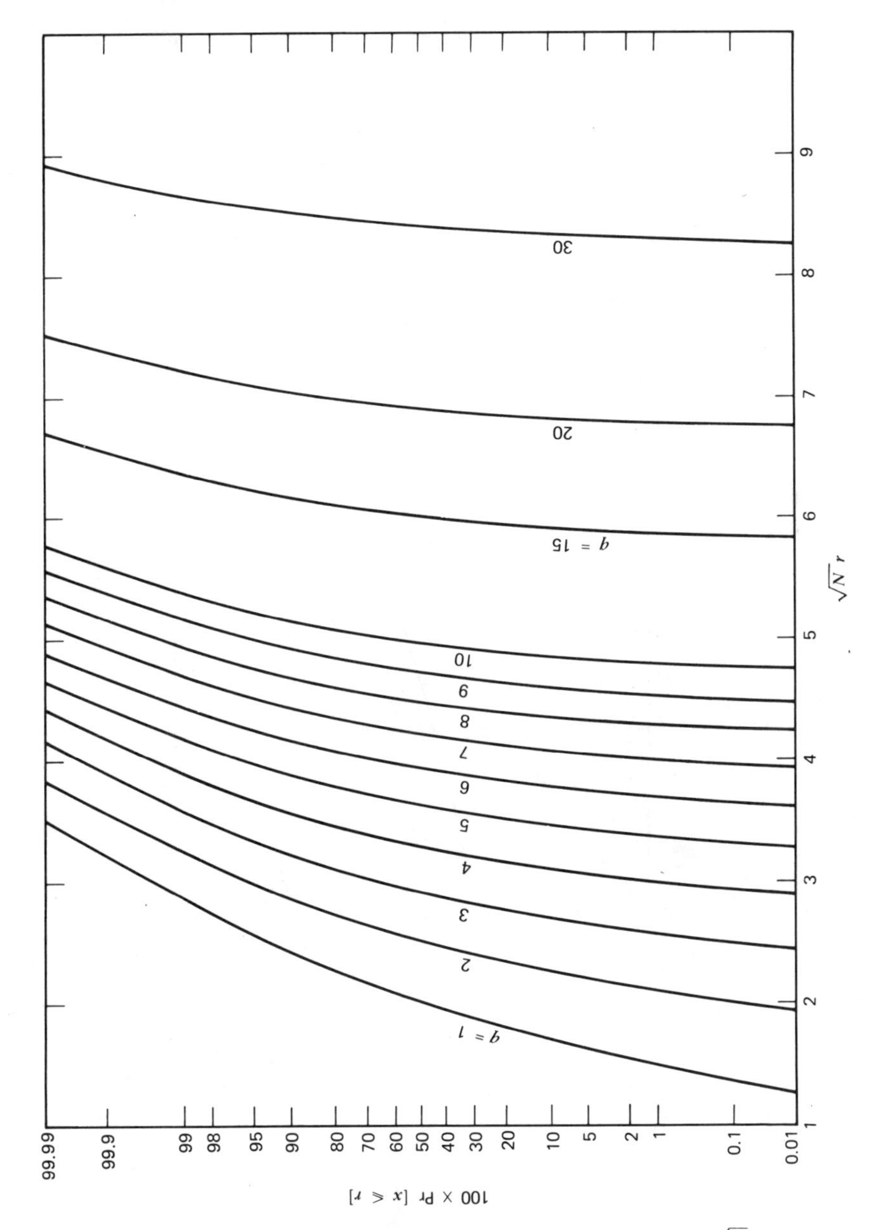


**Figure 2.11** Cumulative probability that sidelobe level ≤ $r$ versus $\sqrt{N}\, r$.

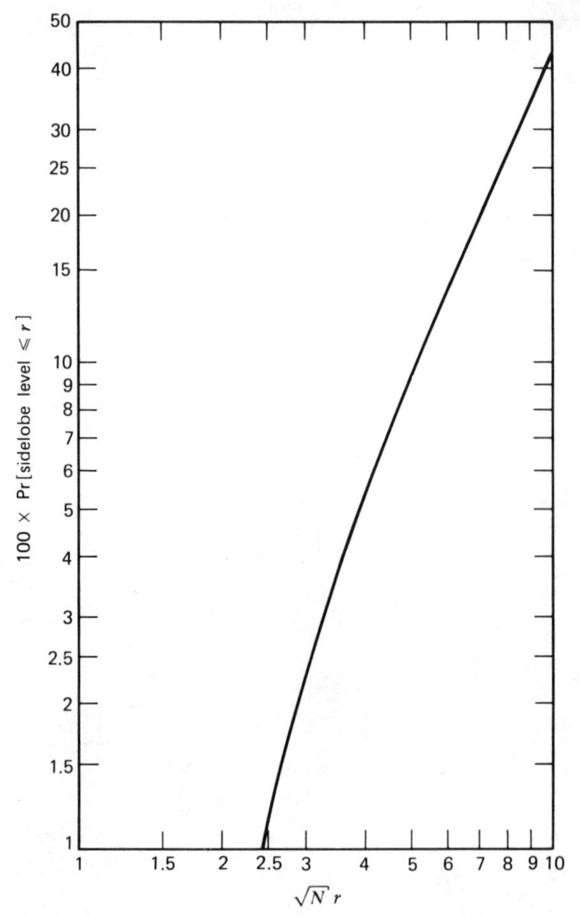


**Figure 2.12** Pr[sidelobe level $\leqslant r$] versus $\sqrt{N}\, r$ for $\Pr[x \leqslant r]=0.9$.

then

$$\Pr_2 = erf(x) \tag{2.38}$$

where

$$erf(x) \triangleq \frac{2}{\sqrt{\pi}} \int_0^x e^{-t^2} dt \tag{2.39}$$

The resolution of an array composed of randomly spaced elements can be described in terms of the expected value of the half-power beamwidth $2\theta_0$, where [18]

$$E\{2\theta_0\} = \frac{0.73}{a} \text{(radians)} \tag{2.40}$$

The standard deviation to be associated with the mean value given by (2.40) depends on the number of elements contained in the array. This standard deviation as a function of the number of array elements is plotted in Figure 2.13.

For a two-dimensional array having dimension $a = b = 50\lambda$ (so that $p = q \cong 1.7$), employing $\lambda/2$ element spacing requires $10^4$ array elements. Now suppose it is desired to use randomly spaced elements while maintaining the sidelobe level less than $r = 0.245$ (which is comparable to the large $\lambda/2$ element spaced array) with a 90% probability. The corresponding cumulative probability in Figure 2.11 is $\Pr[x \leqslant .245] = 0.968$, and the value of $\sqrt{N}\ r$ corresponding to $q = 3.4$ is 3.55. Consequently the required number of elements to maintain the same sidelobe levels is $N = (3.55)^2/(.245)^2 \cong 210$. This number of elements required for the random array is less than the number of elements required for the $\lambda/2$ uniformly spaced array by a factor of nearly 50. For 20 dB sidelobes ($r = 0.1$), the required number of elements in the random array increases to approximately 1260, which is still a factor of 8 reduction. It can also be shown that the beamwidth for the thinned random array is comparable to that for the $\lambda/2$ uniformly spaced array.

Finally, it is worth noting that should a compelling reason exist to employ triangular or rectangular element spacing in a planar array, then a simple analytical technique developed by Kmetzo [20] can be exploited to minimize the number of elements required in an array while avoiding the formation of end-fire grating lobes. A thorough discussion of array optimization techniques may be found in reference [21].

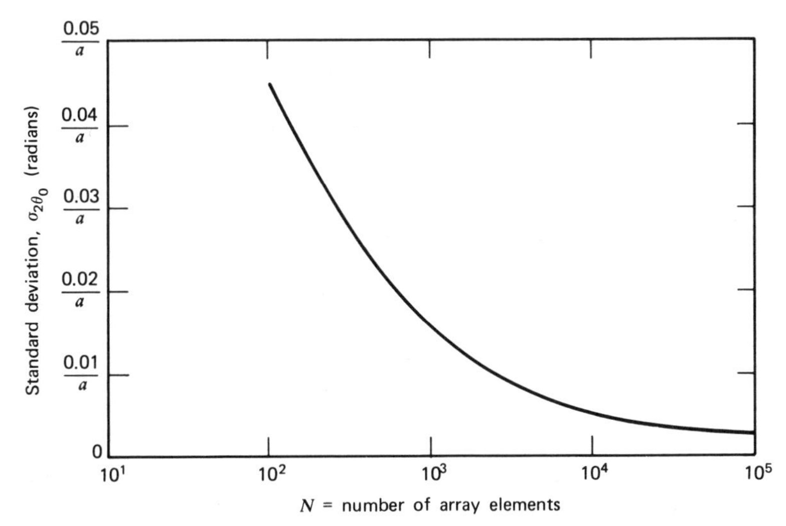


**Figure 2.13** Standard deviation of half-power beamwidth $\sigma_{2\theta_0}$ versus number of elements for a random array.

## 2.4 NULLING LIMITATIONS DUE TO MISCELLANEOUS ARRAY EFFECTS

A principal measure of the effectiveness of an adaptive array in cancelling an undesired interference signal is the ratio of the total output noise power $P_0$ to internal noise power $P_N$ as a function of frequency. To obtain this ratio for a simple example, consider the two-element adaptive array of Figure 2.14. If the interference signal arriving at element 1 from the direction $\theta$ is $s(t)$, then the signal appearing at element 2 is $s(t+\tau)$ where $\tau=(d/\mathfrak{v})\sin\theta$ and $\mathfrak{v}=$ wavefront propagation velocity.

The signals in the two element channels are weighted and summed to form the array output $y(t)$ so that

$$y(t)=w_1 s(t)+w_2 s(t+\tau) \tag{2.41}$$

which can be written in the frequency domain as

$$Y(\omega)=S(\omega)\left[w_1+w_2 e^{-j\omega\tau}\right] \tag{2.42}$$

In order to exactly cancel the interference signal at the array output at a particular frequency, $f_0$ (termed the center frequency of the jamming signal), it is necessary that

$$w_2=-w_1 e^{j\omega_0\tau} \tag{2.43}$$

If we select the two adaptive weights to satisfy (2.43), it follows that the interference signal component of the array output at any frequency is given by

$$Y(\omega)=S(\omega)w_1\left[1-e^{-j\tau(\omega-\omega_0)}\right] \tag{2.44}$$

and consequently the interference component of output power at any

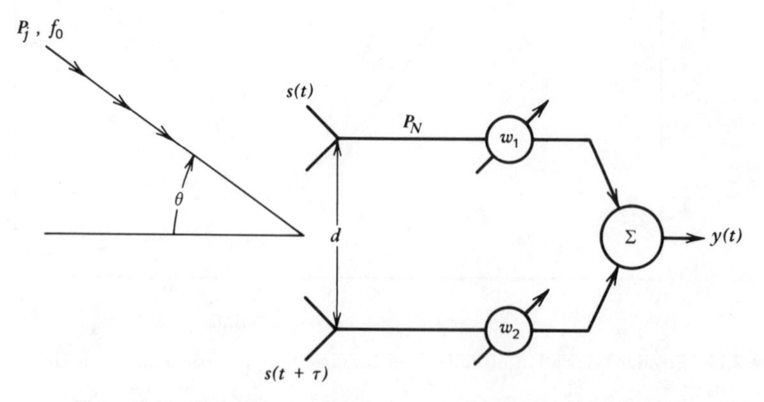


**Figure 2.14** Two-element adaptive array with jammer located at $\theta$.

frequency may be expressed as (assuming momentarily for simplicity that $|S(\omega)|^2 = 1$)

$$|Y(\omega)|^2 = |w_1|^2\{2 - 2\cos[\tau(\omega - \omega_0)]\} \tag{2.45}$$

Let $|S(\omega)|^2 = P_J$ and denote the constant spectral density of the internal noise power in each channel by $P_N$, it follows that the total output noise power spectral density $P_o(\omega)$ can be written as

$$P_o(\omega) = 2|w_1|^2[1 - \cos(\tau(\omega - \omega_0))]P_J + 2|w_1|^2 P_N \tag{2.46}$$

If we recognize that the output thermal noise power spectral density is just $P_n = 2|w_1|^2 P_N$ and normalize the above expression to $P_n$, it follows that the ratio of the total output interference plus noise power spectral density $P_0$ to the output noise power spectral density $P_n$ is then given by

$$\frac{P_0}{P_n}(\omega) = 1 + \frac{\{1 - \cos[\tau(\omega - \omega_0)]\}P_J}{P_N} \tag{2.47}$$

where $P_J/P_N$ = jammer power spectral density to internal noise power spectral density per channel
$\tau = (d/v)\sin\theta$
$d$ = sensor element spacing
$\theta$ = angular location of jammer from array broadside
$\omega_0$ = center frequency of jamming signal.

At the center frequency, a null is placed exactly in the direction of the jammer so that $P_0/P_n = 1$. If, however, the jamming signal has a nonzero bandwidth then the other frequencies present in the jamming signal will not be completely attenuated. Figure 2.15 is a plot of (2.47) and shows the

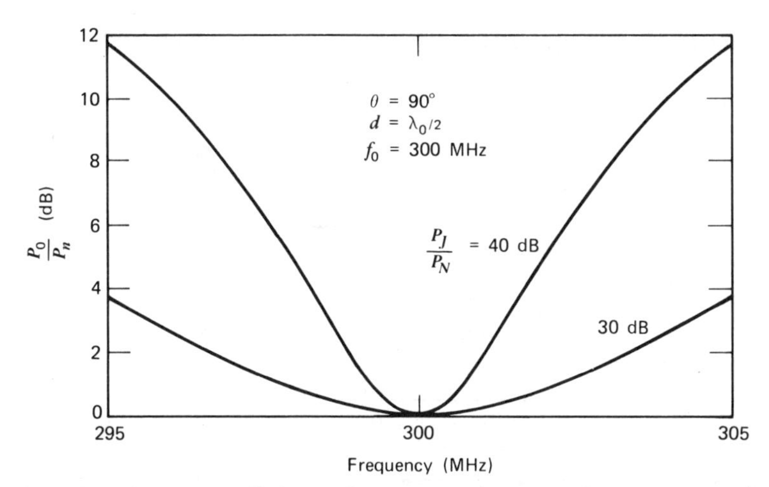


**Figure 2.15** Two-element cancellation performance: $P_0/P_n$ versus frequency for $P_J/P_N = 30$ and 40 dB.

resulting output interference power density in decibels as a function of frequency for a 10 MHz bandwidth jamming signal located at $\theta = 90°$ and an element spacing of $\lambda_0/2$ for two values of $P_J/P_N$. The results shown in Figure 2.15 indicate that for $P_J/P_N = 40$ dB, about 12 dB of uncancelled output interference power remains at the edges of the 10 MHz band (with $f_0 = 300$ MHz). Consequently the output residue power at that point is about $40 - 12 = 28$ dB below the interfering signal that would otherwise be present if no attempt at cancellation were made.

It can also be noted from (2.47) and Figure 2.15 that the null bandwidth decreases as the element spacing increases and as the angle of the interference signal from array broadside increases. Specifically, the null bandwidth is inversely proportional to the element spacing and inversely proportional to the sine of the angle of the interference signal location.

The degree of interference signal cancellation that can be achieved depends principally on three array characteristics: (1) element spacing, (2) interference signal bandwidth, and (3) frequency dependent interelement channel mismatch across the cancellation bandwidth. The effect of sensor element spacing on the overall array sensitivity pattern has already been discussed. Yet another effect of element spacing is the propagation delay across the array aperture.

For the two-element array of Figure 2.16, the propagation delay $\tau$ is given by (2.48). Assume that a single jammer having power $P_J$ is located at $\theta$ and that the jamming signal has a flat power spectral density with (double-sided) bandwidth $B$ Hz. It may be shown that the cancellation ratio $P_0/P_J$, where $P_0$ is the (cancelled) output residue power in this case, is given by

$$\frac{P_0}{P_J} = 1 - \frac{\sin^2(\pi B\tau)}{(\pi B\tau)^2} \tag{2.49}$$

Note that (2.49) is quite different from (2.47), a result that reflects the fact

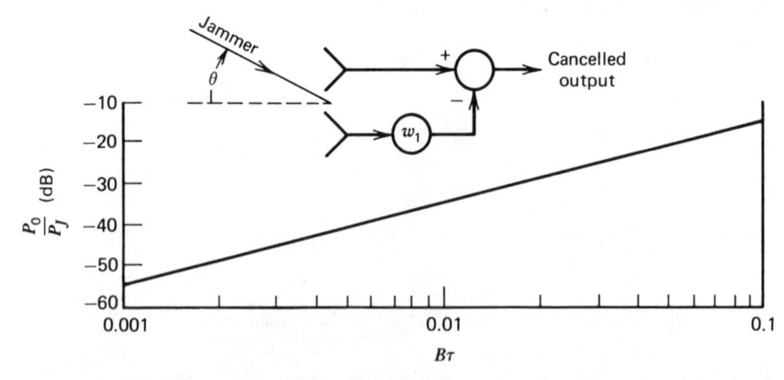


**Figure 2.16** Cancellation ratio versus jammer signal bandwidth—propagation delay product $B\tau$.

that the weight value that minimizes the output jammer power is different from the weight value that yields a null at the center frequency. Equation (2.49) is plotted in Figure 2.16, which shows how the interference signal cancellation decreases as the signal bandwidth-propagation delay product $B\tau$ increases. It is seen from (2.49) that the cancellation ratio $P_0/P_J$ is inversely proportional to the element spacing and the signal bandwidth.

An elementary interchannel amplitude and phase mismatch model for a two-element array is shown in Figure 2.17. Ignoring the effect of any propagation delay, the output signal may be expressed as

$$y(t) = 1 - ae^{j\phi} \tag{2.50}$$

From (2.50) it follows that the cancelled output power is given by

$$|y(t)|^2 = |1 - ae^{j\phi}|^2 = 1 + a^2 - 2a\cos\phi \tag{2.51}$$

Figure 2.17 also shows a plot of the cancellation ratio $P_0/P_J$ based on (2.51) for amplitude errors only ($\phi = 0$) and phase errors only ($a = 1$). In Part Three, where adaptive array compensation is discussed, more realistic interchannel mismatch models are introduced and means for compensating such interchannel mismatch effects are studied. The results obtained from Figure 2.17 indicates that the two-sensor channels must be matched to within about 0.5 dB in amplitude and to within approximately 2.8° in phase in order to obtain 25 dB of interference signal cancellation.

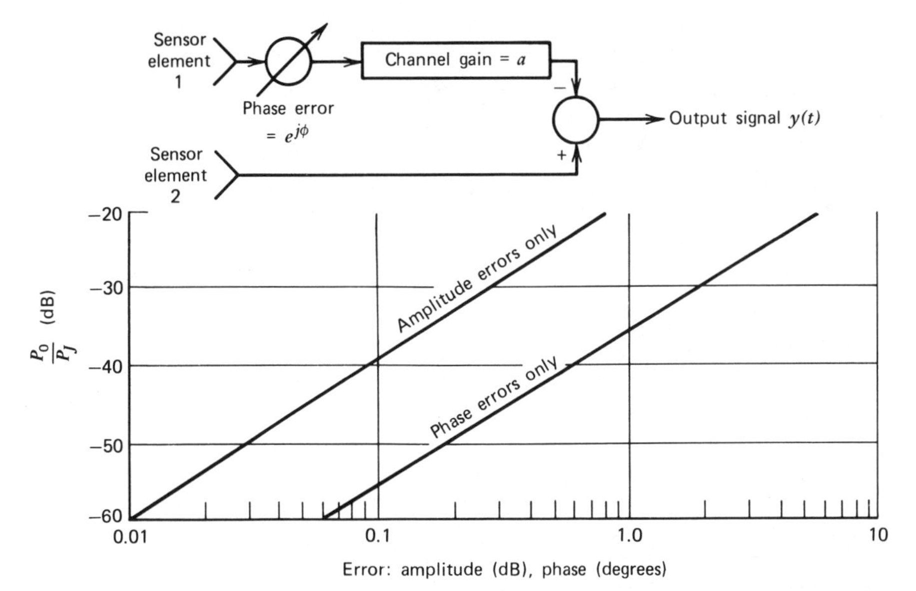


**Figure 2.17** Cancellation ratio $P_0/P_J$ versus amplitude and phase channel mismatch for a two-element array.

## 2.5 NARROWBAND AND BROADBAND SIGNAL PROCESSING CONSIDERATIONS

We have seen that by adjusting the complex weights in the two-element array of Figure 2.10 it is possible to modify the overall array pattern so as to place a pattern null in the direction of an undesired interference signal. So long as the interference signal is "narrowband" (that is, can be adequately characterized by a single frequency $\omega_0$) the resulting array pattern will succeed in suppressing the undesired interference signal. The most common way of realizing complex weighting of an array element output is by means of a quadrature hybrid circuit, which is illustrated in Figure 2.18. Each individual sensor element is divided into an in-phase and quadrature-phase component by means of a 90° phase shifter. Each component is then applied to a variable weight whose output is summed to form the output signal. The resulting complex gain factor is then $Ae^{j\phi}$ where $\phi = -\tan^{-1}(w_2/w_1)$ and $A = \sqrt{w_1^2 + w_2^2}$ . The weights $w_1, w_2$ can assume a continuum of both positive and negative values, and the range of magnitude $A$ is limited only by the range of limitations of the two individual weights. It should be recognized that it is not absolutely necessary to maintain a phase shift of exactly 90° to obtain quite satisfactory processing with quadrature hybrid circuitry.

In the event that the interference signal cannot be adequately characterized by a single frequency and has a frequency content encompassing a significant spectrum, then the complex weight selection appropriate for one frequency $\omega_1$ will not be appropriate for a different frequency $\omega_2$, since the array pattern nulls shift as the value of $\lambda_0$ changes; this fact leads to the conclusion that different complex weights are required at different frequencies if an array null is to be maintained in the same direction for all frequencies of interest. A simple and effective way of obtaining different amplitude and phase weightings at a number of frequencies over the band of interest is to replace the quadrature hybrid circuitry of Figure 2.18 by a transversal filter having the transfer function $\mathfrak{h}(\omega)$. Such a transversal filter can be realized by a tapped-delay line having $L$ complex weights as shown in Figure 2.19 [22], [23]. A tapped-delay line has a transfer function that is periodic, repeating over consecutive filter bandwidths $B_f$, as shown in Appendix A. If the tap spacing is sufficiently close and the number of taps is large, this network approxi-

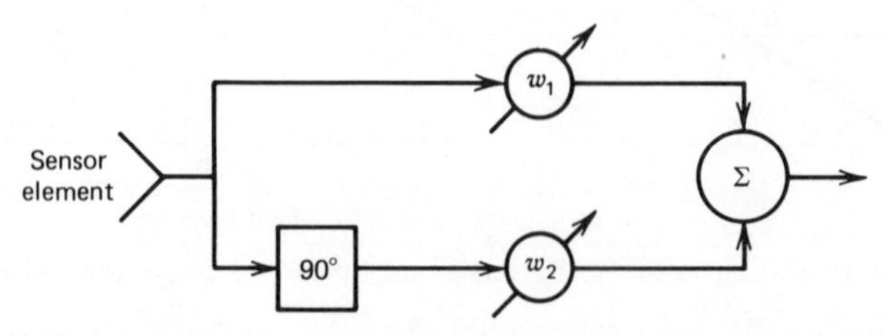


**Figure 2.18** Realization of a complex weight by means of a quadrature hybrid circuit.

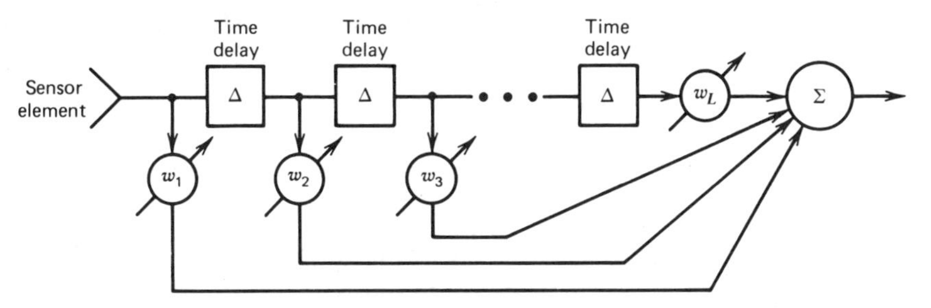


**Figure 2.19** Transversal filter realized with a tapped-delay line having $L$ complex weights.

mates an ideal filter that allows exact control of gain and phase at each frequency within the band of interest. An upper limit on the tap spacing is given by the desired array cancellation bandwidth $B_a$, since $B_f \geqslant B_a$ and $B_f = 1/\Delta$ for uniform tap spacing. The transversal filter not only is useful for providing the desired adjustment of gain and phase over the frequency band of interest for wideband signals, but is also well suited for providing array compensation for the effects of multipath, finite array propagation delay, and interchannel mismatch effects; these additional uses of transversal filters are explored further in Part Three. Transversal filters are also being exploited to enhance the ability of an AMTI (airborne moving target indication) radar system to reject clutter, compensate for platform motion, and compensate for near-field scattering effects and element excitation errors [24].

To model a complete multichannel processor (in a manner analogous to that of Section 1.4 for the narrowband processing case), consider the tapped-delay line multichannel processor depicted in Figure 2.20. The multichannel wideband signal processor consists of $N$ sensor element channels in which each channel contains one tapped-delay line like that of Figure 2.19 consisting of $L$ tap points, $(L-1)$ time delays of $\Delta$ seconds each, and $L$ complex weights. On comparing Figure 2.20 with Figure 1.1, it is seen that $x_1(t), x_2(t), \cdots x_N(t)$ in the former correspond exactly with $x_1(t), x_2(t) \ldots x_N(t)$ in the latter which were defined in (1.6) to form the elements of the vector $\mathbf{x}(t)$. In like manner, therefore, define a complex vector $\mathbf{x}_1'(t)$ such that

$$(\mathbf{x}_1')^T = (\mathbf{x}')^T = [x_1 x_2 \cdots x_N] \tag{2.52}$$

The signals appearing at the second tap point in all channels are merely a time-delayed version of the signals appearing at the first tap point, so define a second complex signal vector $\mathbf{x}_2'(t)$ where

$$\begin{aligned}[\mathbf{x}_2'(t)]^T &= [\mathbf{x}'(t-\Delta)]^T \\ &= [x_1(t-\Delta) x_2(t-\Delta) \cdots x_N(t-\Delta)]\end{aligned} \tag{2.53}$$

Continuing in the above manner for all $L$ tap points, a complete signal vector

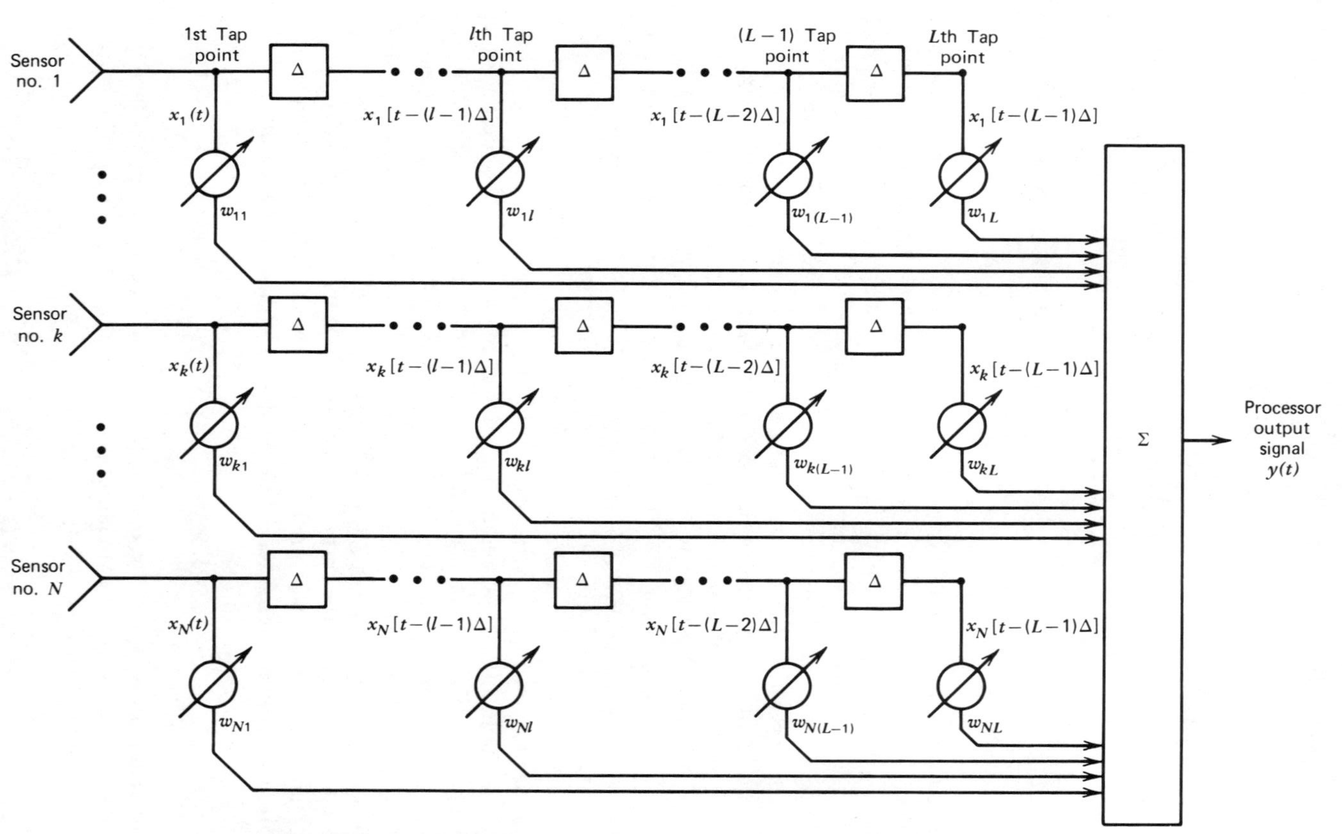


**Figure 2.20** Tapped-delay line multichannel processor for wideband signal processing.

for the entire multichannel processor can be defined as

$$\mathbf{x}(t) \triangleq \begin{bmatrix} \mathbf{x}'_1(t) \\ \mathbf{x}'_2(t) \\ \vdots \\ \mathbf{x}'_L(t) \end{bmatrix} = \begin{bmatrix} \mathbf{x}'(t) \\ \mathbf{x}'(t-\Delta) \\ \vdots \\ \mathbf{x}'[t-(L-1)\Delta] \end{bmatrix} \tag{2.54}$$

It is seen that the signal vector $\mathbf{x}(t)$ contains $L$ component vectors each having dimension $N$.

In an exactly similar manner define the weight vector

$$(\mathbf{w}'_1)^T \triangleq [w_{11} w_{21} \cdots w_{N1}] \tag{2.55}$$

The weight vector for the entire multichannel processor is then given by

$$\mathbf{w} \triangleq \begin{bmatrix} \mathbf{w}'_1 \\ \mathbf{w}'_2 \\ \vdots \\ \mathbf{w}'_L \end{bmatrix} \tag{2.56}$$

It is seen that the weight vector $\mathbf{w}$ so defined is a $NL$-dimensional vector containing $L$ component vectors, each having dimension $N$.

As a consequence of the signal and weight vector definitions introduced above, the output of the multichannel tapped-delay line processor can be written in the form

$$\begin{aligned} y(t) &= \sum_{l=1}^{L} (\mathbf{w}'_l)^T \mathbf{x}'_l(t) = \sum_{l=1}^{L} (\mathbf{w}'_l)^T \mathbf{x}'[t-(l-1)\Delta] \\ &= \mathbf{w}^T \mathbf{x}(t) \end{aligned} \tag{2.57}$$

which is of exactly the same form as (1.4). Yet another reason for expressing the signal vector in the form of (2.54) is that this construction leads to a Toeplitz form for the correlation matrix of the input signals, as shown in Chapter 3.

The array processors of Figures 2.19 and 2.20 are examples of classical time domain processors that involve the use of complex weights and delay lines. More recently fast Fourier transform (FFT) techniques have been exploited to replace a conventional time domain processor by an equivalent frequency domain processor using the frequency domain equivalent of time delay. The advantage of such an approach lies in alleviating the hardware problem. The use of phase shifters or delay lines to form a directional beam in the time domain can quickly become cumbersome from a hardware standpoint as the number of delay elements increase, whereas using the frequency domain equivalent of time delay permits beamforming to be accomplished with a digital computer, thereby simplifying the hardware. The notion of the frequency domain equivalent of time delay is introduced in the problems section, and more detailed presentations of the techniques of digital beamforming may be found in references [25]–[27].

The time domain processor of Figure 2.20 is depicted operating on the outputs of the individual sensor elements within an array; a processor that operates on the sensor element outputs is referred to as an "element space processor" [29]. It follows that there are both time domain and frequency domain element space processors. Frequency domain processors provide for adaptation in the space dimension by resolving and processing signals in contiguous frequency bins, and enjoy a performance advantage in terms of cancelling continuous wave (cw) interference signals. Time domain processors, on the other hand, provide adaptivity in both the space and time (frequency) dimensions and are therefore quite useful for cancelling angularly spread interference. Consequently frequency domain processors are most suitable for narrowband signal detection problems while time domain processors are most suitable for broadband signal detection applications; it should be noted, however, that the actual selection of a processor type for a specific application must also involve questions of sensitivity to perturbations as well as suitability for the specified signal environment [28].

Rather than apply the output of each individual sensor element of an array to a corresponding adaptive processor channel, the number of processors required can be reduced by first combining the outputs of sensor groups within the array to form beams, and then weighting the output of each beam signal channel by an adaptive processor. When designed to process a beam channel output signal rather than a sensor element output, an adaptive processor is referred to as a "beam space processor" [29]. For partially adaptive array applications (a subject considered in Chapter 12) beam space processors yield performance characteristics that are superior to element space processors.

## 2.6 SUMMARY AND CONCLUSIONS

The signal environment in which an adaptive array system is expected to operate is introduced by discussing the different types of interference and modeling of the desired signal as either a random or nonrandom process. Signal propagation effects are also important because of the distortion introduced into the desired signal, and the rate of change of variations in the ambient noise environment directly affects the ability of an adaptive system to cope with a nonstationary signal environment by adjusting the pattern-forming network elements.

Sensor element spacing within an array directly affects its ability to respond appropriately to any given signal environment and yield enhanced reception performance. The potential of an array to offer enhanced reception performance is demonstrated with a simple example. The presence of undesirable pattern grating lobes may be avoided either by using a "filled array" for systems having a small number of elements, or by using a "thinned array" by exploiting the theory of randomly spaced elements for large arrays to achieve an efficient design. In addition to element spacing, interference signal nulling limitations may also derive from:

1. Jammer signal bandwidth.
2. Signal propagation delay across the array aperture.
3. Interchannel mismatch effects.

Narrowband and broadband signal processing requirements differ in that narrowband processing can be achieved by means of a single constant complex weight in each element channel, whereas a broadband signal requires frequency-dependent weighting that leads to the introduction of a transversal filter (tapped-delay line). Time domain processors can be replaced by equivalent frequency domain processors whose functions are realized by a digital computer thereby alleviating the need for cumbersome hardware.

The best possible steady-state performance that can be achieved by an adaptive array system can be computed theoretically, without explicit consideration of the array factors affecting such performance. The theoretical performance limits for steady-state array operation are addressed in Chapter 3.

## PROBLEMS

**1** For the circular array of Figure 2.21 having $N$ equally spaced elements, select as a phase reference the center of the circle. For a signal whose direction of arrival is $\theta$ with respect to the selected angle reference, define the angle $\phi_k$ for which $\phi_k = \theta - \psi_k$.

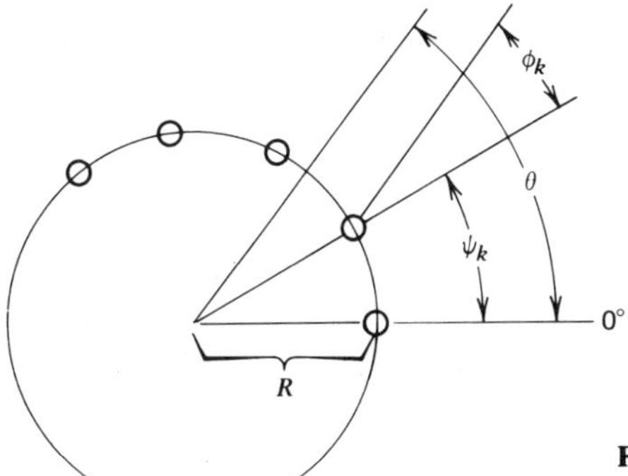


**Figure 2.21** Azimuth plane of circular array having $N$ equally spaced elements.

(a) Show that $\psi_k=(2\pi/N)(k-1)$ for $k=1,2,\dots,N$.

(b) Show that the path length phase difference for any element with respect to the center of the circle is given by $u_k=\pi(R/\lambda)\cos\phi_k$.

(c) Show that an interelement spacing of $\lambda/2$ is maintained by choosing the radius of the array circle to be

$$R=\frac{\lambda}{4\sin(\pi/N)}$$

**2** Consider the linear array geometry of Figure 2.22 in which $\mathbf{v}$ is the vector from the array center to the signal source, $\theta$ is the angle between $\mathbf{v}$ and the array normal, and $\mathfrak{v}$ is the propagation velocity.

(a) Show that for a spherical signal wavefront the time delay experienced by an element located at position $x$ in the array with respect to the array center is given by

$$\tau(x)=\frac{\|\mathbf{v}\|}{\mathfrak{v}}\left\{1-\sqrt{1+\left(\frac{x^2}{\|\mathbf{v}\|^2}\right)-2\left(\frac{x}{\|\mathbf{v}\|}\right)\sin\theta}\right\}$$

($\tau$ is positive for time advance and negative for time delay) where $\|\mathbf{v}\|$ denotes the length of the vector $\mathbf{v}$.

(b) Show that for a spherical signal wavefront the attenuation factor experienced by an element located at position $x$ in the array with respect to the array center is given by

$$\rho(x)=\frac{\|\mathbf{v}\|}{\|\mathbf{l}\|}=\frac{1}{\sqrt{1+\left(\frac{x^2}{\|\mathbf{v}\|^2}\right)-2\left(\frac{x}{\|\mathbf{v}\|}\right)\sin\theta}}$$

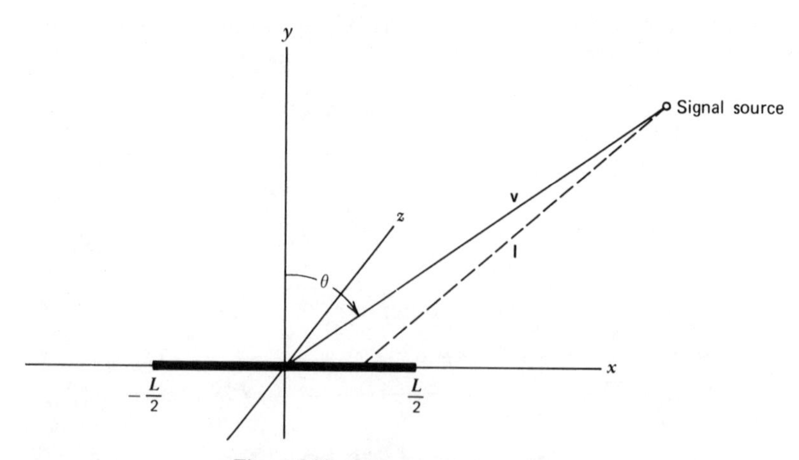


**Figure 2.22** Linear array geometry.

where $\|\mathbf{l}\|$ denotes the distance between the signal source and the array element of concern.

**3** [30] ***Uniformly Spaced Linear Array.*** Show that for a uniformly spaced linear array having $N$ elements, the normalized array factor is given by

$$A(u)=\frac{1}{N}\sum_{i=1}^{N}\cos\left(\frac{iu}{2}\right)=\frac{1}{N}\frac{\sin(Nu/2)}{\sin(u/2)}$$

where $u=2\pi(d/\lambda)\sin\theta$.

**4** [15] ***Representation of Nonuniformly Spaced Arrays***

(a) For nonuniformly spaced elements in a linear array, a convenient "base separation" $d$ can be selected and the element position then represented by

$$d_i=\left(\frac{i}{2}+\epsilon_i\right)d$$

where $\epsilon_i$ is the fractional change from the uniform spacing represented by $d$. The normalized field pattern for a nonuniformly spaced array is then given by

$$A=\frac{1}{N}\sum_{i=1}^{N}\cos u_i=\frac{1}{N}\sum_{i=1}^{N}\cos\left[\left(\frac{i}{2}+\epsilon_i\right)u\right]$$

where $u=2\pi(d/\lambda)\sin\theta$. Show that $A$ above can be put in the form:

$$A=A_u-\frac{1}{N}\sum_i\left[\sin\epsilon_i u\sin i\frac{u}{2}+(1-\cos\epsilon_i u)\cos i\frac{u}{2}\right]$$

where $A_u$ is the pattern of a uniform array having element separation equal to the base separation and is given by the result obtained in Problem 3.

(b) Assuming all $\epsilon_i u$ are small, show that the result obtained in part (a) above reduces to

$$A=A_u-\frac{u}{N}\sum_i\epsilon_i\sin i\left(\frac{u}{2}\right)$$

(c) The result obtained in part (b) can be rearranged to read

$$\sum_i\epsilon_i\sin\left(i\frac{u}{2}\right)=\frac{N}{u}(A_u-A)$$

which may be regarded as a Fourier series representation of the quantity on the right-hand side of the above equation. Consequently the $\epsilon_i$ are given by the formula for Fourier coefficients:

$$\epsilon_i = \frac{2N}{\pi}\int_0^{\pi}\frac{1}{u}(A_u - A)\sin\left(i\frac{u}{2}\right)du$$

Let

$$\frac{A_u - A}{u} = \frac{1}{u}\sum_{k=1}^{L} a_k\,\delta(u - u_k)$$

and show that

$$\epsilon_i = 2\frac{N}{\pi}\sum_{k=1}^{L} a_k \frac{\sin[(i/2)u_k]}{u_k}$$

(d) For uniform arrays the pattern sidelobes have maxima at positions approximately given by

$$u_k = \frac{\pi}{N}(2k+1)$$

Consequently, $u_k$ gives the positions of the necessary impulse functions in the representation of $(A_u - A)/u$ given in part (c). Furthermore since the sidelobes adjacent to the main lobe drop off as $1/u$, $(A_u - A)/u$ is now given by

$$\frac{A_u - A}{u} = A\frac{1}{u^2}\sum_{k=1}^{L}(-1)^k\,\delta\left[u - \frac{\pi}{N}(2k+1)\right]$$

For this representation of $(A_u - A)/u$ show that

$$\epsilon_i = 2A\left(\frac{N}{\pi}\right)^3\sum_{k=1}^{L}(-1)^k\frac{\sin(i\pi/2N)(2k+1)}{(2k+1)^2}$$

5 [14] ***Nonuniformly Spaced Arrays and the Equivalent Uniformly Spaced Array.*** One method of gaining insight into the gross behavior of a nonuniformly spaced array and making it amenable to a linear analysis is provided by the correspondence between a nonuniformly spaced array and its equivalent uniformly spaced array (EUA). The EUA provides a best mean square representation for the original nonuniformly spaced array.

(a) Define the uniformly spaced array factor $A(\theta)$ as the normalized phasor sum of the responses of each of the sensor elements to a narrowband signal in the uniform array of Figure 2.23, where the normalization is taken with respect to the response of the center

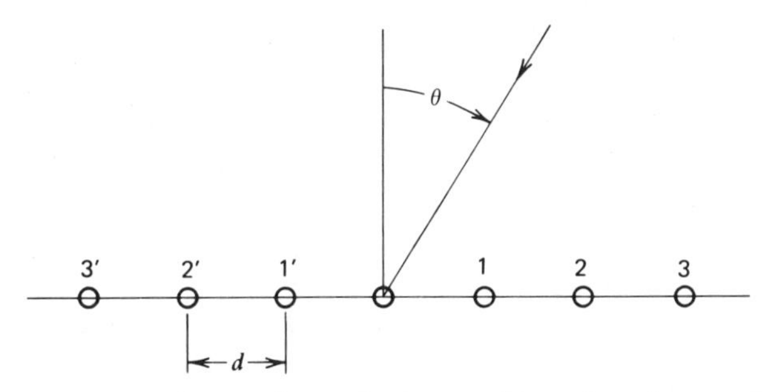


Figure 2.23 Uniformly spaced array having an odd number of sensor elements.

element. Show that

$$A(\theta)=1+2\sum_{i=1}^{(N-1)/2}\cos[2\pi i l\sin\theta]$$

where $l=d/\lambda$. Furthermore show that when $d/\lambda=2$ the array factor has its second principal maxima occurring at $\theta=\pi/6$.

(b) Show that the nonuniformly spaced array factor for the array of Figure 2.24 is given by

$$A(\theta)=1+2\sum_{i=1}^{(N-1)/2}\cos[2\pi l_i\sin\theta]$$

where $l_i=d_i/\lambda$.

(c) The result obtained in part (b) may be regarded as $A(\theta)=1+2\Sigma_i\cos\omega_i$ where

$$\omega_i=\left(\frac{2}{R}l_i\right)\underbrace{(\pi R\sin\theta)}_{\omega_1},\qquad R=\text{arbitrary scale factor}$$

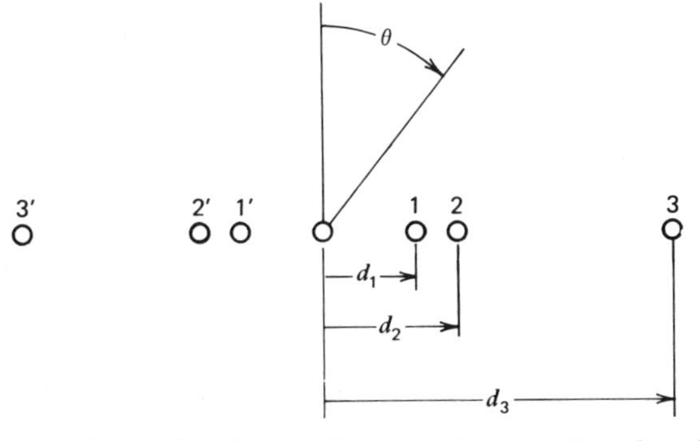


Figure 2.24 Symmetrical nonuniformly spaced array having an odd number of sensor elements.

Regarding $\omega_1 = \pi R \sin\theta$ as the lowest nonzero superficial array frequency in the EUA, the scale factor $R$ determines the element spacing in the EUA. For example, $R=1$ corresponds to half-wavelength spacing, and $R=\frac{1}{2}$ corresponds to quarter wavelength spacing. Each term in $A(\theta)$ above may be expanded (in a Fourier series representation for each higher harmonic) into an infinite number of uniformly spaced equivalent elements. Since the amplitude of the equivalent elements varies considerably across the array, only a few terms of the expansion need be considered for a reasonable equivalent representation in terms of EUA's. The $m$th term in $A(\theta)$ is given by

$$\cos\omega_m = \cos\frac{2}{R} l_m \omega_1 = \cos\mu_m \omega_1$$

where $\mu_m = (2l_m / R)$

The $m$th term may consequently be expanded in an integral harmonic cosine series as follows:

$$\cos\mu_m\omega_1 = \sum_{\nu=0}^{\infty} a_{m\nu}\cos(\nu\omega_1)$$

where

$$a_{m\nu} = \frac{2}{\pi}\int_0^{\pi R} \cos\mu_m\omega_1 \cos\nu\omega_1 \, d\omega_1$$

Show that when $R_1 = 1$, then

$$a_{m\nu} = \frac{2}{\pi}\frac{\mu_m(-1)^\nu \sin\mu_m\pi}{\mu_m^2 - \nu^2} \qquad \text{for } \mu_m \neq 0, \pm 1, \pm 2, \cdots$$

(d) Applying the harmonic cosine series expansion given in part (c) to each component $\cos\omega_i$, show that the resulting amplitude coefficients may be added term by term to give the EUA representation:

$$A(\theta) = 1 + 2\sum_{i=1}^{P}\left\{\sum_{\nu=0}^{\infty} a_{m\nu}\cos(\nu\omega_1)\right\}$$

$$= 1 + 2\sum_{r=0}^{P} A_r \cos r\omega_1$$

where

$$A_0 = a_{10} + a_{20} + a_{30} + \cdots$$

$$A_1 = a_{11} + a_{21} + a_{31} + \cdots$$

Note that the representation given by the above result is not unique. Many choices are possible as a function of the scale factor $R$. From the least mean square error property of the Fourier expansion, each representation will be the best possible for a given value of $\omega_1$.

**6** ***Array Propagation Delay Effects.*** Array propagation delay effects can be investigated by considering the simple two-element array model of Figure 2.25. It is desired to adjust the value of $w_1$ so the output signal power $P_0$ is minimized for a directional interference signal source.

(a) Since $x_2(\omega)=e^{-j\omega\tau}x_1(\omega)$ in the frequency domain, the task of $w_1$ may be regarded as one of providing the best estimate (in the mean square error sense) of $x_1(\omega)e^{-j\omega\tau}$, and the error in this estimate is given by $\epsilon(\omega)=x_1(\omega)(e^{-j\omega\tau}-w_1)$. For the weight $w_1$ to be optimal, the error in the estimate must be orthogonal to the signal $x_1(\omega)$ so it is necessary that

$$E\left\{|x_1(\omega)|^2\left[e^{-j\omega\tau}-w_1\right]\right\}=0$$

where the expectation $E\{\cdot\}$ is taken over frequency. If $|x_1(\omega)|^2$ is a constant independent of frequency, it follows from the above orthogonality condition that $w_1$ must satisfy

$$E\left\{e^{-j\omega\tau}-w_1\right\}=0$$

If the signal has a rectangular spectrum over the bandwidth $-\pi B \leqslant \omega \leqslant \pi B$, show that the above equation results in

$$w_1=\frac{\sin(\pi B\tau)}{\pi B\tau}$$

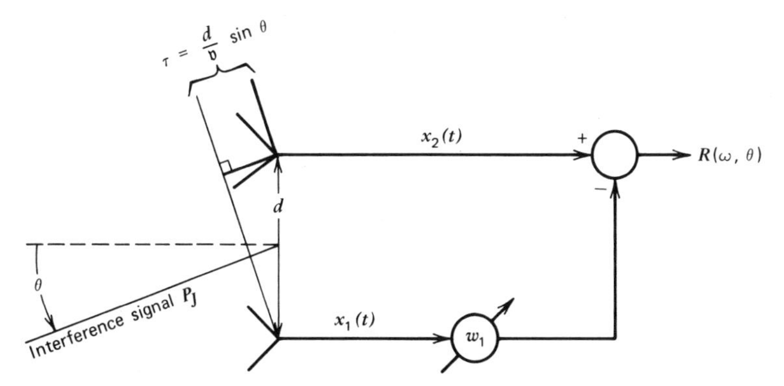


**Figure 2.25** Two-element linear array having a single complex weight with interelement propagation delay, $\tau$.

(b) The output power may be expressed as

$$P_0 = \int_{-\pi B}^{\pi B} \phi_{RR}(\omega, \theta)\, d\omega$$

where $\phi_{RR}(\omega,\theta)$ is the output power spectral density. The output power spectral density is in turn given by

$$\phi_{RR}(\omega,\theta) = |e^{-j\omega\tau} - w_1|^2 \phi_{ss}(\omega)$$

where $\phi_{ss}(\omega)$ is the power spectral density of the directional interference signal. Assuming that $\phi_{ss}(\omega)$ is unity over the signal bandwidth (so that $P_J = 2\pi B$), then from part (a) it follows that

$$P_0 = \int_{-\pi B}^{\pi B} \left| e^{-j\omega\tau} - \frac{\sin(\pi B\tau)}{\pi B\tau} \right|^2 d\omega$$

Show that $P_0$ above can be expressed as

$$\frac{P_0}{P_J} = 1 - \left[ \frac{\sin(\pi B\tau)}{\pi B\tau} \right]^2$$

where $P_J$ is the directional interference signal power.

7 [31] ***Open-Loop Scanning of Adaptively Formed Beams.*** To scan an adaptively formed beam about the direction of a received signal requires open-loop scanning. Open-loop scanning, however, requires knowledge of the array element locations. An airborne thinned array may contain elements distributed over the entire air frame. Although element coordinates may be measured accurately on the ground, an airframe is flexible, and consequently the relative element positions may vary considerably in flight, with rms deviations of $10^{-1}$ ft easily occurring.

For open-loop scanning, four theoretical results have been obtained as follows [31]:

1. The rms array element coordinate tolerance is

$$\sigma_x = \sigma_y = \sigma_z = \frac{\lambda}{4\pi\theta_s}$$

where $\lambda$ is the radiated wavelength and $\theta_s$ is the maximum scan angle from the initial pointing direction of the array (i.e., $\theta_s$ is half the field of view).

2. The rms tolerance in the estimate of initial pointing direction is

$$\sigma_\theta \cong \frac{\Delta\theta}{2\theta_s}$$

where $\Delta\theta$ is the beamwidth $\cong \lambda/D$ and where $D$ is the linear dimension of the array.

3. The distance beyond which far-field scanning may be employed is given by

$$R_m \approx \frac{\theta_s D^2}{2\lambda}$$

where $D$ is the linear dimension of the array.

4. The rms tolerance in the range estimate of the target is

$$\sigma_{R_0} \cong \frac{R_0 \Delta\theta}{2\theta_s}$$

where $R_0 < R_m$ is the target range.

(a) Assuming $\lambda = \sigma_x = \sigma_y = \sigma_z = 10^{-1}$ ft, what field of view can successfully be scanned with open-loop techniques?

(b) What is the minimum range at which far-field scanning may be used?

(c) For the field of view found in part (a), determine the rms tolerance in the initial pointing error.

**8** [27] ***Frequency Domain Equivalent of Time Delay.*** The conventional time domain processor for beamforming involving sum and delay techniques can be replaced by an equivalent frequency domain processor using the fast Fourier transform (FFT).

The wavefront of a plane wave arriving at a line of sensors having a spacing $d$ between elements will be delayed in time by an amount $\tau_1$ between adjacent sensors given by

$$\tau_1 = \left(\frac{d}{\mathfrak{v}}\right)\sin\theta_1$$

where $\theta_1$ is the angle between the direction of arrival and the array normal. Consequently for a separation of $n$ elements,

$$\tau_{n1} = n\tau_1$$

In classical time-domain array processing, a spatial beam is formed at an angle $\theta_1$ by introducing appropriate delay lines at the output of each sensor. Let $x_n(t)$ denote the input signal from the plane wave at the $n$th sensor, and let the time delay present in the $n$th sensor channel be $\tau_n$. Since the array output is formed by coherent addition of each of the sensor outputs, the output $y(t)$ is given by

$$y(t) = \sum_{n=0}^{N-1} x_n(t - \tau_n)$$

for an array having $N$ sensors

(a) Show the following time-frequency domain relationships based on the Fourier transformation:

| Time Domain | Frequency Domain |
|---|---|
| $x_n(t)$ | $X_n(\omega)$ |
| $x_n(t-\tau_n)$ | $X_n(\omega)e^{-j\phi_n}$ |
| $\sum_{n=0}^{N-1} x_n(t-\tau_n)$ | $\sum_{n=0}^{N-1} X_n(\omega)e^{-j\phi_n}$ |

Note that $\phi_n$ represents the phase shift equivalent to the time delay $\tau_n$.

(b) The plane wave arriving at the array with arrival angle $\theta_1$ can be represented by

$$x(t,z) = \cos\left[\omega\left(t - \frac{z}{\upsilon}\right)\right] = \cos[\omega t - kz]$$

where $k$ is the wavenumber of the plane wave and $z$ is the axis of propagation at an angle $\theta_1$ with respect to the array normal. Show that the phase shift $\phi_n$ associated with a time delay $\tau_n$ is given by

$$\phi_n = \omega\tau_n$$

where

$$\tau_n = n\left(\frac{d}{\upsilon}\right)\sin\theta_1$$

(c) It is convenient to index $\theta$ and $\omega$ to account for different combinations of the angle $\theta_l$ and frequency $\omega$ so that

$$\theta = \theta_l = l\,\Delta\theta, \qquad l = 0, 1, \ldots, N-1$$

$$\omega = \omega_m = m\,\Delta\omega, \qquad m = 0, 1, \ldots, M-1$$

Furthermore, index the input time samples in $i$ so that

$$t = t_i = i\,\Delta t, \qquad i = 0, 1, \ldots, M-1$$

Let $X_n(\omega)$ be represented at discrete frequencies (a discrete Fourier transform) by

$$X_n(m\,\Delta\omega) = X_n(\omega_m) = X_{mn}$$

Likewise the sampled time waveform from the $n$th array element can be represented by

$$x_n(i\,\Delta t) = x_n(t_i) = x_{in}$$

The time-frequency domain equivalence between $x_{in}$ and $X_{mn}$ is then given by:

| Time Domain | | Frequency Domain |
|---|---|---|
| $x_{in} = x_n(i\,\Delta t)$ | $\leftrightarrow$ | $X_n(m\,\Delta\omega) = X_{mn}$ |

Consequently the summed expression from part (a) can be written as

$$\sum_{n=0}^{N-1} X_n(\omega) e^{-j\phi_n} = \sum_{n=0}^{N-1} X_n(m\,\Delta\omega) e^{-j\phi_n}$$

Show that $\phi_{nl} = nkd \sin\theta_l$. Consequently show that the above summed expressions can be rewritten in terms of both frequency and angle as

$$X_m(l\,\Delta\theta) = \sum_{n=1}^{N-1} x_n(m\,\Delta\omega) e^{-j\phi_{nl}}$$

(d) The phase shift $\phi_{nl}$ can be expressed as

$$\phi_{nl} = \frac{nl}{N}$$

implying that $\Delta\theta = 1/Nd$, so that

$$x_{ml} = x_m(l\,\Delta\theta) = \sum_{n=0}^{N-1} x_n(m\,\Delta\omega) e^{-jnl/N}$$

which is of the form of a discrete Fourier transform (DFT).

Show that $X_{mn} = X_n(m\,\Delta\omega)$ can also be expressed as a DFT,

$$X_{mn} = \sum_{i=0}^{M-1} x_n(i\,\Delta t) e^{-jim/N}$$

Now expressing $x_n(i\,\Delta t)$ as

$$x_n(i\,\Delta t) = x_{ni}$$

a double DFT can now be written as:

$$X_{ml} = \sum_{n=0}^{N-1} \sum_{i=0}^{M-1} x_{ni} e^{-jim/N} e^{-jnl/N}$$

The array output has therefore been obtained in the form of spectral sample versus beam number, which results after a two-dimensional DFT transformation of array input. The required DFT's can then be implemented using the FFT algorithm.

## REFERENCES


[1] P. W. Howells, "Explorations in Fixed and Adaptive Resolution at GE and SURC," *IEEE Trans. Antennas Propag.*, Vol. AP-24, No. 5, September 1976, pp. 575–584.

[2] R. L. Deavenport, "The Influence of the Ocean Environment on Sonar System Design," IEEE 1975 EASCON Record, Electronics and Aerospace Systems Convention, September 29–October 1, pp. 66.A–66.E.

[3] A. A. Winder, "Underwater Sound—A Review: II. Sonar System Technology," *IEEE Trans. Sonics and Ultrason.*, Vol. SU-22, No. 5, September 1975, pp. 291–332.

[4] L. W. Nolte and W. S. Hodgkiss, "Directivity or Adaptivity?," IEEE 1975 EASCON Record, Electronics and Aerospace Systems Convention, September 29–October 1, pp. 35.A–35.H.

[5] G. M. Wenz, "Review of Underwater Acoustics Research: Noise," *J. Acoust. Soc. Am.*, Vol. 51, No. 3 (Part 2), March 1972, pp. 1010–1024.

[6] B. W. Lindgren, *Statistical Theory*, MacMillan, New York, 1962, Ch. 4.

[7] F. Bryn, "Optimum Structures of Sonar Systems Employing Spatially Distributed Receiving Elements," in NATO Advanced Study Institute on Signal Processing with Emphasis on Underwater Acoustics, Vol. 2, Paper No. 30, Enschede, the Netherlands, August 1968.

[8] H. Cox, "Optimum Arrays and the Schwartz Inequality," *J. Acoust. Soc. Am.*, Vol. 45, No. 1, January 1969, pp. 228–232.

[9] R. S. Elliott, "The Theory of Antenna Arrays," in *Microwave Scanning Antennas, Vol. 2, Array Theory and Practice*, edited by R. C. Hansen, Academic Press, New York and London, 1966; Ch. 1.

[10] M. T. Ma, *Theory and Application of Antenna Arrays*, Wiley, New York, 1974, Ch. 3.

[11] B. D. Steinberg, *Principles of Aperture and Array System Design*, Wiley, New York, 1976.

[12] C. A. Lewis, "On the Grating Lobes of Arrays of Large Subapertures," Paper C3-4, 1978 National Radio Science Meeting, International Union of Radio Science, January, Boulder, CO.

[13] F. Hodjat and S. A. Hovanessian, "Nonuniformly Spaced Linear and Planar Array Antennas for Sidelobe Reduction," *IEEE Trans. Antennas Propag.*, Vol. AP-26, No. 2, March 1978, pp. 198–204.

[14] S. S. Sandler, "Some Equivalences Between Equally and Unequally Spaced Arrays," *IRE Trans. Antennas and Progag.*, Vol. AP-8, No. 5, September 1960, pp. 496–500.

[15] R. F. Harrington, "Sidelobe Reduction by Nonuniform Element Spacing," *IRE Trans. Antennas Propag.*, Vol. AP-9, No. 2, March 1961, pp. 187–192.

[16] D. D. King, R. F. Packard, and R. K. Thomas, "Unequally Spaced, Broadband Antenna Arrays," *IRE Trans. Antennas Propag.*, Vol. AP-8, No. 4, July 1960, pp. 380–385.

[17] M. I. Skolnik, "Nonuniform Arrays," in *Antenna Theory, Part I*, edited by R. E. Collin and F. J. Zucker, McGraw-Hill, New York, 1969, Ch. 6.

[18] Y. T. Lo, "A Mathematical Theory of Antenna Arrays with Randomly Spaced Elements," *IEEE Trans. Antennas Propag.*, Vol. AP-12, May 1964, pp. 257–268.

[19] Y. T. Lo and R. J. Simcoe, "An Experiment on Antenna Arrays with Randomly Spaced Elements," *IEEE Trans. Antennas Propag.*, Vol. AP-15, March 1967, pp. 231–235.

[20] J. L. Kmetzo, "An Analytical Approach to the Coverage of a Hemisphere by *N* Planar Phased Arrays," *IEEE Trans. Antennas Propag.*, Vol. AP-15, May 1967, pp. 367–371.

[21] D. K. Cheng, "Optimization Techniques for Antenna Arrays," *Proc. IEEE*, Vol. 59, No. 12, December 1971, pp. 1664–1674.

[22] B. Widrow and M. E. Hoff, Jr., "Adaptive Switching Circuits," IRE 1960 WESCON Convention Record, Part 4, pp. 96–104.

[23] J. H. Chang and F. B. Tuteur, "A New Class of Adaptive Array Processors," *J. Acoust. Soc. Am.*, Vol. 49, No. 3, March 1971, pp. 639–649.

[24] L. E. Brennan, J. D. Mallet, and I. S. Reed, "Adaptive Arrays in Airborne MTI Radar," *IEEE Trans. Antennas Propag.*, Special Issue Adaptive Antennas, Vol. AP-24, No. 5, September 1976, pp. 607–615.

[25] P. Rudnick, "Digital Beamforming in the Frequency Domain," *J. Acoust. Soc. Am.*, Vol. 46, No. 5, 1969.

[26] D. T. Deihl, "Digital Bandpass Beamforming with the Discrete Fourier Transform," Naval Research Laboratory Report 7359, AD-737-191, 1972.

[27] L. Armijo, K. W. Daniel, and W. M. Labuda, "Applications of the FFT to Antenna Array Beamforming," EASCON 1974 Record, IEEE Electronics and Aerospace Systems Convention, October 7–9, Washington, DC, pp. 381–383.

[28] A. M. Vural, "A Comparative Performance Study of Adaptive Array Processors," Proceedings of the 1977 International Conference on Acoustics, Speech, and Signal Processing, May, pp. 695–700.

[29] _____ "Effects of Perturbations on the Performance of Optimum/Adaptive Arrays," *IEEE Trans. Aerosp. Electron. Syst.*, Vol. AEC-15, No. 1, January 1979, pp. 76–87.

[30] R. J. Urick, *Principles of Underwater Sound*, McGraw-Hill, New York, 1975, p. 52.

[31] S. H. Taheri and B. D. Steinberg, "Tolerances in Self-Cohering Antenna Arrays of Arbitrary Geometry," *IEEE Trans. Antennas Propag.*, Vol. AP-24, No. 5, September 1976, pp. 733–739.

# Chapter 3. Optimum Array Processing: Steady-State Performance Limits and the Wiener Solution

The problem of optimum space-time processing of array signal data has received a great deal of attention during the last two decades and there is consequently an abundance of literature in this area [1]–[12]. Optimum array processing may be regarded as basically an optimum multichannel filtering problem. The objective of array processing is to enhance the reception (or detection) of a desired signal that may be either random or deterministic in a signal environment containing numerous interference signal sources. The desired signal itself may also contain one or several uncertain parameters (such as spatial location, signal energy, and phase) that it may be desired to estimate.

Since an array of sensors can be utilized to obtain some degree of spatial filtering or directional sensitivity, much of the early literature on array processing was concerned with desirable beam patterns that could be achieved through appropriate weighting of the individual sensor element outputs. While such an approach seems reasonable, it nevertheless inherently assumes that array beamforming is the best way to facilitate the overall objective of enhanced signal detection.

A more global approach to the signal detection problem makes no assumptions concerning the desirability of beamforming and offers the promise of substantial performance improvement over more conventional approaches. Based upon a selected measure of optimality, the array processor structure can be found to emerge freely from the mathematics of the problem under consideration. In this manner, beamforming turns out to be a part of the optimization problem solution, but this structure is not imposed from the beginning, and the resulting processor will be as nearly optimum as the mathematical models used in obtaining the solution are valid in representing actual operating conditions.

Optimum array processing techniques may be broadly classified as: (1) processing appropriate for ideal propagation conditions and (2) processing appropriate for perturbed propagation conditions. Ideal propagation conditions refer to the situation that exists when propagation takes place in an ideal nonrandom, nondispersive medium where the desired signal is a plane (or spherical) wave and the receiving sensors are distortionless. In this case the optimum processor is said to be matched to a plane wave signal. Any performance degradation resulting from deviation of the actual operating conditions from the assumed ideal conditions is minimized by the use of complementary methods (such as the introduction of constraints). When operating under the above ideal conditions, vector weighting of the input data succeeds in accomplishing the desired signal matching.

When perturbations in either the propagating medium or the receiving mechanism occur, the plane wavefront signal assumption no longer holds, and vector weighting of the input data will not produce the desired signal matching. Likewise if it is desired to match the array processor to a signal of arbitrary characteristics, matched array techniques [17] involving matrix weighting of the input data become appropriate. The principal advantage of such an element space-matched array processor is realized when it operates on noncoherent wavefront signals where matching can only be performed in a statistical sense.

The determination of steady-state performance limits is important because such limits provide a measure of how well any selected design can ever be expected to perform. Quite naturally the problem of optimum array processing that assumes ideal propagation conditions has received by far the greatest share of attention, and various approaches to this problem have been proposed for both narrowband and broadband signal applications. By far the most popular and widely reported approaches involve the adoption of an array performance measure that is to be optimized by appropriate selection of an optimum weight vector. Such performance measure optimization approaches have been widely used for radar, communication, and sonar systems, and therefore are discussed at some length. Recently, approaches to array processing involving the use of the maximum entropy technique (which is particularly applicable to the passive sonar problem) [13], [14] and eigenvalue resolution techniques [15], [16] have also been reported.

To determine the optimal weighting for an array processor and its associated performance limits, some mathematical preliminaries are first discussed, and the signal descriptions for conventional and signal-aligned arrays are introduced. It is well known that when all elements in an array are uniformly weighted, then the maximum SNR is obtained if the noise contributions from the various element channels have equal power and are uncorrelated [18]. When there is any directional interference, however, the noise from the various element channels will be correlated. Consequently the problem of selecting an optimum set of weights may be regarded as a problem of attempting to cancel out correlated noise components. Signal environment descriptions in terms of correlation matrices therefore play a fundamental role in determining the optimum solution for the complex weight vector.

The problem of formulating some popular array performance measures in terms of complex envelope signal characterizations is discussed. It is a remarkable fact that the different performance measures considered here all converge (to within a constant scalar factor) toward the same steady-state solution— the optimum Wiener solution. Passive detection systems face the problem of designing an array processor for optimum detection performance. Classical statistical detection theory yields an array processor based on a likelihood ratio test that leads to a canonical structure for the array processor. This canonical structure contains a weighting network that is closely related to the steady-state weighting solutions found for the selected array performance measures. It therefore turns out that what at first appear to be quite different optimization problems actually possess an underlying unity that is revealed by introducing the more general array processing concepts discussed here.

## 3.1 MATHEMATICAL PRELIMINARIES

In deriving the optimal weight settings it is convenient to represent the signal envelopes as well as the adaptive weights in their complex envelope form. The meaning of such complex representations therefore is briefly reviewed.

In any system only real signals actually appear, and the relationship between these real signals and their corresponding complex envelope representations is

$$\text{actual signal} = \text{Re}\{(\text{complex envelope representation})e^{j\omega_0 t}\} \tag{3.1}$$

where Re{ } denotes "real part of." This result is further discussed in Appendix B. The meaning of (1.4) when complex representations are used is that (boldface lower case symbols denote vectors while boldface upper case symbols denote matrices)

$$\text{actual } y(t) = \text{Re}\{\mathbf{x}^T\mathbf{w}^* e^{j\omega_0 t}\} = \text{Re}\{\mathbf{w}^\dagger \mathbf{x} e^{j\omega_0 t}\} \tag{3.2}$$

where $\mathbf{w}^\dagger\mathbf{x}$ is the complex envelope representation of $y(t)$, * denotes complex conjugate, and † denotes complex conjugate transpose $[(\ )^*]^T$. A closely related complex representation is the analytic signal $\psi(t)$ for which $y(t) \triangleq \text{Re}\{\psi(t)\}$ and

$$\psi(t) \triangleq y(t) + j\check{y}(t) \tag{3.3}$$

where $\check{y}(t)$ denotes the Hilbert transform of $y(t)$. It follows that for a complex weight representation $w = w_1 + jw_2$ having an input signal with analytic representation $x_1+jx_2$, the resulting actual output signal is given by

$$\text{actual output} = \text{Re}\{(w_1-jw_2)(x_1(t)+jx_2(t))\} = w_1x_1(t)+w_2x_2(t) \quad (3.4)$$

where $x_2(t)=\check{x}_1(t)$. Note that (complex envelope) $e^{j\omega_0 t}=\psi(t)$. As a consequence of the above meaning of complex signal and weight representations, two alternate approaches to the optimization problems for the optimal weight solutions may be taken as follows:

1. Reformulate the problem involving complex quantities in terms of completely real quantities, so that familiar mathematical concepts and operations can be conveniently carried out.
2. Revise the definitions of certain concepts and operations (covariance matrices, gradients, etc.) so that all complex quantities are handled directly in an appropriate manner.

Numerous examples of both approaches can be found in the adaptive array literature. It is therefore appropriate to consider how to use both approaches since there are no compelling advantages favoring one approach over the other.

### *3.1.1 Problem Formulation in Terms of Real Variables*

Let $z_1$ be the first component of a complex vector $\mathbf{z}$ having $n$ components ($\mathbf{z}$ may represent a weight vector, signal vector, etc.). Furthermore let the real part of $z_1$ be denoted by $x_1$ and the imaginary part of $z_1$ be denoted by $x_2$. Continuing in this manner we have

$$z_k = x_{2k-1} + jx_{2k} \quad (3.5)$$

It follows that the $n$-component complex vector $\mathbf{z}$ can be completely represented by the $2n$-component real vector $\mathbf{x}$. By representing all complex quantities in terms of corresponding real vectors, the adaptive array system problem can then be solved using familiar definitions of mathematical concepts and procedures. If we carry out the foregoing procedure for the single complex weight $w_1+jw_2$ and the complex signal $x_1+jx_2$, there results

$$\mathbf{w}^T\mathbf{x} = [w_1 w_2]\begin{bmatrix} x_1 \\ x_2 \end{bmatrix} = w_1x_1 + w_2x_2 \quad (3.6)$$

which is in agreement with the result expressed by (3.4). If the foregoing approach is adopted, then certain off-diagonal elements of the corresponding correlation matrix (which is defined below) will be zero. This fact does not affect the results to be obtained in this chapter and will not be pursued now; the point is discussed more fully in connection with an example presented in Chapter 11.

### *3.1.2 Correlation Matrices for Real Signals*

Once the signal and noise processes in an adaptive array system have been described in terms of statistical properties, the system performance can be conveniently evaluated in terms of its average behavior. The evaluation of average behavior leads directly to an interest in quantities related to the second statistical moment such as autocorrelation and crosscorrelation matrices. The degree of correlation that exists between the components of two random vectors is given by the elements of the correlation matrix between the two vectors [19]. For example, the crosscorrelation matrix between the vectors $\mathbf{x}(t)$ and $\mathbf{y}(t)$ having stationary statistical properties is defined by

$$\mathbf{R}_{xy}(\tau) \triangleq E\{\mathbf{x}(t)\mathbf{y}^T(t-\tau)\} \tag{3.7}$$

where $E\{\ \}$ denotes the expected value and $\tau$ is a running time-delay variable. Likewise the autocorrelation matrix for the vector $\mathbf{x}(t)$ is defined by

$$\mathbf{R}_{xx}(\tau) \triangleq E\{\mathbf{x}(t)\mathbf{x}^T(t-\tau)\} \tag{3.8}$$

If a signal vector $\mathbf{x}(t)$ is composed of uncorrelated desired signal and noise components so that

$$\mathbf{x}(t)=\mathbf{s}(t)+\mathbf{n}(t) \tag{3.9}$$

then

$$\mathbf{R}_{xx}(\tau)=\mathbf{R}_{ss}(\tau)+\mathbf{R}_{nn}(\tau) \tag{3.10}$$

The correlation matrices of principal concern in the analysis of adaptive array system behavior are those for which the time-delay variable $\tau$ is zero. Rather than write the correlation matrix argument explicitly as $\mathbf{R}_{xy}(0)$ and $\mathbf{R}_{xx}(0)$, it will prove more convenient to simply define

$$\mathbf{R}_{xx}=\mathbf{R}_{xx}(0) \tag{3.11}$$

and

$$\mathbf{R}_{xy}=\mathbf{R}_{xy}(0) \tag{3.12}$$

It follows that for an $N$-dimensional vector $\mathbf{x}(t)$, the autocorrelation matrix is simply written as $\mathbf{R}_{xx}$ where

$$\mathbf{R}_{xx} \triangleq E\{\mathbf{x}\mathbf{x}^T\} = \begin{bmatrix} \overline{x_1(t)x_1(t)} & \overline{x_1(t)x_2(t)} & \cdots & \overline{x_1(t)x_N(t)} \\ \overline{x_2(t)x_1(t)} & \overline{x_2(t)x_2(t)} & \cdots & \\ \vdots & & & \\ \overline{x_N(t)x_1(t)} & & \cdots & \overline{x_N(t)x_N(t)} \end{bmatrix} \tag{3.13}$$

where $\overline{x_i(t)x_k(t)}$ denotes $E\{x_i(t)x_k(t)\}$.

From the above definitions, it immediately follows that

$$\mathbf{R}_{xx}^T = \mathbf{R}_{xx} \quad \text{and} \quad \mathbf{R}_{xy}^T = \mathbf{R}_{yx} \tag{3.14}$$

so the autocorrelation matrix is symmetric. In general the autocorrelation matrix $\mathbf{R}_{xx}$ is only positive semidefinite, and the inverse matrix $\mathbf{R}_{xx}^{-1}$ may not exist. Since in practice only estimates of the signal environment correlation matrices are available and such estimates are based on distinct time samples, then a sufficient number of such time samples for a nonzero bandwidth signal process may guarantee positive definiteness so the inverse matrix will exist. The signal vectors $\mathbf{x}(t)$ and $\mathbf{n}(t)$ of (3.9), for example, often contain uncorrelated amplifier self-noise and therefore may be regarded as having positive definite correlation matrices. The desired signal vector $\mathbf{s}(t)$ may, however, contain correlated components, as when a uniform plane wave arrives simultaneously at two or more array sensor elements thereby producing identical, inphase signal components. Consequently the autocorrelation matrix $\mathbf{R}_{ss}$ is only positive semidefinite, and its inverse $\mathbf{R}_{ss}^{-1}$ may not exist.

Now consider the correlation matrix that results for the tapped-delay line processor of Section 2.5 by examining the $(N \times N)$-dimensional autocorrelation matrix given by

$$\mathbf{R}'_{xx}(\tau) \triangleq E\{\mathbf{x}'(t)\mathbf{x}'^T(t-\tau)\} \tag{3.15}$$

where $\mathbf{x}'(t)$ is the (real) signal vector defined in (2.32) for a tapped-delay line multichannel processor. Likewise for the $NL$-dimensional vector of all signals observed at the tap points, the autocorrelation matrix is given by

$$\mathbf{R}_{xx}(\tau) \triangleq E\{\mathbf{x}(t)\mathbf{x}^T(t-\tau)\} \tag{3.16}$$

Substituting (2.34) for $x(t)$ into (3.16) then yields

$$\mathbf{R}_{xx}(\tau) = E\left\{\begin{bmatrix} \mathbf{x}'(t) \\ \mathbf{x}'(t-\Delta) \\ \vdots \\ \mathbf{x}'[t-(L-1)\Delta] \end{bmatrix} \left[\mathbf{x}'^T(t-\tau)\mathbf{x}'^T(t-\tau-\Delta)\cdots\mathbf{x}'^T[t-\tau-(L-1)\Delta]\right]\right\} \tag{3.17}$$

and using (3.15), $R_{xx}(\tau)$ then becomes:

$$\mathbf{R}_{xx}(\tau) = \begin{bmatrix} \mathbf{R}'_{xx}(\tau) & \mathbf{R}'_{xx}(\tau+\Delta) & \cdots & \mathbf{R}'_{xx}[\tau+(L-1)\Delta] \\ \mathbf{R}'^T_{xx}(\tau-\Delta) & \mathbf{R}'_{xx}(\tau) & & \vdots \\ \vdots & & & \vdots \\ \mathbf{R}'^T_{xx}[\tau-(L-1)\Delta] & \cdots & & \mathbf{R}'_{xx}(\tau) \end{bmatrix} \tag{3.18}$$

The $(NL\times NL)$-dimensional matrix $\mathbf{R}_{xx}(\tau)$ given by (3.18) has the form of a Toeplitz matrix—that is, a matrix having equal valued matrix elements along any diagonal [20]. The desirability of the Toeplitz form lies in the fact that the entire matrix can be constructed from the first row of submatrices, that is, $\mathbf{R}'_{xx}(\tau)$, $\mathbf{R}'_{xx}(\tau+\Delta),\ldots,\mathbf{R}'_{xx}[\tau+(L-1)\Delta]$. Consequently only a $(N\times NL)$ dimensional matrix need be stored to have all the information contained in $\mathbf{R}_{xx}(\tau)$.

Covariance matrices are closely related to correlation matrices since the covariance matrix between the vectors $\mathbf{x}(t)$ and $\mathbf{y}(t)$ is defined by

$$\operatorname{cov}[\mathbf{x}(t),\mathbf{y}(t)] \triangleq E\{(\mathbf{x}(t)-\bar{\mathbf{x}})(\mathbf{y}(t)-\bar{\mathbf{y}})^T\} \tag{3.19}$$

where

$$\bar{\mathbf{x}}=E\{\mathbf{x}(t)\} \qquad \text{and} \qquad \bar{\mathbf{y}}=E\{\mathbf{y}(t)\}$$

Consequently for zero-mean processes and with $\tau=0$, correlation matrices and covariance matrices are identical, and the adaptive array literature frequently uses the two terms interchangeably.

Frequency domain signal descriptions as well as time domain signal descriptions will be found to be valuable in considering array processors for broadband applications. The frequency domain equivalent of time domain descriptions may be found by taking the Fourier transform of the time domain quantities, $\mathfrak{f}(\omega)=\mathfrak{F}\{f(t)\}$. The Fourier transform of a signal correlation matrix yields the signal cross-spectral density matrix,

$$\mathbf{\Phi}_{xx}(\omega)=\mathfrak{F}\{\mathbf{R}_{xx}(\tau)\} \tag{3.20}$$

Cross-spectral density matrices therefore present the signal information contained in correlation matrices in the frequency domain.

### *3.1.3 Revised Definitions Required for Complex Vector Quantities*

The optimization problems that must be solved to yield the most desirable complex weight vector choices for different performance measures involve the use of norms, gradients, and covariance (or correlation) matrices of complex vector quantities. It is therefore useful to consider the definitions of these terms for both real and complex vectors.

The norm of a vector in Hilbert space, denoted by $\|\mathbf{x}\|$, represents the length of the vector. For a real vector,

$$\|\mathbf{x}\| \triangleq \sqrt{\mathbf{x}^T\mathbf{x}}$$

while for a complex vector,

$$\|\mathbf{x}\| \triangleq \sqrt{\mathbf{x}^\dagger\mathbf{x}} \tag{3.22}$$

The gradient operator $\nabla_y$ is applied to a scalar-valued function of the vector $\mathbf{y}$, $f(\mathbf{y})$, to obtain an indication of the partial derivatives of $f(\cdot)$ along the various component directions of $\mathbf{y}$. In the case of real variables, the gradient operator is therefore a vector operator given by

$$\nabla_y \triangleq \left[ \frac{\partial}{\partial y_1} \cdots \frac{\partial}{\partial y_n} \right]^T \tag{3.23}$$

so that

$$\nabla_y f(\mathbf{y}) = \frac{\partial f}{\partial y_1}\mathbf{e}_1 + \frac{\partial f}{\partial y_2}\mathbf{e}_2 + \cdots + \frac{\partial f}{\partial y_n}\mathbf{e}_n \tag{3.24}$$

where $\mathbf{e}_1, \mathbf{e}_2, \ldots, \mathbf{e}_n$ is the set of unit basis vectors for the vector $\mathbf{y}$.

For a complex vector $\mathbf{y}$ each element $y_k$ has a real and an imaginary component:

$$y_k = x_k + jz_k \tag{3.25}$$

Consequently, each partial derivative in (3.24) now has a real and an imaginary component so that [21]

$$\frac{\partial f}{\partial y_k} = \frac{\partial f}{\partial x_k} + \frac{(j)\partial f}{\partial z_k} \tag{3.26}$$

In the optimization problems to be encountered in this chapter, it is frequently desired to obtain the gradient with respect to the vector $\mathbf{x}$ of the scalar quantity $\mathbf{x}^T\mathbf{a} = \mathbf{a}^T\mathbf{x}$ and of the quadratic form $\mathbf{x}^T\mathbf{A}\mathbf{x}$ (which is also a scalar) where $\mathbf{A}$ is a symmetric matrix. Note that $\mathbf{x}^T\mathbf{a}$ is an inner product,

$$(\mathbf{x}, \mathbf{a}) = \mathbf{x}^T\mathbf{a} = \mathbf{a}^T\mathbf{x} \tag{3.27}$$

If $\mathbf{A}$ has a dyadic structure, the quadratic form $\mathbf{x}^T\mathbf{A}\mathbf{x}$ can be regarded as an inner product squared, or

$$\mathbf{x}^T\mathbf{A}\mathbf{x} = (\mathbf{x}, \mathbf{v})^2 \tag{3.28}$$

where

$$\mathbf{A} = \mathbf{v}\mathbf{v}^T \tag{3.29}$$

The trace of a matrix product given by

$$\text{trace}\left[\mathbf{A}\mathbf{B}^T\right] = \sum_i \sum_k a_{ik} b_{ik} \tag{3.30}$$

has all the properties of an inner product, so formulas for the differentiation of the trace of various matrix products are of interest in obtaining solutions to optimization problems. A partial list of convenient differentiation formulas is given in Appendix C. From these formulas it follows for real variables that

$$\nabla_x(\mathbf{y}^T\mathbf{A}\mathbf{x}) = \mathbf{y}^T\mathbf{A} \tag{3.31}$$

and

$$\nabla_x(\mathbf{x}^T\mathbf{A}\mathbf{x}) = 2\mathbf{A}\mathbf{x} \tag{3.32}$$

whereas for complex variables the corresponding gradients are given by

$$\nabla_x(\mathbf{y}^\dagger\mathbf{A}\mathbf{x}) = \mathbf{y}^\dagger\mathbf{A} \tag{3.33}$$

and

$$\nabla_x(\mathbf{x}^\dagger\mathbf{A}\mathbf{x}) = 2\mathbf{A}\mathbf{x} \tag{3.34}$$

### *3.1.4 Correlation Matrices for Complex Signals*

For complex vector quantities the corresponding correlation matrix definitions must be revised from (3.7) and (3.8) so that

$$\mathbf{R}_{xx} \triangleq E\{\mathbf{x}^*\mathbf{x}^T\} \quad \text{and} \quad \mathbf{R}_{xy} \triangleq E\{\mathbf{x}^*\mathbf{y}^T\} \tag{3.35}$$

An alternative correlation matrix definition that is also found in the literature is given by

$$\mathbf{R}_{xx} \triangleq E\{\mathbf{x}\mathbf{x}^\dagger\} \quad \text{and} \quad \mathbf{R}_{xy} \triangleq E\{\mathbf{x}\mathbf{y}^\dagger\} \tag{3.36}$$

The definition of (3.36) yields a matrix that is just the complex conjugate (or the transpose) of the definition given by (3.35). So long as one adheres consistently to either one definition or the other, the results obtained (in terms of the selected performance measure) will turn out to be the same; so it is immaterial which definition is elected for use. With either of the above definitions it immediately follows that

$$\mathbf{R}_{xx}^\dagger = \mathbf{R}_{xx} \quad \text{and} \quad \mathbf{R}_{xy}^\dagger = \mathbf{R}_{yx} \tag{3.37}$$

So the autocorrelation matrix $\mathbf{R}_{xx}$ is Hermitian, whereas the crosscorrelation matrix $\mathbf{R}_{xy}$ is in general not Hermitian (since $\mathbf{R}_{xy}^\dagger \neq \mathbf{R}_{xy}$). Whether or not the autocorrelation matrix is positive definite or positive semidefinite once again is determined by whether the signal vector of concern is $\mathbf{x}(t)$, $\mathbf{s}(t)$, or $\mathbf{n}(t)$.

## 3.2 SIGNAL DESCRIPTIONS FOR CONVENTIONAL AND SIGNAL ALIGNED ARRAYS

There are two types of arrays that have proven useful in different applications: the conventional array illustrated in Figure 3.1 and the signal aligned array illustrated in Figure 3.2. The signal aligned array is useful where the direction of arrival of the desired signal is known a priori, and this information can be used to obtain time-coincident desired signals in each channel. One advantage of the signal aligned array structure is that a set of processor weights independent of the desired signal time structure can be determined, which provides a distortionless output for any waveform incident on the array from the (assumed) known desired signal direction [22]. Such a processor is useful for extracting impulse "bursts," which are only present during relatively short time intervals.

The outputs of the arrays illustrated in Figures 3.1 and 3.2 can be expressed respectively as

$$y(t)=\mathbf{w}^{\dagger}\mathbf{x}(t) \qquad \text{and} \qquad y(t)=\mathbf{w}^{\dagger}\mathbf{z}(t) \tag{3.38}$$

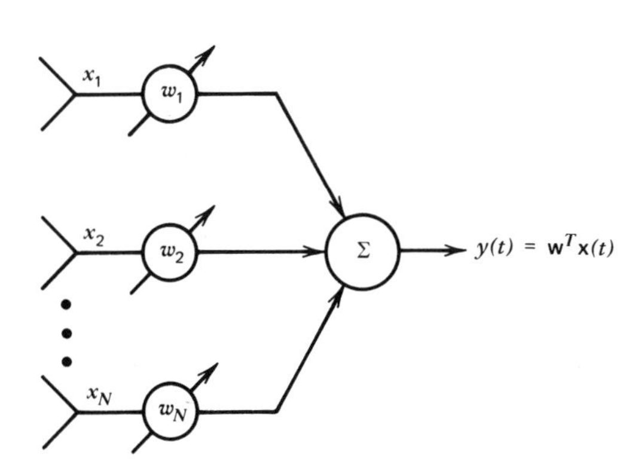


**Figure 3.1** Conventional narrowband array.

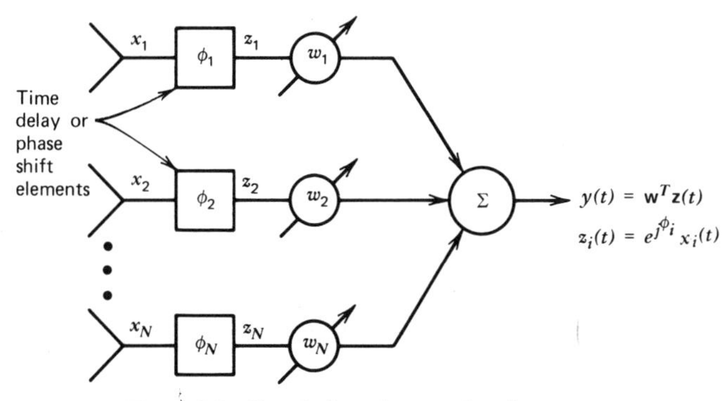


**Figure 3.2** Signal aligned narrowband array.

where $\mathbf{x}(t)=\mathbf{s}(t)+\mathbf{n}(t)$ is the vector of received signals that are complex valued functions. The noise vector $\mathbf{n}(t)$ may be assumed to be stationary and ergodic and to have both directional and thermal noise components that are independent of the signal. The signal vector $\mathbf{s}(t)$ is induced within the individual array sensor elements from a single directional signal source assumed to be

$$s(t)=\sqrt{S}\, e^{j\omega_0 t} \tag{3.39}$$

where $\omega_0$ is the (radian) carrier frequency and $S$ represents the signal power. Assuming identical antenna elements, the resulting signal component in each array element is just a phase shifted version (due to propagation along the array) of the signal appearing at the first array element encountered by the directional signal source. It therefore follows that the signal vector $\mathbf{s}(t)$ can be written as

$$\begin{aligned}\mathbf{s}^T(t)&=\left[\sqrt{S}\, e^{j\omega_0 t}, \sqrt{S}\, e^{j\omega_0 t+\theta_1},\ldots,\sqrt{S}\, e^{j\omega_0 t+\theta_{N-1}}\right]\\ &=s(t)\mathbf{v}^T\end{aligned} \tag{3.40}$$

where $\mathbf{v}$ is defined to be the array propagation vector,

$$\mathbf{v}^T=\left[1,e^{j\theta_1},\ldots,e^{j\theta_{N-1}}\right] \tag{3.41}$$

Consequently, the received signal vector for the conventional array of Figure 3.1 can be written as

$$\mathbf{x}(t)=s(t)\mathbf{v}+\mathbf{n}(t) \tag{3.42}$$

The received signal vector (after the time delay or phase shift elements) for the signal aligned array of Figure 3.2 can be written as

$$\mathbf{z}(t)=s(t)\mathbf{1}+\mathbf{n}'(t) \tag{3.43}$$

where now $\mathbf{v}$ of (3.42) has been replaced by $\mathbf{1}=(1,1,\ldots,1)^T$ since the desired signal terms in each channel are time aligned and therefore identical. The components of the noise vector then become

$$n_i'(t)=n_i(t)e^{j\phi_i} \tag{3.44}$$

In developing the optimal solutions for selected performance measures, four correlation matrices will be required. These correlation matrices are defined as follows for narrowband uncorrelated signal processes:

$$\mathbf{R}_{ss}\triangleq E\left\{\mathbf{s}^*(t)\mathbf{s}^T(t)\right\}=S\mathbf{v}^*\mathbf{v}^T \tag{3.45}$$

where $S$ denotes the signal power,

$$\mathbf{R}_{nn} \triangleq E\{\mathbf{n}^*(t)\mathbf{n}^T(t)\} \tag{3.46}$$

$$\mathbf{r}_{xs} \triangleq E\{\mathbf{x}^*(t)s(t)\} = S\mathbf{v}^* \tag{3.47}$$

and

$$\mathbf{R}_{xx} \triangleq E\{\mathbf{x}^*(t)\mathbf{x}^T(t)\} = \mathbf{R}_{ss} + \mathbf{R}_{nn} \tag{3.48}$$

## 3.3 OPTIMUM ARRAY PROCESSING FOR NARROWBAND APPLICATIONS

It was previously noted that several different performance measures can be adopted to govern the operation of the adaptive processor that adjusts the weighting for each of the sensor element outputs of Figure 1.1. Now consider the problem of formulating four popular performance measures in terms of complex envelope signal characterizations and determining the optimum steady-state solution for the adaptive weight vector.

The performance measures to be considered are the following:

1. Mean square error (MSE) criterion.
2. Signal to noise ratio (SNR) criterion.
3. Maximum likelihood (ML) criterion.
4. Minimum noise variance (MV) criterion.

The use of the above criteria will be considered for the design of adaptive processors in narrowband systems. It will be recalled that for narrowband signals, the adaptive processor may be regarded as consisting of complex weights in each channel; for wideband signals, however, the adaptive processor consists of a set of linear filters that are usually approximated by tapped-delay lines in each channel of the array. Consequently, consideration of the narrowband processing case is slightly easier, although conceptually the wideband processor can be regarded as a set of frequency dependent complex weights.

The optimum solutions for the complex weight vector for each of the above four performance measures are derived first. Since it is desirable to be equally familiar with real and complex notation, the derivations are carried out using real notation; the corresponding complex solutions are then given so the student may develop the complex results using the derivations for real quantities but employing complex notation. Following the derivation of the desired results for each performance measure, it is shown that each solution is closely related to a single optimal solution by means of factorization of the results: this factorization provides the connection between the various solutions obtained and the form known as the Wiener solution.

### *3.3.1 The Mean Square Error (MSE) Performance Measure*

The MSE performance measure was considered by Widrow et. al. [5] for the conventional array configuration, and additional procedures based on this criterion have been developed and further extended [23]–[25]. Suppose the desired directional signal $s(t)$ is known and represented by a reference signal $d(t)$. This assumption is never strictly met in practice because a communication signal cannot possibly be known a priori if it is to convey information, and hence the desired signal must be unknown in some respect. Nevertheless, it turns out that in practice enough is usually known about the desired signal that a suitable reference signal $d(t)$ can be obtained to approximate $s(t)$ in some sense by appropriately processing the array output signal. For example, when $s(t)$ is an amplitude modulated signal, it is possible to use the carrier component of $s(t)$ for $d(t)$ and still obtain suitable operation. Consequently the desired or "reference" signal concept is a valuable tool, and one can proceed with the analysis as though the adaptive processor had a complete desired signal characterization.

The difference between the desired array response and the actual array output signal defines an error signal as shown in Figure 3.3:

$$\epsilon(t) = d(t) - \mathbf{w}^T\mathbf{x}(t) \tag{3.49}$$

The squared error can therefore be written as

$$\epsilon^2(t) = d^2(t) - 2d(t)\mathbf{w}^T\mathbf{x}(t) + \mathbf{w}^T\mathbf{x}(t)\mathbf{x}^T(t)\mathbf{w} \tag{3.50}$$

Taking expected values of both sides of (3.50) then yields

$$E\{\epsilon^2(t)\} = \overline{d^2(t)} - 2\mathbf{w}^T\mathbf{r}_{xd} + \mathbf{w}^T\mathbf{R}_{xx}\mathbf{w} \tag{3.51}$$

where

$$\mathbf{r}_{xd} = \begin{bmatrix} \overline{x_1(t)d(t)} \\ \overline{x_2(t)d(t)} \\ \vdots \\ \overline{x_N(t)d(t)} \end{bmatrix} \tag{3.52}$$

Since $d(t) = s(t)$, from (3.39) it follows that $\overline{d^2(t)} = S$ so that

$$E\{\epsilon^2(t)\} = S - 2\mathbf{w}^T\mathbf{r}_{xd} + \mathbf{w}^T\mathbf{R}_{xx}\mathbf{w} \tag{3.53}$$

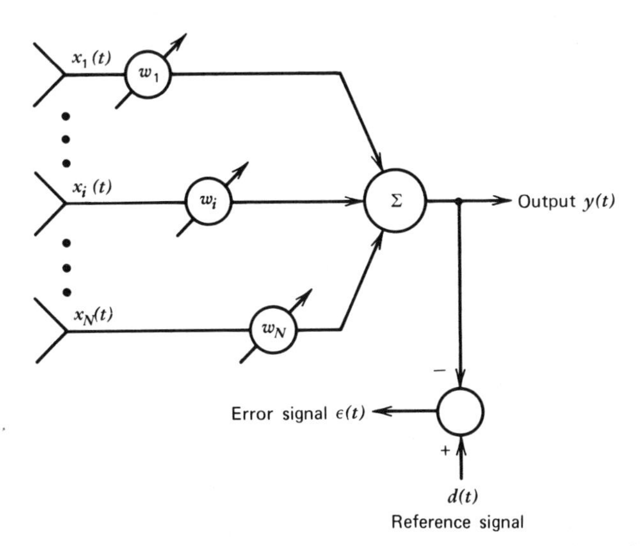


**Figure 3.3** Basic adaptive array structure with known desired signal.

It is desired to minimize (3.53) by appropriately selecting the weight vector **w**. Since (3.53) is a quadratic function of **w**, the extremum of this function is a minimum. Therefore the value of **w** that gives the minimum of $E\{\epsilon^2(t)\}$ may be found by setting the gradient of (3.53) with respect to the weight vector equal to zero, that is,

$$\nabla_w \overline{(\epsilon^2)} = 0 \tag{3.54}$$

Since

$$\nabla_w \overline{(\epsilon^2)} = -2\mathbf{r}_{xd} + 2\mathbf{R}_{xx}\mathbf{w} \tag{3.55}$$

it follows that the optimum choice for the weight vector must satisfy

$$\mathbf{R}_{xx}\mathbf{w}_{\text{opt}} = \mathbf{r}_{xd} \qquad \text{or} \qquad \mathbf{w}_{\text{opt}} = \mathbf{R}_{xx}^{-1}\mathbf{r}_{xd} \tag{3.56}$$

Equation (3.56) for $\mathbf{w}_{\text{opt}}$ is the Wiener-Hopf equation in matrix form and is consequently referred to as the optimum "Wiener solution."

If we use $d(t) = s(t)$ and (3.39) and (3.42), it then follows that

$$\mathbf{r}_{xd} = E\{\mathbf{x}d\} = S\mathbf{v} \tag{3.57}$$

so that

$$\mathbf{w}_{\text{MSE}} = S\mathbf{R}_{xx}^{-1}\mathbf{v} \tag{3.58}$$

where it is assumed that $\mathbf{R}_{xx}$ is nonsingular so that $\mathbf{R}_{xx}^{-1}$ exists. Setting the weight vector equal to $\mathbf{w}_{\text{MSE}}$, the resulting minimum MSE is found from (3.53) to be

$$\overline{\epsilon_{\min}^2} = S - \mathbf{r}_{xd}^T \mathbf{R}_{xx}^{-1} \mathbf{r}_{xd} \tag{3.59}$$

For complex quantities (3.53) becomes

$$\overline{|\epsilon(t)|^2} = S - 2\,\text{Re}\left\{\mathbf{w}^\dagger \mathbf{r}_{xd}\right\} + \mathbf{w}^\dagger \mathbf{R}_{xx} \mathbf{w} \tag{3.60}$$

(3.58) becomes

$$w_{\text{MSE}} = S \mathbf{R}_{xx}^{-1} \mathbf{v}^* \tag{3.61}$$

and (3.59) becomes

$$\overline{|\epsilon|_{\min}^2} = S - \mathbf{r}_{xd}^\dagger \mathbf{R}_{xx}^{-1} \mathbf{r}_{xd} \tag{3.62}$$

### *3.3.2 The Signal-to-Noise Ratio (SNR) Performance Measure*

Adaptive processors whose operation is based on maximizing the SNR have been developed for the conventional array configuration for use in communication and detection systems [26]–[28]. The output signal from the adaptive array of Figure 3.1 can be written as

$$y(t) = \mathbf{w}^T \mathbf{x}(t) \tag{3.63}$$

where the input signal vector may be regarded as comprised of a signal component $\mathbf{s}(t)$ and a noise component $\mathbf{n}(t)$ so that

$$\mathbf{x}(t) = \mathbf{s}(t) + \mathbf{n}(t) \tag{3.64}$$

The signal and noise components of the array output signal may therefore be written as

$$y_s(t) = \mathbf{w}^T \mathbf{s}(t) = \mathbf{s}^T(t)\mathbf{w} \tag{3.65}$$

and

$$y_n(t) = \mathbf{w}^T \mathbf{n}(t) = \mathbf{n}^T(t)\mathbf{w} \tag{3.66}$$

where

$$\mathbf{s}(t) = \begin{bmatrix} s_1(t) \\ s_2(t) \\ \vdots \\ s_N(t) \end{bmatrix} \quad \text{and} \quad \mathbf{n}(t) = \begin{bmatrix} n_1(t) \\ n_2(t) \\ \vdots \\ n_N(t) \end{bmatrix} \tag{3.67}$$

Consequently the output signal power may be written as

$$E\{|y_s(t)|^2\} = |\overline{\mathbf{w}^T\mathbf{s}}|^2 \tag{3.68}$$

and the output noise power is

$$E\{|y_n(t)|^2\} = |\overline{\mathbf{w}^T\mathbf{n}}|^2 \tag{3.69}$$

The output SNR is therefore given by

$$\left(\frac{s}{n}\right) = \frac{|\overline{\mathbf{w}^T\mathbf{s}}|^2}{|\overline{\mathbf{w}^T\mathbf{n}}|^2} = \frac{\mathbf{w}^T\left[\overline{\mathbf{s}\mathbf{s}^T}\right]\mathbf{w}}{\mathbf{w}^T\left[\overline{\mathbf{n}\mathbf{n}^T}\right]\mathbf{w}} = \frac{\mathbf{w}^T\mathbf{R}_{ss}\mathbf{w}}{\mathbf{w}^T\mathbf{R}_{nn}\mathbf{w}} \tag{3.70}$$

The ratio given by (3.70) can be rewritten as

$$\left(\frac{s}{n}\right) = \frac{\mathbf{z}^T\mathbf{R}_{nn}^{-1/2}\mathbf{R}_{ss}\mathbf{R}_{nn}^{-1/2}\mathbf{z}}{\mathbf{z}^T\mathbf{z}} \tag{3.71}$$

where

$$\mathbf{z} \triangleq \mathbf{R}_{nn}^{1/2}\mathbf{w} \tag{3.72}$$

Equation (3.70) may be recognized as a standard quadratic form and is bounded by the minimum and maximum eigenvalues of the symmetric matrix $\mathbf{R}_{nn}^{-1/2}\mathbf{R}_{ss}\mathbf{R}_{nn}^{-1/2}$ (or more conveniently $\mathbf{R}_{nn}^{-1}\mathbf{R}_{ss}$) [29]. The optimization of (3.70) by appropriately selecting the weight vector $\mathbf{w}$ consequently results in an eigenvalue problem where the ratio $(s/n)$ must satisfy the relationship [30]

$$\mathbf{R}_{ss}\mathbf{w} = (s/n)\mathbf{R}_{nn}\mathbf{w} \tag{3.73}$$

in which $(s/n)$ now represents an eigenvalue of the symmetric matrix noted above. The maximum eigenvalue satisfying (3.73) is denoted by $(s/n)_{\text{opt}}$. Corresponding to $(s/n)_{\text{opt}}$ there is a unique eigenvector $\mathbf{w}_{\text{opt}}$ which represents the optimum element weights. Therefore

$$\mathbf{R}_{ss}\mathbf{w}_{\text{opt}} = \left(\frac{s}{n}\right)_{\text{opt}}\mathbf{R}_{nn}\mathbf{w}_{\text{opt}} \tag{3.74}$$

Substitution of (3.70) corresponding to $(s/n)_{\text{opt}}$ into (3.74) yields

$$\mathbf{R}_{ss}\mathbf{w}_{\text{opt}} = \frac{\mathbf{w}_{\text{opt}}^T\mathbf{R}_{ss}\mathbf{w}_{\text{opt}}}{\mathbf{w}_{\text{opt}}^T\mathbf{R}_{nn}\mathbf{w}_{\text{opt}}}\mathbf{R}_{nn}\mathbf{w}_{\text{opt}} \tag{3.75}$$

Substituting $\mathbf{R}_{ss} = [\overline{\mathbf{s}\mathbf{s}^T}]$ and noting that $\mathbf{s}^T\mathbf{w}_{\text{opt}}$ is a scalar quantity occurring on both sides of (3.75) that may be cancelled, we get the results

$$\mathbf{s} = \frac{\mathbf{w}_{\text{opt}}^T \mathbf{s}}{\mathbf{w}_{\text{opt}}^T \mathbf{R}_{nn} \mathbf{w}_{\text{opt}}} \cdot \mathbf{R}_{nn} \mathbf{w}_{\text{opt}} \tag{3.76}$$

The ratio $(\mathbf{w}_{\text{opt}}^T \mathbf{s} / \mathbf{w}_{\text{opt}}^T \mathbf{R}_{nn} \mathbf{w}_{\text{opt}})$ may be seen to be just a complex (scalar) number, denoted here by $c$. It therefore follows that

$$\mathbf{w}_{\text{opt}} = \left(\frac{1}{c}\right) \mathbf{R}_{nn}^{-1} \mathbf{s} \tag{3.77}$$

Since from (3.39) the envelope of $\mathbf{s}$ is just $\sqrt{S}\,\mathbf{v}$, it follows that

$$\mathbf{w}_{\text{SNR}} = \alpha \mathbf{R}_{nn}^{-1} \mathbf{v} \tag{3.78}$$

where

$$\alpha = \frac{\sqrt{S}}{c}$$

The maximum possible value that $(s/n)_{\text{opt}}$ can attain is easily derived by converting the original system into orthonormal system variables. Since $\mathbf{R}_{nn}$ is a positive definite Hermitian matrix, it can be diagonalized by a nonsingular coordinate transformation as shown in Figure 3.4. Such a transformation can be selected so that all element channels have equal noise power components that are uncorrelated. Denote the transformation matrix that accomplishes this diagonalization by $\mathbf{A}$, so that

$$\mathbf{s}' = \mathbf{A}\mathbf{s} \tag{3.79}$$

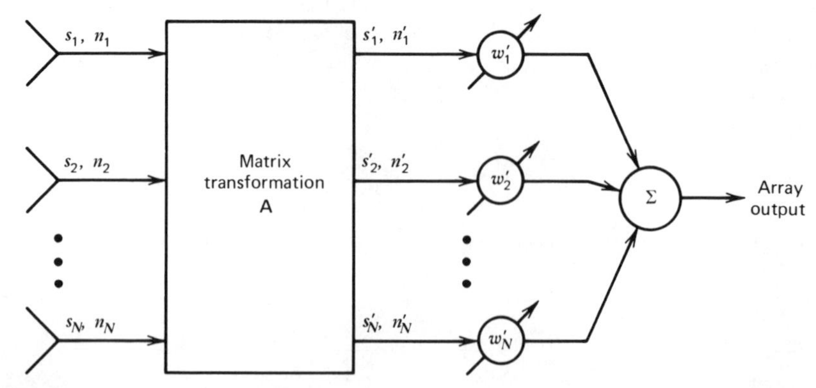


**Figure 3.4** Functional representation of orthonormal transformation and adaptive weight combiner equivalent to system of Figure 3.1.

and

$$\mathbf{n}' = \mathbf{A}\mathbf{n} \tag{3.80}$$

where a prime (′) denotes quantities after the transformation.

The signal component of the array output now becomes

$$y_s = \mathbf{w}'^T\mathbf{s}' = \mathbf{w}'^T\mathbf{A}\mathbf{s} \tag{3.81}$$

and the noise component becomes

$$y_n = \mathbf{w}'^T\mathbf{n}' = \mathbf{w}'^T\mathbf{A}\mathbf{n} \tag{3.82}$$

For the array output of the system in Figure 3.4 to be equivalent to the output of the system in Figure 3.1, it is necessary that

$$\mathbf{w} = \mathbf{A}^T\mathbf{w}' \tag{3.83}$$

The output noise power of the orthonormal system is given by

$$E\{|y_n(t)|^2\} = E\{|\mathbf{w}'^T\mathbf{n}'|^2 = \mathbf{w}'^T E\{\mathbf{n}'\mathbf{n}'^T\}\mathbf{w}' \tag{3.84}$$

Since the transformation matrix $\mathbf{A}$ decorrelates the various noise components and equalizes their powers, the covariance matrix of the noise process $\mathbf{n}'(t)$ is just the identity matrix, that is,

$$E\{\mathbf{n}'\mathbf{n}'^T\} = \mathbf{I}_N \tag{3.85}$$

It immediately follows from (3.84) and (3.85) that

$$E\{|y_n(t)|^2\} = \mathbf{w}'^T\mathbf{w}' = \|\mathbf{w}'\|^2 \tag{3.86}$$

The output noise power of the original system of Figure 3.1 is given by

$$E\{|y_n(t)|^2\} = \mathbf{w}^T\mathbf{R}_{nn}\mathbf{w} \tag{3.87}$$

Substituting (3.78) into (3.82), we find it follows that

$$E\{|y_n(t)|^2\} = \mathbf{w}'^T\mathbf{A}\mathbf{R}_{nn}\mathbf{A}^T\mathbf{w}' \tag{3.88}$$

For the output noise power of the two systems to be equivalent, it is necessary that

$$\mathbf{A}\mathbf{R}_{nn}\mathbf{A}^T = \mathbf{I}_N \tag{3.89}$$

or

$$\mathbf{R}_{nn} = [\mathbf{A}^T\mathbf{A}]^{-1} \tag{3.90}$$

Equation (3.90) simply expresses the fact that the transformation **A** diagonalizes and normalizes the matrix $\mathbf{R}_{nn}$.

The signal output of the orthonormal array system is given by (3.81). Applying the Cauchy-Schwartz inequality (see Appendix D and reference [4]) to this expression immediately yields an upper bound on the array output signal power as

$$|y_S(t)|^2 \leqslant \|\mathbf{w}'\|^2 \|\mathbf{s}'\|^2 \tag{3.91}$$

where

$$\|\mathbf{s}'\|^2 = \mathbf{s}'^T \mathbf{s}' \qquad \text{and} \qquad \|\mathbf{w}'\|^2 = \mathbf{w}'^T \mathbf{w}' \tag{3.92}$$

From (3.86) and (3.91) it follows that the maximum possible value of the SNR is given by

$$\text{SNR}_{\text{max}} = \|\mathbf{s}'\|^2 \tag{3.93}$$

Substituting (3.79) into (3.83) and using (3.93) and (3.90), we find it then follows that

$$\text{SNR}_{\text{opt}} = \mathbf{s}^T \mathbf{R}_{nn}^{-1} \mathbf{s} \tag{3.94}$$

For complex quantities, (3.70) becomes

$$\left(\frac{s}{n}\right) = \frac{\mathbf{w}^\dagger \left[\, \overline{\mathbf{s}^* \mathbf{s}^T} \,\right] \mathbf{w}}{\mathbf{w}^\dagger \left[\, \overline{\mathbf{n}^* \mathbf{n}^T} \,\right] \mathbf{w}} = \frac{\mathbf{w}^\dagger \mathbf{R}_{ss} \mathbf{w}}{\mathbf{w}^\dagger \mathbf{R}_{nn} \mathbf{w}} \tag{3.95}$$

Equation (3.78) is replaced by

$$\mathbf{w}_{\text{SNR}} = \alpha \mathbf{R}_{nn}^{-1} \mathbf{v}^* \tag{3.96}$$

(3.90) is now

$$\mathbf{R}_{nn} = \left[\mathbf{A}^T \mathbf{A}^*\right]^{-1} \tag{3.97}$$

while (3.94) becomes

$$\text{SNR}_{\text{opt}} = \mathbf{s}^T \mathbf{R}_{nn}^{-1} \mathbf{s}^* \tag{3.98}$$

Designing the adaptive processor so that the weights satisfy $\mathbf{R}_{nn}\mathbf{w} = \alpha \mathbf{v}^*$ means that the output SNR is the governing performance criterion, even for the quiescent environment (when no jamming signal and no desired signal is present). It is usually desirable, however, to compromise the output SNR in order to exercise some control over the array beam pattern sidelobe levels. An alternative performance measure that yields more flexibility in beam shaping can be introduced in the manner described in the following material.

Suppose in the normal quiescent signal environment that the most desirable array weight vector selection is given by $\mathbf{w}_q$ (where now $\mathbf{w}_q$ represents the designer's most desirable compromise between gain, sidelobe levels, etc.). For this quiescent environment the signal covariance matrix is $\mathbf{R}_{nn_q}$. Define the column vector $\mathbf{t}$ by the equation

$$\mathbf{R}_{nn_q}\mathbf{w}_q = \alpha \mathbf{t}^* \tag{3.99}$$

On comparing (3.99) with (3.96), we see that the ratio being maximized is no longer (3.95), but the modified ratio given by

$$\frac{|\mathbf{w}^\dagger \mathbf{t}|^2}{\mathbf{w}^\dagger \mathbf{R}_{nn} \mathbf{w}} \tag{3.100}$$

This ratio is a more general criterion than the SNR and is often used for practical applications. The vector $\mathbf{t}$ is referred to as a generalized signal vector, and the ratio (3.100) is called the generalized signal-to-noise ratio (GSNR). For tutorial purposes the ordinary SNR is usually employed, although the GSNR is often used in practice.

The adaptive array of Figure 3.1 may be considered as a generalization of a coherent sidelobe canceller (CSLC). As an illustration of the application of the foregoing SNR performance measure concepts, it is useful to show how sidelobe cancellation may be regarded as a special case of an adaptive array.

The functional diagram of a standard sidelobe cancellation system is shown in Figure 3.5. A sidelobe cancellation system consists of a main antenna with high gain designated as channel $o$ and $N$ auxiliary antenna elements with their associated channels. The auxiliary antennas are designed so their gain patterns approximate the average sidelobe level of the main antenna gain pattern [18], [31]. It is also sometimes convenient to constrain the weights in the auxiliary channels to have a magnitude less than unity, in which case the auxiliary antennas must be designed so their gain patterns always exceed the peak sidelobe level of the main antenna gain pattern. With properly designed auxiliary antennas, it is then possible for the auxiliary channels to provide replicas of the jamming signals appearing in the sidelobes of the main antenna pattern. These replica jamming signals may then be used to provide coherent cancellation in the main channel output signal; thereby providing an interference free array output response as far as the sidelobes are concerned. Any jamming signal output occurring as a result of main beam response to jamming signal presence of course cannot be cancelled.

In applying the GSNR performance measure, it is first necessary to select a $\mathbf{t}$ column vector. Since any desired signal component contributed by the auxiliary channels to the total array desired signal output is negligible in comparison to the main channel contribution, and the main antenna has a

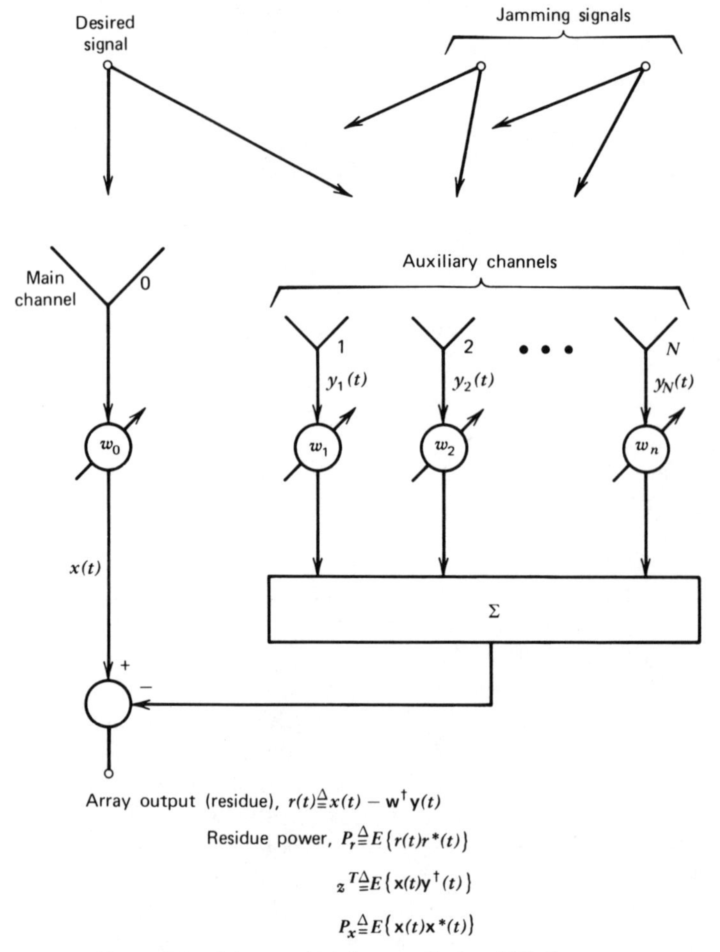


**Figure 3.5** Coherent sidelobe cancellation (CSLC) system.

carefully designed pattern, a reasonable choice for $t$ is the $N+1$ component vector:

$$\mathbf{t} = \begin{bmatrix} 1 \\ 0 \\ 0 \\ \vdots \\ 0 \end{bmatrix} \tag{3.101}$$

which preserves the main channel response signal. From (3.99) the optimum weight vector must satisfy

$$\mathbf{R}'_{nn}\mathbf{w}' = \alpha \mathbf{t} \tag{3.102}$$

where $\mathbf{R}'_{nn}$ is the $(N+1)\times(N+1)$ covariance matrix of all channel input signals (in the absence of the desired signal), and $\mathbf{w}'$ is the $N+1$ column vector of all channel weights.

Let the $N \times N$ covariance matrix of the auxiliary channel input signals (again in the absence of the desired signal) be denoted by $\mathbf{R}_{nn}$ and let $\mathbf{w}$ be the $N$-component column vector of the auxiliary channel weights. Equation (3.102) may now be partitioned to yield:

$$\left[\begin{array}{c|c} P_0 & \boldsymbol{\Psi}^\dagger \\ \hline \boldsymbol{\Psi} & \mathbf{R}_{nn} \end{array}\right]\left[\begin{array}{c} w_0 \\ \hline \mathbf{w} \end{array}\right]=\left[\begin{array}{c} \alpha \\ \hline \mathbf{0} \end{array}\right] \tag{3.103}$$

where $P_0$ = noise power output of the main channel and

$$\boldsymbol{\Psi} \triangleq E\begin{bmatrix} x_1^* x_0 \\ x_2^* x_0 \\ \vdots \\ x_N^* x_0 \end{bmatrix} = \begin{array}{l}\text{cross correlation vector between the} \\ \text{main channel output and the output of} \\ \text{each auxiliary channel}\end{array}$$

Equation (3.103) may in turn be written as two separate equations:

$$\text{(scalar equation) } P_0 w_0 + \boldsymbol{\Psi}^\dagger \mathbf{w} = \alpha \tag{3.104}$$

and

$$\text{(matrix equation) } \mathbf{R}_{nn}\mathbf{w} = -w_0 \boldsymbol{\Psi} \tag{3.105}$$

The solution represented by (3.102) can be implemented using the $N+1$ weights $w_0, w_1, \ldots, w_N$ shown in Figure 3.5. It is also possible to attain the same optimum SNR using only the $N$ weights in the auxiliary channels, however. To see this, note that if $\mathbf{w}'$ is optimum for a given noise environment (a given $\mathbf{R}'_{nn}$), then any scalar multiple of $\mathbf{w}'$ will also yield the optimum SNR since the output SNR does not depend on the absolute level of $\mathbf{w}'$. Consequently the scalar weight $w_0$ in (3.104) can be fixed at any convenient nonzero value. Note that since $(\mathbf{w}')^T\mathbf{t} = w_0$, the GSNR represented by (3.100) will never be optimized by $w_0 = 0$. Consequently with $w_0$ fixed at some convenient $\hat{w}_0$, it is only necessary to adjust the weights in the auxiliary channels so that

$$\mathbf{R}_{nn}\mathbf{w}_{\text{opt}} = -\hat{w}_0 \boldsymbol{\Psi} \tag{3.106}$$

Since $\hat{w}_0$ is fixed and (3.106) optimizes the GSNR ratio of $\hat{w}_0$ to the output noise power, it follows that $\mathbf{w}_{\text{opt}}$ must be minimizing the output noise power resulting from sidelobe response.

### 3.3.3 The Maximum Likelihood (ML) Performance Measure

When the desired signal waveform is completely unknown (as with seismic waves, for example), then the desired signal may be regarded as a time function which is to be estimated. The derivation of the maximum likelihood estimator of the desired signal requires the assumption that the noise components have a multidimensional Gaussian distribution [22], [32].

The input signal vector may once again be written as

$$\mathbf{x}(t)=\mathbf{s}(t)+\mathbf{n}(t) \tag{3.107}$$

where

$$\mathbf{s}(t)=s(t)\mathbf{v} \tag{3.108}$$

for the conventional array of Figure 3.1, and it is desired to obtain an estimate of $s(t)$. Define the likelihood function of the input signal vector as

$$\mathfrak{L}[\mathbf{x}(t)]=-\ln[P\{\mathbf{x}(t)/\mathbf{x}(t)=\mathbf{s}(t)+\mathbf{n}(t)\}] \tag{3.109}$$

where $P\{z/y\}$ is the probability density function for $z$ given the event $y$. Thus, the likelihood function defined by (3.109) is the negative natural logarithm of the probability density function for the input signal vector $\mathbf{x}(t)$ given that $\mathbf{x}(t)$ contains both the desired signal and interfering noise.

Now assume that the noise vector $\mathbf{n}(t)$ can be modeled as a stationary, zero-mean Gaussian random vector with a covariance matrix $\mathbf{R}_{nn}$. Furthermore, assume that $\mathbf{x}(t)$ is also a stationary Gaussian random vector having the mean $s(t)\mathbf{v}$, where $s(t)$ is a deterministic but unknown quantity. With these assumptions, the likelihood function can be written as

$$\mathfrak{L}[\mathbf{x}(t)]=c[\mathbf{x}(t)-s(t)\mathbf{v}]^T\mathbf{R}_{nn}^{-1}[\mathbf{x}(t)-s(t)\mathbf{v}] \tag{3.110}$$

where $c$ is a scalar constant independent of $\mathbf{x}(t)$ and $s(t)$.

The maximum likelihood processor is obtained by solving for the estimate of $s(t)$, denoted by $\hat{s}(t)$, which maximizes (3.110). Taking the partial derivative of $\mathfrak{L}[\mathbf{x}(t)]$ with respect to $s(t)$ and setting the result equal to zero yields

$$0=\frac{\partial\mathfrak{L}[\mathbf{x}(t)]}{\partial s(t)}=-2\mathbf{v}^T\mathbf{R}_{nn}^{-1}\mathbf{x}+2\hat{s}(t)\mathbf{v}^T\mathbf{R}_{nn}^{-1}\mathbf{v} \tag{3.111}$$

It immediately follows that the estimate $\hat{s}(t)$ that maximizes $\mathfrak{L}[\mathbf{x}(t)]$ is given by

$$\hat{s}(t)\mathbf{v}^T\mathbf{R}_{nn}^{-1}\mathbf{v}=\mathbf{v}^T\mathbf{R}_{nn}^{-1}\mathbf{x} \tag{3.112}$$

Since the quantity $\mathbf{v}^T\mathbf{R}_{nn}^{-1}\mathbf{v}$ is a scalar, (3.112) can be rewritten as

$$\hat{s}(t)=\frac{\mathbf{v}^T\mathbf{R}_{nn}^{-1}}{\mathbf{v}^T\mathbf{R}_{nn}^{-1}\mathbf{v}}\mathbf{x}(t) \tag{3.113}$$

which is of the form $\hat{s}(t)=\mathbf{w}_{\text{ML}}^T\mathbf{x}(t)$. Consequently the maximum likelihood weight vector is given by

$$\mathbf{w}_{\text{ML}}=\frac{\mathbf{R}_{nn}^{-1}\mathbf{v}}{\mathbf{v}^T\mathbf{R}_{nn}^{-1}\mathbf{v}} \tag{3.114}$$

For complex quantities (3.110) becomes

$$\mathfrak{L}[\mathbf{x}(t)]=c[\mathbf{x}(t)-s(t)\mathbf{v}]^{\dagger}\,\mathbf{R}_{nn}^{-1}[\mathbf{x}(t)-s(t)\mathbf{v}] \tag{3.115}$$

Likewise (3.112) is replaced by

$$\hat{s}(t)\mathbf{v}^{\dagger}\mathbf{R}_{nn}^{-1}\mathbf{v}=\mathbf{v}^{\dagger}\mathbf{R}_{nn}^{-1}\mathbf{x}(t) \tag{3.116}$$

and the result expressed by (3.114) then becomes

$$\mathbf{w}_{\text{ML}}=\frac{\mathbf{R}_{nn}^{-1}\mathbf{v}}{\mathbf{v}^{\dagger}\mathbf{R}_{nn}^{-1}\mathbf{v}} \tag{3.117}$$

### *3.3.4 The Minimum Noise Variance (MV) Performance Measure*

When both the desired signal $s(t)$ and the desired signal direction are known (as with a signal aligned array), then minimizing the output noise variance provides a means of ensuring good signal reception, and methods based on this performance measure have been developed [33], [34]. For the signal aligned array of Figure 3.2, the array output is given by

$$y(t)=\mathbf{w}^T\mathbf{z}(t)=s(t)\sum_{i=1}^{N}w_i+\sum_{i=1}^{N}w_i n_i' \tag{3.118}$$

where the $n_i'$ represent the noise components after the signal aligning phase shifts. When we constrain the sum of the array weights to be unity, then the output signal becomes

$$y(t)=s(t)+\mathbf{w}^T\mathbf{n}'(t) \tag{3.119}$$

which represents an unbiased output signal since

$$E\{y(t)\}=s(t) \tag{3.120}$$

The variance of the array output may therefore be expressed as

$$\text{var}[y(t)] = E\{\mathbf{w}^T\mathbf{n}'(t)\mathbf{n}'^T(t)\mathbf{w}\} = \mathbf{w}^T\mathbf{R}_{n'n'}\mathbf{w} \tag{3.121}$$

The relationship between $\mathbf{n}(t)$, the noise vector appearing before the signal aligning phase shifts, and $\mathbf{n}'(t)$ is given by

$$\mathbf{n}'(t) = \mathbf{\Phi}\mathbf{n}(t) \tag{3.122}$$

where $\mathbf{\Phi}$ is the diagonal unitary transformation described by

$$\mathbf{\Phi} = \begin{bmatrix} e^{j\phi_1} & & & 0 \\ & e^{j\phi_2} & & \\ & & \ddots & \\ 0 & & & e^{j\phi_N} \end{bmatrix} \tag{3.123}$$

The variance of the array output remains unaffected by such a unitary transformation so that

$$\text{var}[y(t)] = \mathbf{w}^T\mathbf{R}_{n'n'}\mathbf{w} = \mathbf{w}^T\mathbf{R}_{nn}\mathbf{w} \tag{3.124}$$

It is now desired to minimize (3.124) subject to the constraint

$$\mathbf{w}^T\mathbf{1} = 1 \tag{3.125}$$

where

$$\mathbf{1} = [1, 1, \ldots, 1]^T \tag{3.126}$$

To solve this minimization problem, form the modified performance criterion:

$$\mathfrak{P}_{\text{MV}} = \tfrac{1}{2}\mathbf{w}^T\mathbf{R}_{nn}\mathbf{w} + \lambda[1 - \mathbf{w}^T\mathbf{1}] \tag{3.127}$$

where the factor $\lambda$ is a Lagrange multiplier. Since $\mathfrak{P}_{\text{MV}}$ above is a quadratic function of $\mathbf{w}$, it follows that the optimal choice for $\mathbf{w}$ may be found by setting the gradient $\nabla_w\mathfrak{P}_{\text{MV}}$ equal to zero. The gradient is given by

$$\nabla_w\mathfrak{P}_{\text{MV}} = \mathbf{R}_{nn}\mathbf{w} - \lambda\mathbf{1} \tag{3.128}$$

so that

$$\mathbf{w}_{\text{MV}} = \mathbf{R}_{nn}^{-1}\mathbf{1}\lambda \tag{3.129}$$

The optimum solution $\mathbf{w}_{\text{MV}}$ must satisfy the constraint condition so that

$$\mathbf{w}_{\text{MV}}^T\mathbf{1} = 1 \tag{3.130}$$

and on substituting (3.130) into (3.129), there results

$$\lambda = \frac{1}{\mathbf{1}^T \mathbf{R}_{nn}^{-1} \mathbf{1}} \tag{3.131}$$

It follows immediately that

$$\mathbf{w}_{\text{MV}} = \frac{\mathbf{R}_{nn}^{-1} \mathbf{1}}{\mathbf{1}^T \mathbf{R}_{nn}^{-1} \mathbf{1}} \tag{3.132}$$

where $\mathbf{w}_{\text{MV}}$ satisfies (3.130).

On substituting (3.132) into (3.124), the minimum value of the output noise variance is found to be

$$\text{var}_{\min}[y(t)] = \frac{1}{\mathbf{1}^T \mathbf{R}_{nn}^{-1} \mathbf{1}} \tag{3.133}$$

If complex quantities are introduced, all the foregoing expressions remain unchanged, since the only modification required is the definition of the covariance matrix $\mathbf{R}_{nn}$ appropriate for complex vectors.

### *3.3.5 Factorization of the Optimum Solutions*

The solutions obtained in the preceding sections all applied to the conventional array, with the single exception of the minimum variance weights where it was necessary to use the signal aligned array in order to define a well posed problem. Each of the solutions obtained is closely related to one another since (as will be shown) they differ only by a scalar gain factor. Hence the different solutions possess identical output SNR's. This relationship can easily be shown by factoring the various solutions into a linear matrix filter followed by a scalar processor as described in the following paragraphs [35].

The optimum weight vector obtained for the minimum MSE performance measure can be written from (3.61) as

$$\mathbf{w}_{\text{MSE}} = \mathbf{R}_{xx}^{-1} S \mathbf{v}^* = \left[ S \mathbf{v}^* \mathbf{v}^T + \mathbf{R}_{nn} \right]^{-1} S \mathbf{v}^* \tag{3.134}$$

Applying the matrix identity (D.10) of Appendix D to (3.134) there results

$$\begin{aligned} \mathbf{w}_{\text{MSE}} &= \left[ S \mathbf{R}_{nn}^{-1} - \frac{S^2 \mathbf{R}_{nn}^{-1} \mathbf{v}^* \mathbf{v}^T \mathbf{R}_{nn}^{-1}}{1 + S \mathbf{v}^T \mathbf{R}_{nn}^{-1} \mathbf{v}^*} \right] \mathbf{v}^* \\ &= \left[ \frac{S}{1 + S \mathbf{v}^T \mathbf{R}_{nn}^{-1} \mathbf{v}^*} \right] \mathbf{R}_{nn}^{-1} \mathbf{v}^* \end{aligned} \tag{3.135}$$

From (3.135) it is seen that the minimum MSE weights are given by the product of a matrix filter $\mathbf{R}_{nn}^{-1}\mathbf{v}^*$ (which is also common to the other weight vector solutions) and a scalar factor. Since the MV solution only applied to a signal aligned array, for the other solutions to pertain to the signal aligned array it is only necessary to replace the $\mathbf{v}$ vector wherever it occurs by the $\mathbf{1}$ vector.

Now consider the noise power $N_0$ and the signal power $S_0$ that appear at the output of the linear matrix filter corresponding to $\mathbf{w}=\mathbf{R}_{nn}^{-1}\mathbf{v}^*$ as follows:

$$N_0=\mathbf{w}^{\dagger}\mathbf{R}_{nn}\mathbf{w}=\mathbf{v}^T\mathbf{R}_{nn}^{-1}\mathbf{v}^* \tag{3.136}$$

and

$$S_0=\mathbf{w}^{\dagger}\mathbf{R}_{ss}\mathbf{w}=SN_0^2 \tag{3.137}$$

The optimal weight vector solution given in Sections 3.3.1 and 3.3.3 for the MSE and the ML ratio performance measures can now be written in terms of the above quantities as

$$\mathbf{w}_{\text{MSE}}=\frac{1}{N_0}\cdot\frac{S_0}{N_0+S_0}\cdot\mathbf{R}_{nn}^{-1}\mathbf{v}^* \tag{3.138}$$

and

$$\mathbf{w}_{\text{ML}}=\frac{1}{N_0}\cdot\mathbf{R}_{nn}^{-1}\mathbf{v}^* \tag{3.139}$$

Likewise for the signal aligned array (where $\mathbf{v}=\mathbf{1}$) the ML weights reduce to the unbiased, MV weights, that is,

$$\mathbf{w}_{\text{ML}}\big|_{\mathbf{v}=\mathbf{1}}=\frac{\mathbf{R}_{nn}^{-1}\mathbf{1}}{\mathbf{1}^T\mathbf{R}_{nn}^{-1}\mathbf{1}}=\mathbf{w}_{\text{MV}} \tag{3.140}$$

The above expressions show that the minimum MSE processor can be regarded as factoring into a linear matrix filter followed by a scalar processor that contains the estimates corresponding to the other performance measures, as shown in Figure 3.6. The different solutions obtained above for the optimum weight vector can therefore all be expressed in the form

$$\mathbf{w}=\beta\mathbf{R}_{nn}^{-1}\mathbf{v}^* \tag{3.141}$$

where $\beta$ is an appropriate scalar gain, and hence they all yield the same SNR, which can then be expressed as

$$\left(\frac{s}{n}\right)=\frac{\mathbf{w}^{\dagger}\mathbf{R}_{ss}\mathbf{w}}{\mathbf{w}^{\dagger}\mathbf{R}_{nn}\mathbf{w}}=\frac{\beta^2S\mathbf{v}^T(\mathbf{R}_{nn}^{-1})\mathbf{v}^*\mathbf{v}^T\mathbf{R}_{nn}^{-1}\mathbf{v}^*}{\beta^2\mathbf{v}^T(\mathbf{R}_{nn}^{-1})\mathbf{v}^*}=S\mathbf{v}^TR_{nn}^{-1}\mathbf{v}^* \tag{3.142}$$

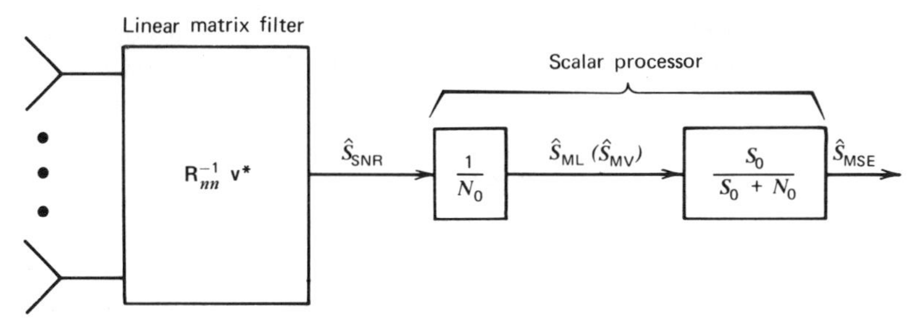


**Figure 3.6** The factored minimum MSE processor showing relationships to the maximum SNR, ML, and MV processors.

For the case of a wideband processor it can similarly be shown that the solutions to various estimation and detection problems can be factored into a linear matrix filter followed by appropriate scalar processing. This development is undertaken in the next section.

The fact that the optimum weight vector solutions for an adaptive array using the different performance criteria indicated in the preceding section are all given (to within a constant factor) by the Wiener solution underscores the fundamental importance of the Wiener-Hopf equation in establishing theoretical adaptive array steady-state performance limits. These theoretical performance limits provide the designer with a standard for determining how much any improvement in array implementation can result in enhanced array steady-state performance, and they are a valuable tool for judging the merit of alternate designs.

## 3.4 OPTIMUM ARRAY PROCESSING FOR BROADBAND APPLICATIONS

One way of regarding the task of an array processor that is particularly appropriate for passive sensing devices is to note that the processor must decide whether the random processes observed at the array element outputs consist of a desired signal obscured by noise or consist of noise alone. For a system where such a decision must be made, the adoption of some performance measure like maximum output SNR or minimum MSE will not by itself be adequate when decision errors are important, since the performance measure considers no mechanism for making decisions concerning the presence or absence of a signal. It is well known from statistical detection theory, however, that decisions based upon the likelihood ratio test in some sense minimize the risk associated with an incorrect decision. In addition, the likelihood ratio test yields decisions that are also optimum for a wide range of performance criteria [36]–[38], and one of the concerns of this section is to establish the relationship between the likelihood ratio test and some other popular performance measures for broadband signal applications.

Let the observation vector $\mathbf{x}$ consist of elements representing the outputs from each of the array sensors $x_i(t)$, $i=1,2,\ldots,N$. The likelihood ratio is then given by the ratio of conditional probability density functions [39]

$$\Lambda(\mathbf{x}) \triangleq \frac{p[\mathbf{x}/\text{signal present}]}{p[\mathbf{x}/\text{signal absent}]} \tag{3.143}$$

If $\Lambda(\mathbf{x})$ exceeds a certain threshold $\eta$ then the signal present assumption is accepted as true, while if $\Lambda(\mathbf{x})$ is less than this threshold, the signal absent assumption is accepted. The ratio (3.143) therefore represents the likelihood that the sample $\mathbf{x}$ was observed, given that the signal is present relative to the likelihood that it was observed given that the signal is absent. Such an approach certainly comes far closer to determining the "best" system for a given class of decisions since it assumes at the outset that the system must make such decisions and obtains the processor design accordingly.

It is worthwhile to mention briefly some extentions of the likelihood ratio test represented by (3.143). In many practical cases one or several signal parameters (such as spatial location, phase, or signal energy) may be uncertain. When uncertain signal parameters (denoted by $\boldsymbol{\theta}$) are present, one intuitively appealing approach is explicitly to form an estimate of $\boldsymbol{\theta}$ (denoted by $\hat{\boldsymbol{\theta}}$) and use this estimate to form a classical generalized likelihood ratio (GLR) [40],

$$\Lambda_G(\mathbf{x}/\hat{\boldsymbol{\theta}}) \triangleq \frac{p[\mathbf{x}/\hat{\boldsymbol{\theta}}, \text{signal present}]}{p[\mathbf{x}/\text{signal absent}]} \tag{3.144}$$

Uncertain signal parameters can also be modeled as random variables with any prior information about them summarized in the form of an *a priori* density function $p(\boldsymbol{\theta})$. The likelihood ratio can then be written as the ratio of marginal density functions and is referred to as the Bayes likelihood ratio [40]

$$\Lambda_B(\mathbf{x}) \triangleq \frac{\int_{\theta} p[\mathbf{x}/\boldsymbol{\theta}, \text{signal present}]\, p(\boldsymbol{\theta})\, d\boldsymbol{\theta}}{p[\mathbf{x}/\text{signal absent}]} \tag{3.145}$$

where $\boldsymbol{\theta} \in \theta$.

It is well known that likelihood ratio tests can be regarded as completely equivalent to matched filter processing in which the output SNR is maximized at a certain observation time $t=T$ [41]–[43]. Another goal of this section is to determine the role that matched filtering plays in optimum array processing.

Another important concept in likelihood ratio tests is that of a "sufficient statistic" [44]. A brute force approach to the tests represented by the ratios

(3.143) to (3.145) is to process the observation vector $\mathbf{x}$ so as to actually construct the ratio $\Lambda(\mathbf{x})$ and compare this ratio to a threshold $\eta$ in order to make a decision. In many cases, although the observation vector is $N$-dimensional, the threshold decision can be regarded as occurring along a one-dimensional coordinate in the $N$-dimensional space, so that choosing a sufficient statistic simply amounts to picking a coordinate system in which one coordinate contains all the information required to make a decision. If such a sufficient statistic is available and is denoted by the scalar $l(\mathbf{x})$, then it is not necessary to construct an unwieldy ratio $\Lambda(\mathbf{x})$, since only the relatively simple quantity $l(\mathbf{x})$ need be considered in arriving at the required decision.

Detection performance is commonly measured in terms of the receiver operating characteristic (ROC). The ROC is a figure displaying the probability of detection (given by $\Pr[\Lambda(\mathbf{x})>\eta/\text{signal present}]$) versus the probability of false alarm (given by $\Pr[\Lambda(\mathbf{x})>\eta/\text{signal absent}]$) with the SNR as an independent parameter [38], [44]. Solutions for the optimum detection processor may also be obtained very simply by working instead with the "detection index" $d$ at a specified observation time $T$ where

$$d \triangleq \frac{E\{y(T)/\text{signal present}\} - E\{y(T)/\text{signal absent}\}}{\{\text{var}[y(T)/\text{signal absent}]\}^{1/2}} \tag{3.146}$$

and $y(T)$ is the processor output at $t=T$. By maximizing the detection index the detection performance is also optimized, since this corresponds to obtaining the greatest possible expected normalized difference of the processor output between the two signal conditions it is desired to discriminate between.

The unification of a variety of problems occurring in estimation and detection theory can be achieved by considering an observation vector described by

$$\mathbf{x}(t) = \mathbf{M}\mathbf{s}(t) + \mathbf{n}(t) \tag{3.147}$$

where $\mathbf{x}$ = observation vector
$\mathbf{n}$ = additive random noise vector having zero mean and noise covariance matrix $\mathbf{R}_{nn}$
$\mathbf{s}$ = signal vector (known, unknown, or random)
$\mathbf{M}$ = known transformation matrix

The treatment of (3.147) for different classes of problems given here generally follows the development presented by Cox [10]. In most cases results are obtained here using only arguments by analogy, which sacrifice the mathematical rigor of techniques which transform continuous time functions into discrete functions and use arguments based on Fourier series expansions of stationary random processes, but indicate how important results come about with very little effort. Equation (3.147) can be interpreted as a complex equation, which makes it necessary to define the meaning of a complex random vector. For purposes of this chapter, a complex random vector,

$$\mathbf{z}=\boldsymbol{\zeta}+j\boldsymbol{\gamma} \tag{3.148}$$

is required to have the following two properties, which remain invariant under any linear transformation:

1. The real part $\boldsymbol{\zeta}$ and the imaginary part $\boldsymbol{\gamma}$ are both real random vectors having the same covariance matrix.
2. All components of $\boldsymbol{\zeta}$ and $\boldsymbol{\gamma}$ satisfy

$$E\{\zeta_l\gamma_m\}=-E\{\zeta_m\gamma_l\} \qquad \text{for all } l \text{ and } m \tag{3.149}$$

The development that follows involves some useful matrix properties and generalizations of the Schwartz inequality that are summarized for convenience in Appendix D. Many of the results to be obtained will apply to Gaussian random vectors, and some useful properties of both real and complex Gaussian random vectors are given in Appendix E.

### *3.4.1 Estimation of a Random Signal*

Assuming that **s** in (3.147) is random, then the problem is posed of estimating **s** given the observation vector **x**. Let the mean value of **s** be given by

$$E\{\mathbf{s}\}=\mathbf{u} \tag{3.150}$$

and the associated covariance matrix be given by

$$E\{(\mathbf{s}-\mathbf{u})(\mathbf{s}-\mathbf{u})^{\dagger}\}=\mathbf{R}_{ss} \tag{3.151}$$

where $\mathbf{u}$ and $\mathbf{R}_{ss}$ are both known. The best estimate of **s**, given **x**, for a quadratic cost criterion, is just the conditional mean $E\{\mathbf{s}/\mathbf{x}\}$.

***Gaussian Random Signal.*** When **s** and **n** are both Gaussian and independent, then **x** and **s** are jointly Gaussian, and the conditional mean $E\{\mathbf{s}/\mathbf{x}\}$ can be obtained by considering the vector formed by combining **s** and **x** into a single vector. From (3.147), it follows that

$$E\left\{\begin{bmatrix}\mathbf{s}\\ \mathbf{x}\end{bmatrix}\right\}=\begin{bmatrix}\mathbf{u}\\ \mathbf{Mu}\end{bmatrix} \tag{3.152}$$

and

$$\text{cov}\begin{bmatrix}\mathbf{s}\\ \mathbf{x}\end{bmatrix}=\begin{bmatrix}\mathbf{R}_{ss} & \mathbf{R}_{ss}\mathbf{M}^{\dagger}\\ \mathbf{MR}_{ss} & \mathbf{MR}_{ss}\mathbf{M}^{\dagger}+\mathbf{R}_{nn}\end{bmatrix} \tag{3.153}$$

Applying (E.14) or (E.42) from Appendix E it follows immediately that

$$\hat{\mathbf{s}} = E\{\mathbf{s}/\mathbf{x}\} = \mathbf{u} + \mathbf{R}_{ss}\mathbf{M}^{\dagger}\left[\mathbf{M}\mathbf{R}_{ss}\mathbf{M}^{\dagger} + \mathbf{R}_{nn}\right]^{-1}(\mathbf{x} - \mathbf{M}\mathbf{u}) \tag{3.154}$$

The associated covariance matrix of $\hat{\mathbf{s}}$ can be obtained from (E.15) or (E.43) from Appendix E so that

$$\text{cov}(\hat{\mathbf{s}}) = \mathbf{R}_{ss} - \mathbf{R}_{ss}\mathbf{M}^{\dagger}\left[\mathbf{M}\mathbf{R}_{ss}\mathbf{M}^{\dagger} + \mathbf{R}_{nn}\right]^{-1}\mathbf{M}\mathbf{R}_{ss} \tag{3.155}$$

Some useful equivalent forms of (3.154) and (3.155) may be obtained by applying the matrix identities (D.10) and (D.11) of Appendix D to yield

$$\hat{\mathbf{s}} = \mathbf{u} + \left[\mathbf{R}_{ss}^{-1} + \mathbf{M}^{\dagger}\mathbf{R}_{nn}^{-1}\mathbf{M}\right]^{-1}\mathbf{M}^{\dagger}\mathbf{R}_{nn}^{-1}(\mathbf{x} - \mathbf{M}\mathbf{u}) \tag{3.156}$$

or

$$\hat{\mathbf{s}} = \left[\mathbf{I} + \mathbf{R}_{ss}\mathbf{M}^{\dagger}\mathbf{R}_{nn}^{-1}\mathbf{M}\right]^{-1}\left[\mathbf{u} + \mathbf{R}_{ss}\mathbf{M}^{\dagger}\mathbf{R}_{nn}^{-1}\mathbf{x}\right] \tag{3.157}$$

and

$$\text{cov}(\hat{\mathbf{s}}) = \left[\mathbf{R}_{ss}^{-1} + \mathbf{M}^{\dagger}\mathbf{R}_{nn}\mathbf{M}\right]^{-1} \tag{3.158}$$

or

$$\text{cov}(\hat{\mathbf{s}}) = \left[\mathbf{I} + \mathbf{R}_{ss}\mathbf{M}^{\dagger}\mathbf{R}_{nn}^{-1}\mathbf{M}\right]^{-1}\mathbf{R}_{ss} \tag{3.159}$$

Assuming $\mathbf{u} = \mathbf{0}$ yields the result for the interesting and practical zero mean case. Setting $\mathbf{u} = \mathbf{0}$, then (3.156) yields

$$\hat{\mathbf{s}} = \left[\mathbf{R}_{ss}^{-1} + \mathbf{M}^{\dagger}\mathbf{R}_{nn}\mathbf{M}\right]^{-1}\mathbf{M}^{\dagger}\mathbf{R}_{nn}^{-1}\mathbf{x} \tag{3.160}$$

which is a result that will be used later.

Since the mean, the mode, and the maximum likelihood of a Gaussian density function all coincide, the best estimate of $\mathbf{s}$ corresponds to the maximum of the *a posteriori* density function. In other words, the value of $\mathbf{s}$ is chosen that maximizes

$$p(\mathbf{s}/\mathbf{x}) = \frac{p(\mathbf{x}/\mathbf{s})p(\mathbf{s})}{p(\mathbf{x})} \tag{3.161}$$

From (3.147) it is seen that $p(\mathbf{x}/\mathbf{s})$ is Gaussian with mean $\mathbf{M}\mathbf{s}$ and covariance matrix $\mathbf{R}_{nn}$ and it is known that $p(\mathbf{s})$ is Gaussian with mean $\mathbf{u}$ and covariance matrix $\mathbf{R}_{ss}$. Consequently from (3.161) and (E.49) of Appendix E it follows that

$$p(\mathbf{s}/\mathbf{x})=(\text{const.})\exp\left\{-\frac{\alpha}{2}\left[(\mathbf{x}-\mathbf{Ms})^{\dagger}\mathbf{R}_{nn}^{-1}(\mathbf{x}-\mathbf{Ms})+(\mathbf{s}-\mathbf{u})^{\dagger}\mathbf{R}_{ss}^{-1}(\mathbf{s}-\mathbf{u})\right]\right\} \tag{3.162}$$

where $\alpha=2$ for a complex random vector, $\alpha=1$ for a real random vector, and "const." denotes a constant of proportionality. Choosing $\mathbf{s}$ to maximize (3.162) is completely equivalent to choosing $\mathbf{s}$ to minimize the following part of the exponent of (3.162):

$$J=(\mathbf{x}-\mathbf{Ms})^{\dagger}\mathbf{R}_{nn}^{-1}(\mathbf{x}-\mathbf{Ms})+(\mathbf{s}-\mathbf{u})^{\dagger}\mathbf{R}_{ss}^{-1}(\mathbf{s}-\mathbf{u}) \tag{3.163}$$

Either maximizing (3.162) or minimizing (3.163) then immediately leads to (3.154) as the appropriate equation for the estimate $\hat{\mathbf{s}}$. This procedure of minimizing the exponent of (3.162) is sometimes used in nonlinear estimation problems where the quantity $\mathbf{Ms}$ is replaced by a nonlinear function $\mathbf{m}(\mathbf{s})$ in which case the conditional mean is not so easily computed.

***Non-Gaussian Random Signal.*** In the event that $\mathbf{s}$ and $\mathbf{n}$ are non-Gaussian random vectors, the conditional mean $E\{\mathbf{s}/\mathbf{x}\}$ still remains the best estimate for a quadratic cost criterion. This fact may not be very helpful, however, since in many cases convenient expressions for computing the conditional mean do not exist, and, furthermore, available information may be limited to second-order statistics.

One approach to this problem that is sometimes adopted is to still use the estimate given by (3.154) or its equivalent forms even though such an estimate no longer represents the conditional mean of $p(\mathbf{s}/\mathbf{x})$.

Another approach often used in non-Gaussian problems is to find the linear estimate that minimizes the MSE. A linear estimate is one that has the form

$$\hat{\mathbf{s}}=\mathbf{a}+\mathbf{Bx} \tag{3.164}$$

The values assigned to $\mathbf{a}$ and $\mathbf{B}$ are determined by minimizing the expression

$$e=\text{trace}\left\{E\left[(\mathbf{s}-\hat{\mathbf{s}})(\mathbf{s}-\hat{\mathbf{s}})^{\dagger}\right]\right\}=E\left\{(\mathbf{s}-\hat{\mathbf{s}})^{\dagger}(\mathbf{s}-\hat{\mathbf{s}})\right\} \tag{3.165}$$

On combining (3.147) and (3.164) it follows that

$$(\mathbf{s}-\hat{\mathbf{s}})=\left[\mathbf{I}-\mathbf{BM}\right](\mathbf{s}-\mathbf{u})+\left\{\left[\mathbf{I}-\mathbf{BM}\right]\mathbf{u}-\mathbf{a}\right\}-\mathbf{Bn} \tag{3.166}$$

Consequently

$$E\{(\mathbf{s}-\hat{\mathbf{s}})(\mathbf{s}-\hat{\mathbf{s}})^{\dagger}\}=[\mathbf{I}-\mathbf{BM}]\mathbf{R}_{ss}[\mathbf{I}-\mathbf{M}^{\dagger}\mathbf{B}^{\dagger}]+\mathbf{BR}_{nn}\mathbf{B}^{\dagger}$$
$$+\{[\mathbf{I}-\mathbf{BM}]\mathbf{u}-\mathbf{a}\}\{[\mathbf{I}-\mathbf{BM}]\mathbf{u}-\mathbf{a}\}^{\dagger} \quad (3.167)$$

By setting the gradient of (3.167) with respect to $\mathbf{a}$ equal to zero, it is easily shown that the value of $\mathbf{a}$ that minimizes (3.165) is given by

$$\mathbf{a}=[\mathbf{I}-\mathbf{BM}]\mathbf{u} \quad (3.168)$$

After we substitute (3.168) into (3.167) and complete the square in the manner of (D.9) of Appendix D, it follows that

$$E\{(\mathbf{s}-\hat{\mathbf{s}})(\mathbf{s}-\hat{\mathbf{s}})^{\dagger}\}=\mathbf{R}_{ss}-\mathbf{R}_{ss}\mathbf{M}^{\dagger}[\mathbf{MR}_{ss}\mathbf{M}^{\dagger}+\mathbf{R}_{nn}]^{-1}\mathbf{MR}_{ss}$$
$$+\{\mathbf{B}-\mathbf{R}_{ss}\mathbf{M}^{\dagger}[\mathbf{MR}_{ss}\mathbf{M}^{\dagger}+\mathbf{R}_{nn}]^{-1}\}[\mathbf{MR}_{ss}\mathbf{M}^{\dagger}+\mathbf{R}_{nn}]$$
$$\cdot\{\mathbf{B}^{\dagger}-[\mathbf{MR}_{ss}\mathbf{M}^{\dagger}+\mathbf{R}_{nn}]^{-1}\mathbf{MR}_{ss}\} \quad (3.169)$$

The value of $\mathbf{B}$ that minimizes (3.165) may easily be found from (3.169) to be given by

$$\mathbf{B}=\mathbf{R}_{ss}\mathbf{M}^{\dagger}[\mathbf{MR}_{ss}\mathbf{M}^{\dagger}+\mathbf{R}_{nn}]^{-1} \quad (3.170)$$

By substituting the results from (3.168) and (3.170) into (3.164), the expression for $\hat{\mathbf{s}}$ corresponding to (3.154) is once again obtained. Likewise, by substituting (3.170) into (3.169) the same expression for the error covariance matrix as appeared in (3.155) also results. It therefore follows that the same estimation equations result from several different approaches. In the Gaussian case the estimate $\hat{s}$ given by (3.154) is the conditional mean, whereas in the non-Gaussian case the same estimate is the "best" linear estimate in the sense that it minimizes the MSE.

***Application of Random Signal Estimation Results to Optimum Array Processing.*** Assume that an array consisting of $N$ sensors is located in some region of space and that the received signal vector can be expressed as

$$\mathbf{x}(t)=\int_{-\infty}^{t}\mathbf{m}(t-v)s(v)\,dv+\mathbf{n}(t) \quad (3.171)$$

where $\mathbf{m}(t)$ is a known linear transformation that represents propagation effects and any signal distortion occurring in the sensor itself. In the case of ideal (nondispersive) propagation and distortion-free electronics, the elements of $\mathbf{m}(t)$ are pure time delays, $\delta(t-T_i)$. The quantity $s(t)$ is a scalar function of time representing the desired signal.

When the signal and noise processes are stationary and the observation interval is long (so that $t\rightarrow\infty$), then the convolution appearing in (3.171) can be circumvented by working in the frequency domain using Fourier transform techniques. Formally taking the Fourier transform of all quantities in (3.171) yields

$$\mathfrak{x}(\omega)=\mathfrak{m}(\omega)\mathfrak{s}(\omega)+\mathfrak{n}(\omega) \tag{3.172}$$

where the German quantities denote Fourier transformed variables. Note that (3.172) is of the same form as (3.147) although it is a frequency domain equation whereas (3.147) is a time domain equation. The fact that (3.172) is a frequency domain equation implies that cross spectral density matrices (which are the Fourier transforms of covariance matrices) now fill the role that covariance matrices formerly played for (3.147).

Now apply the frequency domain equivalent of (3.160) to obtain the solution

$$\hat{\mathfrak{s}}(\omega)=\left[\phi_{ss}^{-1}(\omega)+\mathfrak{m}^{\dagger}(\omega)\mathbf{\Phi}_{nn}^{-1}(\omega)\mathfrak{m}(\omega)\right]^{-1}\mathfrak{m}^{\dagger}(\omega)\mathbf{\Phi}_{nn}^{-1}(\omega)\mathfrak{x}(\omega) \tag{3.173}$$

Note that the quantity $\phi_{ss}^{-1}(\omega)+\mathfrak{m}^{\dagger}(\omega)\mathbf{\Phi}_{nn}^{-1}(\omega)\mathfrak{m}(\omega)$ is just a scalar so that (3.173) can be rewritten as

$$\hat{\mathfrak{s}}(\omega)=|\mathfrak{a}(\omega)|^2\mathfrak{m}^{\dagger}(\omega)\mathbf{\Phi}_{nn}^{-1}(\omega)\mathfrak{x}(\omega) \tag{3.174}$$

where

$$|\mathfrak{a}(\omega)|^2=\frac{\phi_{ss}(\omega)}{1+\phi_{ss}(\omega)\mathfrak{m}^{\dagger}(\omega)\mathbf{\Phi}_{nn}^{-1}(\omega)\mathfrak{m}(\omega)} \tag{3.175}$$

and $\phi_{ss}(\omega)$ is simply the power of $s(t)$ appearing at the frequency $\omega$. Note that in general a frequency response such as that given by (3.174) is not realizable, and it is necessary to introduce a time delay in order to obtain a good approximation to the corresponding time waveform $\hat{s}(t)$. Taking the inverse Fourier transform of (3.174) to obtain $\hat{s}(t)$ then yields

$$\hat{s}(t)=\frac{1}{2\pi}\int_{-\infty}^{\infty}|\mathfrak{a}(\omega)|^2\mathfrak{m}^{\dagger}(\omega)\mathbf{\Phi}_{nn}^{-1}(\omega)\mathfrak{x}(\omega)e^{j\omega t}\,d\omega \tag{3.176}$$

Equation (3.174) is especially interesting in view of the following interpretation of that result. Note that (3.174) can be rewritten as

$$\hat{\mathfrak{s}}(\omega)=|\mathfrak{a}(\omega)|^2\mathfrak{h}(\omega)\mathfrak{x}(\omega) \tag{3.177}$$

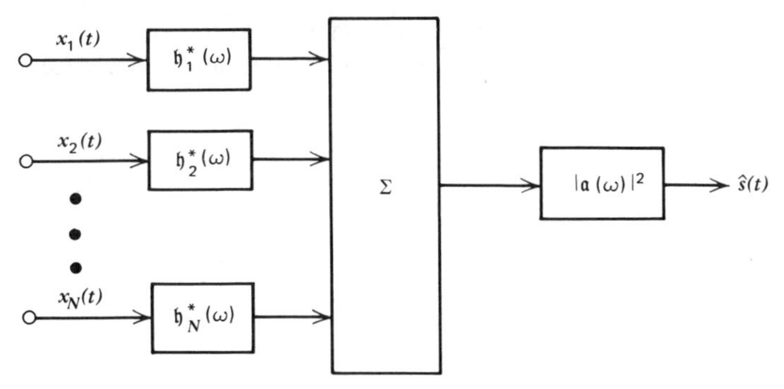


**Figure 3.7** Optimum array processor for estimation of a random signal.

where

$$\mathfrak{h}(\omega)=\mathfrak{m}^{\dagger}(\omega)\mathbf{\Phi}_{nn}^{-1}(\omega) \tag{3.178}$$

is a $1\times n$ row vector of individual filters. The single filter $\mathfrak{h}_j(\omega)$ operates on the received signal component $\mathfrak{x}_j(\omega)$ so that it performs the operation of spacial prewhitening and then matching to the known propagation and distortion effects imprinted on the structure of the signal. The processor obtained for this estimation problem therefore employs the principle: first prewhiten, then match. A block diagram of the processor corresponding to (3.174) is shown in Figure 3.7.

### *3.4.2 Estimation of an Unknown, Nonrandom Signal*

Now assume that $\mathbf{s}$ in (3.147) is a vector of unknown parameters rather than a vector of random variables. The problem then once again arises of estimating $\mathbf{s}$ given the observation vector $\mathbf{x}$. An estimate having desirable optimal properties for nonrandom parameters is the maximum likelihood estimate that corresponds to the value of $\mathbf{s}$ that maximizes the conditional density function $p(\mathbf{x}/\mathbf{s})$ for a particular observation $\mathbf{x}$.

***Gaussian Noise Case.*** When the noise vector $\mathbf{n}$ is Gaussian, the conditional density function takes the form

$$p(\mathbf{x}/\mathbf{s})=\text{const.}\ \exp\left\{-\tfrac{1}{2}\alpha(\mathbf{x}-\mathbf{Ms})^{\dagger}\mathbf{R}_{nn}^{-1}(\mathbf{x}-\mathbf{Ms})\right\} \tag{3.179}$$

Maximizing (3.179) corresponds to minimizing the exponent, and the estimate that minimizes this exponent is easily shown to be given by

$$\hat{\mathbf{s}}=\left[\mathbf{M}^{\dagger}\mathbf{R}_{nn}^{-1}\mathbf{M}\right]^{-1}\mathbf{M}^{\dagger}\mathbf{R}_{nn}^{-1}\mathbf{x} \tag{3.180}$$

It is seen from (3.180) that $\hat{\mathbf{s}}$ is obtained from a linear transformation on $\mathbf{x}$, so the distribution of $\hat{\mathbf{s}}$ is also Gaussian. From (3.180) it immediately follows that since $E\{\mathbf{x}\}=\mathbf{Ms}$, then

$$E\{\hat{\mathbf{s}}\}=\mathbf{s} \tag{3.181}$$

and

$$\text{cov}(\hat{\mathbf{s}})=\left[\mathbf{M}^{\dagger}\mathbf{R}_{nn}^{-1}\mathbf{M}\right]^{-1} \tag{3.182}$$

It is also interesting to note that this same estimate (3.180) is reached in the limit in (3.156) as $\mathbf{u}\rightarrow\mathbf{0}$ and $\mathbf{R}_{ss}^{-1}\rightarrow\mathbf{0}$. In other words, the ML estimate corresponds to the estimate obtained with an *a priori* Gaussian distribution $p(\mathbf{s})$ having zero mean and infinite variance.

***Non-Gaussian Noise Case.*** When $\mathbf{n}$ is not Gaussian, the likelihood function will not have the simple form of (3.179), and consequently the ML estimate may not be so easily obtained. Furthermore, the available information is usually limited to second-order statistics. Consequently, a weighted least square estimate is popular to use in these circumstances, and it can be obtained by minimizing

$$J=(\mathbf{x}-\mathbf{Ms})^{\dagger}\mathbf{R}_{nn}^{-1}(\mathbf{x}-\mathbf{Ms}) \tag{3.183}$$

The value of $\mathbf{s}$ that minimizes (3.183) is easily shown to be given by (3.180).

Yet another approach is to use the minimum variance unbiased linear estimate. A linear estimate will have the form

$$\hat{\mathbf{s}}=\mathbf{a}+\mathbf{Bx} \tag{3.184}$$

and an unbiased estimate requires that

$$E\{\hat{\mathbf{s}}\}=\mathbf{s} \tag{3.185}$$

If an estimate is to have minimum variance, then it is necessary that

$$\text{trace}\left\{E\left[(\hat{\mathbf{s}}-\mathbf{s})(\hat{\mathbf{s}}-\mathbf{s})^{\dagger}\right]\right\}=E\left\{(\hat{\mathbf{s}}-\mathbf{s})^{\dagger}(\hat{\mathbf{s}}-\mathbf{s})\right\} \tag{3.186}$$

be minimized. On combining (3.147) and (3.184) there results

$$\hat{\mathbf{s}}=\mathbf{a}+\mathbf{BMs}+\mathbf{Bn} \tag{3.187}$$

In order for (3.185) to hold, it is necessary that

$$\mathbf{a}=\mathbf{0} \tag{3.188}$$

and

$$\mathbf{BM}=\mathbf{I} \tag{3.189}$$

From (3.187) and (3.189) it follows that

$$E\{\hat{\mathbf{s}}\hat{\mathbf{s}}^\dagger\}=\mathbf{s}\mathbf{s}^\dagger+\mathbf{B}\mathbf{R}_{nn}\mathbf{B}^\dagger \tag{3.190}$$

and (3.186) will therefore be minimized by choosing $\mathbf{B}$ to minimize the quantity trace($\mathbf{B}\mathbf{R}_{nn}\mathbf{B}^\dagger$) subject to the constraint (3.189). This constrained minimization problem can be solved by introducing a matrix Lagrange multiplier $\boldsymbol{\lambda}$ and minimizing the quantity

$$J=\text{trace}\left\{\mathbf{B}\mathbf{R}_{nn}\mathbf{B}^\dagger+\left[\mathbf{I}-\mathbf{BM}\right]\boldsymbol{\lambda}+\boldsymbol{\lambda}^\dagger\left[\mathbf{I}-\mathbf{M}^\dagger\mathbf{B}^\dagger\right]\right\} \tag{3.191}$$

On completing the square of (3.191) using the formula (D.9) in Appendix D, the above expression can be rewritten as

$$\begin{aligned}J=\text{trace}\{&\boldsymbol{\lambda}+\boldsymbol{\lambda}^\dagger-\boldsymbol{\lambda}^\dagger\mathbf{M}^\dagger\mathbf{R}_{nn}^{-1}\mathbf{M}\boldsymbol{\lambda}\\&+\left[\mathbf{B}-\boldsymbol{\lambda}^\dagger\mathbf{M}^\dagger\mathbf{R}_{nn}^{-1}\right]\mathbf{R}_{nn}\left[\mathbf{B}^\dagger-\mathbf{R}_{nn}^{-1}\mathbf{M}\boldsymbol{\lambda}\right]\}\end{aligned} \tag{3.192}$$

From (3.192), it follows that the minimizing value of $\mathbf{B}$ is given by

$$\mathbf{B}=\boldsymbol{\lambda}^\dagger\mathbf{M}^\dagger\mathbf{R}_{nn}^{-1} \tag{3.193}$$

To eliminate $\boldsymbol{\lambda}$, use (3.189) so that

$$\boldsymbol{\lambda}^\dagger\mathbf{M}^\dagger\mathbf{R}_{nn}^{-1}\mathbf{M}=\mathbf{I} \tag{3.194}$$

or

$$\boldsymbol{\lambda}^\dagger=\left[\mathbf{M}^\dagger\mathbf{R}_{nn}^{-1}\mathbf{M}\right]^{-1} \tag{3.195}$$

and hence

$$\mathbf{B}=\left[\mathbf{M}^\dagger\mathbf{R}_{nn}^{-1}\mathbf{M}\right]^{-1}\mathbf{M}^\dagger\mathbf{R}_{nn}^{-1} \tag{3.196}$$

Consequently the estimate (3.184) once again reduces to (3.180), and the mean and covariance of the estimate are given by (3.181) and (3.182), respectively.

***Application of Unknown, Nonrandom Signal Estimation Results to Optimum Array Processing.*** Again using the frequency domain equation (3.172) it follows by analogy with (3.147) and (3.180) that the optimum frequency domain estimator is given by

$$\hat{\mathfrak{s}}(\omega)=\left[\mathfrak{m}^{\dagger}(\omega)\boldsymbol{\Phi}_{nn}^{-1}(\omega)\mathfrak{m}(\omega)\right]^{-1}\mathfrak{m}^{\dagger}(\omega)\boldsymbol{\Phi}_{nn}^{-1}(\omega)\mathfrak{x}(\omega) \tag{3.197}$$

Noting that $\mathfrak{m}^{\dagger}(\omega)\boldsymbol{\Phi}_{nn}^{-1}(\omega)\mathfrak{m}(\omega)$ is a scalar, (3.197) can be written as

$$\hat{\mathfrak{s}}(\omega)=\left[\frac{1}{\phi_{ss}(\omega)\mathfrak{m}^{\dagger}(\omega)\boldsymbol{\Phi}_{nn}^{-1}(\omega)\mathfrak{m}(\omega)}+1\right]\cdot|\mathfrak{a}(\omega)|^{2}\mathfrak{h}(\omega)\mathfrak{x}(\omega) \tag{3.198}$$

where $|\mathfrak{a}(\omega)|^2$ and $\mathfrak{h}(\omega)$ are given by (3.175) and (3.178), respectively. Therefore the only difference between the optimum array processor for random signal estimation and for unknown, nonrandom signal estimation is the presence of an additional scalar transfer function. The characteristic prewhitening and matching operation represented by $\mathfrak{h}(\omega)=\mathfrak{m}^{\dagger}(\omega)\boldsymbol{\Phi}_{nn}^{-1}(\omega)$ is still required.

### 3.4.3 *Detection of a Known Signal*

The binary detection problem in which one must decide whether or not a signal is present leads to a likelihood ratio test in which the value of the likelihood ratio given by (3.143) is compared with a threshold $\eta$.

***Gaussian Noise Case.*** When the noise is Gaussian, then the likelihood ratio (3.143) can be written as

$$\Lambda(\mathbf{x})=\frac{\exp\left\{-\frac{1}{2}\alpha(\mathbf{x}-\mathbf{Ms})^{\dagger}\mathbf{R}_{nn}^{-1}(\mathbf{x}-\mathbf{Ms})\right\}}{\exp\left\{-\frac{1}{2}\alpha(\mathbf{x}^{\dagger}\mathbf{R}_{nn}^{-1}\mathbf{x})\right\}} \tag{3.199}$$

where $\alpha=2$ for complex $\mathbf{x}$ and $\alpha=1$ for real $\mathbf{x}$. Clearly (3.199) can be written in terms of a single exponential function so that

$$\Lambda(\mathbf{x})=\exp\left\{-\tfrac{1}{2}\alpha\left[\mathbf{s}^{\dagger}\mathbf{M}^{\dagger}\mathbf{R}_{nn}^{-1}\mathbf{Ms}-\mathbf{s}^{\dagger}\mathbf{M}^{\dagger}\mathbf{R}_{nn}^{-1}\mathbf{x}-\mathbf{x}^{\dagger}\mathbf{R}_{nn}^{-1}\mathbf{Ms}\right]\right\} \tag{3.200}$$

Since the term $\mathbf{s}^{\dagger}\mathbf{M}^{\dagger}\mathbf{R}_{nn}^{-1}\mathbf{Ms}$ does not depend on any observation of $\mathbf{x}$, a sufficient test statistic for making a decision is the variable

$$y=\tfrac{1}{2}(\mathbf{s}^{\dagger}\mathbf{M}^{\dagger}\mathbf{R}_{nn}^{-1}\mathbf{x}+\mathbf{x}^{\dagger}\mathbf{R}_{nn}^{-1}\mathbf{Ms})=\mathrm{Re}\{\mathbf{s}^{\dagger}\mathbf{M}^{\dagger}\mathbf{R}_{nn}^{-1}\mathbf{x}\} \tag{3.201}$$

The factor $\alpha$ is assumed to be incorporated into the threshold level setting.

The distribution of the sufficient statistic $y$ is Gaussian since it results from a linear operation on $\mathbf{x}$, which in turn is Gaussian both when the signal is present and when it is absent.

When the signal is absent, then

$$E\{y\}=0 \tag{3.202}$$

and from (E.7) and (E.54) of Appendix E, it follows that

$$\mathrm{var}(y) = \frac{\mathbf{s}^\dagger \mathbf{M}^\dagger \mathbf{R}_{nn}^{-1} \mathbf{M}\mathbf{s}}{\alpha} \tag{3.203}$$

Likewise when the signal is present, then

$$E\{y\} = \mathbf{s}^\dagger \mathbf{M}^\dagger \mathbf{R}_{nn}^{-1} \mathbf{M}\mathbf{s} \tag{3.204}$$

and the variance of $y$ is the same as when the signal is absent.

***Non-Gaussian Noise Case.*** In the event that the noise vector $\mathbf{n}$ is non-Gaussian and only second-order statistics are available, a linear transformation of the observation vector $\mathbf{x}$ is commonly sought that is of the form

$$y = \tfrac{1}{2}\left[\mathbf{k}^\dagger \mathbf{x} + \mathbf{x}^\dagger \mathbf{k}\right] = \mathrm{Re}\{\mathbf{k}^\dagger \mathbf{x}\} \tag{3.205}$$

where the vector $\mathbf{k}$ is selected so the output SNR is maximized. The ratio given by

$$r_0 = \frac{\text{change in mean-squared output due to signal presence}}{\text{mean-squared output for noise alone}}$$

or equivalently

$$r_0 = \frac{E\{y^2/\text{signal present}\} - E\{y^2/\text{signal absent}\}}{E\{y^2/\text{signal absent}\}} \tag{3.206}$$

is referred to as the signal-to-noise ratio. Using (3.205) and (E.54) of Appendix E, the ratio can be written as

$$r_0 = \frac{\alpha\left[\mathrm{Re}\{\mathbf{k}^\dagger \mathbf{M}\mathbf{s}\}\right]^2}{\mathbf{k}^\dagger \mathbf{R}_{nn} \mathbf{k}} \leqslant \frac{\alpha \mathbf{k}^\dagger \mathbf{M}\mathbf{s}\mathbf{s}^\dagger \mathbf{M}^\dagger \mathbf{k}}{\mathbf{k}^\dagger \mathbf{R}_{nn} \mathbf{k}} \tag{3.207}$$

It is convenient to factor $\mathbf{R}_{nn}$ by introducing

$$\mathbf{R}_{nn} = \mathbf{T}^\dagger \mathbf{T} \tag{3.208}$$

in which case the upper bound for $r_0$ given in (3.207) can be written as

$$\frac{\alpha \mathbf{k}^\dagger \mathbf{M}\mathbf{s}\mathbf{s}^\dagger \mathbf{M}^\dagger \mathbf{k}}{\mathbf{k}^\dagger \mathbf{R}_{nn} \mathbf{k}} = \frac{\alpha \mathbf{k}^\dagger \mathbf{T}^\dagger (\mathbf{T}^\dagger)^{-1} \mathbf{M}\mathbf{s}\mathbf{s}^\dagger \mathbf{M}^\dagger \mathbf{T}^{-1} \mathbf{T}\mathbf{k}}{\mathbf{k}^\dagger \mathbf{T}^\dagger \mathbf{T}\mathbf{k}} \tag{3.209}$$

With the upper bound written as in (3.209), (D.16) of Appendix D can be applied and the resulting terms rearranged to yield

$$r_0 = \frac{\alpha\left[\mathrm{Re}\{\mathbf{k}^\dagger \mathbf{M}\mathbf{s}\}\right]^2}{\mathbf{k}^\dagger \mathbf{R}_{nn}\mathbf{k}} \leqslant \alpha \mathbf{s}^\dagger \mathbf{M}^\dagger \mathbf{R}_{nn}^{-1}\mathbf{M}\mathbf{s} \tag{3.210}$$

where equality obtains when

$$\mathbf{k}^\dagger = \mathbf{s}^\dagger \mathbf{M}^\dagger \mathbf{R}_{nn}^{-1} \tag{3.211}$$

On substituting $\mathbf{k}^\dagger$ of (3.211) into (3.205), it is found that the test statistic is once again given by (3.201).

A different approach to the problem of detecting a known signal embedded in non-Gaussian noise is to maximize the detection index given by (3.146). For $y$ given by (3.205), it is easily shown that

$$d = \sqrt{\alpha}\,\frac{\mathrm{Re}\{\mathbf{k}^\dagger \mathbf{M}\mathbf{s}\}}{\sqrt{\mathbf{k}^\dagger \mathbf{R}_{nn}\mathbf{k}}} \leqslant \frac{\sqrt{\alpha \mathbf{k}^\dagger \mathbf{M}\mathbf{s}\mathbf{s}^\dagger \mathbf{M}^\dagger \mathbf{k}}}{\sqrt{\mathbf{k}^\dagger \mathbf{R}_{nn}\mathbf{k}}} \tag{3.212}$$

Now applying (D.17) of Appendix D to the upper bound in (3.212) in the same manner as (D.16) was applied to the upper bound in (3.209), the detection index can be shown to satisfy

$$d = \sqrt{\alpha}\,\frac{\mathrm{Re}\{\mathbf{k}^\dagger \mathbf{M}\mathbf{s}\}}{\sqrt{\mathbf{k}^\dagger \mathbf{R}_{nn}\mathbf{k}}} \leqslant \sqrt{\alpha \mathbf{s}^\dagger \mathbf{M}^\dagger \mathbf{R}_{nn}^{-1}\mathbf{M}\mathbf{s}} \tag{3.213}$$

Equality obtains in (3.213) when (3.211) is satisfied so (3.201) once again results for the test statistic.

***Application of Known Signal Detection Results to Optimum Array Processing.*** With the received signal vector expressed by (3.171) and when the signal is known, it is simplest to once again cast the problem in the frequency domain as in (3.172) and maximize the detection index at a specified observation time $T$. The maximization can be carried out directly by applying an appropriate form of the Schwartz inequality to the detection index. With the processor specified to be linear as in Figure 3.8, then

$$E\{y(T)/\text{signal present}\} = \frac{1}{2\pi}\int_{-\infty}^{\infty} \mathfrak{k}^\dagger(\omega)\mathfrak{m}(\omega)\mathfrak{s}(\omega)e^{j\omega T}\,d\omega \tag{3.214}$$

and

$$\mathrm{var}[y(T)] = \frac{1}{2\pi}\int_{-\infty}^{\infty} \mathfrak{k}^\dagger(\omega)\mathbf{\Phi}_{nn}(\omega)\mathfrak{k}(\omega)\,d\omega \tag{3.215}$$

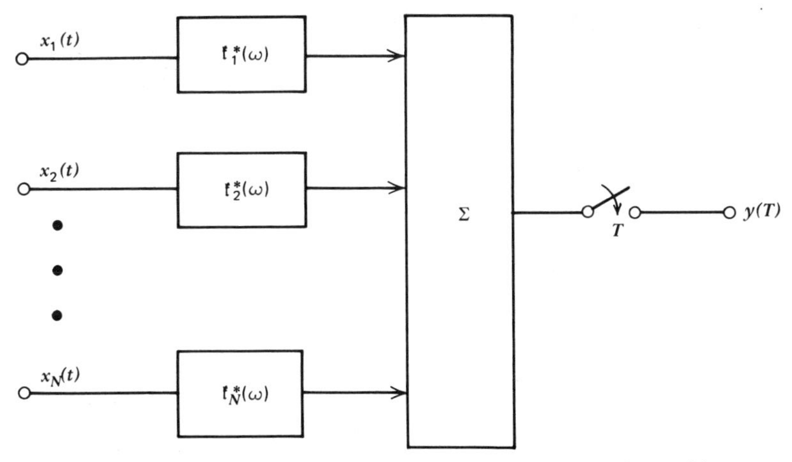


**Figure 3.8** Linear processor structure for known signal detection problem.

Since $E\{y(T)/\text{signal absent}\}=0$, the detection index becomes

$$d=\frac{1/\sqrt{2\pi}\int_{-\infty}^{\infty}\mathfrak{k}^{\dagger}(\omega)\mathfrak{m}(\omega)\mathfrak{s}(\omega)e^{j\omega T}\,d\omega}{\left[\int_{-\infty}^{\infty}\mathfrak{k}^{\dagger}(\omega)\boldsymbol{\Phi}_{nn}(\omega)\mathfrak{k}(\omega)\,d\omega\right]^{1/2}} \tag{3.216}$$

Now let $\boldsymbol{\Phi}_{nn}(\omega)=\mathfrak{T}^{\dagger}(\omega)\mathfrak{T}(\omega)$ so (3.216) can be rewritten as

$$d\sqrt{2\pi}=\frac{\int_{-\infty}^{\infty}\mathfrak{k}^{\dagger}(\omega)\mathfrak{T}^{\dagger}(\omega)\left[\mathfrak{T}^{\dagger}(\omega)\right]^{-1}\mathfrak{m}(\omega)\mathfrak{s}(\omega)e^{j\omega T}\,d\omega}{\left[\int_{-\infty}^{\infty}\mathfrak{k}^{\dagger}(\omega)\mathfrak{T}^{\dagger}(\omega)\mathfrak{T}(\omega)\mathfrak{k}(\omega)\,d\omega\right]^{1/2}} \tag{3.217}$$

The Schwartz inequality in the form of (D.14) of Appendix D may be applied to (3.217) by letting $\mathfrak{k}^{\dagger}(\omega)\mathfrak{T}^{\dagger}(\omega)$ play the role of $\mathbf{f}^{\dagger}$ and $[\mathfrak{T}^{\dagger}(\omega)]^{-1}\mathfrak{m}(\omega)\mathfrak{s}(\omega)e^{j\omega T}$ play the role of $\mathbf{g}$. It then follows that

$$d\sqrt{2\pi}\leqslant\left[\int_{-\infty}^{\infty}\mathfrak{s}^{*}(\omega)\mathfrak{m}^{\dagger}(\omega)\boldsymbol{\Phi}_{nn}^{-1}(\omega)\mathfrak{m}(\omega)\mathfrak{s}(\omega)\,d\omega\right]^{1/2} \tag{3.218}$$

where equality obtains if and only if

$$\mathfrak{k}^{\dagger}(\omega)=e^{-j\omega T}\mathfrak{s}^{*}(\omega)\mathfrak{m}^{\dagger}(\omega)\boldsymbol{\Phi}_{nn}^{-1}(\omega) \tag{3.219}$$

Once again the usual spatial prewhitening and matching operation represented by $\mathfrak{m}^{\dagger}(\omega)\boldsymbol{\Phi}_{nn}^{-1}(\omega)$ appears in the optimum processor.

### *3.4.4 Detection of a Random Signal*

When $\mathbf{s}$ is a random signal vector it will again be assumed that the mean value $\mathbf{u}$ and the covariance matrix $\mathbf{R}_{ss}$ are both known. The optimum processor for the binary detection problem again leads to a likelihood ratio test.

***Gaussian Case.*** When the vectors $\mathbf{s}$ and $\mathbf{n}$ are both Gaussian, the likelihood ratio (3.143) can be written as

$$\Lambda(\mathbf{x}) = \text{const.} \frac{\exp\left\{-(\alpha/2)(\mathbf{x}-\mathbf{Mu})^{\dagger}\left[\mathbf{MR}_{ss}\mathbf{M}^{\dagger}+\mathbf{R}_{nn}\right]^{-1}(\mathbf{x}-\mathbf{Mu})\right\}}{\exp\left\{-(\alpha/2)\mathbf{x}^{\dagger}\mathbf{R}_{nn}^{-1}\mathbf{x}\right\}} \tag{3.220}$$

where "const." represents a constant of proportionality. Expanding (3.220) by carrying out indicated multiplications and taking logarithms yields the following sufficient test statistic:

$$\begin{aligned} y = {} & \tfrac{1}{2}(\mathbf{x}-\mathbf{Mu})^{\dagger}\left\{\mathbf{R}_{nn}^{-1}-\left[\mathbf{MR}_{ss}\mathbf{M}^{\dagger}+\mathbf{R}_{nn}\right]^{-1}\right\}(\mathbf{x}-\mathbf{Mu}) \\ & + \tfrac{1}{2}\mathbf{u}^{\dagger}\mathbf{M}^{\dagger}\mathbf{R}_{nn}^{-1}\mathbf{x} + \tfrac{1}{2}\mathbf{x}^{\dagger}\mathbf{R}_{nn}^{-1}\mathbf{Mu} \end{aligned} \tag{3.221}$$

where the factor $\alpha$ has again been incorporated into the threshold level setting.

Thinking of the known mean value $\mathbf{u}$ as the deterministic part of $\mathbf{s}$ and the deviation of the observation from its mean $\mathbf{x}-\mathbf{Mu}$ as the random part of the observation, then we see the test statistic (3.221) to be comprised of a linear term corresponding to the test statistic in the known signal case [represented by (3.201)], and a quadratic term involving only the random part of the observation. Since the known signal case has been treated in the Section 3.4.3, only the random part of the problem will be considered here by assuming that the mean has been subtracted out or, equivalently, by assuming that $\mathbf{s}$ has zero mean.

When $\mathbf{u}=0$, then (3.221) reduces to

$$y = \tfrac{1}{2}\mathbf{x}^{\dagger}\left\{\mathbf{R}_{nn}^{-1}-\left[\mathbf{MR}_{ss}\mathbf{M}^{\dagger}+\mathbf{R}_{nn}\right]^{-1}\right\}\mathbf{x} \tag{3.222}$$

Applying the matrix identities (D.11) and (D.12) from Appendix D, we may obtain equivalent forms of (3.222) as follows:

$$y = \tfrac{1}{2}\mathbf{x}^{\dagger}\mathbf{R}_{nn}^{-1}\mathbf{M}\left[\mathbf{R}_{ss}^{-1}+\mathbf{M}^{\dagger}\mathbf{R}_{nn}^{-1}\mathbf{M}\right]^{-1}\mathbf{M}^{\dagger}\mathbf{R}_{nn}^{-1}\mathbf{x} \tag{3.223}$$

or

$$y = \tfrac{1}{2}\mathbf{x}^{\dagger}\mathbf{R}_{nn}^{-1}\mathbf{MR}_{ss}\mathbf{M}^{\dagger}\left[\mathbf{MR}_{ss}\mathbf{M}^{\dagger}+\mathbf{R}_{nn}\right]^{-1}\mathbf{x} \tag{3.224}$$

A special case of some practical importance arises when $\mathbf{s}$ is only a scalar $s$ and $\mathbf{M}$ is a column vector $\mathbf{m}$. Then (3.223) becomes

$$2y = \frac{|\mathbf{x}^\dagger \mathbf{R}_{nn}^{-1}\mathbf{m}|^2}{\phi_{ss} + \mathbf{m}^\dagger \mathbf{R}_{nn}^{-1}\mathbf{m}} \tag{3.225}$$

Another special case of some importance is the small signal case where

$$\left[\mathbf{M}\mathbf{R}_{ss}\mathbf{M}^\dagger + \mathbf{R}_{nn}\right]^{-1} \approx \mathbf{R}_{nn}^{-1} \tag{3.226}$$

In this case, (3.224) becomes

$$y = \tfrac{1}{2}\mathbf{x}^\dagger \mathbf{R}_{nn}^{-1}\mathbf{M}\mathbf{R}_{ss}\mathbf{M}^\dagger \mathbf{R}_{nn}^{-1}\mathbf{x} \tag{3.227}$$

***Gaussian Noise, Non-Gaussian Signal.*** Whereas the sufficient test statistic for the known signal detection problem involved only a linear operation on the observation vector, the random signal detection problem of the preceding section involved a test statistic that was quadratic in the observation vector. Likewise when $\mathbf{s}$ is non-Gaussian and only second-order statistics are available, then the "best" quadratic processor is found (in the sense that the detection index is maximized). When we assume $\mathbf{s}$ has zero mean, the test statistic is presumed to have the form:

$$y = \mathbf{x}^\dagger \mathbf{K}\mathbf{K}^\dagger \mathbf{x} \tag{3.228}$$

and the problem is now to select the matrix $\mathbf{K}$ to maximize the detection index defined by (3.146). Note that since $y$ is quadratic in $\mathbf{x}$ and the variance of $y$ in the denominator of (3.146) involves $E\{y^2/\text{signal absent}\}$, then fourth-order moments of the distribution for $\mathbf{x}$ are involved. By assuming the noise field is Gaussian, the required fourth-order moments can be expressed in terms of the covariance matrix by applying (E.51) of Appendix E.

The numerator of (3.146) when $y$ is given by (3.228) can be written as

$$E\{y/\text{signal present}\} - E\{y/\text{signal absent}\} = \text{trace}\left[\mathbf{K}^\dagger \mathbf{M}\mathbf{R}_{ss}\mathbf{M}^\dagger \mathbf{K}\right] \tag{3.229}$$

By use of (E.51), the denominator of (3.146) becomes

$$\sqrt{\text{var}(y/\text{signal absent})} = \sqrt{\text{trace}\left[\left(\mathbf{K}^\dagger \mathbf{R}_{nn}\mathbf{K}\right)^2\right]\cdot\frac{2}{\alpha}} \tag{3.230}$$

and consequently

$$d = \frac{\sqrt{(\alpha/2)}\ \text{trace}(\mathbf{K}^\dagger \mathbf{M}\mathbf{R}_{ss}\mathbf{M}^\dagger \mathbf{K})}{\sqrt{\text{trace}\left[\left(\mathbf{K}^\dagger \mathbf{R}_{nn}\mathbf{K}\right)^2\right]}} \tag{3.231}$$

It is now desired to maximize $d$ by applying the Schwartz inequality to obtain an upper bound. It is again convenient to introduce $\mathbf{R}_{nn} = \mathbf{T}^\dagger\mathbf{T}$ before applying (D.16) so that

$$d = \frac{\sqrt{(\alpha/2)}\ \mathrm{trace}\left[(\mathbf{T}^\dagger)^{-1}\mathbf{M}\mathbf{R}_{ss}\mathbf{M}^\dagger\mathbf{T}^{-1}\mathbf{T}\mathbf{K}\mathbf{K}^\dagger\mathbf{T}^\dagger\right]}{\sqrt{\mathrm{trace}\left\{(\mathbf{T}\mathbf{K}\mathbf{K}^\dagger\mathbf{T}^\dagger)^2\right\}}}$$

$$\leqslant \sqrt{(\alpha/2)\mathrm{trace}\left\{(\mathbf{M}\mathbf{R}_{ss}\mathbf{M}^\dagger\mathbf{R}_{nn}^{-1})^2\right\}} \tag{3.232}$$

Equality obtains in (3.232) when

$$\mathbf{K}^\dagger = \mathbf{A}^\dagger\mathbf{M}^\dagger\mathbf{R}_{nn}^{-1} \tag{3.233}$$

where

$$\mathbf{R}_{ss} = \mathbf{A}\mathbf{A}^\dagger \tag{3.234}$$

Substituting (3.233) into (3.228) then yields the test statistic

$$y = \mathbf{x}^\dagger\mathbf{R}_{nn}^{-1}\mathbf{M}\mathbf{R}_{ss}\mathbf{M}^\dagger\mathbf{R}_{nn}^{-1}\mathbf{x} \tag{3.235}$$

which is precisely the same as the test statistic (3.227) obtained from the likelihood ratio in the small signal case.

***Application of Random Signal Detection Results to Optimum Array Processing.*** Once again using the frequency domain equation (3.172), we find it follows directly from (3.223) that a sufficient test statistic is given by

$$\mathfrak{y}(\omega) = \tfrac{1}{2}|\mathfrak{a}(\omega)|^2\left[\mathfrak{m}^\dagger(\omega)\mathbf{\Phi}_{nn}^{-1}(\omega)\mathfrak{x}(\omega)\right]^\dagger\left[\mathfrak{m}^\dagger(\omega)\mathbf{\Phi}_{nn}^{-1}(\omega)\mathfrak{x}(\omega)\right]$$

$$= \tfrac{1}{2}|\mathfrak{a}(\omega)\mathfrak{m}^\dagger(\omega)\mathbf{\Phi}_{nn}^{-1}(\omega)\mathfrak{x}(\omega)|^2 \tag{3.236}$$

where $|\mathfrak{a}(\omega)|^2$ is given by (3.175). The structure of the optimum processor corresponding to (3.176) is shown in Figure 3.9 where Parseval's theorem has been invoked to write (3.236) in the time domain:

$$y(T) = \frac{1}{2T}\int_0^T |z(t)|^2\,dt \tag{3.237}$$

If the small signal assumption is applicable, then using (3.227) instead of (3.223) leads to

$$\mathfrak{y}(\omega) = \tfrac{1}{2}\phi_{ss}(\omega)|\mathfrak{m}^\dagger(\omega)\mathbf{\Phi}_{nn}^{-1}(\omega)\mathfrak{x}(\omega)|^2 \tag{3.238}$$

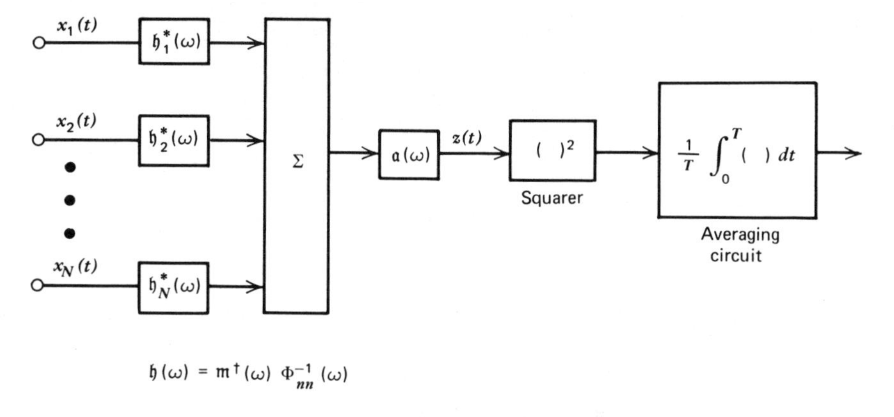


$$|\mathfrak{a}(\omega)|^2 = \frac{\phi_{ss}(\omega)}{1+\phi_{ss}(\omega)\,\mathfrak{m}^\dagger(\omega)\,\Phi_{nn}^{-1}(\omega)\,\mathfrak{m}^\dagger(\omega)}$$

**Figure 3.9** Optimum array processor for detection of a random signal.

The result expressed by (3.238) is different from (3.236) only in that the $\phi_{ss}(\omega)$ has replaced the scalar factor $|\mathfrak{a}(\omega)|^2$ in (3.236).

### *3.4.5 Detection of an Unknown, Nonrandom Signal*

When $\mathbf{s}$ is unknown but nonrandom, an approach that often yields good results is to use the generalized likelihood ratio (GLR) test. The procedure is to form the ratio

$$\Lambda_G(\mathbf{x}) = \frac{p(\mathbf{x}/\hat{\mathbf{s}}, \text{signal present})}{p(\mathbf{x}/\text{signal absent})} \tag{3.239}$$

where $\hat{\mathbf{s}}$ is selected to maximize the conditional density function $p(\mathbf{x}/\mathbf{s}$, signal present) and is therefore just the maximum likelihood estimate of $\mathbf{s}$.

***Gaussian Noise Case.*** When $\mathbf{n}$ is Gaussian the likelihood ratio is given by

$$\frac{p(\mathbf{x}/\mathbf{s}, \text{signal present})}{p(\mathbf{x}/\text{signal absent})} = \frac{\exp\{-(\alpha/2)(\mathbf{x}-\mathbf{Ms})^\dagger \mathbf{R}_{nn}^{-1}(\mathbf{x}-\mathbf{Ms})\}}{\exp\{-(\alpha/2)\mathbf{x}^\dagger \mathbf{R}_{nn}^{-1}\mathbf{x}\}} \tag{3.240}$$

It is now desired to substitute the maximum likelihood estimate of $\mathbf{s}$ into (3.240). The maximum likelihood estimate for $\mathbf{s}$ is given by (3.180). Substituting $\hat{\mathbf{s}}$ of (3.180) into (3.240) then yields the generalized likelihood ratio:

$$\Lambda_G(\mathbf{x}) = \exp\left\{\frac{\alpha}{2}\mathbf{x}^\dagger \mathbf{R}_{nn}^{-1}\mathbf{M}\left[\mathbf{M}^\dagger \mathbf{R}_{nn}^{-1}\mathbf{M}\right]^{-1}\mathbf{M}^\dagger \mathbf{R}_{nn}^{-1}\mathbf{x}\right\} \tag{3.241}$$

The sufficient test statistic for (3.241) is obviously

$$y=\tfrac{1}{2}\mathbf{x}^{\dagger}\mathbf{R}_{nn}^{-1}\mathbf{M}\left[\mathbf{M}^{\dagger}\mathbf{R}_{nn}^{-1}\mathbf{M}\right]^{-1}\mathbf{M}^{\dagger}\mathbf{R}_{nn}^{-1}\mathbf{x} \tag{3.242}$$

or

$$y=\tfrac{1}{2}\hat{\mathbf{s}}_1^{\dagger}\mathbf{M}^{\dagger}\mathbf{R}_{nn}^{-1}\mathbf{x} \tag{3.243}$$

where

$$\hat{\mathbf{s}}_1=\left[\mathbf{M}^{\dagger}\mathbf{R}_{nn}^{-1}\mathbf{M}\right]^{-1}\mathbf{M}^{\dagger}\mathbf{R}_{nn}^{-1}\mathbf{x} \tag{3.244}$$

It is worthwhile noting that by setting $\mathbf{R}_{ss}^{-1}=\mathbf{0}$ in (3.223) for the case of a random signal vector, then (3.223) reduces to (3.242).

***Application of Unknown, Nonrandom Signal Detection Results to Optimum Array Processing.*** Again using the frequency domain equation (3.172), we find it follows from (3.242) that a sufficient test statistic is given by

$$\begin{aligned}\mathfrak{y}(\omega)&=\tfrac{1}{2}\left[\mathfrak{m}^{\dagger}(\omega)\mathbf{\Phi}_{nn}^{-1}(\omega)\mathfrak{m}(\omega)\right]^{-1}\left[\mathfrak{m}^{\dagger}(\omega)\mathbf{\Phi}_{nn}^{-1}(\omega)\mathfrak{x}(\omega)\right]^{\dagger}\\&\quad\cdot\left[\mathfrak{m}^{\dagger}(\omega)\mathbf{\Phi}_{nn}^{-1}(\omega)\mathfrak{x}(\omega)\right]\\&=\tfrac{1}{2}\left[\mathfrak{m}^{\dagger}(\omega)\mathbf{\Phi}_{nn}^{-1}(\omega)\mathfrak{m}(\omega)\right]^{-1}\left|\mathfrak{m}^{\dagger}(\omega)\mathbf{\Phi}_{nn}^{-1}(\omega)\mathfrak{x}(\omega)\right|^{2}\end{aligned} \tag{3.245}$$

The result expressed by (3.245) again reveals the characteristic prewhitening and matching operator $\mathfrak{m}^{\dagger}(\omega)\mathbf{\Phi}_{nn}^{-1}(\omega)$.

### *3.4.6 Array Gain*

The array gain is defined to be the ratio of the output signal-to-noise spectral ratio to the input signal-to-noise spectral ratio. The linear processor structure of Figure 3.8 consists of a set of filters (one for each sensor), followed by a summation device for which the array gain is given by [17]

$$G(\omega)=\frac{\mathfrak{k}^{\dagger}(\omega)\tilde{\mathbf{\Phi}}_{ss}(\omega)\mathfrak{k}(\omega)}{\mathfrak{k}^{\dagger}(\omega)\tilde{\mathbf{\Phi}}_{nn}(\omega)\mathfrak{k}(\omega)} \tag{3.246}$$

where $\tilde{\mathbf{\Phi}}_{ss}(\omega)$ and $\tilde{\mathbf{\Phi}}_{nn}(\omega)$ represent the normalized cross spectral density matrices of the signal vector and noise vector, that is,

$$\mathbf{\Phi}_{ss}(\omega)=\sigma_s^2(\omega)\tilde{\mathbf{\Phi}}_{ss}(\omega) \tag{3.247}$$

and

$$\mathbf{\Phi}_{nn}(\omega)=\sigma_n^2(\omega)\tilde{\mathbf{\Phi}}_{nn}(\omega) \tag{3.248}$$

where $\sigma_n^2(\omega)$ is the noise power spectral density averaged over the $N$ sensors so that

$$\sigma_n^2(\omega) = \frac{1}{N}\operatorname{trace}\left[\mathbf{\Phi}_{nn}(\omega)\right] \tag{3.249}$$

and

$$\sigma_s^2(\omega) = \frac{1}{N}\operatorname{trace}\left[\mathbf{\Phi}_{ss}(\omega)\right] \tag{3.250}$$

The array gain corresponding to (3.246) can be recognized as the ratio of the output signal-to-noise spectral ratio to the input signal-to-noise spectral ratio.

Whenever the signal vector $\mathbf{s}(t)$ is related to a scalar signal $s(t)$ by a known transformation $\mathbf{m}(t)$ such that

$$\mathbf{s}(t) = \int_{-\infty}^{t} \mathbf{m}(t-\tau)s(\tau)\,d\tau \tag{3.251}$$

Then $\tilde{\mathbf{\Phi}}_{ss}(\omega)$ is simply given by the dyad [17]

$$\tilde{\mathbf{\Phi}}_{ss}(\omega) = \tilde{\mathfrak{m}}(\omega)\tilde{\mathfrak{m}}^{\dagger}(\omega) \tag{3.252}$$

where $\tilde{\mathfrak{m}}(\omega)$ denotes the normalized Fourier transform of $\mathbf{m}(t)$ so that

$$\tilde{\mathfrak{m}}^{\dagger}(\omega)\tilde{\mathfrak{m}}(\omega) = N \tag{3.253}$$

When $\tilde{\mathbf{\Phi}}_{ss}(\omega)$ is given by (3.252), then the array gain becomes

$$G(\omega) = \frac{|\mathfrak{f}^{\dagger}(\omega)\tilde{\mathfrak{m}}(\omega)|^2}{\mathfrak{f}^{\dagger}(\omega)\tilde{\mathbf{\Phi}}_{nn}(\omega)\mathfrak{f}(\omega)} \tag{3.254}$$

The quantity $\mathfrak{f}^{\dagger}(\omega)\tilde{\mathfrak{m}}(\omega)$ may be regarded as the inner product of the beam-steering vector $\mathfrak{f}$ and the signal direction $\tilde{\mathfrak{m}}$ for plane wave propagation. Define a generalized angle $\gamma$ between $\mathfrak{f}$ and $\tilde{\mathfrak{m}}$ as described in Appendix F such that

$$\cos^2(\gamma) = \frac{|\mathfrak{f}^{\dagger}(\omega)\tilde{\mathfrak{m}}(\omega)|^2}{(\mathfrak{f}^{\dagger}(\omega)\mathfrak{f}(\omega))(\tilde{\mathfrak{m}}^{\dagger}(\omega)\tilde{\mathfrak{m}}(\omega))} \tag{3.255}$$

In a conventional beamformer the vector $\mathfrak{f}$ is chosen to be proportional to $\tilde{\mathfrak{m}}$, thus making $\gamma$ equal to zero and "matching to the signal direction." This operation also maximizes the array gain against spatially white noise as shown below.

Substituting (3.252) into (3.246) and using $\mathfrak{z} = \tilde{\mathbf{\Phi}}_{nn}^{\frac{1}{2}}\mathfrak{f}$ yields

$$G(\omega) = \frac{|\mathfrak{z}^{\dagger}(\omega)\tilde{\mathbf{\Phi}}_{nn}^{-\frac{1}{2}}(\omega)\tilde{\mathfrak{m}}(\omega)|^2}{\mathfrak{z}^{\dagger}(\omega)\mathfrak{z}(\omega)} \tag{3.256}$$

Applying the Schwartz inequality (D.14) to (3.256) then gives

$$G(\omega) \leqslant \tilde{\mathfrak{m}}^{\dagger}(\omega)\tilde{\Phi}_{nn}^{-1}(\omega)\tilde{\mathfrak{m}}(\omega) \tag{3.257}$$

where equality obtains when the vector $\mathfrak{f}^{\dagger}$ is chosen to be a scalar multiple of $\tilde{\mathfrak{m}}^{\dagger}\tilde{\Phi}_{nn}^{-1}$. Therefore maximizing the array gain with no constraints yields the same prewhitening and matching operation found in the preceding sections for a variety of detection and estimation problems. Note also that this processor reduces to the conventional direction matching beamformer when the noise field is uncorrelated from sensor to sensor so that $\tilde{\Phi}_{nn} = \mathbf{I}$, the identity matrix [45].

A close relationship exists between maximizing the array gain and minimizing the array output signal variance under a desired signal response constraint because this is completely equivalent to minimizing the denominator of (3.246) subject to a constraint on the numerator. Since $G(\omega)$ is not changed by any scaling of the vector $\mathfrak{f}$, introducing a constraint like $\mathfrak{f}^{\dagger}\tilde{\mathfrak{m}} = \beta$ does not affect $G(\omega)$ and merely determines the scalar multiple used in selecting $\mathfrak{f}$. Consequently selecting $\mathfrak{f} = \beta\tilde{\Phi}_{nn}^{-1}\tilde{\mathfrak{m}}/(\tilde{\mathfrak{m}}^{\dagger}\tilde{\Phi}_{nn}^{-1}\tilde{\mathfrak{m}})$ both maximizes $G(\omega)$ and satisfies the constraint $\mathfrak{f}^{\dagger}\tilde{\mathfrak{m}} = \beta$. It has also been shown that maximizing the array gain and minimizing the output signal distortion yield exactly the same filters for monochromatic signals [3]. It can also be shown that the filter that yields the maximum likelihood estimate and the Wiener filter are different by only a scalar transfer function [46]. Furthermore, the likelihood ratio processor, the maximum SNR filter, and the Wiener filter are known to be equivalent in the case of a narrowband signal that is known except for phase and corrupted by additive, Gaussian noise [47].

### 3.4.7 *Criterion Invariant Array Processor*

Various detection and estimation techniques and performance measures related to optimum array processing were treated in the preceding sections by means of a unified theory. The likelihood ratio test for optimum signal detection, several signal estimation problems, and various performance criteria have been found to be all related by a prewhitening and matching operation that defines a vector operator for the received signal vector that has a scalar output. This scalar output is a single waveform which is then processed by different scalar operators depending on the problem of concern.

The results obtained for the various classes of problems may be conveniently summarized in the criterion invariant processor shown in Figure 3.10. This figure illustrates that the principle "*first prewhiten, then match,*" represented by the operator $\mathfrak{m}^{\dagger}\Phi_{nn}^{-1}$, is common for a wide variety of different optimization problems. Since the optimum processor depends on the inverse of the noise cross spectral density matrix but in practice the only measurable quantity is the cross spectral matrix of the sensor outputs (which in general

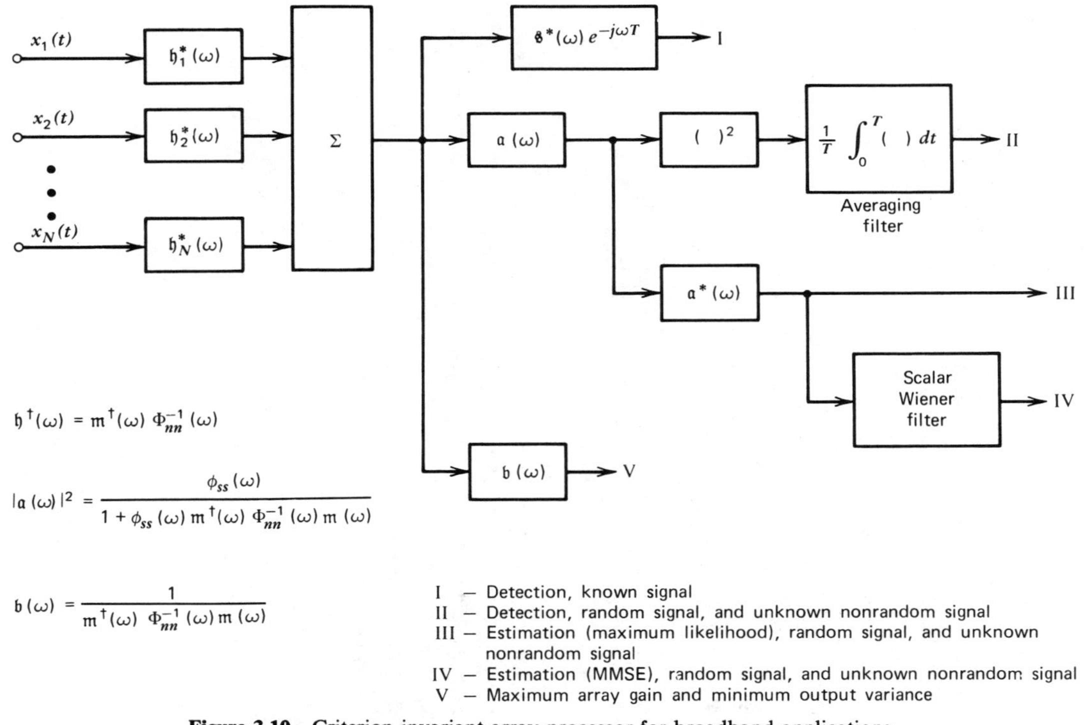


$$\mathfrak{h}^{\dagger}(\omega) = \mathfrak{m}^{\dagger}(\omega)\, \Phi_{nn}^{-1}(\omega)$$

$$|\mathfrak{a}(\omega)|^2 = \frac{\phi_{ss}(\omega)}{1 + \phi_{ss}(\omega)\, \mathfrak{m}^{\dagger}(\omega)\, \Phi_{nn}^{-1}(\omega)\, \mathfrak{m}(\omega)}$$

$$\mathfrak{b}(\omega) = \frac{1}{\mathfrak{m}^{\dagger}(\omega)\, \Phi_{nn}^{-1}(\omega)\, \mathfrak{m}(\omega)}$$

I — Detection, known signal
II — Detection, random signal, and unknown nonrandom signal
III — Estimation (maximum likelihood), random signal, and unknown nonrandom signal
IV — Estimation (MMSE), random signal, and unknown nonrandom signal
V — Maximum array gain and minimum output variance

**Figure 3.10** Criterion invariant array processor for broadband applications.

contains desired signal plus noise components), the use of the signal-plus-noise spectral matrix may result in performance degradation unless provision is made to obtain a signal-free estimate of the noise cross spectral matrix or the use of the signal-plus-noise spectral matrix is specifically provided for. The consequences involved for providing an inexact prewhitening and matching operation are discussed in reference [45]. It may be further noted that the minimum mean square error (MMSE) signal estimate is different from the maximum likelihood (distortionless) estimate (or any other estimate) only by a scalar Wiener filter. A Wiener processor may therefore be regarded as forming the undistorted signal estimate for observational purposes before introducing the signal distortion resulting from the scalar Wiener filter. The nature of the scalar Wiener filter is further considered in the problems section.

## 3.5 OPTIMUM ARRAY PROCESSING FOR PERTURBED PROPAGATION CONDITIONS

The processor structure of Figure 3.8 is known to be optimum only when the signal cross spectral density matrix is a simple dyad [17], that is

$$\mathbf{\Phi}_{ss}(\omega)=\mathbf{m}(\omega)\mathbf{m}^{\dagger}(\omega) \tag{3.258}$$

The effect of any perturbations occurring in the propagation process is to destroy the dyad nature of $\mathbf{\Phi}_{ss}(\omega)$, and consideration must be given to a processor structure more general than that of Figure 3.8. To determine the nature of a more general processor structure, the optimum processor for a noncoherent narrowband signal will be found.

Consider the problem of detecting a signal that is narrowband with unknown amplitude and phase. Such conditions frequently occur when the received signal has undergone unknown amplitude and phase changes during propagation. When the signal is present, the received waveform can be expressed as

$$x(t)=\mathrm{Re}\{ae^{j\theta}r(t)e^{j\phi(t)}e^{j\omega_0 t}\}+n(t) \tag{3.259}$$

where

$a$ = unknown amplitude
$\theta$ = unknown phase
$r(t)$ = known amplitude modulation
$\phi(t)$ = known phase modulation
$\omega_0$ = known carrier frequency

Using real notation we can rewrite (3.259) as

$$x(t)=m_1(t)s_1+m_2(t)s_2+n(t) \tag{3.260}$$

where

$$s_1 = a\cos\theta$$
$$s_2 = -a\sin\theta$$
$$m_1(t) = r(t)\cos[\omega_0 t + \phi(t)] = f(t)\cos(\omega_0 t) - g(t)\sin(\omega_0 t) \tag{3.261}$$
$$m_2(t) = r(t)\sin[\omega_0 t + \phi(t)] = f(t)\sin(\omega_0 t) + g(t)\cos(\omega_0 t) \tag{3.262}$$

and

$$f(t) = r(t)\cos[\phi(t)] \tag{3.263}$$
$$g(t) = r(t)\sin[\phi(t)] \tag{3.264}$$

where

$$r(t) = \sqrt{f^2(t) + g^2(t)}$$
$$\phi(t) = \tan^{-1}\left\{\frac{g(t)}{f(t)}\right\}$$

Equation (3.259) may now be rewritten as

$$x = \mathbf{ms} + n \tag{3.265}$$

where

$$\mathbf{m} = [m_1 m_2], \qquad \mathbf{s} = \begin{bmatrix} s_1 \\ s_2 \end{bmatrix}, \qquad \text{and } \operatorname{var}(n) = \sigma_n^2$$

The results of Section 3.4.5 for the detection of an unknown, nonrandom signal may now be applied, for which the sufficient test statistic is given by (3.242). For real variables (3.242) becomes

$$y = \tfrac{1}{2}\mathbf{x}^T\mathbf{R}_{nn}^{-1}\mathbf{M}[\mathbf{M}^T\mathbf{R}_{nn}^{-1}\mathbf{M}]^{-1}\mathbf{M}^T\mathbf{R}_{nn}^{-1}\mathbf{x} \tag{3.266}$$

Note that $m_1(t)$ and $m_2(t)$ are orthogonal functions having equal energy so that $|m_1|^2 = |m_2|^2$, and $m_1(t)\cdot m_2(t) = 0$. Consequently when the functions are sampled in time to form $\mathbf{x}^T = [x(t_1), x(t_2), \ldots]$, $\mathbf{m}_i^T = [m_i(t_1), m_i(t_2), \ldots]$, then (3.266) for the test statistic can be rewritten as

$$y = \tfrac{1}{2}\mathbf{x}^2[\mathbf{m}_1^T\mathbf{m}_2^T]\begin{bmatrix} \mathbf{m}_1 \\ \mathbf{m}_2 \end{bmatrix}(|\mathbf{m}_1|^2\sigma_n^2)^{-1}$$
$$= \tfrac{1}{2}(\sigma_n^2|\mathbf{m}_1|^2)^{-1}\{(\mathbf{m}_1^T\mathbf{x})^2 + (\mathbf{m}_2^T\mathbf{x})^2\} \tag{3.267}$$

Since the scalar factor $\frac{1}{2}(\sigma_n^2|\mathbf{m}_1|^2)^{-1}$ in (3.267) can be incorporated into the threshold setting for the likelihood ratio test, a suitable test statistic is

$$z=(\mathbf{m}_1^T\mathbf{x})^2+(\mathbf{m}_2^T\mathbf{x})^2 \tag{3.268}$$

For a continuous time observation, the test statistic given by (3.268) can be rewritten as

$$z=\left[\int_0^T m_1(t)x(t)dt\right]^2+\left[\int_0^T m_2(t)x(t)dt\right]^2 \tag{3.269}$$

The test statistic represented by (3.269) is conveniently expressed in terms of the "sine" and "cosine" components, $f(t)$ and $g(t)$. Using (3.261) and (3.262), we then can rewrite (3.269) as shown.

$$z=\left[\int_0^T f(t)\cos(\omega_0 t)x(t)dt-\int_0^T g(t)\sin(\omega_0 t)x(t)dt\right]^2$$
$$+\left[\int_0^T f(t)\sin(\omega_0 t)x(t)dt+\int_0^T g(t)\cos(\omega_0 t)x(t)dt\right]^2 \tag{3.270}$$

The above test statistic suggests the processor shown in Figure 3.11.

The optimum detector structure for a noncoherent signal shown in Figure 3.11 suggests the more general array processor structure illustrated in Figure 3.12. This more general processor structure is appropriate when propagating medium and/or receiving mechanism perturbations cause the plane wave desired signal assumption to hold no longer, or when it is desired to match the processor to a signal of arbitrary covariance structure. A matched array processor like that of Figure 3.12 utilizing matrix weighting is referred to as an element space matched array processor [48].

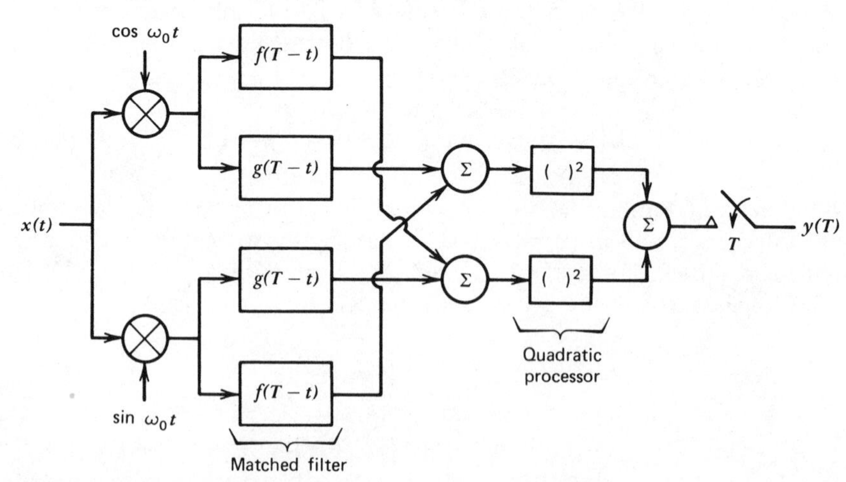


**Figure 3.11** Quadrature matched filter and envelope detector for received waveform having unknown amplitude and phase.

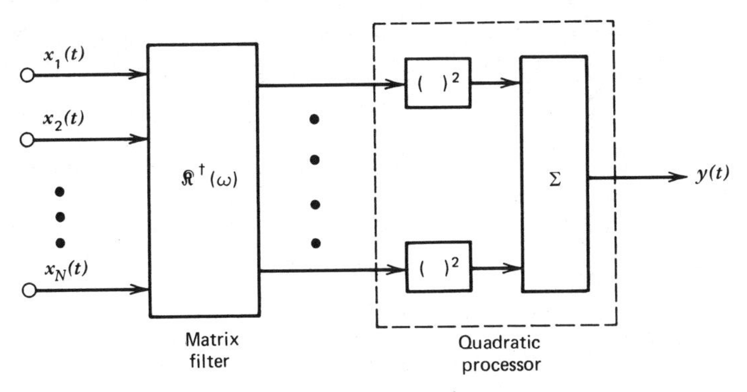


**Figure 3.12** Structure for general array processor.

Regarding the quantity **Ms** in (3.147) as the signal having arbitrary characteristics, then it is appropriate to define an array gain in terms of the detection index at the output of a general quadratic processor [17]. Where the noise is Gaussian, the results summarized in (3.231) may be utilized to give

$$G = \frac{\text{trace}\left[\mathfrak{K}^\dagger(\omega)\mathbf{\Phi}_{ss}(\omega)\mathfrak{K}(\omega)\right]}{\left\{\text{trace}\left[\left(\mathfrak{K}^\dagger(\omega)\mathbf{\Phi}_{nn}(\omega)\mathfrak{K}(\omega)\right)^2\right]\right\}^{1/2}} \tag{3.271}$$

It may be seen that (3.271) reduces to (3.246) when $\mathfrak{K}$ is a column vector. Under perturbed propagation conditions, $\mathbf{\Phi}_{ss}$ is given by [17]

$$\mathbf{\Phi}_{ss}(\omega) = \mathfrak{A}(\omega)\mathfrak{A}^\dagger(\omega) \tag{3.272}$$

where the matrix $\mathfrak{A}$ has $N$ rows and $r$ columns, and $r$ denotes the rank of the matrix $\mathbf{\Phi}_{ss}$. For a plane wave signal, the cross spectral density matrix $\mathbf{\Phi}_{ss}$ has rank one, and the dyad structure of (3.258) holds. The array gain given by (3.271) may be maximized in the same manner as (3.246), now using (D.16) instead of (D.14), to yield [17]

$$G \leqslant \left\{\text{trace}\left(\left[\mathbf{\Phi}_{ss}(\omega)\mathbf{\Phi}_{nn}^{-1}(\omega)\right]^2\right)\right\}^{1/2} \tag{3.273}$$

where equality obtains when the matrix $\mathfrak{K}$ is chosen to be a scalar multiple of $\mathfrak{A}^\dagger\mathbf{\Phi}_{nn}^{-1}$. It follows that the maximum gain cannot be achieved by any $\mathfrak{K}$ having less than $r$ columns.

## 3.6 SUMMARY AND CONCLUSIONS

Optimum array processing techniques for several classes of detection, estimation, and array performance optimization problems have been shown to be closely related to one another and, in fact, are different only by virtue of some scalar processing that follows a common matrix filter and combiner operator. This matrix filter operator embodies the principle of "first prewhiten, then match" in its realization. For narrowband signals the matching operation provides for time-delay (or phase) steering to obtain an undistorted representation of the signal. For broadband signals it is necessary to provide for signal spectral matching, which is not equivalent to time-delay steering. Perturbed signal propagation conditions resulting in noncoherent wave-front signals require element space matched processing in which the signal matching can only be performed in a statistical sense. The optimum processors derived in this chapter have, for the most part, involved solving unconstrained optimization problems. Various classes of constrained optimum processors are introduced in the problems section of this chapter.

The optimum array processor solutions resulting from the prewhitening and matching operation are intimately related to the optimal Wiener solution, which thereby provides a convenient theoretical performance limit against which to compare actual system performance. Direct implementation of an optimum processor solution is impractical because the signal environment statistics are usually unknown. Furthermore, even if the signal environment statistics were known, an optimal processor designed for a particular noise field would have limited utility—consequently adaptive techniques for realizing an optimum processor are extremely important. Since the spatial properties of the steady-state solutions resulting from various different performance measures are either identical or very similar, the question of which array performance measure to select for a given application is usually not very significant; rather, it is the temporal properties of the adaptive control algorithm to be used to reach the steady-state solutions that are of principal concern to the designer. The characteristics of adaptive control algorithms to be used for controlling the weights in the array pattern-forming network are therefore highly important, and it is to these concerns that Part Two of this book is addressed.

## PROBLEMS

**1** [18] ***Use of the Maximum SNR Performance Measure in a Single Jammer Encounter.*** The behavior of a linear adaptive array controlled by the maximum SNR performance measure when the noise environment consists of a single jammer added to the quiescent environment is of some practical interest. Consider a linear, uniformly spaced array whose weights are determined according to (3.96). Assume that the quiescent noise environment is characterized by the covariance matrix

$$\mathbf{R}_{nn_q} = \begin{bmatrix} p_q & & & 0 \\ & p_q & & \\ & & \ddots & \\ 0 & & & p_q \end{bmatrix} = p_q \mathbf{I}_k$$

where $p_q$ = noise power output of each of the $K$ array elements
$\mathbf{I}_k$ = identity matrix of order $K$
In general, the various weight amplitudes are not equal, and the array beam pattern is given by

$$G_q(\beta)=\sum_{k=1}^{K} a_k e^{j(k-1)(\beta-\beta_s)}$$

where

$$\beta=\frac{2\pi d}{\lambda}\sin\theta$$

or in matrix notation, $G_q(\beta)=\mathbf{b}^T\mathbf{w}_q$ where

$$\mathbf{b}=\begin{bmatrix}1\\ e^{j\beta}\\ e^{j2\beta}\\ \vdots\\ e^{j(K-1)\beta}\end{bmatrix},\qquad \mathbf{w}_q=\begin{bmatrix}a_1\\ a_2e^{-j\beta_s}\\ a_3e^{-j2\beta_s}\\ \vdots\\ a_ke^{-j(K-1)\beta_s}\end{bmatrix}$$

In view of the expressions for $\mathbf{w}_q$ and $G_q(\beta)$ given above, it follows that

$$\mathbf{UHw}_q=G_q(\beta_J)\begin{bmatrix}1\\1\\ \vdots\\1\end{bmatrix}$$

where $\mathbf{U}$ and $\mathbf{H}$ are described below.

Consider a single narrowband jammer located at an angle $\theta_J$ from mechanical boresight. Denote the jamming signal appearing in the first element channel by $J(t)$, then the jamming signal appearing in the $k$th channel will be $J(t)\exp[-j(k-1)\beta_J]$ where $\beta_J=(2\pi d/\lambda)\sin\theta_J$. Let $p_J$ represent the envelope jamming power in each channel, then the covariance of the jamming signals in the $k$th and $l$th channels is $p_J\exp[-j(k-l)\beta_J]$, which represents the $kl$th element of the jammer covariance matrix $\mathbf{R}_{JJ}$.

(a) Show that $\mathbf{R}_{JJ}$ can be expressed as

$$\mathbf{R}_{JJ}=p_J\mathbf{H}^*\mathbf{UH}$$

where $\mathbf{H}$ is the diagonal matrix

$$\mathbf{H}=\begin{bmatrix}1 & & & & 0\\ & e^{j\beta_J} & & & \\ & & e^{j2\beta_J} & & \\ & & & \ddots & \\ 0 & & & & e^{j(K-1)\beta_J}\end{bmatrix}$$

and $\mathbf{U}$ is a $K\times K$ matrix of ones.

(b) By use of the control law (3.99), it follows that the weight vector for the quiescent noise environment should satisfy $\mathbf{R}_{nn_q}\mathbf{w}_q=\mu\mathbf{t}^*$ where $\mathbf{R}_{nn_q}\mathbf{w}_q=p_q\mathbf{w}_q$. If the noise environment changes so the noise covariance matrix is now $\mathbf{R}_{nn}$ instead of $\mathbf{R}_{nn_q}$, it follows that the optimum weight vector should then satisfy $\mathbf{R}_{nn}\mathbf{w}=\mu\mathbf{t}^*$ or $\mathbf{w}=p_q\mathbf{R}_{nn}^{-1}\mathbf{w}_q$. The covariance matrix corresponding to the total noise environment (consisting of the quiescent noise plus jammer) is the sum $\mathbf{R}_{nn}=\mathbf{R}_{nn_q}+\mathbf{R}_{JJ}$ or $\mathbf{R}_{nn}=p_q\mathbf{I}_k+p_J\mathbf{H}^*\mathbf{U}\mathbf{H}$. To obtain the optimum weight vector for the total noise environment requires the inverse of $\mathbf{R}_{nn}$. Using the fact that $\mathbf{H}^*=\mathbf{H}^{-1}$, show that

$$\mathbf{R}_{nn}^{-1}=\frac{1}{p_q}\left\{\mathbf{I}_k-\left(\frac{p_J}{p_q+Kp_J}\right)\mathbf{H}^*\mathbf{U}\mathbf{H}\right\}$$

(c) Using the result from part (b), show that the optimum weight vector is given by

$$\mathbf{w}=\mathbf{w}_q-\left(\frac{p_J}{p_q+Kp_J}\right)\mathbf{H}^*\mathbf{U}\mathbf{H}\mathbf{w}_q$$

(d) Let the optimum quiescent weight vector for a desired signal located in the direction $\theta_s$ from mechanical boresight be expressed as

$$\mathbf{w}_q=\begin{bmatrix} a_1 \\ a_2e^{-j\beta_s} \\ a_3e^{-j2\beta_s} \\ \vdots \\ a_Ke^{-j(K-1)\beta_s} \end{bmatrix}$$

where

$$\beta_s=\frac{2\pi d}{\lambda}\sin\theta_s$$

and the $a_k$ represent weight amplitudes. If the various $a_k$ are all equal, the resulting array beam pattern is of the form $(\sin Kx)/(\sin x)$.

It then follows from the definition of $\mathbf{H}$ in part (a) that

$$\mathbf{H}\mathbf{w}_q=\begin{bmatrix} a_1 \\ a_2e^{j(\beta_J-\beta_s)} \\ \vdots \\ a_Ke^{j(K-1)(\beta_J-\beta_s)} \end{bmatrix}$$

From the foregoing expressions and the results of part (c), show that the pattern obtained with the optimum weight vector may be expressed as

$$G(\beta)=\mathbf{b}^T\mathbf{w}=\mathbf{b}^T\mathbf{w}_q-\left(\frac{P_J}{P_q+KP_J}\right)G_q(\beta_J)\mathbf{b}_J^*$$

where $\mathbf{b}_J$ is just $\mathbf{b}$ with the variable $\beta_J$ replacing $\beta$.

(e) Recalling from part (d) that $\mathbf{b}^T\mathbf{w}_q=G_q(\beta)$, show that $\mathbf{b}^T\mathbf{b}_J^*=C(\beta-\beta_J)$ where

$$C(x)=\exp\left\{j\left[\frac{(K-1)x}{2}\right]\frac{\sin Kx/2}{\sin x/2}\right\}$$

(f) Using the fact that $\mathbf{b}^T\mathbf{w}_q=G_q(\beta)$ and the results of parts (d) and (e), show that

$$G(\beta)=G_q(\beta)-\left(\frac{P_J}{P_q+KP_J}\right)G_q(\beta_J)C(\beta-\beta_J)$$

This result expresses the fact that the array beam pattern of an adaptively controlled linear array in the presence of one jammer consists of two parts. The first part is the quiescent pattern $G_q(\beta)$, while the second part (which is subtracted from the first part) is a $(\sin Kx)/(\sin x)$ shaped cancellation beam centered on the jammer.

(g) From the results of part (e) it may be seen that $C(x)|_{x=0}=K$. Using this fact show that the gain of the array in the direction of the jammer is given by

$$G(\beta_J)=\left(\frac{P_q}{P_q+KP_J}\right)G_q(\beta_J)$$

With the array weights fixed at $\mathbf{w}_q$, the array pattern gain in the direction of the jammer would be $G_q(\beta_J)$. The foregoing result therefore shows that the adaptive control reduces the gain in the direction of the jammer by the factor

$$\frac{P_q}{P_q+KP_J}=\frac{1}{1+K(P_J/P_q)}$$

(h) The proper measure of performance improvement against the jammer realized by the introduction of adaptively controlled weights is the cancellation ratio $\Gamma$ defined by

$$\Gamma\triangleq\frac{\gamma}{1-\gamma\cdot(J/N)_q}$$

where

$$\gamma \triangleq \frac{\mathbf{w}_q^\dagger \mathbf{R}_{nn_q} \mathbf{R}_{nn}^{-1} \mathbf{R}_{JJ} \mathbf{w}_q}{\mathbf{w}_q^\dagger \mathbf{R}_{JJ} \mathbf{w}_q}$$

and $(J/N)_q$ denotes the ratio of the jammer noise output to the quiescent noise output when the weights are fixed at $\mathbf{w}_q$, that is,

$$\left(\frac{J}{N}\right)_q \triangleq \frac{\mathbf{w}_q^\dagger \mathbf{R}_{JJ} \mathbf{w}_q}{\mathbf{w}_q^\dagger \mathbf{R}_{nn_q} \mathbf{w}_q}$$

It is now desired to evaluate $\gamma$ and $\Gamma$ for the single jammer environment. First show that

$$\mathbf{R}_{nn_q} \mathbf{R}_{nn}^{-1} = I_k - \left(\frac{1}{P_q + KP_J}\right) \mathbf{R}_{JJ}$$

Then show that

$$\mathbf{R}_{JJ} \mathbf{R}_{JJ} = KP_J \mathbf{R}_{JJ}$$

to obtain the result

$$\mathbf{R}_{nn_q} \mathbf{R}_{nn}^{-1} \mathbf{R}_{JJ} = \left(\frac{P_q}{P_q + KP_J}\right) \mathbf{R}_{JJ}$$

Substituting these results into the expression for $\gamma$ immediately yields

$$\gamma = \frac{1}{1 + (KP_J / P_q)}$$

For a main beam of the form $(\sin Kx)/(\sin x)$, the maximum possible value of output jammer noise power is $Kp_J$, which will only occur at the peak of the main beam. Consequently the maximum possible value of $(J/N)_q$ is $KP_J/P_q$, in which case $\Gamma = 1$ (thereby indicating no performance improvement against the jammer). For jammers located in the sidelobe region, however, $(J/N)_q \ll KP_J/P_q$ so that $\gamma \cdot (J/N)_q \ll 1$ and hence, $\Gamma \approx \gamma$. For a jammer located in the sidelobe region, the adaptive control "cancels' the jammer power by approximately the jammer-to-noise ratio in the cancellation beam.

**2** [49] ***Effect of Adaptive Weight Errors***

(a) For the CSLC system depicted in Figure 3.5 with $w_0 = 1$, show that $P_r = P_x - \mathbf{w}^\dagger \mathbf{z} - \mathbf{z}^\dagger \mathbf{w} + \mathbf{w}^\dagger \mathbf{R}_{yy} \mathbf{w}$.

(b) Since $\mathbf{w}_{opt} = \mathbf{R}_{yy}^{-1}\mathbf{z}^*$ and $P_{r_{min}} = P_x - \mathbf{z}^\dagger \mathbf{R}_{yy}^{-1}\mathbf{z}$, show that when $\mathbf{w} = \mathbf{w}_{opt} + \boldsymbol{\Delta}$, then $P_r = P_{r_{min}} + P_{add}$ where $P_{add} = \boldsymbol{\Delta}^\dagger \mathbf{R}_{yy} \boldsymbol{\Delta}$. Note that since $P_{add}$ is given by a Hermitian quadratic form, it is bounded by the smallest and largest eigenvalues of $\mathbf{R}_{yy}$ as $\lambda_{min}\|\boldsymbol{\Delta}\|^2 \leqslant P_{add} \leqslant \lambda_{max}\|\boldsymbol{\Delta}\|^2$ where $\|\boldsymbol{\Delta}\|^2 \triangleq \boldsymbol{\Delta}^\dagger \boldsymbol{\Delta}$.

(c) Since $\lambda_{min}$ and $\lambda_{max}$ are both geometry dependent, it is instructive to use a worst case design approach. Assume each auxiliary array element receives the same interference power $P_{I_a}$. Furthermore assume that the receiver of each array element has the same effective received noise power $P_n$. Under these conditions, each diagonal entry of $\mathbf{R}_{yy}$ is equal to $P_n + P_{I_a}$, and hence trace $(\mathbf{R}_{yy}) = N(P_n + P_{I_a})$. Since the largest eigenvalue of $\mathbf{R}_{yy}$ is less than the trace of $\mathbf{R}_{yy}$, it follows that $P_{add} \leqslant N(P_n + P_{I_a})\|\boldsymbol{\Delta}\|^2$. Assume that the weight errors are due to quantization errors where the quanta size of the in-phase and quadrature channel is $q$. Under worst case conditions, each complex weight component quantization error is identical and equal to $(q/2)(i \pm j)$. Hence show that

$$P_{add} \leqslant \frac{N^2 q^2}{2}(P_n + P_{I_a})$$

(d) The interference-to-noise ratio $(P_I/P_n)_{main}$ for the main element can be related to the ratio $(P_{I_a}/P_n)_{aux}$ for each auxiliary element by

$$\left(\frac{P_I}{P_n}\right)_{main} = |\alpha|^2 \left(\frac{P_{I_a}}{P_n}\right)_{aux}$$

where $\alpha$ is the average voltage gain of the main antenna in the sidelobe region. Assume that $\alpha$ is given by

$$\alpha = q \cdot 2^{B-1}$$

where $B$ represents the number of bits available to implement the quantizer. Show that by assuming $(P_I/P_n)_{main} \gg 1$ and using the results of part (c), then

$$\frac{P_{add}}{P_{r_{min}}} = \left(\frac{N^2}{2^{2B-1}}\right) R_0$$

where

$$R_0 \triangleq \frac{P_{r_{min}}}{P_n}$$

(e) The principle use of a worst case analysis is to determine the minimum number of bits required to avoid any degradation of the signal-to-interference ratio (SIR) performance. When the number of available quantization bits is significantly less than the minimum number predicted by worst case analysis, a better prediction of SIR performance can be obtained using a MSE analysis. The only change required for a MSE analysis is to treat $\|\boldsymbol{\Delta}\|^2$ in an appropriate manner. Assuming the weight errors of the in-phase and quadrature channels are independent and uniformly distributed, show that

$$\|\boldsymbol{\Delta}\|^2_{\text{average}} = \tfrac{1}{6}\|\boldsymbol{\Delta}\|^2_{\max}$$

and develop corresponding expressions for $P_{\text{add}}$ and $P_{\text{add}}/P_{r_{\min}}$.

3 [50] ***Wiener Linear Minimum Mean Square Error (MMSE) Filtering for Broadband Signals.*** Consider the scalar signal $x(t)$,

$$x(t) = s(t) + n(t)$$

where the desired signal $s(x)$ and the noise $n(t)$ are uncorrelated. The signal $x(t)$ is to be passed through a linear filter $h(t)$ so that the output $y(t)$ will best approximate $s(t)$ in the MMSE sense. The error in the estimate $y(t)$ is given by

$$e(t) = y(t) - s(t)$$

and it is desired to minimize $E\{e^2(t)\}$ where

$$y(t) = \int_{-\infty}^{\infty} h(\tau)s(t-\tau)\,d\tau = \underbrace{\int_{-\infty}^{\infty} h(\tau)s(t-\tau)\,dt}_{\triangleq\, s'(t)} + \underbrace{\int_{-\infty}^{\infty} h(\tau)n(t-\tau)\,d\tau}_{\triangleq\, n'(t)}$$

(a) Show that

$$E\{e^2(t)\} = E\{\underbrace{[s'(t) - s(t)]^2}_{\substack{\text{signal}\\ \text{distortion}\\ \text{component}}} + \underbrace{n'^2(t)}_{\substack{\text{noise}\\ \text{component}}}\}$$

(b) Show that

$$E\{n'^2(t)\}=\int_{-\infty}^{\infty}\int_{-\infty}^{\infty} r_{nn}(\nu)h(\tau)h(\tau+\nu)\,d\nu\,d\tau$$

where $r_{nn}(\nu)$ = autocorrelation function of $n(t)$.

(c) By not imposing realizability or finite memory requirements on $h(t)$, then $H(\omega)$ may be introduced using

$$H(\omega)=\int_{-\infty}^{\infty} h(t)e^{-j\omega t}\,dt=\mathfrak{F}\{h(t)\}$$

Show that

$$E\{n'^2(t)\}=\int_{-\infty}^{\infty} H(\omega)H^*(\omega)\phi_{nn}(\omega)\,d\omega$$

where $\phi_{nn}(\omega)=\mathfrak{F}\{r_{nn}(\tau)\}$, the spectral density function of $n(t)$. Likewise show that

$$E\{[s'(t)-s(t)]^2\}=\int_{-\infty}^{\infty}[1-H(\omega)][1-H^*(\omega)]\cdot\phi_{ss}(\omega)\,d\omega$$

so that

$$E\{e^2(t)\}=\int_{-\infty}^{\infty}\{H(\omega)H^*(\omega)\phi_{nn}(\omega)+[1-H(\omega)][1-H^*(\omega)]\phi_{ss}(\omega)\}\,d\omega$$

(d) It is desired to minimize $E\{e^2(t)\}$ by appropriately choosing $H(\omega)$. Since the integrand of the expression for $E\{e^2(t)\}$ appearing in part (c) is positive for all choices of $H(\omega)$, it is only necessary to minimize the integrand by choosing $H(\omega)$. Show that by setting the gradient of the integrand of $E\{e^2(t)\}$ with respect to $H(\omega)$ equal to zero, there results

$$H_{\text{opt}}(\omega)=\frac{\phi_{ss}(\omega)}{\phi_{ss}(\omega)+\phi_{nn}(\omega)}$$

which is the optimum scalar Wiener filter. Therefore to obtain the scalar Wiener filter, it is only necessary to know the signal spectral density and the noise spectral density at the point in question. For scalar processes, the foregoing result may be used to obtain the MMSE signal estimate by introducing the appropriate scalar Wiener filter at the point where a MMSE signal estimate is desired. It is easy to show that the corresponding result for vector processes is given by

$$\mathfrak{h}_{\text{opt}}(\omega)=\left[\boldsymbol{\Phi}_{ss}(\omega)+\boldsymbol{\Phi}_{nn}(\omega)\right]^{-1}\boldsymbol{\Phi}_{ss}(\omega)$$

and when $\mathbf{x}(t)=\mathbf{s}(t)+\mathbf{n}(t)$ where $\mathbf{s}(t)=\mathbf{v}d(t)$ then $\mathfrak{h}_{\text{opt}}(\omega)$ can be expressed as

$$\mathfrak{h}_{\text{opt}}(\omega)=\boldsymbol{\Phi}_{xx}^{-1}(\omega)\boldsymbol{\phi}_{xd}(\omega)$$

which represents the broadband signal generalization of (3.56).

(e) Determine the appropriate scalar Wiener filter to insert at point III in Figure 3.10 to obtain the MMSE estimate of $s(t)$. Likewise determine the appropriate scalar Wiener filter to insert at point V to obtain the MMSE estimate of $s(t)$.

**4** [17] ***The Use of Multiple Linear Constraints in Array Gain Maximization.*** It was found that in the process of maximizing array gain with a single linear constraint of the form, $\mathbf{k}^\dagger\mathbf{m}=\beta$, the constraint could be handled by using the degree of freedom in specifying the magnitude of $\mathbf{k}$. In the event that a null constraint of the form $\mathbf{k}^\dagger\mathbf{d}=0$ is introduced, however, the situation must be handled as a special case of the more general problem involving multiple linear constraints. Multiple linear constraints may be introduced to keep the array gain relatively constant over a range of signal perturbations thereby reducing any sensitivity to such perturbations. Likewise, multiple linear constraints may be used to control sidelobes in a specific neighborhood of directions. The basic problem can be stated as that of minimizing the output power $z$ subject to a constraint of the form $\mathbf{H}^\dagger\mathbf{k}=\mathbf{g}$ where each row vector of the constraint matrix $\mathbf{H}^\dagger$ imposes a constraint of the form $\mathbf{h}_i^\dagger\mathbf{k}=g_i$. Clearly, then, the constraint matrix $\mathbf{H}^\dagger$ has a row for each constraint, and the total number of rows must be less than the number of sensors, or the problem will be overspecified. Introducing the multiple linear constraints into the expression for output power by using a vector Lagrange multiplier $\boldsymbol{\lambda}$, it is therefore desired to minimize

$$z=\mathbf{k}^\dagger\boldsymbol{\Phi}_{xx}\mathbf{k}+\boldsymbol{\lambda}^\dagger\left[\mathbf{H}^\dagger\mathbf{k}-\mathbf{g}\right]+\left[\mathbf{k}^\dagger\mathbf{H}-\mathbf{g}^\dagger\right]\boldsymbol{\lambda}$$

(a) Complete the square of the above expression for $z$ to obtain the equivalent result

$$z=\left[\mathbf{k}^\dagger+\boldsymbol{\lambda}^\dagger\mathbf{H}^\dagger\boldsymbol{\Phi}_{xx}^{-1}\right]\boldsymbol{\Phi}_{xx}\left[\boldsymbol{\Phi}_{xx}^{-1}\mathbf{H}\boldsymbol{\lambda}+\mathbf{k}\right]-\boldsymbol{\lambda}^\dagger\mathbf{H}_{xx}^{\dagger-1}\mathbf{H}\boldsymbol{\lambda}-\boldsymbol{\lambda}^\dagger\mathbf{g}-\mathbf{g}^\dagger\boldsymbol{\lambda}$$

Since $\mathbf{k}$ appears only in the first quadratic term above, the minimizing value of $\mathbf{k}$ is obviously that for which the quadratic term is zero or

$$\mathbf{k}_{\text{opt}}=-\boldsymbol{\Phi}_{xx}^{-1}\mathbf{H}\boldsymbol{\lambda}$$

(b) Use the constraint equation $\mathbf{H}^{\dagger}\mathbf{k}=\mathbf{g}$ to eliminate $\boldsymbol{\lambda}$ from the result for $\mathbf{k}_{\text{opt}}$ obtained in part (a) and thereby show that

$$\mathbf{k}_{\text{opt}}=\boldsymbol{\Phi}_{xx}^{-1}\mathbf{H}\left[\mathbf{H}^{\dagger}\boldsymbol{\Phi}_{xx}^{-1}\mathbf{H}\right]^{-1}\mathbf{g}$$

With this value of $\mathbf{k}$ the output power then becomes

$$z=\mathbf{g}^{\dagger}\left[\mathbf{H}\boldsymbol{\Phi}_{xx}^{-1}\mathbf{H}\right]^{-1}\mathbf{g}$$

(c) For the general array processor of Figure 3.12, the output power is given by $z=\text{trace}(\mathfrak{R}^{\dagger}\boldsymbol{\Phi}_{xx}\mathfrak{R})$, and the problem of minimizing $z$ subject to multiple linear constraints of the form $\mathbf{H}^{\dagger}\mathfrak{R}=\mathbf{L}$ is handled by using a matrix Lagrange multiplier, $\boldsymbol{\Lambda}$, and considering

$$z=\text{trace}\left(\mathfrak{R}^{\dagger}\boldsymbol{\Phi}_{xx}\mathfrak{R}+\boldsymbol{\Lambda}^{\dagger}\left[\mathbf{H}^{\dagger}\mathfrak{R}-\mathbf{L}\right]+\left[\mathfrak{R}^{\dagger}\mathbf{H}-\mathbf{L}^{\dagger}\right]\boldsymbol{\Lambda}\right)$$

Complete the square of the above expression to show that the optimum solution is

$$\mathfrak{R}_{\text{opt}}=\boldsymbol{\Phi}_{xx}^{-1}\mathbf{H}\left[\mathbf{H}^{\dagger}\boldsymbol{\Phi}_{xx}^{-1}\mathbf{H}\right]^{-1}\mathbf{L}$$

for which $\mathfrak{R}_{\text{opt}}^{\dagger}\boldsymbol{\Phi}_{xx}\mathfrak{R}_{\text{opt}}=\mathbf{L}^{\dagger}[\mathbf{H}^{\dagger}\boldsymbol{\Phi}_{xx}^{-1}\mathbf{H}]^{-1}\mathbf{L}$

**5** [15] ***Introduction of Linear Constraints to the MSE Performance Measure.*** The theoretically optimum array processor structure for maximizing (or minimizing) some performance measure may be too complex or costly to fully implement. This fact leads to the consideration of suboptimal array processors for which the processor structure is properly constrained within the context of the signal and interference environment.

The $K$-component weight vector $\mathbf{w}$ is said to be linearly constrained if

$$f=\mathbf{c}^{\dagger}\mathbf{w}$$

The number of linear, orthonormal constraints on $\mathbf{w}$ must be less than $K$ if any remaining degrees of freedom are to be available for adaptation.

(a) The MSE can be expressed as

$$E\left\{|y-y_A|^2\right\}$$

where $y_A=\mathbf{w}^{\dagger}\mathbf{x}$.
Consequently

$$E\left\{|y-y_A|^2\right\}=E\left\{|y|^2\right\}-2\,\text{Re}\left\{\mathbf{w}^{\dagger}\mathbf{r}_{xy}\right\}+\mathbf{w}^{\dagger}\mathbf{R}_{xx}\mathbf{w}$$

Append the constraint equation to the MSE by means of a complex

Lagrange multiplier to form

$$J=E\{|y|^2\}-2\,\mathrm{Re}\{\mathbf{w}^\dagger\mathbf{r}_{xy}\}+\mathbf{w}^\dagger\mathbf{R}_{xx}\mathbf{w}+\lambda[f-\mathbf{c}^\dagger\mathbf{w}]+[f^*-\mathbf{w}^\dagger\mathbf{c}]\lambda^*$$

Take the gradient of the foregoing expression with respect to $\mathbf{w}$ and set the result equal to zero to obtain

$$\mathbf{w}_{\mathrm{opt}}=\mathbf{R}_{xx}^{-1}[\mathbf{r}_{xy}+\lambda^*\mathbf{c}]$$

(b) Apply the constraint $f=\mathbf{c}^\dagger\mathbf{w}$ to the result in part (a) thereby obtaining a solution for $\lambda$:

$$\lambda=\frac{(f^*-\mathbf{r}_{xy}^\dagger\mathbf{R}_{xx}^{-1}\mathbf{c})}{(\mathbf{c}^\dagger\mathbf{R}_{xx}^{-1}\mathbf{c})}$$

This solution may be substituted into the result of part (a) to obtain the resulting solution for the constrained suboptimal array processor.

**6** [15] ***Introduction of Multiple Linear Constraints.*** A multiple linear constraint on the array processor takes the form

$$\mathbf{f}=\mathbf{C}^\dagger\mathbf{w}$$

where the matrix $\mathbf{C}$ has the vector $\mathbf{c}_i$ as its $i$th column, and the set $\{\mathbf{c}_i/i=1,2,\ldots,M\}$ must be a set of orthonormal constraint vectors.

Append the multiple linear constraint to the expected output power of the array output signal to determine the optimum constrained solution for $\mathbf{w}$.

**7** [15] ***Introduction of Quadratic Constraints to an Expected Output Power Performance Measure.*** The weight vector $\mathbf{w}$ is said to be quadratically constrained if

$$g=\mathbf{w}^\dagger\mathbf{Q}\mathbf{w}$$

where $\mathbf{Q}$ is a $K\times K$ Hermitian matrix.

(a) The expected output power of an array output signal is

$$E\{|y|^2\}=E\{\mathbf{w}^\dagger\mathbf{x}\mathbf{x}^\dagger\mathbf{w}\}=\mathbf{w}^\dagger E\{\mathbf{x}\mathbf{x}^\dagger\}\mathbf{w}=\mathbf{w}^\dagger\mathbf{R}_{xx}\mathbf{w}$$

Appending the constraint equation to the expected output power with a complex Lagrange multiplier yields

$$J=\mathbf{w}^\dagger\mathbf{R}_{xx}\mathbf{w}+\lambda[g-\mathbf{w}^\dagger\mathbf{Q}\mathbf{w}]$$

Take the gradient of the above expression with respect to $\mathbf{w}$ and set the result equal to zero (to obtain the extremum value of $\mathbf{w}$) to obtain

$$\mathbf{R}_{xx}^{-1}\mathbf{Q}\mathbf{w}=\lambda^{-1}\mathbf{w}$$

(b) Note that the above result for $\mathbf{w}$ is satisfied when $\mathbf{w}$ is an eigenvector of $(\mathbf{R}_{xx}^{-1}\mathbf{Q})$ and $\lambda$ is the corresponding eigenvalue. Therefore maximizing (minimizing) $J$ corresponds to selecting the largest (smallest) eigenvalue of $(\mathbf{R}_{xx}^{-1}\mathbf{Q})$. It follows that a quadratic constraint can be regarded simply as a means of scaling the weight vector $\mathbf{w}$ in the array processor.

(c) Append both a multiple linear constraint and a quadratic constraint to the expected output power of the array to determine the optimum constrained solution for $\mathbf{w}$. Note that once again the quadratic constraint merely results in scaling the weight vector $\mathbf{w}$.

**8** [48] ***Introduction of Single-Point, Multiple-Point, and Derivative Constraints to a Minimum Power Output Criterion for a Signal-Aligned Array.***

(a) Consider the problem of minimizing the array output power given by

$$P_0=\mathbf{w}^{\dagger}\mathbf{R}_{xx}\mathbf{w}$$

subject to the constraint $\mathbf{C}^{\dagger}\mathbf{w}=\mathbf{f}$. Show that the solution to this problem is given by

$$\mathbf{w}_{\text{opt}}=\mathbf{R}_{xx}^{-1}\mathbf{C}\left[\mathbf{C}^{\dagger}\mathbf{R}_{xx}^{-1}\mathbf{C}\right]^{-1}\mathbf{f}$$

(b) A single-point constraint corresponds to the case when the following conditions hold:

$\mathbf{C}=\mathbf{1}$, the $N\times 1$ vector of one's where $N$ is the number of array elements

so the constraint equation becomes

$$\mathbf{w}^{\dagger}\mathbf{1}=N$$

Show that the corresponding optimal weight vector is given by

$$\mathbf{w}_{\text{opt}}=\frac{N\mathbf{R}_{xx}^{-1}\mathbf{1}}{(\mathbf{1}^{T}\mathbf{R}_{xx}^{-1}\mathbf{1})}$$

Under the single-point constraint, the weight vector minimizes the output power in all directions except in the look (presteered) direction.

(c) A multiple-point constraint can be introduced by specifying that $\mathbf{C}$ is a matrix of dimension $N \times 3$ (for a three-point constraint) given by

$$\mathbf{C} = \mathbf{P} = [\mathbf{e}_1, \mathbf{e}, \mathbf{e}_2]$$

where $\mathbf{e}_1$ and $\mathbf{e}_2$ are direction vectors referenced to the beam axis on either side, and

$$\mathbf{f} = \mathbf{P}^\dagger \mathbf{e} = \begin{bmatrix} \mathbf{e}_1^\dagger \mathbf{e} \\ N \\ \mathbf{e}_2^\dagger \mathbf{e} \end{bmatrix}$$

where $\mathbf{C}^\dagger \mathbf{w} = \mathbf{f}$. Show that the corresponding optimal constrained weight vector is given by

$$\mathbf{w}_{\text{opt}} = \mathbf{R}_{xx}^{-1} \mathbf{P} [\mathbf{P}^\dagger \mathbf{R}_{xx}^{-1} \mathbf{P}]^{-1} \mathbf{P}^\dagger \mathbf{e}$$

(d) Derivative constraints can be used to maintain a broader region of the main lobe by specifying both the response on the beam axis and the derivatives of the response on the beam axis. The constraint matrix now has dimension $N \times k$ and is given by

$$\mathbf{C} = \mathbf{D} = [\mathbf{e}_0, \mathbf{e}_0', \mathbf{e}_0'', \ldots]$$

where $\mathbf{e}_0$ = the array steering vector

$\mathbf{e}_0'$ = the derivative of $\mathbf{e}_0$ with respect to $\sin\theta_0$ where $\theta_0$ is the look direction

$\mathbf{e}_0''$ = the second derivative of $\mathbf{e}_0$ with respect to $\sin\theta_0$

and $\mathbf{f}^T = \boldsymbol{\delta}_{10}^T = [N, 0, 0, \ldots]$ has dimension $1 \times k$. Show that the corresponding optimal weight vector is given by

$$\mathbf{w}_{\text{opt}} = \mathbf{R}_{xx}^{-1} \mathbf{D}^\dagger [\mathbf{D} \mathbf{R}_{xx}^{-1} \mathbf{D}]^{-1} \boldsymbol{\delta}_{10}$$

**9** [39] ***Maximum Likelihood (ML) Estimates of Target Range and Bearing.*** Optimal array processors designed for the detection and estimation of desired signals may only partially satisfy the requirements placed on the system signal processor since the second stage of the signal processing problem often involves the extraction of information concerning parameters such as target range and bearing.

A maximum likelihood estimator selects the parameter $\alpha$ which maximizes the conditional probability density function (or "likelihood function") $p(\mathbf{x}/\alpha)$. Since it is usually preferable to work with $\ln p(\cdot)$, solutions are sought to the equation

$$\frac{\partial}{\partial \alpha} \ln p(\mathbf{x}/\alpha) = 0$$

By forming the likelihood functional

$$y(\alpha) = \frac{\partial}{\partial \alpha} \ln p(\mathbf{x}/\alpha)$$

the maximum likelihood estimate $\hat{\alpha}_{\text{ML}}$ can be found by requiring $y(\alpha)$ to be equal to zero, and the statistics of $\hat{\alpha}_{\text{ML}}$ can be related to those of $y(\alpha)$.

When the components of the received signal vector $\mathbf{x}$ (using a Fourier coefficients representation) are complex Gaussian random processes, then

$$p(\mathbf{x}/\alpha) = \frac{1}{Det(\pi \mathbf{M})} \exp(-\mathbf{x}^{\dagger} M^{-1} x)$$

where $\mathbf{M} = \mathbf{\Phi}_{ss}(\omega) + \mathbf{\Phi}_{nn}(\omega)$.

(a) Using the expression for $p(\mathbf{x}/\alpha)$ given above, show that the likelihood functional can be expressed as

$$y(\alpha) = \mathbf{x}^{\dagger} \mathbf{M}^{-1} \frac{\partial \mathbf{M}}{\partial \alpha} \mathbf{M}^{-1} \mathbf{x} - \text{trace}\left( \mathbf{M}^{-1} \frac{\partial \mathbf{M}}{\partial \alpha} \right)$$

Note that $\mathbf{\Phi}_{ss}(\omega)$ is a function of $\alpha$ whereas $\mathbf{\Phi}_{nn}(\omega)$ is not.

(b) It can be shown [39] that the variance associated with the ML estimate $\hat{\alpha}_{\text{ML}}$ for the case of spatially incoherent noise is proportional to the quantity

$$\text{var}(\hat{\alpha}_{\text{ML}}) \propto \left[ \text{trace}(\mathbf{T}\mathbf{T}^{\dagger}) \right]^{-1}$$

where $\mathbf{T}$ is a weighting matrix that incorporates all geometrical properties of the array. In particular, for linear arrays the $\mathbf{T}$ matrix has elements given by the following:

$$\text{For bearing estimation: } t_{ij} = \frac{\sin \theta}{\mathfrak{v}} (z_i - z_j)$$

$$\text{For range estimation: } t_{ij} = -\frac{\sin^2 \theta}{2 \mathfrak{v} r^2} \left( z_i^2 - z_j^2 \right)$$

where $\theta$ = signal bearing with respect to the array normal,
$z_n$ = position of $n$th sensor along the array axis (the $z$-axis),
$\mathfrak{v}$ = velocity of signal propagation, and
$r$ = true target range.

Show that for an array having length $L$, and with $K \gg 1$ equally spaced sensors then

$$\text{For bearing estimation: } \left[\text{trace}(\mathbf{T}\mathbf{T}^{\dagger})\right]^{-1} = \frac{6\mathfrak{v}^2}{K^2 L^2 \sin^2\theta}$$

$$\text{For range estimation: } \left[\text{trace}(\mathbf{T}\mathbf{T}^{\dagger})\right]^{-1} = \frac{45\mathfrak{v}^2 r^4}{2L^4 K^2 \sin^4\theta}$$

The foregoing results show that the range estimate accuracy is critically dependent on the true range, while the bearing estimate is not (except for the range dependence of the SNR). The range estimate is also more critically dependent on the array aperture $L$ than the bearing estimate.

**10** [9] ***Suboptimal Bayes Estimate of Target Angular Location.*** The signal processing task of extracting information concerning parameters such as target angular location can also be carried out by means of Bayes estimation. A Bayes estimator is just the expectation of the variable being estimated conditioned on the observed data, that is

$$\hat{u} = E\{u_k/\mathbf{x}\} = \int_{-\infty}^{\infty} u_k p(u_k/\mathbf{x})\, du$$

where $u_k = \sin\theta$ denotes the angular location of the $k$th target, $\mathbf{x}$ denotes the observed data vector, and the a posteriori probability density function $p(u/\mathbf{x})$ can be rewritten by application of Bayes' rule as

$$p(u/\mathbf{x}) = \frac{p(\mathbf{x}/u)p(u)}{p(\mathbf{x})}$$

The optimum estimators that result using the foregoing approach are quite complex, requiring the evaluation of multiple integrals that may well be too lengthy for many portable applications. Consequently the development of a simple suboptimum estimator that approximates the optimum Bayes estimator is of practical importance.

(a) The lowest order nonlinear approximation to the optimum Bayes location estimator is given by [9]

$$\hat{\mathbf{u}} = \mathbf{x}^T \mathbf{B}$$

where the matrix of $N$ target locations is a $1 \times N$ row vector

$$\hat{\mathbf{u}} = \left[\hat{u}_1, \hat{u}_2, \ldots, \hat{u}_N\right]$$

Assuming a $(2K+1)$ element array where each element ouput is amplified and detected with synchronous quadrature detectors, then the output of the $m$th pair of quadrature detectors is

$$x_{ym}(t)=s_{ym}(t)+n_{ym}(t)$$

and

$$x_{zm}(t)=s_{zm}(t)+n_{zm}(t)$$

The data vector $\mathbf{x}$ can then be defined by a $2K(2K+1)\times 1$ column vector:

$$\mathbf{x}=\left[x_{ym}x_{zn}\right]$$

in which each index pair $m,n$ occupies a separate row. Define a $2K(2K+1)\times 1$ column vector of coefficients

$$\mathbf{b}_k=\left[b_{mn}^{(k)}\right], \qquad m\neq n$$

where again each index pair $m,n$ occupies a separate row. The full matrix of coefficients $\mathbf{B}$ is then given by the $2K(2K+1)\times N$ matrix

$$\mathbf{B}=\left[\mathbf{b}_1,\mathbf{b}_2,\ldots,\mathbf{b}_N\right]$$

Show that by choosing the coefficient matrix $\mathbf{B}$ so that the resulting estimator is orthogonal to the estimation error, that is $E[\hat{\mathbf{u}}^T(\mathbf{u}-\hat{\mathbf{u}})]=0$, then $\mathbf{B}$ is given by

$$\mathbf{B}=\left[E\{\mathbf{x}\mathbf{x}^T\}\right]^{-1}E\{\mathbf{x}\mathbf{u}\}$$

With $\mathbf{B}$ selected as indicated above, the MSE in the location estimator for the $k$th target is then given by

$$\mathcal{E}_k=E\left[u_k(u_k-\hat{u}_k)\right]=E\left\{u_k^2\right\}-\left[E\{\mathbf{x}u_k\}\right]^T$$
$$\cdot\left[E\{\mathbf{x}\mathbf{x}^T\}\right]^{-1}\cdot\left[E\{\mathbf{x}u_k\}\right]$$

(b) Consider a one-target location estimation problem using a two-element array. From the results of part (a), it follows that

$$\hat{u}=b_{12}x_{y1}x_{z2}+b_{21}x_{y2}x_{y1}$$

Let the signal components of the quadrature detector outputs be given by

$$s_{ym}=\alpha\cos m\pi u+\beta\sin m\pi u$$

and

$$s_{zm} = \beta \cos m\pi u - \alpha \sin m\pi u$$

where the joint probability density function (PDF) for $\alpha, \beta$ is given by

$$p(\alpha,\beta) = \frac{1}{2\pi\sigma^2}\exp\left\{-\frac{(\alpha^2+\beta^2)}{2\sigma^2}\right\}$$

Furthermore assume the quadrature noise components are independent so the PDF for the noise originating at the $m$th antenna element is

$$p(n_{ym}, n_{zm}) = \frac{1}{2\pi\sigma_n^2}\exp\left\{-\frac{(n_{ym}^2+n_{zm}^2)}{2\sigma_n^2}\right\}$$

The above signal and noise models correspond to a Rayleigh fading environment and additive noise due to scintillating clutter. Show that the optimum coefficients for determining $\hat{u}$ are given by

$$b_{12} = -b_{21} = \frac{(1/\sigma^2)E\{u \sin \pi u\}}{4E\{\sin^2 \pi u\} + 2/\gamma + 1/\gamma^2}$$

where $\gamma \triangleq \sigma^2/\sigma_n^2$, the SNR. Finally show that the MSE for this estimator is given by

$$\text{MSE} = E\{u^2\} - \frac{E^2\{u \sin \pi u\}}{2E\{\sin^2 \pi u\} + 1/\gamma + 2/2\gamma^2}$$

and therefore depends only on the SNR and $p(u)$.

**11** [48] ***Constrained Minimum Power Criterion for Element Space Matched Array Processing in Narrowband Application.*** A minimum variance estimate of the signal power $P_s$ for an element space matched array processor may be obtained by solving the following problem:

$$\text{minimize: } \operatorname{var}\left[\mathbf{x}^\dagger \mathbf{K} \mathbf{x}\right] = \operatorname{tr}\left[(\mathbf{K}\mathbf{R}_{xx})^2\right]$$

subject to the constraint $\operatorname{tr}[\mathbf{K}\mathbf{R}_{ss}] = 1$.

(a) Show that the solution of the above problem is given by

$$\mathbf{K}_{\text{opt}} = \frac{\mathbf{R}_{xx}^{-1}\mathbf{R}_{ss}\mathbf{R}_{xx}^{-1}}{\operatorname{tr}\left[(\mathbf{R}_{xx}^{-1}\mathbf{R}_{ss})^2\right]}$$

so the expected value of the processor output power is then

$$P_0 = \frac{\operatorname{tr}\left[\mathbf{R}_{xx}^{-1}\mathbf{R}_{ss}\right]}{\operatorname{tr}\left[\left(\mathbf{R}_{xx}^{-1}\mathbf{R}_{ss}\right)^2\right]}$$

(b) Show that when $\mathbf{R}_{ss}$ has rank one and is consequently given by the dyad $\mathbf{R}_{ss} = \mathbf{v}\mathbf{v}^{\dagger}$, then

$$\mathbf{K}_{\text{opt}} = \mathbf{h}(\omega)\mathbf{h}^{\dagger}(\omega)$$

where

$$\mathbf{h} = \text{scalar} \cdot \mathbf{v}^{\dagger}\mathbf{R}_{nn}^{-1}$$

This result demonstrates that under plane wave signal assumptions, the element space matched array processor degenerates into a plane wave matched processor with a quadratic detector.

## REFERENCES


[1] P. L. Stocklin, "Space-Time Sampling and Likelihood Ratio Processing in Acoustic Pressure Fields," *J. Br. IRE*, July 1963, pp. 79–90.

[2] F. Bryn, "Optimum Signal Processing of Three-Dimensional Arrays Operating on Gaussian Signals and Noise," *J. Acoust. Soc. Am.*, Vol. 34, No. 3, March 1962, pp. 289–297.

[3] D. J. Edelblute, J. M. Fisk, and G. L. Kinneson, "Criteria for Optimum-Signal-Detection Theory for Arrays," *J. Acoust. Soc. Am.*, Vol. 41, January 1967, pp. 199–206.

[4] H. Cox, "Optimum Arrays and the Schwartz Inequality," *J. Acoust. Soc. Am.*, Vol. 45, No. 1, January 1969, pp. 228–232.

[5] B. Widrow, P. E. Mantey, L. J. Griffiths, and B. B. Goode, "Adaptive Antenna Systems," *Proc. IEEE*, Vol. 55, December 1967, pp. 2143–2159.

[6] L. J. Griffiths, "A Simple Algorithm for Real-Time Processing in Antenna Arrays," *Proc. IEEE*, Vol. 57, October 1969, pp. 1696–1707.

[7] N. Owsley, "Source Location with an Adaptive Antenna Array," Naval Underwater Systems Center, Rept. NL-3015, January 1971.

[8] A. H. Nuttall and D. W. Hyde, "A Unified Approach to Optimum and Suboptimum Processing for Arrays," U.S. Navy Underwater Sound Laboratory Report 992, April 1969, pp. 64–68.

[9] G. W. Young and J. E. Howard, "Applications of Spacetime Decision and Estimation Theory to Antenna Processing System Design," *Proc. IEEE*, Vol. 58, May 1970, pp. 771–778.

[10] H. Cox, "Interrelated Problems in Estimation and Detection I and II," Vol. 2, NATO Advanced Study Institute on Signal Processing with Emphasis on Underwater Acoustics, Enchede, The Netherlands, August 12–23, 1968, pp. 23-1 to 23-64.

[11] N. T. Gaarder, "The Design of Point Detector Arrays," *IEEE Trans. Inf. Theory*, part 1, Vol. IT-13, January 1967, pp. 42–50; part 2, Vol. IT-12, April 1966, pp. 112–120.

[12] J. Capon, "Applications of Detection and Estimation Theory to Large Array Seismology," *Proc. IEEE*, Vol. 58, May 1970, pp. 760–770.

[13] A. M. Vural, "An Overview of Adaptive Array Processing for Sonar Applications," IEEE 1975 EASCON Record, Electronics and Aerospace Systems Convention, September 29–October 1, pp. 34.A–34 M.

[14] J. P. Burg, "Maximum Entropy Spectral Analysis," Proceedings of NATO Advanced Study Institute on Signal Processing, August 1968.

[15] N. L. Owsley, "A Recent Trend in Adaptive Spatial Processing for Sensor Arrays: Constrained Adaptation," Proceedings of NATO Advanced Study Institute on Signal Processing, Loughborough, England, August 1972, pp. 591–603.

[16] R. R. Kneiper, et al., "An Eigenvector Interpretation of an Array's Bearing Response Pattern," Naval Underwater Systems Center, Rept. No. 1098, May 1970.

[17] H. Cox, "Sensitivity Considerations in Adaptive Beamforming," Proceedings of NATO Advanced Study Institute on Signal Processing, Loughborough, England, August 1972, pp. 621–644.

[18] S. P. Applebaum, "Adaptive Arrays," *IEEE Trans. Antennas Propag.*, Vol. AP-24, No. 5, September 1976, pp. 585–598.

[19] A. Papoulis, *Probability, Random Variables, and Stochastic Processes*, McGraw-Hill, New York, 1965, Ch. 8.

[20] R. A. Wiggins and E. A. Robinson, "Recursive Solution to the Multichannel Filtering Problem," *J. Geophys. Res.*, Vol. 70, No. 8, April 1965, pp. 1885–1891.

[21] R. V. Churchill, *Introduction to Complex Variables and Applications*, McGraw-Hill, New York, 1948, Ch. 8.

[22] M. J. Levin, "Maximum-Likelihood Array Processing," Lincoln Laboratories, Massachusetts Institute of Technology, Lexington, MA, Semiannual Technical Summary Report on Seismic Discrimination, December 31, 1964.

[23] J. Chang and F. Tuteur, Symposium on Common Practices in Communications, Brooklyn Polytechnic Institute, 1969.

[24] P. E. Mantey and L. J. Griffiths, "Iterative Least-Squares Algorithms for Signal Extraction," Proceedings of the 2nd Hawaii Conference on System Sciences, 1969, pp. 767–770.

[25] R. L. Riegler and R. T. Compton, Jr., "An Adaptive Array for Interference Rejection," *Proc. IEEE*, Vol. 61, No. 6, June 1973, pp. 748–758.

[26] S. P. Applebaum, Syracuse University Research Corporation Report SPL TR 66-1, Syracuse, NY, August 1966.

[27] S. W. W. Shor, "Adaptive Technique to Discriminate Against Coherent Noise in a Narrow-Band System," *J. Acous. Soc. Am.*, Vol. 34, No. 1, pp. 74–78.

[28] R. T. Adams, "An Adaptive Antenna System for Maximizing Signal-to-Noise Ratio," WESCON Conference Proceedings, Session 24, 1966, pp. 1–4.

[29] R. Bellman, *Introduction to Matrix Analysis*, McGraw-Hill, New York, 1960.

[30] R. F. Harrington, *Field Computation by Moment Methods*, Macmillan, New York, 1968, Ch. 10.

[31] B. D. Steinberg, *Principles of Aperture and Array System Design*, Wiley, New York, 1976, Ch. 12.

[32] L. J. Griffiths, "Signal Extraction Using Real-Time Adaptation of a Linear Multichannel Filter," SEL-68-017, Tech. Rep. No. 6788-1, System Theory Laboratory, Stanford University, February 1968.

[33] R. T. Lacoss, "Adaptive Combining of Wideband Array Data for Optimal Reception," *IEEE Trans. Geosci. Electron.*, Vol. GE-6, No. 2, May 1968, pp. 78–86.

[34] C. A. Baird, Jr., and J. T. Rickard, "Recursive Estimation in Array Processing," Proceedings of the Fifth Asilomar Conference on Circuits and Systems, 1971, pp. 509–513.

[35] C. A. Baird, Jr., and C. L. Zahm, "Performance Criteria for Narrowband Array Processing," 1971 IEEE Conference on Decision and Control, December 15–17, Miami Beach, FL, pp. 564–565.

[36] W. W. Peterson, T. G. Birdsall, and W. C. Fox, "The Theory of Signal Detectability," *IRE Trans.*, PGIT-4, 1954, pp. 171–211.

[37] D. Middleton and D. Van Meter, "Modern Statistical Approaches to Reception in Communication Theory," *IRE Trans.*, PGIT-4, 1954, pp. 119–145.

[38] T. G. Birdsall, "The Theory of Signal Detectability: ROC Curves and Their Character," Tech. Rept. No. 177, Cooley Electronics Laboratory, University of Michigan, Ann Arbor, MI, 1973.

[39] W. J. Bangs and P. M. Schultheiss, "Space-Time Processing for Optimal Parameter Estimation," Proceedings of NATO Advanced Study Institute on Signal Processing, Loughborough, England, August 1972, pp. 577–589.

[40] L. W. Nolte, "Adaptive Processing: Time-Varying Parameters, Proceedings of NATO Advanced Study Institute on Signal Processing, Loughborough, England, August 1972, pp. 647–655.

[41] D. O. North, "Analysis of the Factors which Determine Signal/Noise Discrimination in Radar," Rept. PTR-6C, RCA Laboratories, 1943.

[42] L. A. Zadeh and J. R. Ragazzini, "Optimum Filters for the Detection of Signals in Noise," *IRE Proc.*, Vol. 40, 1952, pp. 1223–1231.

[43] G. L. Turin, "An Introduction to Matched Filters," *IRE Trans.*, Vol. IT-6, June 1960, pp. 311–330.

[44] H. L. Van Trees, *Detection, Estimation, and Modulation Theory, Part I*, Wiley, New York, 1968, Ch. 2.

[45] H. Cox, "Resolving Power and Sensitivity to Mismatch of Optimum Array Processors," *J. Acoust. Soc. Am.*, Vol., 54, No. 3, September 1973, pp. 771–785.

[46] H. L. Van Trees, "Optimum Processing for Passive Sonar Arrays," IEEE 1966 Ocean Electronics Symposium, August, Hawaii, pp. 41–65.

[47] L. W. Brooks and I. S. Reed, "Equivalence of the Likelihood Ratio Processor, The Maximum Signal-to-Noise Ratio Filter, and the Wiener Filter," *IEEE Trans. Aerosp. Electron. Syst.*, Vol. AES-8, No. 5, September 1972, pp. 690–691.

[48] A. M. Vural, "Effects of Perturbations on the Performance of Optimum/Adaptive Arrays," *IEEE Trans. Aerosp. Electron. Syst.*, Vol. AES-15, No. 1, January 1979, pp. 76–87.

[49] R. Nitzberg, "Effect of Errors in Adaptive Weights," *IEEE Trans. Aerosp. Electron. Syst.*, Vol. AES-12, No. 3, May 1976, pp. 369–373.

[50] J. P. Burg, "Three-Dimensional Filtering with an Array of Seismometers," *Geophysics*, Vol. 29, No. 5, October 1964, pp. 693–713.

# Part II. Adaptive Algorithms

In Chapter 3 it was shown that a variety of popular performance measures lead to optimum weight vector solutions that are all closely related. Consequently the choice of a specific performance measure usually is not a critically important issue; rather the adaptive control algorithm that is selected to adjust the array beam pattern is highly important since this choice directly influences both the speed of the array transient response and the complexity of the circuitry required to implement the algorithm. The adaptive processor is the heart of any adaptive array system since it automatically adjusts the weights to achieve the desired spatial and frequency filtering. Part Two presents a survey of different classes of popular adaptation algorithms and discusses the important performance characteristics of each class. In some cases algorithms are specifically tailored to particular signal conditions whereas in other cases algorithms can be easily modified to handle a variety of signal environments. This information on algorithm performance and applicability to different signal environments provides the designer with a means for determining those candidate algorithms that are most likely to satisfy the system requirements imposed by: (1) required speed of response, (2) operational signal environment (including any available *a priori* information) in which the array response must occur, and (3) circuit complexity for algorithm implementation.

# Chapter 4. Gradient-Based Algorithms

The gradient approach to solving the control problem posed by adaptive array weight adjustment is very popular since it is a relatively simple and generally well understood method that permits the solution of a large class of problems. The selected performance measure and the parameters to be adjusted (which are usually, but not always, complex weights) determine the nature of the performance surface over which the adaptive processor must operate. The performance surface is determined by the nature of $\mathfrak{P}(\mathbf{w})$, where $\mathfrak{P}(\cdot)$ is the performance measure of concern and $\mathbf{w}$ is the complex weight vector whose components it is desired to adjust. When the selected performance measure is a quadratic function of the weight settings, then the performance measure can be visualized as a bowl-shaped surface, so the adaptive processor has the task of continually seeking the "bottom of the bowl." In this case optimization of the performance measure can be accomplished by "hill climbing" methods of which the various gradient methods are representative. In the event that the performance surface is irregular, having several relative optima or saddle points, then the transient response of the gradient-based hill climbing procedures may suffer in comparison with other methods. The various gradient-based algorithms to be considered here include the following:

1. Least mean square (LMS).
2. Differential steepest descent (DSD).
3. Accelerated gradient (AG).

Variations of the above algorithms can easily be derived by introducing constraints into the adjustment rule, and one section develops the procedure for deriving such variations. Finally, changes in the modes of adaptation are discussed, illustrating how two-mode adaptation can enhance the convergence of the adaptation procedure.

## 4.1 INTRODUCTORY CONCEPTS

The properties of gradient-based algorithms are most easily introduced by first considering the method of steepest descent and its applicability to quadratic performance surfaces. Any quadratic performance surface has a unique minimum point that can be sought by a gradient-based algorithm. In the event that the gradient at any point on the performance surface can be determined exactly, then the method of steepest descent can be exploited to find the bottom of the bowl. The introduction of a feedback model of steepest descent then yields insight into the nature of the adjustment process that characterizes gradient-based algorithms.

### *4.1.1 The Quadratic Performance Surface*

To illustrate how a bowl-shaped quadratic performance surface can arise, consider the MSE performance measure for the adaptive array of Figure 3.3. It will be recalled from Chapter 3 that the array output signal is given by

$$y(t)=\mathbf{w}^T(t)\mathbf{x}(t) \tag{4.1}$$

Denoting the desired array response by $d(t)$, we may express the error signal as

$$e(t)=d(t)-y(t)=d(t)-\mathbf{w}^T(t)\mathbf{x}(t) \tag{4.2}$$

The square of the foregoing error signal is then

$$e^2(t)=d^2(t)-2d(t)\mathbf{x}^T(t)\mathbf{w}(t)+\mathbf{w}^T(t)\mathbf{x}(t)\mathbf{x}^T(t)\mathbf{x}(t) \tag{4.3}$$

The MSE is just the expected value of $e^2(t)$, or

$$\begin{aligned} E\{e^2(t)\} &= \xi[\mathbf{w}(t)] \\ &= \overline{d^2}(t)-2\mathbf{r}_{xd}^T(t)\mathbf{w}(t)+\mathbf{w}^T(t)\mathbf{R}_{xx}(t)\mathbf{w}(t) \end{aligned} \tag{4.4}$$

where the overbar denotes expected value, $\mathbf{r}_{xd}(t)$ is given by (3.52), and $\mathbf{R}_{xx}(t)$ is given by (3.13). When the input signals are statistically stationary, then $\mathbf{r}_{xd}$ and $\mathbf{R}_{xx}$ are also stationary and there is no need to write these quantities with the argument $t$. In nonstationary signal environments, however, the notation $\mathbf{r}_{xd}(t)$ and $\mathbf{R}_{xx}(t)$ is required.

From (4.4) it may be seen that the MSE is just a quadratic function of the weight vector $\mathbf{w}(t)$. This bowl-shaped performance surface may migrate in the nonstationary case, so the location of the bottom of the bowl may be moving while the curvature and orientation of the bowl may be changing. The analysis of adaptive processor behavior for time-varying signal statistics is beyond the scope of this book, although it is a subject of current research interest. The concern here is to determine the transient behavior of the adaptation process with the assumption that the signal statistics are stationary but unknown.

### 4.1.2 *The Method of Steepest Descent*

To illustrate the philosophy underlying gradient-based algorithms, it is useful to assume initially that the statistics describing the signal environment are perfectly known so the gradient at any point on the performance surface can be determined exactly. The method of steepest descent may then be exploited to find the bottom of the quadratic performance surface bowl. For the MSE function of (4.4), the gradient is obtained by differentiating with respect to the weight vector to yield [1]

$$\nabla\{\xi[\mathbf{w}(t)]\} = -2\mathbf{r}_{xd} + 2\mathbf{R}_{xx}\mathbf{w}(t) \tag{4.5}$$

It was shown in Chapter 3 that the bottom of the bowl is reached when the weight vector is set equal to the Wiener solution:

$$\mathbf{w}_{\text{opt}} = \mathbf{R}_{xx}^{-1}\mathbf{r}_{xd} \tag{4.6}$$

On substituting (4.6) into (4.4), the minimum MSE is then found to be

$$\xi_{\min} = \overline{d^2(t)} - \mathbf{w}_{\text{opt}}^T\mathbf{r}_{xd} \tag{4.7}$$

The method of steepest descent begins with an initial guess of where the performance surface minimum point may be. This initial guess consists of a set of initial values for each of the weight vector components. Having selected a starting point, we then determine the gradient vector, and the next guess is obtained by making an appropriate change in the current guess. This appropriate change is determined by perturbing the weight vector in the opposite direction from the gradient (that is, in the direction of the steepest downward slope of the wall of the bowl-shaped performance surface).

Two-dimensional plan views of a quadratic performance surface (corresponding to a two-weight adjustment problem) are shown in Figures 4.1 and 4.2. In these figures the MSE is measured along a coordinate normal to the plane of the paper. The ellipses in these figures represent contours of constant MSE having equal increment spacing. The gradient is orthogonal to these constant value contours (pointing in the steepest direction) at every point on the performance surface. Figure 4.1 illustrates what happens with the method of steepest descent when a very small step size is used to update the weight vector starting from an initial guess; the sequence of steps is so small that each individual step is imperceptible, and these steps appear to form a

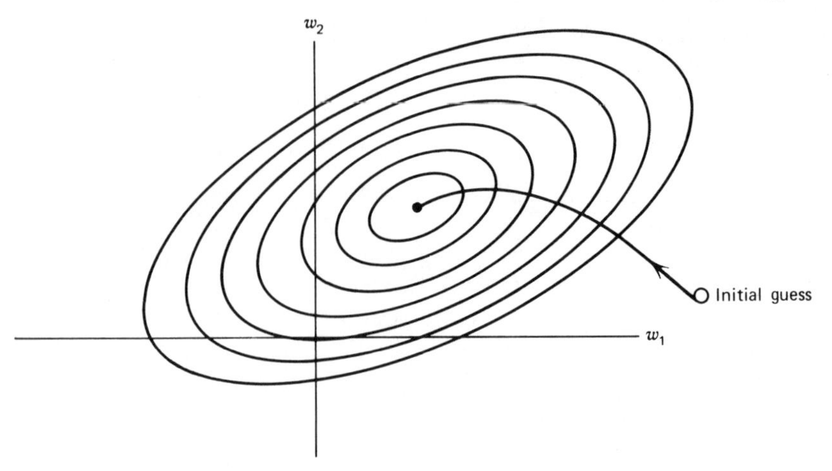


**Figure 4.1** Steepest descent with very small step size (overdamped case).

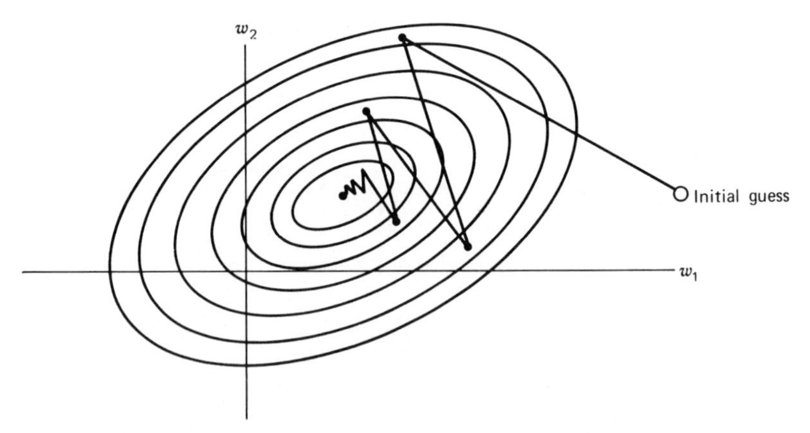


**Figure 4.2** Steepest descent with large step size (underdamped case).

continuous path converging at the performance surface minimum. With the larger step size selection illustrated in Figure 4.2, each individual step can be distinctly seen and is taken in a direction normal to the error contour from which it begins. The precise nature of the transients the weights undergo in reaching the performance surface minimum is discussed in the analysis of steepest descent adaptation that follows. The two adaptation cases illustrated in Figures 4.1 and 4.2 are referred to as "overdamped" and "underdamped," respectively, for reasons that become clear in the next section.

The method of steepest descent can be expressed in discrete mathematical form by the following iterative relation [1]:

$$\mathbf{w}(k+1)=\mathbf{w}(k)-\Delta_s \nabla\left[\overline{e^2(k)}\right] \tag{4.8}$$

where $\mathbf{w}(k)$ = old weight vector guess at time $kT$,
$\mathbf{w}(k+1)$ = new weight vector guess at time $(k+1)T$,
$\nabla[e^2(k)]$ = gradient vector of the MSE determining the direction in which to move from $\mathbf{w}(k)$,
$\Delta_s$ = a constant that determines the amount by which to move (i.e., the "step size") from the old guess.

Substituting the gradient of (4.5) into (4.8) then yields

$$\mathbf{w}(k+1)=\mathbf{w}(k)-2\Delta_s(\mathbf{R}_{xx}\mathbf{w}(k)-\mathbf{r}_{xd}) \tag{4.9}$$

### *4.1.3 Feedback Model of Steepest Descent*

The transient behavior of the method of steepest descent will yield valuable insight into the behavior of the LMS algorithm since the only difference between the two weight adjustment algorithms lies in the fact that with steepest descent the signal environment statistics are perfectly known (so the gradient at any point can be exactly determined), whereas with the LMS algorithm the signal statistics are unknown (although here they are assumed to be stationary) and therefore must be estimated. The first step in determining the transient behavior of the method of steepest descent is to formulate a feedback model of the weight adjustment relationship.

A feedback flow graph of the relationships expressed by (4.8) and (4.9) can be drawn in which the gradient plays the role of an error signal in an $N$-dimensional servomechanism that controls the adjustments of the various weight values. This flow graph is shown in Figure 4.3 where the "signals" in the graph are taken as row vectors rather than column vectors. The symbol $Z^{-1}$ is the $Z$-transform representation [2]–[5] of a unit (one iteration cycle) time delay, and $Z^{-1}\mathbf{I}$ is the matrix transfer function of a unit delay branch. This flow graph represents a first-order multidimensional sampled-data control loop.

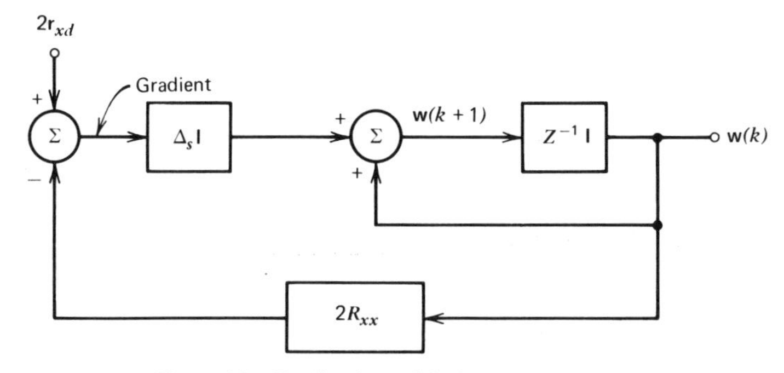


**Figure 4.3** Feedback model of steepest descent.

By setting the initial weight vector $\mathbf{w}(0)$ equal to the initial guess in the flow graph, the resulting sequence of $\mathbf{w}(k)$ will behave exactly as in the actual hill-climbing weight adjustment algorithm. Since the "output" of the flow graph model is the current weight vector $\mathbf{w}(k)$, the transient behavior of the weight adjustment algorithm can be determined by considering the behavior of the flow graph model.

Each transfer function appearing in the flow graph of Figure 4.3 is a diagonal matrix except for the feedback branch denoted by $2\mathbf{R}_{xx}$. This branch matrix will in general have off-diagonal elements since the input signals will usually be mutually correlated. Consequently, transients will cross-couple from one component of the weight vector to the next thereby complicating the study of transient behavior. To alleviate this complication, it is convenient to diagonalize the flow graph and eliminate such cross-coupling effects from consideration: the diagonalization then enables one to consider the natural modes of behavior of the flow graph by merely introducing a coordinate transformation.

To diagonalize the flow graph of Figure 4.3, consider the expression for the MSE given by (4.4). Recalling that $\mathbf{w}_{\text{opt}}$ and $\xi_{\min}$ are given by (4.6) and (4.7), it follows that we can rewrite the MSE as

$$E\{e^2(k)\} = \xi(k) = \xi_{\min} + \left[\mathbf{w}(k) - \mathbf{w}_{\text{opt}}\right]^T \mathbf{R}_{xx}\left[\mathbf{w}(k) - \mathbf{w}_{\text{opt}}\right] \tag{4.10}$$

Since the matrix $\mathbf{R}_{xx}$ is real, symmetric, and positive definite (for real variables), it may be diagonalized by means of a unitary transformation matrix $\mathbf{Q}$ so that

$$\mathbf{R}_{xx} = \mathbf{Q}^{-1}\mathbf{\Lambda}\mathbf{Q} \tag{4.11}$$

where $\mathbf{\Lambda}$ is the diagonal matrix of eigenvalues and $\mathbf{Q}$ is the modal square matrix of eigenvectors. If $\mathbf{Q}$ is constructed from normalized eigenvectors then it is orthonormal so that $\mathbf{Q}^{-1} = \mathbf{Q}^T$, and the MSE now becomes

$$\xi(k) = \xi_{\min} + \left[\mathbf{w}(k) - \mathbf{w}_{\text{opt}}\right]^T \mathbf{Q}^T\mathbf{\Lambda}\mathbf{Q}\left[\mathbf{w}(k) - \mathbf{w}_{\text{opt}}\right] \tag{4.12}$$

Now define

$$\mathbf{Q}\mathbf{w}(k) \triangleq \mathbf{w}'(k) \tag{4.13}$$

and

$$\mathbf{Q}\mathbf{w}_{\text{opt}} \triangleq \mathbf{w}'_{\text{opt}} \tag{4.14}$$

Equation (4.12) can then be rewritten as

$$\xi(k) = \xi_{\min} + \left[\mathbf{w}'(k) - \mathbf{w}'_{\text{opt}}\right]^T \mathbf{\Lambda}\left[\mathbf{w}'(k) - \mathbf{w}'_{\text{opt}}\right] \tag{4.15}$$

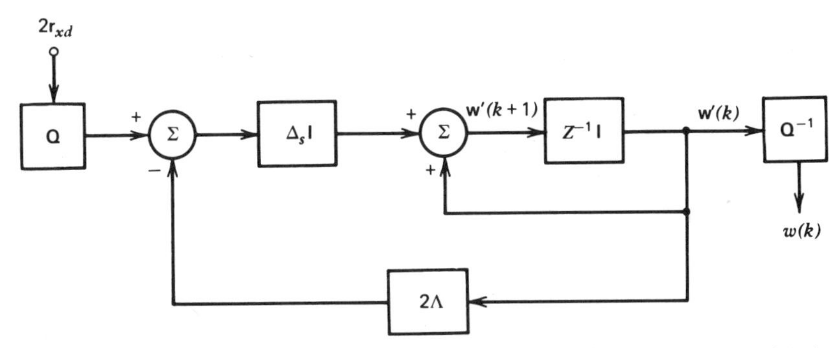


**Figure 4.4** Diagonalized version of feedback model of steepest descent using normal (primed) coordinates.

The transformation $\mathbf{Q}$ projects $\mathbf{w}(k)$ into the primed coordinates [resulting in $\mathbf{w}'(k)$]. Since $\boldsymbol{\Lambda}$ is a diagonal matrix and $\xi(k)$ is a quadratic performance surface, the primed coordinates must comprise the principal axes of the quadratic performance surface. The feedback model of Figure 4.3 can now be simplified by expressing all quantities in terms of the primed coordinate system, resulting in the equivalent feedback model of Figure 4.4 in which all cross-couplings that previously existed within the feedback paths have been eliminated.

The feedback model of steepest descent given by the flow diagram of Figure 4.4 is comprised of the natural modes of the flow graph. The transients of each mode are now isolated (since each of the primed coordinates now has its own natural mode), and the natural behavior of steepest descent can be completely explored by considering the behavior of a single primed coordinate.

An isolated one-dimensional feedback model for the $p$th normal coordinate is shown in Figure 4.5. The pulse transfer function of this closed-loop feedback system is then [1]

$$\frac{\mathbf{w}_p'(z)}{r_p'(z)} = \frac{\Delta_s Z^{-1}}{1+(1-2\Delta_s\lambda_p)Z^{-1}} \tag{4.16}$$

where $\lambda_p$ is the $p$th eigenvalue of the covariance matrix $\mathbf{R}_{xx}$. The impulse response of (4.16) can be found by letting $r_p'(z)=1$ and taking the inverse $Z$-transform of the resulting output $\mathfrak{z}^{-1}\{w_p'(z)\}$. It follows that the impulse response is of the form

$$w_p'(kT) = \text{constant} \times e^{-\alpha_p(kT)}$$

where

$$\alpha_p = -\frac{1}{T}\, ln(1-2\Delta_s\lambda_p) \tag{4.17}$$

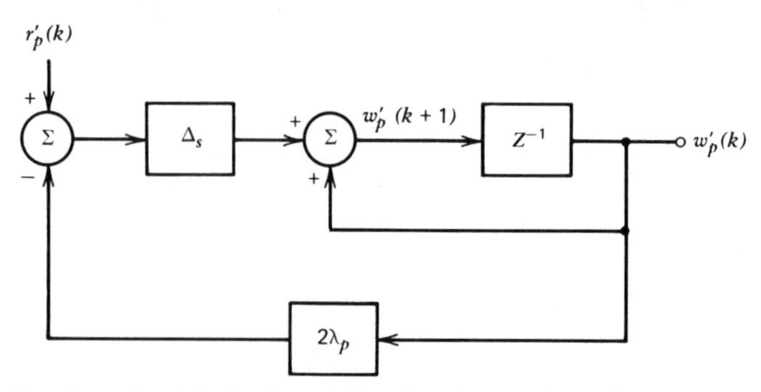


**Figure 4.5** One-dimensional feedback model for the $p$th normal coordinate of steepest descent.

and $T$ = one iteration period. It follows immediately that the time response of (4.17) represents a stable system so long as

$$|1-2\Delta_s\lambda_p|<1 \tag{4.18}$$

The only way that (4.18) can be satisfied is for

$$\left.\begin{aligned}\Delta_s&>0\\ \text{and}\qquad |\Delta_s\lambda_p|&<1\end{aligned}\right\} \tag{4.19}$$

Since the correlation matrix $\mathbf{R}_{xx}$ is positive definite, the eigenvalues of $\mathbf{R}_{xx}$ are such that $\lambda_p \geqslant 0$ for all $p$. Consequently the stability of the multidimensional flow graph of Figure 4.4 is guaranteed if and only if

$$\left.\begin{aligned}\Delta_s&>0\\ \text{and}\qquad |\Delta_s\lambda_{\max}|&<1\end{aligned}\right\} \tag{4.20}$$

where $\lambda_{\max}$ is the maximum eigenvalue of $\mathbf{R}_{xx}$. The stability of the steepest descent adaptation process is therefore guaranteed so long as

$$\frac{1}{\lambda_{\max}}>\Delta_s>0 \tag{4.21}$$

## 4.2 THE LEAST MEAN SQUARE (LMS) ALGORITHM

When signal statistics describing the operational environment are stationary but unknown (a condition that accurately reflects many practical situations), then one is no longer able to compute exactly the gradient of the performance surface at any point since the signal statistics must now be estimated. Under these conditions the gradient must itself be estimated; for this class of problems (having a quadratic performance function) the LMS algorithm introduced by Widrow has proven particularly useful [6]–[10]. It is worthwhile noting that the LMS algorithm requires that an error signal given by (4.2) be generated. Such an error signal in turn requires a reference signal model $d(t)$ to represent the signal it is desired to receive. In communication systems where the desired signal is usually present, this requirement can be met by generating an approximation of the actual signal. In those systems where the desired signal is usually not present (as in radar or sonar systems), it is pointless to try to generate a fictitious desired signal. Consequently the LMS algorithm described here is usually employed to improve communications system performance. The LMS algorithm is exactly like the method of steepest descent except that now changes in the weight vector are made in the direction given by an estimated gradient vector instead of the actual gradient vector. In other words, changes in the weight vector may be expressed as

$$\mathbf{w}(k+1)=\mathbf{w}(k)-\Delta_s \hat{\nabla}[\xi(k)] \tag{4.22}$$

where $\mathbf{w}(k)$ = weight vector before adaptation step,
$\mathbf{w}(k+1)$ = weight vector after adaptation step
$\Delta_s$ = scalar constant determining step size and controlling rate of convergence and stability,
$\hat{\nabla}[\xi(k)]$ = estimated gradient vector of $\xi$ with respect to $\mathbf{w}$.

The adaptation process described by (4.22) attempts to find a solution as close as possible to the Wiener solution given by (4.6). It is tempting to try to solve (4.6) directly since this certainly provides a straightforward way to obtain the desired solution, but such an approach has several drawbacks as follows:

1. Serious computational problems could arise from the necessity of computing and inverting an $N \times N$ matrix when the number of weights $N$ is large and when input data rates are high.
2. This method may require up to $[N(N+3)]/2$ autocorrelation and cross-correlation measurements to be made in order to obtain the elements of $\mathbf{R}_{xx}$ and $\mathbf{r}_{xd}$. Such measurements must be repeated whenever the input signal statistics change with time (as they would in many practical situations).
3. Implementing a direct solution requires setting weight values with high accuracy in open loop fashion, whereas a feedback approach provides self-correction of inaccurate settings thereby giving tolerance to hardware errors.

To obtain the estimated gradient of the MSE performance measure, take the gradient of a single time sample of the squared error as follows:

$$\hat{\nabla}_k=\nabla[\xi(k)]=2e(k)\nabla[e(k)] \tag{4.23}$$

Since

$$e(k) = d(k) - \mathbf{x}^T(k)\mathbf{w} \tag{4.24}$$

it follows that

$$\nabla[e(k)] = \nabla[d(k) - \mathbf{x}^T(k)\mathbf{w}] = -\mathbf{x}(k) \tag{4.25}$$

so that

$$\hat{\nabla}_k = -2e(k)\mathbf{x}(k) \tag{4.26}$$

It is easy to show that the gradient estimate given by (4.26) is unbiased by considering the expected value of the estimate and comparing it with the gradient of the actual MSE. The expected value of the estimate is given by

$$E\{\hat{\nabla}_k\} = -2E\{\mathbf{x}(k)[d(k) - \mathbf{x}^T(k)\mathbf{w}(k)]\} \tag{4.27}$$

$$= -2[\mathbf{r}_{xd}(k) - \mathbf{R}_{xx}(k)\mathbf{w}(k)] \tag{4.28}$$

Now consider the MSE:

$$\xi[\mathbf{x}(k)] = E[d^2(k)] + \mathbf{w}^T\mathbf{R}_{xx}(k)\mathbf{w} - 2\mathbf{w}^T\mathbf{r}_{xd}(k) \tag{4.29}$$

Differentiating (4.29) with respect to $\mathbf{w}$ yields the gradient $\nabla\{\xi[\mathbf{w}(k)]\}$ as

$$\nabla\{\xi[\mathbf{w}(k)]\} = 2\mathbf{R}_{xx}(k)\mathbf{w}(k) - 2\mathbf{r}_{xd}(k) \tag{4.30}$$

Comparing (4.28) and (4.30) reveals that

$$E\{\hat{\nabla}_k\} = \nabla\{\xi[\mathbf{w}(k)]\} \tag{4.31}$$

so the expected value of the estimated gradient is equal to the true value of the gradient of the MSE.

Substituting the estimated gradient of (4.26) into the weight adjustment rule of (4.22) then yields the weight control rule

$$\mathbf{w}(k+1) = \mathbf{w}(k) + 2\Delta_s e(k)\mathbf{x}(k) \tag{4.32}$$

in which the new weight vector is determined by adding to the old weight vector the input signal vector scaled by the error value. Equation (4.32) represents the discrete time form of the LMS algorithm.

The LMS algorithm given by (4.32) can be rewritten for complex quantities as

$$\frac{\mathbf{w}(k+1) - \mathbf{w}(k)}{\Delta t} = 2k_s e(k)\mathbf{x}^*(k) \tag{4.33}$$

where $\Delta t$ is the elapsed time between successive iterations and $\Delta_s = k_s \Delta t$. In the limit as $\Delta t \to 0$, (4.33) yields an equivalent differential equation representation of the LMS algorithm that is appropriate for use in continuous systems as

$$\frac{d\mathbf{w}(t)}{dt} = 2k_s e(t)\mathbf{x}^*(t) \tag{4.34}$$

Equation (4.34) can also be written as

$$\mathbf{w}(t) = 2k_s \int_0^t e(\tau)\mathbf{x}^*(\tau)\,d\tau + \mathbf{w}(0) \tag{4.35}$$

A block diagram representation of the weight adjustment rule represented by (4.35) is shown in Figure 4.6.

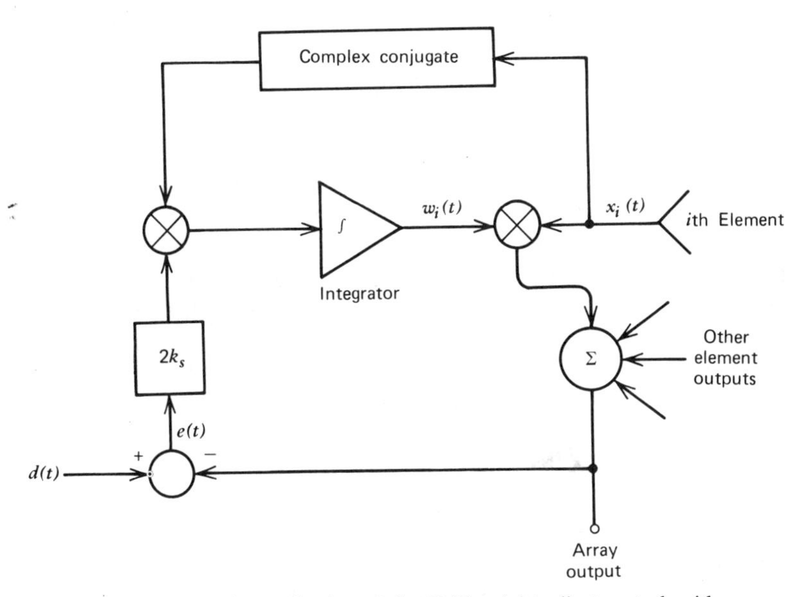


**Figure 4.6** Analog realization of the LMS weight adjustment algorithm.

The discrete version of (4.34) is given by (4.33) and is more commonly written as

$$\mathbf{w}(k+1) = \mathbf{w}(k) + 2k_s \Delta t\, e(k)\mathbf{x}^*(k) \tag{4.36}$$

A block diagram representation of the weight adjustment rule represented by (4.36) is illustrated in Figure 4.7.

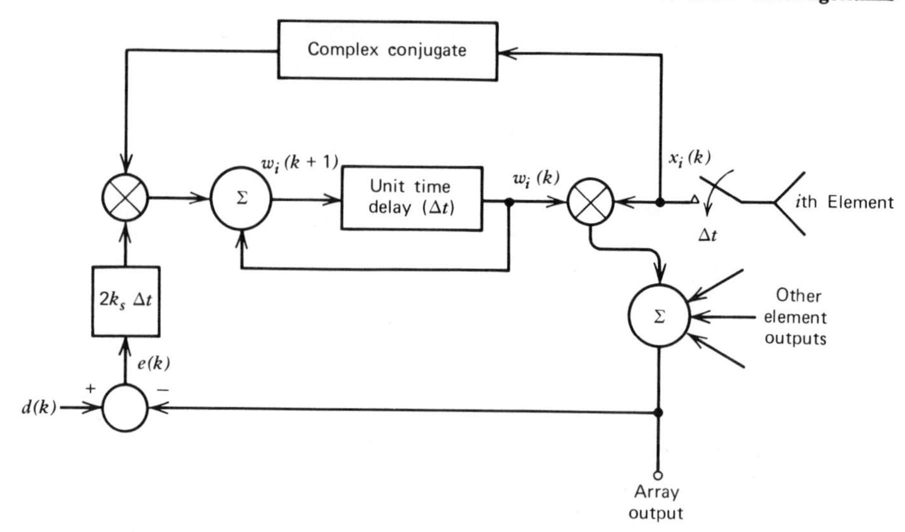


**Figure 4.7** Digital realization of the LMS weight adjustment algorithm.

### *4.2.1 Convergence to the Wiener Solution*

To simplify the discussion, assume that the time between successive iterations of the LMS algorithm is long enough so that the signal vectors $\mathbf{x}(k)$ and $\mathbf{x}(k+1)$ are uncorrelated. From (4.32) it follows that $\mathbf{w}(k)$ is a function only of $\mathbf{x}(k-1)$, $\mathbf{x}(k-2),\ldots,\mathbf{x}(0)$, and $\mathbf{w}(0)$, where the successive input signal vectors are uncorrelated so that $\mathbf{w}(k)$ is independent of $\mathbf{x}(k)$. It will now be shown that for a stationary input signal process meeting the above condition, the expected value of the weight vector $E\{\mathbf{w}(k)\}$ converges (after a sufficient number of iterations) to the Wiener solution given by (4.6).

If we take the expected value of both sides of (4.36) there results

$$E\{\mathbf{w}(k+1)\}=E\{\mathbf{w}(k)\}+2k_s\Delta t \cdot E\{\mathbf{x}^*(k)[d(k)-\mathbf{x}^T(k)\mathbf{w}(k)]\} \tag{4.37}$$

Now let

$$E\{\mathbf{x}^*(k)d(k)\}=\mathbf{r}_{xd} \tag{4.38}$$

and

$$E\{\mathbf{x}^*(k)\mathbf{x}^T(k)\}=\mathbf{R}_{xx} \tag{4.39}$$

Consequently, (4.37) can be rewritten as

$$\begin{aligned}E\{\mathbf{w}(k+1)\}&=E\{\mathbf{w}(k)\}-2k_s\Delta t\mathbf{R}_{xx}E\{\mathbf{w}(k)\}+2k_s\Delta t\mathbf{r}_{xd}\\ &=[\mathbf{I}-2k_s\Delta t\mathbf{R}_{xx}]E\{\mathbf{w}(k)\}+2k_s\Delta t\mathbf{r}_{xd}\end{aligned} \tag{4.40}$$

Starting with an initial guess $\mathbf{w}(0)$, the $(k+1)$ iteration of (4.40) yields

$$E\{\mathbf{w}(k+1)\}=\left[\mathbf{I}-2k_s\Delta t\mathbf{R}_{xx}\right]^{(k+1)}\mathbf{w}(0)$$
$$+2k_s\Delta t\sum_{i=0}^{k}\left[\mathbf{I}-2k_s\Delta t\mathbf{R}_{xx}\right]^{i}\mathbf{r}_{xd} \tag{4.41}$$

Diagonalizing (4.41) by using (4.11) to obtain the normal form results in

$$E\{\mathbf{w}(k+1)\}=\mathbf{Q}^{-1}\left[\mathbf{I}-2k_s\Delta t\mathbf{\Lambda}\right]^{(k+1)}\mathbf{Q}\mathbf{w}(0)$$
$$+2k_s\Delta t\mathbf{Q}^{-1}\sum_{i=0}^{k}\left[\mathbf{I}-2k_s\Delta t\mathbf{\Lambda}\right]^{i}\mathbf{Q}\mathbf{r}_{xd} \tag{4.42}$$

As long as all the terms in the diagonal matrix $[\mathbf{I}-2k_s\Delta t\mathbf{\Lambda}]$ have a magnitude less than unity, then

$$\lim_{k\to\infty}\left[\mathbf{I}-2k_s\Delta t\mathbf{\Lambda}\right]^{(k+1)}\to 0 \tag{4.43}$$

and consequently the first term of (4.42) vanishes after a sufficient number of iterations. The summation factor in the second term of (4.42) likewise becomes (using the formula for the sum of a geometric series):

$$\lim_{k\to\infty}\sum_{i=0}^{k}\left[\mathbf{I}-2k_s\Delta t\mathbf{\Lambda}\right]^{i}=\frac{1}{2k_s\Delta t}\mathbf{\Lambda}^{-1} \tag{4.44}$$

Therefore, after a sufficient number of iterations, (4.42) yields

$$\lim_{k\to\infty}E\{\mathbf{w}(k+1)\}=2k_s\Delta t\mathbf{Q}^{-1}\left(\frac{1}{2k_s\Delta t}\mathbf{\Lambda}^{-1}\right)\mathbf{Q}\mathbf{r}_{xd}$$
$$=\mathbf{R}_{xx}^{-1}\mathbf{r}_{xd} \tag{4.45}$$

The result of (4.45) shows that after a sufficient number of iterations the expected value of the weight vector in the LMS algorithm does in fact converge to the Wiener solution.

In obtaining the result of (4.45), it was assumed that all the terms in the diagonal matrix $[\mathbf{I}-2k_s\Delta t\mathbf{\Lambda}]$ had a magnitude less than unity. Since all the eigenvalues appearing in $\mathbf{\Lambda}$ are positive, it follows that all the terms in the above diagonal matrix will have a magnitude less than unity provided that

$$|1-2k_s\Delta t\lambda_{\max}|<1$$

or

$$\frac{1}{\lambda_{\max}}>k_s\Delta t>0 \tag{4.46}$$

where $\lambda_{max}$ is the maximum eigenvalue of $\mathbf{R}_{xx}$. The convergence condition (4.46) is exactly the same as the stability condition (4.21) for the noise-free steepest descent feedback model.

The foregoing condition on $k_s$ for convergence of the mean value of the LMS algorithm can be related to the total input signal power $P_{\text{IN}}$, as described below. Since $\lambda_{max}$ satisfies the inequality,

$$\lambda_{max} \leqslant \text{trace}[\mathbf{R}_{xx}] \tag{4.47}$$

where

$$\text{trace}[\mathbf{R}_{xx}] = E\{\mathbf{x}^{\dagger}(k)\mathbf{x}(k)\} = \sum_{i=1}^{N} E\{|x_i|^2\} \triangleq P_{\text{IN}} \tag{4.48}$$

then the convergence condition (4.46) is assured if

$$\frac{1}{P_{\text{IN}}} > k_s \Delta t > 0 \tag{4.49}$$

The foregoing convergence results for the LMS algorithm were obtained using the assumption that successive input signal samples are independent. There is good reason to believe that this independence assumption is overly restrictive, since Griffiths [11] has presented experimental results that show that adaptation using highly correlated successive samples also converges to the Wiener solution, although the resulting steady-state MSE is slightly higher than that which results for statistically independent successive samples. For some applications, mean-squared convergence and its associated stability properties may be of concern, in which case more stringent conditions on $k_s$ must be satisfied [12].

### *4.2.2 Transient Response Characteristics for LMS Adaptation*

As in deriving the foregoing convergence results, it is also convenient to use the system normal coordinates to consider the detailed transient response of the adaptation network. In normal coordinates, the adaptive weight transients consist of sums of exponentials with time constants given by

$$\tau_p = \frac{1}{2(k_s \Delta t)\lambda_p}, \qquad p = 1, 2, \ldots, N \tag{4.50}$$

where $\lambda_p$ is the $p$th eigenvalue of the correlation matrix $\mathbf{R}_{xx}$. Since $\tau_p$ is inversely proportional to $\lambda_p$ (where $\lambda_p$ is the signal power associated with the $p$th normal signal coordinate), it follows that transient response is fastest for strong signals and slowest for weak signals. This result reveals the susceptibility of the LMS algorithm convergence time to eigenvalue spread in the $\mathbf{R}_{xx}$

matrix: an $\mathbf{R}_{xx}$ matrix having widely diverse eigenvalues. The transient response of the LMS algorithm will not be complete until the exponential having the longest time constant (corresponding to the smallest normal coordinate signal power) has decayed. Often there is no choice for the value of the constant $k_s$ that represents a good compromise between the various eigenvalues that will yield a desirably short transient period of operation.

The nature of the problem of eigenvalue spread can be seen graphically in Figure 4.8, where a two-dimensional plan view of a quadratic performance surface is shown corresponding to two widely diverse eigenvalues. The constant MSE ellipses in Figure 4.8 are highly elongated, with the result that many adaptive iterations are required before weight values become acceptably close to the desired Wiener solution.

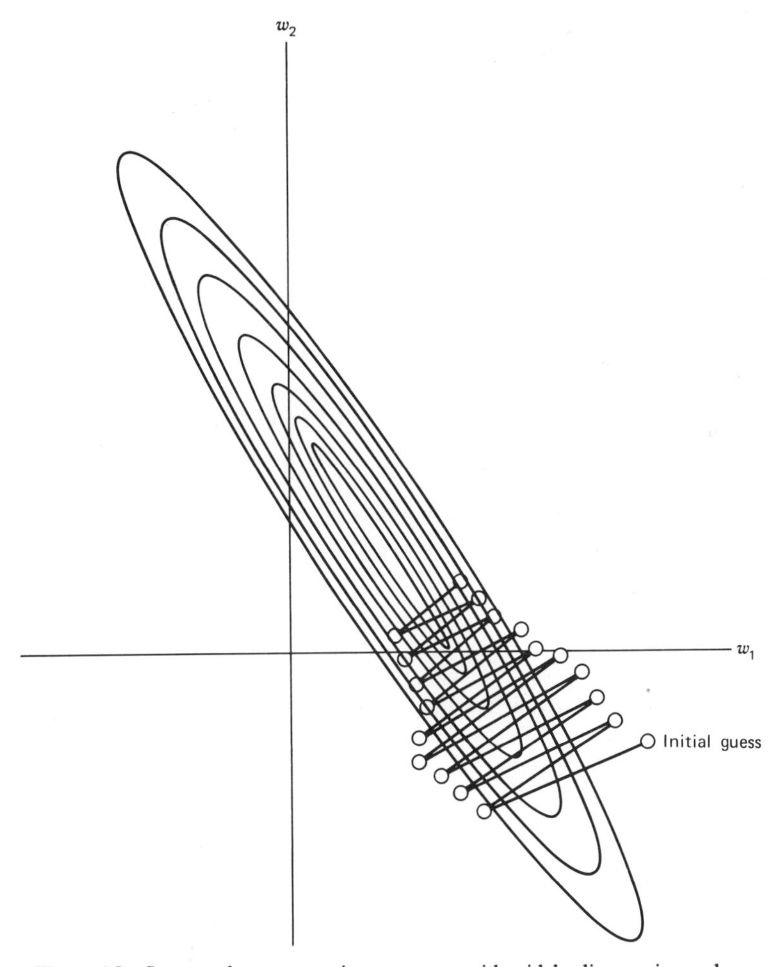


**Figure 4.8** Steepest descent transient response with widely diverse eigenvalues.

In the event that all the eigenvalues are equal, then all the time constants are equal, and

$$\tau = \frac{1}{2(k_s \Delta t)\lambda} \tag{4.51}$$

A "learning curve" that plots the expected value of the performance measure at each stage of the learning process as a function of the number of adaptation iterations provides a convenient way of monitoring the progress of an adaptive process. It has just been shown that the underlying transient behavior of the adaptive weights has an exponential nature, and since the MSE is a quadratic function of the weight values, the transients in the MSE function are also exponential in nature.

Since the square of an exponential function is an exponential having half the time constant of the original exponential function, it follows that when all the time constants are equal, then the MSE learning curve is an exponential having the time constant

$$\tau_{\text{MSE}} = \frac{\tau}{2} = \frac{1}{4(k_s \Delta t)\lambda} \tag{4.52}$$

In general, of course, the eigenvalues of $\mathbf{R}_{xx}$ are unequal so that

$$\tau_{p_{\text{MSE}}} = \frac{\tau_p}{2} = \frac{1}{4(k_s \Delta t)\lambda_p} \tag{4.53}$$

where $\tau_{p_{\text{MSE}}}$ is the time constant for the MSE learning curve, $\tau_p$ is the time constant of the $p$th normal mode in the weights, and $\lambda_p$ is the eigenvalue corresponding to the $p$th normal mode. Since one signal data sample per iteration is used in the adaptation process, the time constant expressed in terms of the number of data samples is simply

$$T_{p_{\text{MSE}}} = \tau_{p_{\text{MSE}}} \tag{4.54}$$

Plots of actual experimental learning curves have the appearance of noisy exponentials—an effect that results from the inherent noise that is present in the adaptation process. The slower the rate of adaptation (i.e., the smaller the magnitude of $k_s$) the smaller is the amplitude of the apparent noise that corrupts the learning curve.

### *4.2.3 Weight Misadjustment During LMS Adaptation*

The LMS adaptation process bases the adjustment of the weights on a gradient-estimation technique. The gradient-estimation process may be thought of as a statistical average taken with a limited sample size. Even if perfect gradient estimates could be taken, the resulting transient response would conform with that predicted for the method of steepest descent. It follows that when the adaptation process is based on noisy gradient estimates, then losses in performance compared to the noise-free case must occur. How rapidly a system adapts therefore depends on two factors: (1) the step size adopted for the weight adjustment and (2) the sample size used to compute the required statistical averages. If a large step size is selected for the weight adjustment process, then the excursions in successive weight values will also tend to be large, resulting in transient behavior like the underdamped case shown in Figure 4.2. If a small number of samples is chosen on which to base the estimate of the desired statistical averages, then the time elapsed in obtaining such averages will be small, but the quality of the resulting estimates will not be high. In general, then, the faster an adaptive algorithm adapts, the poorer its expected steady-state performance will be.

Since the input signal statistics are not known *a priori*, the minimum MSE corresponding to the Wiener solution is not achieved. The degree by which the actual MSE exceeds the minimum MSE is a measure of the extent to which the adaptive weights are misadjusted by comparison to the optimum Wiener solution. Define the "misadjustment" $M$ as the ratio

$$M \triangleq \frac{[\xi_{\text{actual}} - \xi_{\min}]}{\xi_{\min}} \tag{4.55}$$

where

$$\xi = E\{e^2\}$$

The misadjustment that obtains with the LMS algorithm can be evaluated for a specified value of $k_s \Delta t$ by considering the noise associated with the gradient-estimation process.

Assume that the adaptive process has converged to a steady state in the neighborhood of the MSE surface minimum point. The gradient-estimation noise of the adjustment algorithm at the minimum point (where the true gradient is zero) is just the gradient estimate itself, so the gradient noise vector $\mathbf{g}$ is given by

$$\mathbf{g}(k) = \hat{\nabla}(k) = -2e(k)\mathbf{x}(k) \tag{4.56}$$

The covariance of this estimation noise is given by

$$\text{cov}[\mathbf{g}(k)] = E\{\mathbf{g}(k)\mathbf{g}^T(k)\} = 4E\{e^2(k)\mathbf{x}(k)\mathbf{x}^T(k)\} \tag{4.57}$$

When the weight vector is optimized (so that $\mathbf{w}(k) = \mathbf{w}_{\text{opt}}$) then it is well known from Wiener filter theory that the error $e(k)$ is uncorrelated with the input vector $\mathbf{x}(k)$. If it is furthermore assumed that $e(k)$ and $\mathbf{x}(k)$ are Gaussian processes, then not only are they uncorrelated at the minimum point of the MSE surface, but they are also statistically independent. With these conditions (4.57) becomes

$$\text{cov}[\mathbf{g}(k)] = 4E\{e^2(k)\}E\{\mathbf{x}(k)\mathbf{x}^T(k)\} = 4\xi_{\min}\mathbf{R}_{xx} \tag{4.58}$$

In the primed normal coordinates, the above covariance can be written as

$$\text{cov}[\mathbf{g}'(k)] = \mathbf{Q}\text{cov}[\mathbf{g}(k)]\mathbf{Q}^{-1} = 4\xi_{\min}\mathbf{\Lambda} \tag{4.59}$$

Adaptation based on noisy gradient estimates results in noise in the weight vector. Recall that the noise-free method of steepest descent can be described by the iterative relation

$$\mathbf{w}(k+1) = \mathbf{w}(k) + \Delta_s[-\nabla(k)] \tag{4.60}$$

where $\Delta_s$ is the constant that controls stability and rate of convergence, and $\nabla(k)$ is the gradient at the point on the performance surface corresponding to $\mathbf{w} = \mathbf{w}(k)$. Following Widrow and McCool [15], subtract $\mathbf{w}_{\text{opt}}$ from both sides of (4.60) and define $\mathbf{v}(k) \triangleq \mathbf{w}(k) - \mathbf{w}_{\text{opt}}$ to obtain

$$\mathbf{v}(k+1) = \mathbf{v}(k) + \Delta_s[-\nabla(k)] \tag{4.61}$$

With estimated gradients instead of exact gradients, (4.61) can be rewritten as

$$\mathbf{v}(k+1) = \mathbf{v}(k) + \Delta_s(-\hat{\nabla}(k)) = \mathbf{v}(k) + \Delta_s[-\nabla(k) - \mathbf{g}(k)] \tag{4.62}$$

Now since $\nabla(k)$ is given by (4.5), it follows that

$$\nabla(k) = -2\mathbf{r}_{xd} + 2\mathbf{R}_{xx}[\mathbf{w}_{\text{opt}} + \mathbf{v}(k)] = 2\mathbf{R}_{xx}\mathbf{v}(k) \tag{4.63}$$

Consequently (4.62) can be written as

$$\mathbf{v}(k+1) = (\mathbf{I} - 2\Delta_s\mathbf{R}_{xx})\mathbf{v}(k) - \Delta_s\mathbf{g}(k) \tag{4.64}$$

which represents a first-order vector difference equation with a stochastic driving function $-\Delta_s\mathbf{g}(k)$. To simplify the analysis it is convenient to introduce the primed normal coordinate system by premultiplying both sides of (4.64) by $\mathbf{Q}$:

$$\mathbf{v}'(k+1) = (\mathbf{I} - 2\Delta_s\mathbf{\Lambda})\mathbf{v}'(k) - \Delta_s\mathbf{g}'(k) \tag{4.65}$$

After initial transients have died out and the steady state is reached, $\mathbf{v}'(k)$ responds to the stationary driving function $-\Delta_s\mathbf{g}'(k)$ in the manner of a stationary random process. The absence of any cross-coupling in the primed normal coordinate system means that the components of both $\mathbf{g}'(k)$ and $\mathbf{v}'(k)$ are mutually uncorrelated, and the covariance matrix of $\mathbf{g}'(k)$ is consequently diagonal. To find the covariance matrix of $\mathbf{v}'(k)$ consider

$$\begin{aligned}\mathbf{v}'(k+1)\mathbf{v}'^T(k+1) = &(\mathbf{I}-2\Delta_s\mathbf{\Lambda})\mathbf{v}'(k)\mathbf{v}'^T(k)(\mathbf{I}-2\Delta_s\mathbf{\Lambda}) \\ &+\Delta_s^2\mathbf{g}'(k)\mathbf{g}'^T(k)-\Delta_s(\mathbf{I}-2\Delta_s\mathbf{\Lambda})\mathbf{v}'(k)\mathbf{g}'^T(k) \\ &-\Delta_s\mathbf{g}'(k)\mathbf{v}'^T(k)(\mathbf{I}-2\Delta_s\mathbf{\Lambda})\end{aligned} \tag{4.66}$$

Taking expected values of both sides of (4.66) (and noting that $\mathbf{v}'(k)$ and $\mathbf{g}'(k)$ are uncorrelated since $\mathbf{v}'(k)$ is affected only by gradient noise from previous iterations), we find there results

$$\begin{aligned}\text{cov}[\mathbf{v}'(k)] &= (\mathbf{I}-2\Delta_s\mathbf{\Lambda})\text{cov}[\mathbf{v}'(k)](\mathbf{I}-2\Delta_s\mathbf{\Lambda})+\Delta_s^2\text{cov}[\mathbf{g}'(k)] \\ &= \Delta_s^2[4\Delta_s\mathbf{\Lambda}-4\Delta_s^2\mathbf{\Lambda}^2]^{-1}\text{cov}[\mathbf{g}'(k)]\end{aligned} \tag{4.67}$$

Invariably in practical applications the LMS algorithm is implemented with a small value for $\Delta_s$, so that

$$\Delta_s\mathbf{\Lambda} \ll \mathbf{I} \tag{4.68}$$

With (4.68) satisfied, the squared terms involving $\Delta_s\mathbf{\Lambda}$ in (4.67) may be neglected, with the result that

$$\text{cov}[\mathbf{v}'(k)] = \frac{\Delta_s}{4}\mathbf{\Lambda}^{-1}\text{cov}[\mathbf{g}'(k)] \tag{4.69}$$

Using the result given by (4.59), we find it then follows that

$$\text{cov}[\mathbf{v}'(k)] = \frac{\Delta_s}{4}\mathbf{\Lambda}^{-1}(4\xi_{\min}\mathbf{\Lambda}) = \Delta_s\xi_{\min}\mathbf{I} \tag{4.70}$$

Therefore the covariance of the steady-state noise in the weight vector (near the minimum point of the MSE surface) is

$$\text{cov}[\mathbf{v}(k)] = \Delta_s\xi_{\min}\mathbf{I} \tag{4.71}$$

Without noise in the weight vector, the actual MSE experienced would be $\xi_{\min}$. The presence of noise in the weight vector causes the steady-state weight vector solution to meander randomly about the minimum point. This random meandering results in an "excess" MSE—that is, a MSE that is greater than $\xi_{\min}$. Since

$$\xi(k) = \overline{d^2}(k) - 2\mathbf{r}_{xd}^T\mathbf{w}(k) + \mathbf{w}^T(k)\mathbf{R}_{xx}\mathbf{w}(k) \tag{4.72}$$

where

$$\xi_{\min} = \overline{d^2}(k) - \mathbf{w}_{\text{opt}}^T\mathbf{r}_{xd} \tag{4.73}$$

and

$$\mathbf{w}_{\text{opt}} = \mathbf{R}_{xx}^{-1}\mathbf{r}_{xd} \tag{4.74}$$

it follows that (4.72) can be rewritten as (also see (4.10))

$$\xi(k) = \xi_{\min} + \mathbf{v}^T(k)\mathbf{R}_{xx}\mathbf{v}(k) \tag{4.75}$$

In terms of the primed normal coordinates, (4.75) can be rewritten as

$$\xi(k) = \xi_{\min} + \mathbf{v}'^T(k)\mathbf{\Lambda}\mathbf{v}'(k) \tag{4.76}$$

It immediately follows from (4.76) that the average excess MSE is

$$E\{\mathbf{v}'^T(k)\mathbf{\Lambda}\mathbf{v}'(k)\} = \sum_{p=1}^{N} \lambda_p E\{[\mathbf{v}'_p(k)]^2\} \tag{4.77}$$

Using (4.70) to recognize that $E\{[\mathbf{v}'_p(k)]^2\}$ is just $\Delta_s\xi_{\min}$ for each $p$, we see it then follows that

$$\begin{aligned} E\{\mathbf{v}'^T(k)\mathbf{\Lambda}\mathbf{v}'(k)\} &= \Delta_s\xi_{\min}\sum_{p=1}^{N}\lambda_p \\ &= \Delta_s\xi_{\min}tr(\mathbf{R}_{xx}) \end{aligned} \tag{4.78}$$

The misadjustment in the LMS algorithm is therefore given by

$$M = \frac{E\{\mathbf{v}'^T(k)\mathbf{\Lambda}\mathbf{v}'(k)\}}{\xi_{\min}} = \Delta_s tr(\mathbf{R}_{xx}) \tag{4.79}$$

Since $\Delta_s = k_s\Delta t$, the result of (4.79) emphasizes the fact that the degree of misadjustment experienced with the LMS algorithm can be controlled merely by adjusting the step size constant $k_s$. Of course when the step size is decreased, the time required to reach the steady-state condition increases, so there is a fundamental trade-off between the degree of misadjustment and the adaptation speed.

The misadjustment in the LMS algorithm can also be expressed in a manner that gives insight into the relationship between misadjustment and adaptation speed. From (4.53) it follows that

$$\Delta_s\lambda_p = \frac{1}{4\tau_{p_{\text{MSE}}}} \tag{4.80}$$

Furthermore,

$$\Delta_s tr(\mathbf{R}_{xx}) = \Delta_s\sum_{p=1}^{N}\lambda_p = \sum_{p=1}^{N}\left(\frac{1}{4\tau_{p_{\text{MSE}}}}\right) = \frac{N}{4}\left(\frac{1}{\tau_{p_{\text{MSE}}}}\right)_{\text{av}} \tag{4.81}$$

where

$$\left(\frac{1}{\tau_{p_{\text{MSE}}}}\right)_{\text{av}} \triangleq \frac{1}{N}\sum_{p=1}^{N}\left(\frac{1}{\tau_{p_{\text{MSE}}}}\right) \tag{4.82}$$

where $N$ is the number of degrees of freedom of the adaptive processor. Consequently the misadjustment can be written as

$$M=\frac{N}{4}\left(\frac{1}{\tau_{p_{\text{MSE}}}}\right)_{\text{av}}=\frac{N}{4}\left(\frac{1}{T_{p_{\text{MSE}}}}\right)_{\text{av}} \tag{4.83}$$

where $T_{p_{\text{MSE}}}$ is the learning curve time constant in units of the number of data samples. It is worth noting that while the basic time unit in digital systems is the sample period, for analog systems it is the equivalent Nyquist sample period corresponding to the signal bandwidth.

### *4.2.4 Practical Considerations for LMS Adaptation*

Generation of the error signal in LMS adaptation requires an appropriate desired response signal for the adaptive processor. If the desired response signal is taken to be the signal itself, then the adaptive array output reproduces the signal in the best MSE sense and (nearly) eliminates the noise. As a practical matter, the signal is not available for adaptation purposes—indeed, if it were available there would be no need for a receiver and a receiving array.

In practical adaptive antenna systems employing the LMS algorithm, an artificially injected signal that is completely known and termed the "reference signal" or "pilot signal" is used to represent the desired response signal. The pilot signal is designed to have the same (or at least similar) directional and spectral characteristics as those of the incoming signal of interest. These directional and spectral characteristics may sometimes be known *a priori*, but it is more commonly the case that only estimates of these parameters are available. Many practical communication systems derive the reference signal from the array output—a practice that requires a high degree of compatibility between the signaling waveforms and the adaptive array. In general it is not feasible to simply put an adaptive array in any arbitrary communication system because [13]:

1. The adaptive array weights are random processes that modulate the desired signal, and consequently either the desired signal waveforms or the adaptive algorithm must be chosen so this modulation does not impair the communication system effectiveness.
2. The desired signal and interference signal waveforms must be different in some respect so this known difference can be exploited by the designer to enable the adaptive array to distinguish these two signal classes.
3. A practical method for reference-signal generation must be available.

The reference signal need not be a perfect replica of the desired signal, but only must satisfy the following criteria [13]:

1. The reference signal must be highly correlated with the desired signal at the array output.
2. The reference signal must be uncorrelated with any interference signal components appearing at the array output.

If the above correlation properties are satisfied, then the adaptive array will behave in the desired manner, since it is only the correlation between the reference signal and the element signals $x_i(t)$ that affects the adaptive weights. The impact of any phase shift occurring in the network responsible for generating the reference signal (when the reference signal is derived from the array output) is discussed in reference [14].

LMS algorithm adaptation with an injected pilot signal causes the array to form a beam in the pilot-signal direction. This array beam has an essentially flat spectra response and linear phase shift characteristic within the passband defined by the spectral characteristic of the pilot signal. Furthermore, directional noises striking the array face will be manifest as correlated noise components to which the array will respond by producing beam pattern nulls in the noise direction within the array passband.

Since injection of the pilot signal could "block" the receiver (by rendering it insensitive to the actual signal of interest), mode-dependent adaptation schemes have been devised to overcome this difficulty. Two such adaptation algorithms are discussed in the following section.

### *4.2.5 One-Mode and Two-Mode LMS Adaptation*

Figure 4.9 illustrates a practical two-mode method [10] for providing a pilot signal to form the array beam and then switching the pilot signal off to accomplish adaptation on the naturally occurring inputs to eliminate noise. The ideal time delays $\delta_1, \delta_2, \ldots, \delta_N$ are selected to produce a set of input signals that correspond to the set that would appear if the array were actually receiving a radiated plane wave signal from the desired direction. The adaptive processor inputs are either connected to the actual sensor elements outputs (during adaptation to eliminate noise) or to the set of delayed signals obtained from the pilot signal generator and the selected time delay elements (to preserve the main lobe in the desired direction).

During adaptation to eliminate noise, all signals applied to the adaptive processor are sensor element outputs derived from the actual noise field. The adaptation process in this mode tends to eliminate all received signals since the desired response signal has been set equal to zero.

During adaptation to preserve the main beam in a certain desired direction, the input signals to the adaptive processor are derived from the pilot signal.

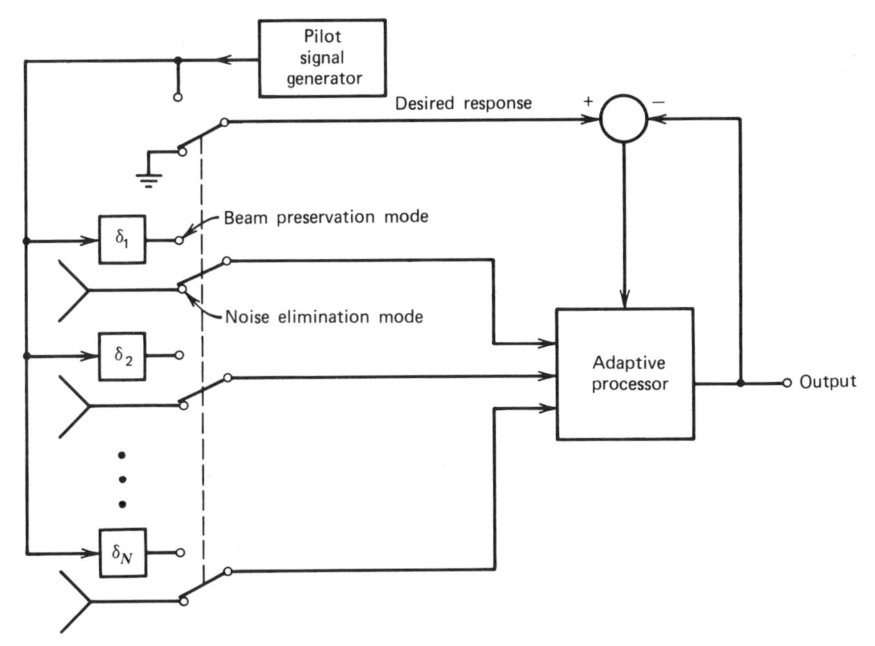


**Figure 4.9** Two-mode LMS adaptation for beam preservation and noise elimination.

For example if a sinusoidal pilot signal having frequency $f_0$ is used, then minimizing the MSE will force the array gain in the desired look direction to have a specific amplitude and phase shift at that frequency. On the other hand, the pilot signal can be chosen to be the sum of several sinusoids having different frequencies; in this case the adaptation process will force the array gain and phase in the desired look direction to have specific values at each one of the pilot-signal frequencies. Finally, if several pilot signals corresponding to different look directions are added together, then the array gain can be constrained simultaneously at the various frequencies and angles corresponding to the different pilot signals selected. In summary, the two-mode adaptation process minimizes the total power of all signals received that are uncorrelated with the pilot signals while constraining the gain and phase of the array beam to values corresponding to the frequencies and angles dictated by the pilot-signal components.

Figure 4.10 illustrates a practical one-mode method for simultaneously eliminating all noises uncorrelated with the pilot signal and forming a desired array beam. The circuitry of Figure 4.10 circumvents the difficulty of being unable to receive the actual signal while the processor is connected to the pilot-signal generator by introducing an auxiliary adaptive processor. For the auxiliary adaptive processor, the desired response is the pilot signal, and both the pilot signal and the actual received signals enter the processor. A second processor performs no adaptation (its weights are slaved to the weights of the adaptive processor) and generates the actual array output signal. The slaved

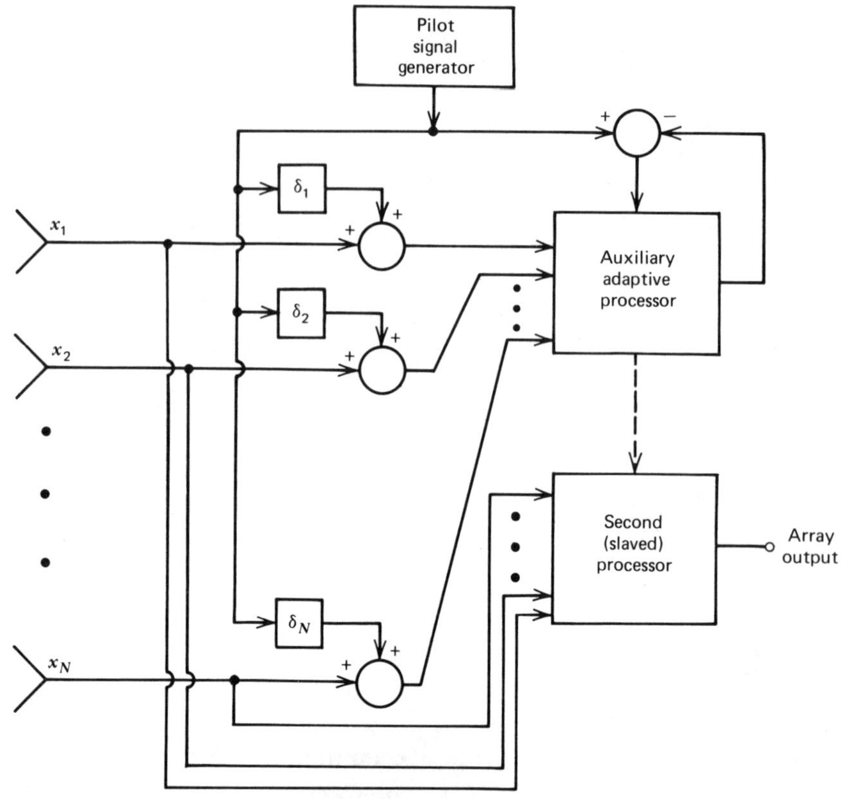


**Figure 4.10** One-mode LMS adaptation for simultaneous beam preservation and noise elimination.

processor inputs do not contain the pilot signal and can therefore receive the transmitted signal at all times.

In the one-mode adaptation method, the pilot signal is on continuously so the adaptive processor that minimizes the MSE forces the adaptive processor output to closely reproduce the pilot signal while rejecting (as best it can) all signals uncorrelated with the pilot signal. The adaptive processor therefore preserves the desired array directivity pattern in the look direction (over the pilot-signal passband) while introducing pattern nulls in the directions of noise sources (over the noise frequency bands).

## 4.3 THE DIFFERENTIAL STEEPEST DESCENT (DSD) ALGORITHM

We have seen that if the gradient could be determined exactly through perfect gradient measurements on each iteration, the adaptive weight vector would converge to the optimal (Wiener) weight vector. In practice, exact gradient measurements are not possible, and gradient vector estimates must be obtained from a limited statistical sample. The differential steepest descent (DSD) algorithm obtains gradient vector estimates by direct measurement and is straightforward and easy to implement [15].

The parabolic performance surface representing the MSE function of a single variable $w$ is defined by

$$\xi[w(k)] \triangleq \xi(k) = \xi_{\min} + \alpha w^2(k) \tag{4.84}$$

Figure 4.11 represents the parabolic performance surface as a function of a single component of the weight vector $\mathbf{w}$. The first and second derivatives of the MSE are:

$$\left[\frac{d\xi(k)}{dw}\right]_{w=w(k)} = 2\alpha w(k) \tag{4.85}$$

and

$$\left[\frac{d^2\xi(k)}{dw}\right]_{w=w(k)} = 2\alpha \tag{4.86}$$

The above derivatives can be numerically estimated by taking the following "symmetric differences":

$$\left[\frac{d\xi(k)}{dw}\right]_{w=w(k)} = \frac{\xi[w(k)+\delta] - \xi[w(k)-\delta]}{2\delta} \tag{4.87}$$

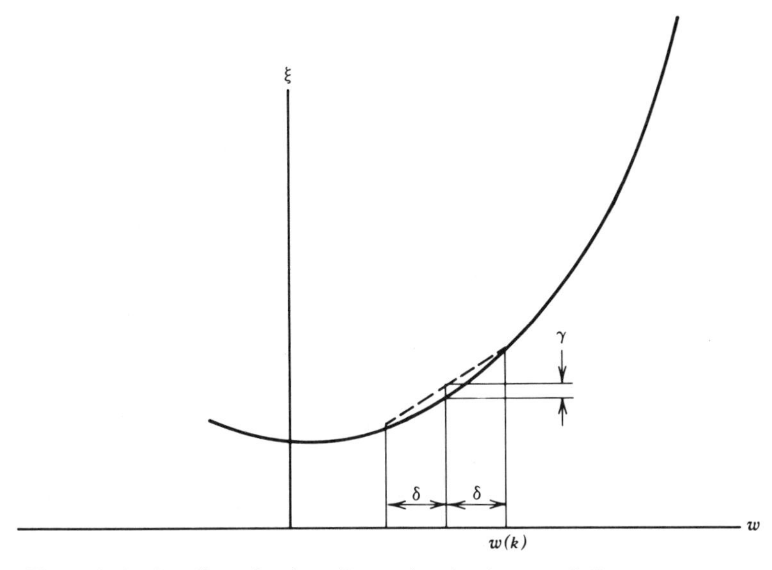


**Figure 4.11** One-dimensional gradient estimation by way of direct measurement.

and

$$\left[\frac{d^2\xi(k)}{dw^2}\right]_{w=w(k)} = \frac{\xi[w(k)+\delta]-2\xi[w(k)]+\xi[w(k)-\delta]}{\delta^2} \tag{4.88}$$

The procedure for estimating the first derivative illustrated in Figure 4.11 requires that the weight adjustment be altered to two distinct settings while the gradient estimate is obtained. If $K$ data samples are taken to estimate the MSE at each of the two weight settings $w(k)+\delta$ and $w(k)-\delta$, then the average MSE experienced (over both settings) will be greater than the MSE at $w(k)$ by an amount $\gamma$. Consequently a performance penalty is incurred that results from the weight alteration used to obtain the derivative estimate.

### 4.3.1 *Performance Penalty Incurred by Gradient Measurement*

From Figure 4.11 it can be deduced that the quantity $\gamma$ can be calculated in the one-dimensional case as

$$\gamma = \frac{\alpha[w(k)+\delta]^2+\alpha[w(k)-\delta]^2+2\xi_{\min}}{2} - \alpha w^2(k) - \xi_{\min} = \alpha\delta^2 \tag{4.89}$$

Consequently the value of $\gamma$ depends only on $\alpha$ and not on $w(k)$. It is convenient to introduce a dimensionless measure of how much the system is perturbed each time the gradient is measured by defining a parameter called the "perturbation" $P$, as follows:

$$P \triangleq \frac{\gamma}{\xi_{\min}} = \frac{\alpha\delta^2}{\xi_{\min}} \tag{4.90}$$

The perturbation is the average increase in the MSE normalized with respect to the minimum achievable MSE.

To consider two-dimensional gradients instead of the foregoing simple one-dimensional derivative, the input signal correlation matrix can be written as

$$\mathbf{R}_{xx} = \begin{bmatrix} r_{11} & r_{12} \\ r_{21} & r_{22} \end{bmatrix} \tag{4.91}$$

The MSE corresponding to this correlation matrix is then

$$\xi = r_{11}w_1^2 + r_{22}w_2^2 + 2r_{12}w_1w_2 + \xi_{\min} \tag{4.92}$$

Measuring the partial derivative of the above performance surface along the coordinate $w_1$ yields a perturbation

$$P = \frac{r_{11}\delta^2}{\xi_{\min}} \tag{4.93}$$

Likewise the perturbation for the measured partial derivative along the coordinate $w_2$ is

$$P = \frac{r_{22}\delta^2}{\xi_{\min}} \tag{4.94}$$

If we allot equal time for the measurement of both partial derivatives (a total of $2K$ data samples are used for both measurements), the average perturbation experienced during the complete measurement process is given by

$$P_{\text{av}} = \frac{\delta^2}{\xi_{\min}} \cdot \frac{r_{11} + r_{22}}{2} \tag{4.95}$$

For $N$ dimensions, define a general perturbation as the average of the perturbations experienced for each of the individual gradient component measurements so that

$$P = \frac{\delta^2}{\xi_{\min}} \cdot \frac{\operatorname{tr}(\mathbf{R}_{xx})}{N} \tag{4.96}$$

where "tr" denotes trace, which is defined as the sum of the diagonal elements of the indicated matrix. When we convert the $\mathbf{R}_{xx}$ matrix to normal coordinates, it is obvious that the trace of $\mathbf{R}_{xx}$ is just the sum of its eigenvalues. Since the sum of the eigenvalues divided by $N$ is just the average of the eigenvalues, the perturbation for the $N$-dimensional case is

$$P = \frac{\delta^2}{\xi_{\min}} \cdot \lambda_{\text{av}} \tag{4.97}$$

Alternative means of measuring the gradient have also been used in practical systems. By perturbing or dithering a single weight sinusoidally, the cross-correlation between the weight value and the performance measure can be measured to determine the derivative of the performance surface. Likewise, all weights can be dithered simultaneously at distinct individual frequencies and the gradient components then obtained by cross-correlation. The procedure for determining the derivative illustrated in Figure 4.11 corresponds to square-wave dithering.

### *4.3.2 Gradient Measurement Noise and Misadjustment in the Weight Vector*

Gradients measured by taking finite differences as in Figure 4.11 are noisy because the MSE measurements on which the differences are based are noisy. Each MSE measurement is an estimate $\hat{\xi}$ of the actual MSE $\xi$ based on $K$ data samples:

$$\hat{\xi} = \frac{1}{K} \sum_{k=1}^{K} e^2(k) \tag{4.98}$$

It is well known [16] that the variance of a sample average estimate of the mean square obtained from $K$ independent samples is given by the difference between the mean fourth and the square of the mean square all divided by $K$. Consequently the variance in $\hat{\xi}$ may be expressed as

$$\text{var}[\hat{\xi}] = \frac{E\{e^4(k)\} - [E\{e^2(k)\}]^2}{K} \tag{4.99}$$

If the random variable $e(k)$ is normally distributed with zero mean and variance $\sigma^2$, then its mean fourth is $3\sigma^4$, and the square of its mean square is $\sigma^4$. Consequently the variance in the estimate of $\xi$ is given by

$$\text{var}[\hat{\xi}] = \frac{1}{K}(3\sigma^4 - \sigma^4) = \frac{2\sigma^4}{K} = \frac{2\xi^2}{K} \tag{4.100}$$

From (4.100) we find that the variance of $\hat{\xi}$ is proportional to the square of $\xi$ and inversely proportional to the number of data samples. In general, the variance can be expressed as

$$\text{var}[\hat{\xi}] = \eta \frac{\xi^2}{K} \tag{4.101}$$

where $\eta$ has the value of 2 for an unbiased Gaussian density function. In the event that the probability density function for $\hat{\xi}$ is not Gaussian, then the value of $\eta$ is generally less than but close to 2. It is therefore convenient to assume that the final result expressed in (4.100) holds for the analysis that follows.

The derivatives required by the DSD algorithm are measured in accordance with (4.87). The measured derivative involves taking finite differences of two MSE estimates, so the error in the measured derivative will involve the sum of two independent components (since the error samples $e(k)$ are assumed to be independent). The variance of each component to the derivative error is given by (4.100). Assume that we are attempting to measure the derivative at a point on the performance surface where the weight vector is near the minimum point of the MSE surface and that the perturbation $P$ is small, then the two components of measured derivative error will have essentially the same variances. The total variance of the measured derivative error will then be the sum of the variances of the two components. From (4.87) and (4.100) it follows that the variance in the estimate of the derivative is given by

$$\begin{aligned} \text{var}\left[\frac{d\xi}{dw}\right]_{w=w(k)} &= \frac{1}{4\delta^2}\left[\frac{2\xi^2[w(k)+\delta]}{K} + \frac{2\xi^2[w(k)-\delta]}{K}\right] \\ &\cong \frac{\xi^2_{\min}}{K\delta^2} \end{aligned} \tag{4.102}$$

When an entire gradient vector is measured, then the errors in each component are independent. It is convenient to define a gradient noise vector $\mathbf{g}(k)$ in terms of the true gradient $\nabla(k)$ and the estimated gradient $\hat{\nabla}(k)$:

$$\hat{\nabla}(k) \triangleq \nabla(k) + \mathbf{g}(k) \tag{4.103}$$

where $\mathbf{g}(k)$ is the gradient noise vector. Under the conditions assumed above, the covariance of the gradient noise vector can be expressed as

$$\text{cov}[\mathbf{g}(k)] = \frac{\xi_{\min}^2}{K\delta^2}\mathbf{I} \tag{4.104}$$

Transforming the gradient noise vector into normal coordinates, we have

$$\mathbf{g}'(k) = \mathbf{Q}\mathbf{g}(k) \tag{4.105}$$

We see from (4.104) that the covariance matrix of $\mathbf{g}(k)$ is a scalar multiplying the identity matrix; so projecting into normal coordinates through the orthonormal transformation $\mathbf{Q}$ yields the same covariance for $\mathbf{g}'(k)$:

$$\text{cov}[\mathbf{g}'(k)] = E\{\mathbf{Q}\mathbf{g}(k)\mathbf{g}^T(k)\mathbf{Q}^{-1}\} = \frac{\xi_{\min}^2}{K\delta^2}\mathbf{I} \tag{4.106}$$

This result merely emphasizes that near the minimum point of the performance surface, the covariance of the gradient noise is essentially a constant and does not depend on $\mathbf{w}(k)$.

The fact that the gradient estimates are noisy means that weight adaptation based on these gradient estimates will also be noisy, and it is consequently of interest to determine the corresponding noise in the weight vector. Using estimated gradients, the method of steepest descent yields the following vector difference equation:

$$\mathbf{v}(k+1) = \mathbf{v}(k) + \Delta_s(-\hat{\nabla}(k)) = \mathbf{v}(k) + \Delta_s[-\nabla(k) - \mathbf{g}(k)] \tag{4.107}$$

where $\mathbf{v}(k) \triangleq \mathbf{w}(k) - \mathbf{w}_{\text{opt}}$. Since the true gradient from (4.63) is given by

$$\nabla(k) = 2\mathbf{R}_{xx}\mathbf{v}(k) \tag{4.108}$$

(4.107) can be rewritten as

$$\mathbf{v}(k+1) = [\mathbf{I} - 2\Delta_s\mathbf{R}_{xx}]\mathbf{v}(k) - \Delta_s\mathbf{g}(k) \tag{4.109}$$

which is a first-order difference equation having a stochastic driving function $-\Delta_s\mathbf{g}(k)$. Projecting the above difference equation into normal coordinates by premultiplying by $\mathbf{Q}$ then yields:

$$\mathbf{v}'(k+1) = [\mathbf{I} - 2\Delta_s\mathbf{\Lambda}]\mathbf{v}'(k) - \Delta_s\mathbf{g}'(k) \tag{4.110}$$

After initial adaptive transients have died out and the steady state is reached, the weight vector $\mathbf{v}'(k)$ behaves like a stationary random process in response to the stochastic driving function $-\Delta_s \mathbf{g}'(k)$. In the normal coordinate system there is no cross-coupling between terms, and the components of $\mathbf{g}'(k)$ are uncorrelated; so the components of $\mathbf{v}'(k)$ are also mutually uncorrelated, and the covariance matrix of $\mathbf{g}'(k)$ will be diagonal. The covariance matrix of $\mathbf{v}'(k)$ describes how noisy the weight vector will be in response to the stochastic driving function, and we now proceed to find this matrix. Since $\text{cov}[\mathbf{v}'(k)] \triangleq E\{\mathbf{v}'(k)\mathbf{v}'^T(k)\}$, it is of interest to determine the quantity $\mathbf{v}'(k+1)\mathbf{v}'^T(k+1)$ by way of (4.110) as follows:

$$\begin{aligned} \mathbf{v}'(k+1)\mathbf{v}'^T(k+1) = &(\mathbf{I}-2\Delta_s\mathbf{\Lambda})\mathbf{v}'(k)\mathbf{v}'^T(k)(\mathbf{I}-2\Delta_s\mathbf{\Lambda}) \\ &+\Delta_s^2\mathbf{g}'(k)\mathbf{g}'^T(k)-\Delta_s(\mathbf{I}-2\Delta_s\mathbf{\Lambda})\mathbf{v}'(k)\mathbf{g}'(k)\mathbf{v}'^T(k) \\ &-\Delta_s\mathbf{g}'(k)\mathbf{v}'^T(k) \end{aligned} \tag{4.111}$$

Taking expected values of both sides of (4.111) and noting that $\mathbf{v}'(k)$ and $\mathbf{g}'(k)$ are uncorrelated since $\mathbf{v}'(k)$ is affected only by gradient noise from previous adaptive cycles, we obtain for the steady state:

$$\begin{aligned} \text{cov}[\mathbf{v}'(k)] &= (\mathbf{I}-2\Delta_s\mathbf{\Lambda})\text{cov}[\mathbf{v}'(k)](\mathbf{I}-2\Delta_s\mathbf{\Lambda}) + \Delta_s^2\text{cov}[\mathbf{g}'(k)] \\ &= (\mathbf{I}-4\Delta_s\mathbf{\Lambda}+4\Delta_s^2\mathbf{\Lambda}^2)\text{cov}[\mathbf{v}'(k)] + \Delta_s^2\text{cov}[\mathbf{g}'(k)] \end{aligned} \tag{4.112}$$

Combining like terms in (4.112) then yields

$$\text{cov}[\mathbf{v}'(k)] = \Delta_s^2[4\Delta_s\mathbf{\Lambda}-4\Delta_s^2\mathbf{\Lambda}^2]^{-1}\text{cov}[\mathbf{g}'(k)] \tag{4.113}$$

In practice, the step size in the method of steepest descent is selected so that

$$\Delta_s\mathbf{\Lambda} \ll \mathbf{I} \tag{4.114}$$

As a result of (4.114), squared terms occurring in (4.113) can be neglected so that

$$\text{cov}[\mathbf{v}'(k)] \cong \frac{\Delta_s}{4}\mathbf{\Lambda}^{-1}\text{cov}[\mathbf{g}'(k)] \tag{4.115}$$

Since $\text{cov}[\mathbf{g}'(k)]$ is given by (4.106), we now have

$$\text{cov}[\mathbf{v}'(k)] \cong \frac{\Delta_s\xi_{\min}^2}{4K\delta^2}\mathbf{\Lambda}^{-1} \tag{4.116}$$

The covariance of the weight vector in the operational coordinate system can be obtained from (4.116) by recalling that $\mathbf{R}_{xx}^{-1} = \mathbf{Q}^{-1}\mathbf{\Lambda}^{-1}\mathbf{Q}$ and $\mathbf{v}' = \mathbf{Q}\mathbf{v}$ so that

$$\begin{aligned}\text{cov}\left[\mathbf{v}(k)\right] &= E\{\mathbf{Q}^{-1}\mathbf{v}'(k)\mathbf{v}'^{T}(k)\mathbf{Q}\}\\ &= \frac{\Delta_s \xi_{\min}^2}{4K\delta^2}\mathbf{R}_{xx}^{-1}\end{aligned} \tag{4.117}$$

Without any noise in the weight vector, the method of steepest descent would adapt the weights so they converge to a steady-state solution at the minimum point of the MSE performance surface (the bottom of the bowl). The MSE would then be $\xi_{\min}$. The noise actually present in the weight vector causes the steady-state solution to randomly wander about the minimum point. The result of this wandering is a steady-state MSE that is greater than $\xi_{\min}$ and hence can be said to have an "excess" MSE. We will now consider how severe this excess MSE is for the noise that is in the weight vector.

We have already seen in Section 4.1.3 that the MSE can be expressed as

$$\xi(k) = \xi_{\min} + \mathbf{v}'^{T}(k)\mathbf{\Lambda}\mathbf{v}'(k) \tag{4.118}$$

where $\mathbf{v}'(k) = \mathbf{w}'(k) - \mathbf{w}'_{\text{opt}}$. Consequently the average excess MSE is

$$E\{\mathbf{v}'^{T}(k)\mathbf{\Lambda}\mathbf{v}'(k)\} = \sum_{p=1}^{N}\lambda_p E\{[v_p'(k)]^2\} \tag{4.119}$$

But from (4.116) we may write

$$E\{[v_p'(k)]^2\} = \frac{\Delta_s \xi_{\min}^2}{4k\delta^2}\left(\frac{1}{\lambda_p}\right) \tag{4.120}$$

Consequently (4.119) can be rewritten as

$$E\{\mathbf{v}'^{T}(k)\mathbf{\Lambda}\mathbf{v}'(k)\} = \frac{N\Delta_s \xi_{\min}^2}{4K\delta^2} \tag{4.121}$$

Recalling that the misadjustment $M$ is defined as the average excess MSE divided by the minimum MSE there results for the DSD algorithm:

$$M = \frac{N\Delta_s \xi_{\min}}{4K\delta^2} \tag{4.122}$$

The foregoing result can be more usefully expressed in terms of time constants of the learning process and the perturbation of the gradient estimation process as developed next.

Each measurement to determine a gradient component uses $2K$ samples of data. Each adaptive weight iteration involves $N$ gradient component measurements, and therefore requires a total of $2KN$ data samples. From Section 4.2.3 it may be recalled that the MSE learning curve has a $p$th mode time constant given by

$$\tau_{p_{\mathrm{MSE}}} = \frac{1}{4\Delta_s \lambda_p} = \frac{\tau_p}{2} \tag{4.123}$$

in time units of the number of iterations. It is useful to define a new time constant $T_{p_{\mathrm{MSE}}}$ whose basic time unit is the data sample and whose value is expressed in terms of the number of data samples. It follows that for the DSD algorithm

$$T_{p_{\mathrm{MSE}}} \triangleq 2KN\tau_{p_{\mathrm{MSE}}} \tag{4.124}$$

The time constant $T_{p_{\mathrm{MSE}}}$ is easily related to real time units (seconds) once the sampling rate is known.

By using the perturbation formula (4.90) to substitute for $\xi_{\min}$ in (4.122), the misadjustment for the DSD algorithm can be rewritten as

$$M = \frac{N\Delta_s \lambda_{\mathrm{av}}}{4KP} \tag{4.125}$$

The time constant defined by (4.124) can also be rewritten using (4.123) as

$$T_{p_{\mathrm{MSE}}} = \frac{NK}{2\Delta_s \lambda_p} \tag{4.126}$$

from which one can conclude that

$$\lambda_p = \frac{NK}{2\Delta_s}\left(\frac{1}{T_{p_{\mathrm{MSE}}}}\right) \tag{4.127}$$

so that

$$\lambda_{\mathrm{av}} = \frac{NK}{2\Delta_s}\left(\frac{1}{T_{\mathrm{MSE}}}\right)_{\mathrm{av}} \tag{4.128}$$

Combining (4.26) and (4.128) then yields the misadjustment as

$$M = \frac{N^2}{8P}\left(\frac{1}{T_{\mathrm{MSE}}}\right)_{\mathrm{av}} \tag{4.129}$$

It may therefore be concluded from (4.129) that for the DSD algorithm, misadjustment is proportional to the square of the number of weights and inversely proportional to the perturbation. In addition, the misadjustment is also inversely proportional to the speed of adaptation (fast adaptation results in high misadjustment). Since the DSD algorithm is based on steepest descent, it also suffers from the disparate eigenvalue problem discussed in Section 4.2.2.

It is appropriate here to compare the misadjustment for the DSD algorithm given by (4.129) with the misadjustment for the LMS algorithm given by (4.83). With a specified level of misadjustment for the LMS algorithm, the adaptive time constants increase linearly with the number of weights rather than with the square of the number of weights as is the case with the DSD algorithm. Furthermore, with the LMS algorithm there is no perturbation. Consequently in typical circumstances much faster adaptation is possible with the LMS algorithm than with the DSD algorithm.

The misadjustment $M$ is defined as a normalized performance penalty that results from noise in the weight vector. In an actual adaptive system employing the DSD algorithm, the weight vector is not only stochastically perturbed due to the presence of noise but is, in addition, deterministically perturbed so the gradient can be measured. As a consequence of the deterministic perturbation, another performance penalty accrues as measured by the perturbation $P$, which is also a normalized ratio of excess MSE. The total excess MSE is therefore the sum of the "stochastic" and "deterministic" perturbation components. The total misadjustment can be expressed as

$$M_{\text{tot}} = M + P \tag{4.130}$$

Adding the above two components then yields

$$M_{\text{tot}} = \frac{N^2}{8P}\left(\frac{1}{T_{\text{MSE}}}\right)_{\text{av}} + P \tag{4.131}$$

Since $P$ is a design parameter given by (4.90), it can be selected by choosing the deterministic perturbation size $\delta$. It is desirable to minimize the total misadjustment $M_{\text{tot}}$ by appropriately selecting $P$. The result of such optimization is to make the two right-hand terms of (4.131) equal so that

$$P_{\text{opt}} = \frac{1}{2} M_{\text{tot}} \tag{4.132}$$

The minimum total misadjustment then becomes

$$(M_{\text{tot}})_{\min} = \frac{N^2}{4P_{\text{opt}}}\left(\frac{1}{T_{\text{MSE}}}\right)_{\text{av}} = \left[\frac{N^2}{2}\left(\frac{1}{T_{\text{MSE}}}\right)_{\text{av}}\right]^{1/2} \tag{4.133}$$

Unlike the LMS algorithm, the DSD algorithm is sensitive to any correlation that may exist between successive samples of the error signal $e(k)$, since such correlation has the effect of making the effective statistical sample size less than the actual number of error samples in computing the estimated gradient vector. As a consequence of such reduced effective sample size, the actual misadjustment experienced will be greater than that predicted by (4.133), which was derived using the assumption of statistical independence between successive error samples.

## 4.4 THE ACCELERATED GRADIENT APPROACH (AG)

Algorithms based on the steepest descent method exhibit an undesirable degree of sensitivity of the convergence speed to the eigenvalue spread in the input signal covariance matrix. A wide variety of methods based on or related to conjugate gradient descent [17]–[22] are available for accelerating the steepest descent approach to obtain faster convergence, and while they have found some applications in practical control problems [23], nevertheless, their application to adaptive array problems has not previously been reported. This reluctance to apply accelerated gradient methods to adaptive array problems is due principally to the following reasons:

1. The increased hardware complexity associated with implementing the algorithm.
2. The enhanced convergence speed realized is not as fast as that which can be obtained with other methods (to be discussed in later chapters).
3. The various accelerated gradient methods are all susceptible (although with different degrees) to signals that are noise corrupted.
4. The required increase in computation and memory space over steepest descent methods.

Despite these objections, it is worthwhile to apply one of the accelerated gradient methods to determine what improvement in convergence speed reasonably might be expected by recourse to such techniques. For this purpose, the descent method introduced by Powell [18] will be considered since this approach is quite tolerant of noise induced errors, although other methods theoretically may yield faster convergence speeds. The method of Powell depends on the fact that if a performance measure $\mathfrak{P}(\mathbf{w})$ is quadratic in the independent variables, then any line that passes through the minimum point of the quadratic performance surface intersects the family of constant performance contours at equal angles. This property is illustrated in Figure 4.12 for two dimensions where it is seen that the line $AC$ connecting point $A$ with the minimum point $C$ intersects the constant performance contours at equal angles. As a consequence of the equal angle property, the line joining the point $A$ with the point $D$ in Figure 4.12 passes through the point $C$ where the derivative of the performance measure $\mathfrak{P}(\mathbf{w})$ with respect to distance along the line $AD$ is zero.

The procedure to follow in finding the minimum value of the performance measure is described below. Given an initial estimate $\mathbf{w}_0$ at point $A$, first find the gradient direction which is normal to the tangent of the constant perfor-

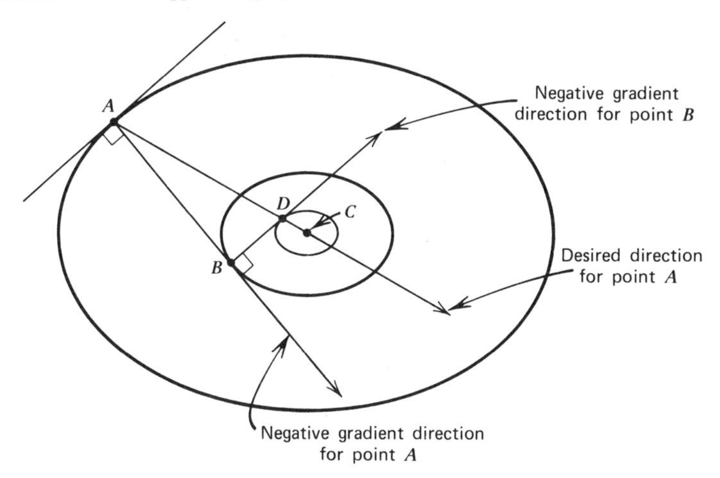


**Figure 4.12** Two-dimensional diagram showing directional relationships for the Powell descent method.

mance measure contour. Proceed along the line defined by the negative gradient direction to the point $B$ where the derivative of $\mathfrak{P}(\mathbf{w})$ with respect to distance along the line is zero. The point $B$ may in fact be any arbitrary point on the line that is a finite distance from $A$; however, by choosing it in the manner described the convergence of the method is assured.

Having found point $B$, we determine the negative gradient direction once again, and this normal direction is parallel to the original tangent at $\mathfrak{P}(\mathbf{w}_0)$. Traveling in this new normal direction we find a point where the derivative of $\mathfrak{P}(\mathbf{w})$ with respect to distance along the line is zero (point $D$ in Figure 4.12). The line passing through the points $A$ and $D$ then also passes through the point $C$. The desired point $C$ may then be found as the point where the derivative of $\mathfrak{P}(\mathbf{w})$ with respect to distance along the line $AD$ is zero.

The generalization of the foregoing procedure to an $N$-dimensional space can be obtained by recognizing that the directional relationships (which depend on the equal angle property) given in Figure 4.12 are valid only in a two-dimensional plane. The first step (moving from point $A$ to point $B$) is simply accomplished by moving in the negative gradient direction in the $N$-dimensional space. Having found point $B$, we can construct $(N-1)$ planes between the original negative gradient direction and $(N-1)$ additional mutually orthogonal vectors, thereby defining points $C, D, E, \ldots,$ until $(N-1)$ additional points have been defined. The last three points in the $N$-dimensional space obtained in the foregoing manner may now be treated in the same fashion as points $A$, $B$, and $D$ of Figure 4.12 by drawing a connecting line between the last point obtained and the point defined two steps earlier. Traveling along the connecting line one may then define a new point corresponding to $C$ of Figure 4.12. This new point may then be considered as point $D$ in Figure 4.12, and a new connecting line drawn between the new point and the point obtained three steps earlier.

The steps corresponding to one complete Powell descent cycle for five dimensions are illustrated in Figure 4.13. The first step from $A$ to $B$ merely involves traveling in the negative gradient direction $\mathbf{v}_1$ with a step size $\alpha_1$ chosen to satisfy the condition

$$\frac{d}{d\alpha_1}\{\mathfrak{P}[\mathbf{w}(0)+\alpha_1\mathbf{v}_1]\}=0 \tag{4.134}$$

so that

$$\mathbf{w}(1)=\mathbf{w}(0)+\alpha_1\mathbf{v}_1 \tag{4.135}$$

Having determined point $B$, we determine point $C$ by traveling in the negative gradient direction $\mathbf{v}_2$ (a direction that is also orthogonal to $\mathbf{v}_1$) from point $B$ with step size $\alpha_2$ selected to satisfy

$$\frac{d}{d\alpha_2}\{\mathfrak{P}[\mathbf{w}(1)+\alpha_2\mathbf{v}_2]\}=0 \tag{4.136}$$

Point $D$ is then determined from point $C$ by using the above procedure, and this process continues until a total of five points ($B$ through $F$ in Figure 4.13) have been defined. A descent direction $\mathbf{v}_6$ is now defined by drawing a

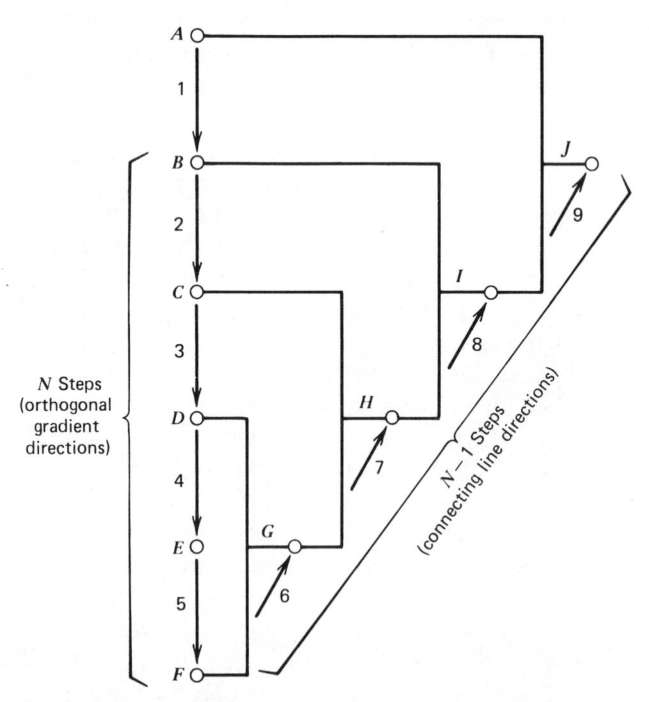


Figure 4.13 Illustration of Powell descent method steps required in five dimensions for one complete cycle.

connecting line between points $D$ and $F$ (analogous to points $A$ and $D$ of Figure 4.12) and traveling along this line with step size $\alpha_6$ selected to satisfy

$$\frac{d}{d\alpha_6}\{\mathfrak{P}[\mathbf{w}(5)+\alpha_6\mathbf{v}_6]\}=0 \tag{4.137}$$

thereby arriving at point $G$. A new descent direction $\mathbf{v}_7$ is now defined by drawing a connecting line between points $C$ and $G$ and traveling along this line with step size $\alpha_7$ selected to satisfy

$$\frac{d}{d\alpha_7}\{\mathfrak{P}[\mathbf{w}(6)+\alpha_7\mathbf{v}_7]\}=0 \tag{4.138}$$

thereby arriving at point $H$. This process continues until the solution point $J$ for the cycle is found from the descent direction $\mathbf{v}_9$ (defined by the connecting line between points $A$ and $I$) and the step size $\alpha_9$. In general, one Powell descent cycle in $N$-dimensional space therefore requires $N+N-1=2N-1$ steps. On completion of one descent cycle, the entire cycle may then be repeated using the last weight vector obtained as the initial weight vector for the new cycle.

### *4.4.1 Algorithm Based on the Powell Accelerated Gradient Cycle*

Each step in a Powell descent cycle involves traveling from a weight vector $\mathbf{w}$ along a direction $\mathbf{v}$ with step size $\alpha$ such that

$$\frac{d}{d\alpha}[\mathfrak{P}(\mathbf{w}+\alpha\mathbf{v})]=0 \tag{4.139}$$

For complex weights the MSE performance measure is given by

$$\xi(\mathbf{w})=E\{d^2\}+\mathbf{w}^\dagger\mathbf{r}_{xd}+\mathbf{r}_{xd}^\dagger\mathbf{w}+\mathbf{w}^\dagger\mathbf{R}_{xx}\mathbf{w} \tag{4.140}$$

The gradient of $\xi(\mathbf{w}+\alpha\mathbf{v})$ with respect to $\alpha$ is then given by

$$\nabla_\alpha[\xi(\mathbf{w}+\alpha\mathbf{v})]=\mathbf{v}^\dagger\mathbf{r}_{xd}+\mathbf{r}_{xd}^\dagger\mathbf{v}+\mathbf{v}^\dagger\mathbf{R}_{xx}\mathbf{w}+\mathbf{w}^\dagger\mathbf{R}_{xx}\mathbf{v}+2\alpha\mathbf{v}^\dagger\mathbf{R}_{xx}\mathbf{v} \tag{4.141}$$

and this gradient is equal to zero when the step size is

$$\alpha=-\frac{\mathbf{v}^\dagger\mathbf{r}_{xd}+\mathbf{v}^\dagger\mathbf{R}_{xx}\mathbf{w}}{\mathbf{v}^\dagger\mathbf{R}_{xx}\mathbf{v}} \tag{4.142}$$

Since $\mathbf{r}_{xd}$ and $\mathbf{R}_{xx}$ are unknown, some estimate of the numerator must be employed to obtain an appropriate step size estimate. Noting that $\mathbf{r}_{xd}+\mathbf{R}_{xx}\mathbf{w}$ is just one-half the gradient of $\xi(\mathbf{w})$, we see it follows that the numerator of (4.142) can be approximated by $\mathbf{v}^\dagger\mathrm{Av}\{e(k)\mathbf{x}(k)\}$. Note that the quantity $\mathbf{v}^\dagger\mathbf{x}$ may be regarded as the output of a processor whose weights correspond to $\mathbf{v}$

and that $\text{Av}\{(\mathbf{v}^{\dagger}\mathbf{x})(\mathbf{x}^{\dagger}\mathbf{v})\}$ may then be taken as an approximation of the quantity $\mathbf{v}^{\dagger}\mathbf{R}_{xx}\mathbf{v}$ where the average Av{ } is taken over a specified number of $K$ data samples. Since the estimates required for the numerator and denominator of the appropriate step size (4.142) should be evaluated simultaneously to accommodate slow variations in the signal environment, the estimates are generated simultaneously. The simultaneous generation of the estimates $\hat{\nabla}_w$ and $\text{Av}\{\mathbf{v}^{\dagger}\mathbf{x}\mathbf{x}^{\dagger}\mathbf{v}\}$ requires parallel processors, one processor with weight values equal to $\mathbf{w}(k)$ and another processor with weight values equal to $\mathbf{v}(k)$. Having described the procedure for determining the appropriate step size along a direction $\mathbf{v}$, we may now consider the steps required to implement an entire Powell descent cycle.

The steps required to generate one complete Powell descent cycle are as follows:

**Step 1** Starting with the initial weight setting $\mathbf{w}(0)$, estimate the negative gradient direction $\mathbf{v}(0)$ using $K$ data samples and then travel in this direction with the appropriate step size to obtain $\mathbf{w}(1)$. The step size determination requires an additional $K$ data samples to obtain by way of (4.142).

**Steps 2→*N*** Estimate the negative gradient direction at $\mathbf{w}(k)$ using $K$ data samples. If the gradient estimates and the preceding step size were error free, the current gradient would automatically be orthogonal to the previous gradient directions. Since the gradient estimate is not error free, determine the new direction of travel $\mathbf{v}(k)$ by requiring it to be orthogonal to all the previous directions $\mathbf{v}(0), \mathbf{v}(1), \ldots, \mathbf{v}(k-1)$ by employing the Schmidt orthogonalization process so that

$$\mathbf{v}(k) = \hat{\nabla}(k) - \sum_{i=0}^{k-1} \frac{\left[\mathbf{v}^{\dagger}(i)\hat{\nabla}(k)\right]}{\left[\mathbf{v}^{\dagger}(i)\mathbf{v}(i)\right]} \cdot \mathbf{v}(i) \tag{4.143}$$

Travel in the direction $-\mathbf{v}(k)$ using the appropriate step size (which requires an additional $K$ data samples to obtain) to arrive at $\mathbf{w}(k+1)$.

**Steps *N*+1→2*N*−1** Determine the new direction of travel at $\mathbf{w}(k)$ by forming

$$\mathbf{v}(k) = \mathbf{w}(k) - \mathbf{w}[2(N-1)-k] \tag{4.144}$$

Travel in the direction $-\mathbf{v}(k)$ from $w[2(N-1)-k]$ using the appropriate step size to arrive at $\mathbf{w}(k+1)$. These steps only require $K$ data samples since now the direction of travel does not require that a gradient estimate be obtained.

## 4.5 GRADIENT ALGORITHM WITH CONSTRAINTS

The early applications of adaptive processing to radar antennas for sidelobe cancellation neglected the effects of signals in the main beam on the adapted response. Such neglect was amply justified because adaptive processors would not respond to low level reflected target signals, and the small number of degrees of freedom then available to the adaptive processor limited the response to large targets or extended clutter. Recently, however, more rigorous demands on radar performance have led to the development of adaptive processors with large numbers of degrees of freedom, and such processors frequently must be explicitly constrained to avoid degradation of the mainlobe response.

With adaptive array systems having large numbers of degrees of freedom and fast response times operating with high energy, long-duration waveforms, then reflected signal returns can be large enough to elicit undesirable responses from the adaptive processor. Such undesirable responses may produce signal cancellation and signal waveform distortion. Furthermore, jammer power level can affect the array response in the mainlobe direction, thereby allowing blinking jammers to modulate the signal response and consequently degrade the performance of any subsequent coherent processing. We therefore consider a constrained optimization procedure introduced by Frost [24] for a gradient type algorithm that imposes constraints on the adaptive weights such that certain main beam properties are preserved during the adaption process. It turns out that the resulting constrained optimization system has two parts: (1) a preprocessing part called a "spatial correction filter," which compensates the signals for the misalignment between the plane wave front and the sensor array geometry and (2) a signal processor that includes the adaptive weights and accounts for the primary function of the adaptive array system.

Suppose it is desired to maintain a certain frequency response characteristic (beam pattern) for the array in a particular direction of interest (called the "look direction"). The algorithm described for this purpose adjusts the adaptive weights of a broadband sensor array depicted in Figure 4.14, and the development below follows that given by Frost [24].

The constrained LMS algorithm to be developed here requires that the direction of arrival and a frequency band of interest be specified *a priori* for the appropriate constraint conditions to be imposed. Because of a simple relation between the look direction frequency response and the adaptive weights, the algorithm is able to maintain a selected frequency response in the look direction while simultaneously minimizing output noise power. Suppose that the look direction is selected as perpendicular to the line of sensors (the appropriate delay settings in the spatial correction preprocessor are then all equal to zero). In that event, identical signal components appear at the first taps [so $x_1(t)=x_2(t)=\cdots=x_N(t)$ in Figure 4.14] and propagate exactly in parallel down the tapped-delay lines following each sensor [so $x_{N+1}(t)=x_{N+2}(t)=\cdots=x_{2N}(t)$, and $x_{(J-1)N+1}(t)=x_{(J-1)N+2}(t)=\cdots=x_{NJ}(t)$]. Noise component waveforms arriving at the sensors from any direction other than the look direction will not usually produce equal voltage components at any vertical column of taps. Consequently as far as the signal is concerned, the adaptive processor appears as an equivalent single tapped-delay line in which

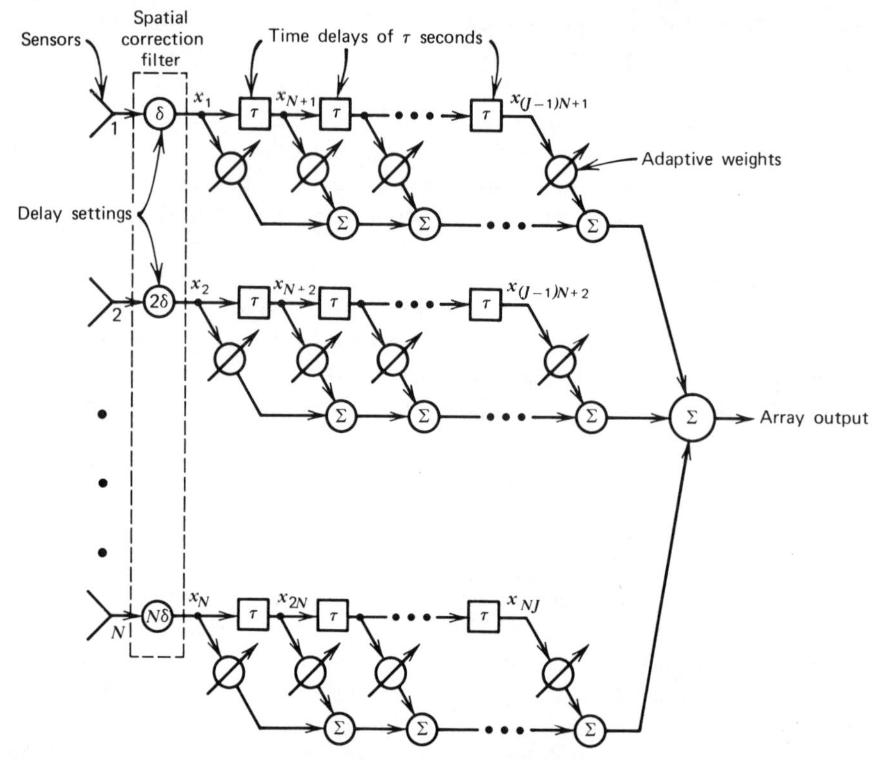


**Figure 4.14** Broadband adaptive array having $N$ sensors and $J$ adjustable weights per sensor.

each adaptive weight is just equal to the sum of the weights in the corresponding vertical column of the original processor. These $J$ summation weights in the equivalent tapped-delay line must each be assigned a value to give the desired frequency response characteristic in the look direction, thereby giving rise to $J$ constraint conditions. In the event that the look direction is selected to be other than that perpendicular to the line of sensors (as the above discussion has assumed), then the various time delays in the spatial correction filter can be adjusted so the signal components of each channel at the output of the preprocessor are in phase.

The adaptive signal processor of Figure 4.14 has $N$ sensors and $J$ taps per sensor for a total of $NJ$ adjustable weights. It was seen above that $J$ constraints are required to determine the look direction frequency response, so the remaining $NJ-J$ degrees of freedom in selecting weight values may be used to minimize the total array output power. Since the $J$ constraints effectively fix the look direction frequency response, minimizing the total output power is equivalent to minimizing the nonlook direction noise power (provided the signal voltages at the taps are uncorrelated with the corresponding noise voltages at these taps). If signal-correlated noise in the array is present, then part or all of the signal component of the array output may be cancelled. Although signal-correlated noise may not occur frequently, sources of such noise include multiple signal-propagation paths, and coherent radar or sonar "clutter."

It is desirable for proper noise cancellation that the noise voltages appearing at the adaptive processor taps be correlated among themselves (although uncorrelated with the signal voltages). Such noise sources may be generated by lightning, "jammers," noise from nearby vehicles, spatially localized incoherent clutter, and self-noise from the structure carrying the array. Noise voltages that are uncorrelated between taps (e.g., amplifier thermal noise) can at least be partially rejected by the adaptive array either as the result of incoherent noise voltage addition at the array output or by reducing the weighting applied to any taps that may have a disproportionately large uncorrelated noise power.

### *4.5.1 Optimum Constrained Weight Vector Solution*

A sampled-data version of the analog signal processor appearing in Figure 4.14 will be considered in developing the optimum weight vector solution by assuming that the voltages appearing at each array tap are sampled every $\Delta$ seconds (where for convenience $\Delta$ may be a multiple of the delay $\tau$ between taps). The vector of tap voltages at the $k$th sample is denoted by $\mathbf{x}(k)$ and is defined by

$$\mathbf{x}^T(k) \triangleq [x_1(k\Delta), x_2(k\Delta), \ldots, x_{NJ}(k\Delta)] \tag{4.145}$$

At any tap the voltages that appear may be regarded as the sums of voltages due to look direction signals $s$ and nonlook direction noises $n$, so that

$$\mathbf{x}(k) = \mathbf{s}(k) + \mathbf{n}(k) \tag{4.146}$$

where the $NJ$-dimensional vector of look direction signals is defined by

$$\mathbf{s}(k) \triangleq \begin{bmatrix} s(k\Delta) \\ \vdots \\ s(k\Delta) \\ s(k\Delta-\tau) \\ \vdots \\ s(k\Delta-\tau) \\ \vdots \\ s[k\Delta-(J-1)\tau] \\ \vdots \\ s[k\Delta-(J-1)\tau] \end{bmatrix} \begin{matrix} \left.\begin{matrix} \\ \\ \\ \end{matrix}\right\} N \text{ taps} \\ \left.\begin{matrix} \\ \\ \\ \end{matrix}\right\} N \text{ taps} \\ \\ \left.\begin{matrix} \\ \\ \\ \end{matrix}\right\} N \text{ taps} \end{matrix} \tag{4.147}$$

and the vector of nonlook direction noises is defined by

$$\mathbf{n}^T(k) \triangleq [n_1(k\Delta), n_2(k\Delta), \ldots, n_{NJ}(k\Delta)] \tag{4.148}$$

The vector of weights appearing at each tap is denoted by $\mathbf{w}$, where

$$\mathbf{w}^T \triangleq [w_1, w_2, \ldots, w_{NJ}] \tag{4.149}$$

We assume that both the signals and noises can be modeled as zero-mean random processes with unknown second-order statistics. The covariance matrices of $\mathbf{x}$, $\mathbf{s}$, and $\mathbf{n}$ are given by:

$$E\{\mathbf{x}(k)\mathbf{x}^T(k)\} = \mathbf{R}_{xx} \tag{4.150}$$

$$E\{\mathbf{s}(k)\mathbf{s}^T(k)\} = \mathbf{R}_{ss} \tag{4.151}$$

$$E\{\mathbf{n}(k)\mathbf{n}^T(k)\} = \mathbf{R}_{nn} \tag{4.152}$$

Since the vector of look direction signals is assumed uncorrelated with the vector of nonlook direction noises,

$$E\{\mathbf{n}(k)\mathbf{s}^T(k)\} = \mathbf{0} \tag{4.153}$$

It is assumed that the noise environment is such that both $\mathbf{R}_{xx}$ and $\mathbf{R}_{nn}$ are positive definite and symmetric.

The adaptive array output (which forms the signal estimate) at the $k$th sample is given by

$$y(k) = \mathbf{w}^T\mathbf{x}(k) = \mathbf{x}^T(k)\mathbf{w} \tag{4.154}$$

From (4.154), it follows that the expected array output power is

$$E\{y^2(k)\} = E\{\mathbf{w}^T\mathbf{x}(k)\mathbf{x}^T(k)\mathbf{w}\} = \mathbf{w}^T\mathbf{R}_{xx}\mathbf{w} \tag{4.155}$$

Now suppose that the weights in the $j$th vertical column of taps should sum to a selected number $f_j$. This constraint may be expressed by the relation

$$\mathbf{c}_j^T\mathbf{w} = f_j, \qquad j = 1, 2, \ldots, J \tag{4.156}$$

where the $NJ$-dimensional vector $\mathfrak{c}_j$ is given by

$$\mathfrak{c}_j = \begin{bmatrix} 0 \\ \vdots \\ 0 \\ \vdots \\ 0 \\ \vdots \\ 0 \\ 1 \\ \vdots \\ 1 \\ 0 \\ \vdots \\ 0 \\ \vdots \\ 0 \\ \vdots \\ 0 \end{bmatrix} \begin{matrix} \left.\begin{matrix} \\ \\ \\ \end{matrix}\right\} N \\ \\ \left.\begin{matrix} \\ \\ \\ \end{matrix}\right\} N \\ \left.\begin{matrix} \\ \\ \\ \end{matrix}\right\} j\text{th column of } N \text{ elements} \\ \left.\begin{matrix} \\ \\ \\ \end{matrix}\right\} N \\ \\ \left.\begin{matrix} \\ \\ \\ \end{matrix}\right\} N \end{matrix} \tag{4.157}$$

Now consider the requirement of constraining the entire weight vector to satisfy all $J$ equations given by (4.156). Define a $J \times NJ$ constraint matrix $\mathbf{C}$ having $\mathfrak{c}_j$ by

$$\mathbf{C} \triangleq [\mathfrak{c}_1 \cdots \mathfrak{c}_j \cdots \mathfrak{c}_J] \tag{4.158}$$

Furthermore define $\mathbf{f}$ as the $J$-dimensional vector of summed weight values for each of the $j$ vertical columns that yield the desired frequency response characteristic in the look direction as

$$\mathbf{f} \triangleq \begin{bmatrix} f_1 \\ f_2 \\ \vdots \\ f_J \end{bmatrix} \tag{4.159}$$

It immediately follows by inspection that the full set of constraints (4.156) can be written in matrix form as

$$\mathbf{C}^T\mathbf{w}=\mathbf{f} \tag{4.160}$$

Now that the look direction frequency response is fixed by the constraint equation (4.160), minimizing the nonlook direction noise power is completely equivalent to minimizing the total output power given by (4.155). The constrained optimization problem is now reduced to the following:

$$\underset{\mathbf{w}}{\text{Minimize}}\, \mathbf{w}^T\mathbf{R}_{xx}\mathbf{w} \tag{4.161}$$

$$\text{subject to } \mathbf{C}^T\mathbf{w}=\mathbf{f} \tag{4.162}$$

Finding $\mathbf{w}_{\text{opt}}$ to satisfy (4.161), (4.162) can be accomplished by the method of Lagrange multipliers [25]. Adjoining the constraint equation (4.162) to the cost function (4.161) by a $J$-dimensional vector $\boldsymbol{\lambda}$, whose elements are undetermined Lagrange multipliers (and including a factor of $\frac{1}{2}$ to simplify the arithmetic), then yields

$$\underset{\mathbf{w}}{\text{Minimize}}\, \mathfrak{P}(\mathbf{w})=\tfrac{1}{2}\mathbf{w}^T\mathbf{R}_{xx}\mathbf{w}+\boldsymbol{\lambda}^T\left[\mathbf{C}^T\mathbf{w}-\mathbf{f}\right] \tag{4.163}$$

The gradient of (4.163) with respect to $\mathbf{w}$ is given by

$$\nabla_{\mathbf{w}}\mathfrak{P}(\mathbf{w})=\mathbf{R}_{xx}\mathbf{w}+\mathbf{C}\boldsymbol{\lambda} \tag{4.164}$$

A necessary condition for (4.163) to be minimized is that the gradient be equal to zero so that

$$\mathbf{R}_{xx}\mathbf{w}+\mathbf{C}\boldsymbol{\lambda}=\mathbf{0} \tag{4.165}$$

Therefore the optimal weight vector is given by

$$\mathbf{w}_{\text{opt}}=-\mathbf{R}_{xx}^{-1}\mathbf{C}\boldsymbol{\lambda} \tag{4.166}$$

where the vector $\boldsymbol{\lambda}$ remains to be determined. The vector of Lagrange multipliers may now be evaluated from the constraint equation:

$$\mathbf{C}^T\mathbf{w}_{\text{opt}}=\mathbf{f}=\mathbf{C}^T\left[-\mathbf{R}_{xx}^{-1}\mathbf{C}\boldsymbol{\lambda}\right] \tag{4.167}$$

It then follows that the vector $\boldsymbol{\lambda}$ is given by

$$\boldsymbol{\lambda}=-\left[\mathbf{C}^T\mathbf{R}_{xx}^{-1}\mathbf{C}\right]^{-1}\mathbf{f} \tag{4.168}$$

where the existence of $[\mathbf{C}^T\mathbf{R}_{xx}^{-1}\mathbf{C}]^{-1}$ is guaranteed by the fact that $\mathbf{R}_{xx}$ is positive definite and $\mathbf{C}$ has full rank. Combining (4.166) and (4.168) then yields the optimum constrained weight vector:

$$\mathbf{w}_{\text{opt}} = \mathbf{R}_{xx}^{-1}\mathbf{C}\left[\mathbf{C}^T\mathbf{R}_{xx}^{-1}\mathbf{C}\right]^{-1}\mathbf{f} \tag{4.169}$$

If we substitute $\mathbf{w}_{\text{opt}}$ into (4.154), it follows that the constrained least squares estimate of the look direction signal provided by the array is

$$y_{\text{opt}}(k) = \mathbf{w}_{\text{opt}}^T\mathbf{x}(k) \tag{4.170}$$

If the vector of summed weight values $\mathbf{f}$ is selected so the frequency response characteristic in the look direction is all-pass and linear phase (distortionless), then the output of the constrained LMS signal processor is the ML estimate of a stationary process in Gaussian noise (provided the angle of arrival is known) [26]. A variety of other optimal processors can also be obtained by a suitable choice of the vector $\mathbf{f}$ [27]. It is also worth noting that the solution (4.169) is sensitive to deviations of the actual signal direction from that specified by $\mathbf{C}$ and to various random errors in the array parameters [28].

### 4.5.2 *The Adaptive Algorithm*

It will be assumed that the correlation matrix $\mathbf{R}_{xx}$ is unknown a priori and must be learned by the adaptive processor. It follows that in stationary environments during learning and in time-varying environments an estimate of the optimum adaptive processor weights must be periodically recomputed. The initial guess of an appropriate weight vector must satisfy (4.162), so a good starting point is

$$\mathbf{w}(0) = \mathbf{C}\left[\mathbf{C}^T\mathbf{C}\right]^{-1}\mathbf{f} \tag{4.171}$$

where the quantity $\mathbf{C}[\mathbf{C}^T\mathbf{C}]^{-1}$ represents the pseudoinverse of the singular matrix $\mathbf{C}^T$ [29]. For a gradient type algorithm, after the $k$th iteration the next weight vector is given by

$$\begin{aligned}\mathbf{w}(k+1) &= \mathbf{w}(k) - \Delta_s \nabla_w \mathfrak{P}\left[\mathbf{w}(k)\right] \\ &= \mathbf{w}(k) - \Delta_s\left[\mathbf{R}_{xx}\mathbf{w}(k) + \mathbf{C}\boldsymbol{\lambda}(k)\right]\end{aligned} \tag{4.172}$$

where $\Delta_s$ is the step size constant and $\mathfrak{P}$ denotes the performance measure. Requiring $\mathbf{w}(k+1)$ to satisfy (4.160) then yields

$$\mathbf{f} = \mathbf{C}^T\mathbf{w}(k+1) = \mathbf{C}^T\left\{\mathbf{w}(k) - \Delta_s\left[\mathbf{R}_{xx}\mathbf{w}(k) + \mathbf{C}\boldsymbol{\lambda}(k)\right]\right\} \tag{4.173}$$

Consequently, the Lagrange multipliers are given by

$$\boldsymbol{\lambda}(k) = -[\mathbf{C}^T\mathbf{C}]^{-1}\mathbf{C}^T\mathbf{R}_{xx}\mathbf{w}(k) - \frac{1}{\Delta_s}[\mathbf{C}^T\mathbf{C}]^{-1} \cdot [\mathbf{f} - \mathbf{C}^T\mathbf{w}(k)] \tag{4.174}$$

Substituting (4.174) into (4.172) then gives the iterative relation

$$\mathbf{w}(k+1) = \mathbf{w}(k) - \Delta_s[I - \mathbf{C}(\mathbf{C}^T\mathbf{C})^{-1}\mathbf{C}^T]\mathbf{R}_{xx}\mathbf{w}(k) + \mathbf{C}(\mathbf{C}^T\mathbf{C})^{-1}[\mathbf{f} - \mathbf{C}^T\mathbf{w}(k)] \tag{4.175}$$

It is convenient to define the $NJ$-dimensional vector,

$$\mathfrak{f} \triangleq \mathbf{C}(\mathbf{C}^T\mathbf{C})^{-1}\mathbf{f} \tag{4.176}$$

and the $NJ \times NJ$ matrix

$$\mathbf{P} \triangleq \mathbf{I} - \mathbf{C}(\mathbf{C}^T\mathbf{C})^{-1}\mathbf{C}^T \tag{4.177}$$

Then the iterative relation (4.175) may be rewritten as

$$\mathbf{w}(k+1) = \mathbf{P}[\mathbf{w}(k) - \Delta_s\mathbf{R}_{xx}\mathbf{w}(k)] + \mathfrak{f} \tag{4.178}$$

In the actual system the input correlation matrix is not known, and it is necessary to adopt some estimate of this matrix to insert in place of $\mathbf{R}_{xx}$ in the iterative weight adjustment equation. An approximation for $\mathbf{R}_{xx}$ at the $k$th iteration is merely the outer product of the tap voltage vector with itself: $\mathbf{x}(k)\mathbf{x}^T(k)$. Substituting this estimate of $\mathbf{R}_{xx}$ into (4.178) and recognizing that $y(k) = \mathbf{x}^T(k)\mathbf{w}(k)$ then yields the constrained LMS algorithm:

$$\left.\begin{aligned} \mathbf{w}(0) &= \mathfrak{f} \\ \mathbf{w}(k+1) &= \mathbf{P}[\mathbf{w}(k) - \Delta_s y(k)\mathbf{x}(k)] + \mathfrak{f} \end{aligned}\right\} \tag{4.179}$$

If it is merely desired to ensure that the complex response of the adaptive array system to a normalized signal input from the look direction is unity, then the spatial correction filter may be dispensed with and the compensation for phase misalignment incorporated directly into the variable weight selection as suggested by Takao et al. [30]. Denote the complex response (amplitude and phase) of the array system by $Y(\theta)$ where $\theta$ is the angle measured from the normal direction to the array face. The appropriate conditions to impose on the adaptive weights are then easily found by requiring that $\Re e\{Y(\theta)\} = 1$ and $Im\{Y(\theta)\} = 0$ when $\theta = \theta_c$, the look direction.

### 4.5.3 Conditions Ensuring Convergence to the Optimum Solution

The weight vector $\mathbf{w}(k)$ obtained by employing (4.179) is a random vector. Convergence of the mean value of the weight vector to the optimum can be shown by considering the length of the difference vector between the mean of the actual weight vector and the optimum weight vector: convergence is assured if the length of the difference vector asymptotically approaches zero.

If we start with the weight adjustment equation,

$$\mathbf{w}(k+1)=\mathbf{P}\left[\mathbf{w}(k)-\Delta_s\mathbf{x}(k)y(k)\right]+\mathfrak{f} \tag{4.180}$$

and recognize that $y(\mathbf{k})=\mathbf{x}^T(k)\mathbf{w}(k)$, then taking the expected value of both sides of (4.180) yields

$$E\left[\mathbf{w}(k+1)\right]=\mathbf{P}\left\{E\left[\mathbf{w}(k)\right]-\Delta_s\mathbf{R}_{xx}E\left[\mathbf{w}(k)\right]\right\}+\mathfrak{f} \tag{4.181}$$

Define the difference vector $\mathbf{v}(k+1)$ by

$$\mathbf{v}(k+1) \triangleq E\left[\mathbf{w}(k+1)\right]-\mathbf{w}_{\text{opt}} \tag{4.182}$$

Substitute (4.181) into (4.182) and use $\mathfrak{f}=(\mathbf{I}-\mathbf{P})\mathbf{w}_{\text{opt}}$ and $\mathbf{P}\mathbf{R}_{xx}\mathbf{w}_{\text{opt}}=0$ [which may be verified by direct substitution of (4.169) and (4.177)], then the difference vector may be shown to satisfy

$$\mathbf{v}(k+1)=\mathbf{P}\mathbf{v}(k)-\Delta_s\mathbf{P}\mathbf{R}_{xx}\mathbf{v}(k) \tag{4.183}$$

Note from (4.177) that $\mathbf{P}$ is idempotent (i.e., $\mathbf{P}^2=\mathbf{P}$), then premultiplying (4.183) by $\mathbf{P}$ reveals that $\mathbf{P}\mathbf{v}(k+1)=\mathbf{v}(k+1)$ for all $k$, so (4.183) can be rewritten as

$$\begin{aligned}\mathbf{v}(k+1)&=\left[\mathbf{I}-\nabla_s\mathbf{P}\mathbf{R}_{xx}\mathbf{P}\right]\mathbf{v}(k)\\&=\left[\mathbf{I}-\nabla_s\mathbf{P}\mathbf{R}_{xx}\mathbf{P}\right]^{(k+1)}\mathbf{v}(0)\end{aligned} \tag{4.184}$$

From (4.184) it follows that the matrix $\mathbf{P}\mathbf{R}_{xx}\mathbf{P}$ determines both the rate of convergence of the mean weight vector to the optimum solution and the steady-state variance of the weight vector about the optimum. The matrix $\mathbf{P}\mathbf{R}_{xx}\mathbf{P}$ has $J$ zero eigenvalues (corresponding to the column vectors of the constraint matrix $\mathbf{C}$) and $NJ-J$ nonzero eigenvalues $\sigma_i$, $i=1,2,\ldots,NJ-J$ [31]. The values of the $NJ-J$ nonzero eigenvalues are bounded by the relation

$$\lambda_{\min}\leqslant\sigma_{\min}\leqslant\sigma_i\leqslant\sigma_{\max}\leqslant\lambda_{\max} \tag{4.185}$$

where $\lambda_{\min}$ and $\lambda_{\max}$ denote the smallest and largest eigenvalues of $R_{xx}$, and $\sigma_{\min}$ and $\sigma_{\max}$ denote the smallest and largest nonzero eigenvalues of $\mathbf{P}\mathbf{R}_{xx}\mathbf{P}$.

The initial difference vector $\mathbf{v}(0)=\mathfrak{f}-\mathbf{w}_{\text{opt}}$ can be expressed as a linear combination of the eigenvectors of $\mathbf{PR}_{xx}\mathbf{P}$ corresponding to the nonzero eigenvalues [31]. Consequently if $\mathbf{v}(0)$ is just equal to an eigenvector $\mathbf{e}_i$ of $\mathbf{PR}_{xx}\mathbf{P}$ corresponding to the nonzero eigenvalue $\sigma_i$, then

$$\begin{aligned}\mathbf{v}(k+1)&=\left[\mathbf{I}-\Delta_s\mathbf{PR}_{xx}\mathbf{P}\right]^{(k+1)}\mathbf{e}_i\\&=\left[1-\Delta_s\sigma_i\right]^{(k+1)}\mathbf{e}_i\end{aligned}\tag{4.186}$$

From (4.186) it follows that along any eigenvector of $\mathbf{PR}_{xx}\mathbf{P}$ the mean weight vector converges to the optimum weight vector geometrically with the geometric ratio $(1-\Delta_s\sigma_i)$. Consequently the time required for the difference vector length to decay to $1/e$ of its initial value is given by the time constant

$$\begin{aligned}\tau_i&=\frac{\Delta t}{\ln(1-\Delta_s\sigma_i)}\\&\cong\frac{\Delta t}{\Delta_s\sigma_i}\qquad\text{if } \Delta_s\sigma_i\ll 1\end{aligned}\tag{4.187}$$

where $\Delta t$ denotes the time interval corresponding to one iteration.

If the step size constant $\Delta_s$ is selected so that

$$0<\Delta_s<\frac{1}{\sigma_{\max}}\tag{4.188}$$

then the length (given by the norm) of any difference vector is bounded by

$$\begin{aligned}(1-\Delta_s\sigma_{\max})^{(k+1)}\|\mathbf{v}(0)\|&\leqslant\|\mathbf{v}(k+1)\|\\&\leqslant(1-\Delta_s\sigma_{\min})^{(k+1)}\|\mathbf{v}(0)\|\end{aligned}\tag{4.189}$$

It immediately follows that if the initial difference vector length is finite, then the mean weight vector converges to the optimum so that

$$\lim_{k\to\infty}\|E\{\mathbf{w}(k)\}-\mathbf{w}_{\text{opt}}\|=0\tag{4.190}$$

where the convergence occurs with the time constants given by (4.187).

The LMS algorithm is designed to cope with nonstationary noise environments by continually adapting the weights in the signal processor. In stationary environments, however, this adaptation results in the weight vector exhibiting an undesirable variance about the optimum solution thereby producing an additional (above the optimum) component of noise to appear at the adaptive array output.

The optimum (minimum) output power level is given by

$$E\{y_{\text{opt}}^2(k)\} = \mathbf{w}_{\text{opt}}\mathbf{R}_{xx}\mathbf{w}_{\text{opt}} = \mathbf{f}^T(\mathbf{C}^T\mathbf{R}_{xx}^{-1}\mathbf{C})^{-1}\mathbf{f} \tag{4.191}$$

The additional noise caused by adaptively adjusting the weights can be compared with (4.191) to determine the penalty incurred by the adaptive algorithm. A direct measure of this penalty is the "misadjustment" $M$ defined by (4.55). For a step size constant satisfying

$$0 < \Delta_s < \frac{1}{\sigma_{\max} + \frac{1}{2}\text{tr}(\mathbf{P}\mathbf{R}_{xx}\mathbf{P})} \tag{4.192}$$

The steady-state misadjustment has been shown to be bounded by [32]

$$\frac{\Delta_s}{2}\cdot\frac{\text{tr}(\mathbf{P}\mathbf{R}_{xx}\mathbf{P})}{1-(\Delta_s/2)[\text{tr}(\mathbf{P}\mathbf{R}_{xx}\mathbf{P})+2\sigma_{\min}]} \leqslant M \leqslant \frac{\Delta_s}{2}\cdot\frac{\text{tr}(\mathbf{P}\mathbf{R}_{xx}\mathbf{P})}{1-(\Delta_s/2)[\text{tr}(\mathbf{P}\mathbf{R}_{xx}\mathbf{P})+2\sigma_{\max}]} \tag{4.193}$$

where "tr" denotes trace.

If $\Delta_s$ is chosen to satisfy

$$0 < \Delta_s < \frac{2}{3\,\text{tr}(\mathbf{R}_{xx})} \tag{4.194}$$

then it will automatically also satisfy (4.192). It is also worth noting that the upper bound in (4.194) can be easily calculated directly from observations since $\text{tr}(\mathbf{R}_{xx}) = E\{\mathbf{x}^T(k)\mathbf{x}(k)\}$, the sum of the powers of the tap voltages.

### *4.5.4 A Useful Geometrical Interpretation*

The constrained LMS algorithm (4.179) has a simple geometrical interpretation [24] that is useful for visualizing the error correcting property that prevents the weight vector from deviating from the constraint condition. Even unavoidable computational errors due to roundoff, truncation, or quantization are prevented from accumulating by the error correcting property, which continuously corrects for any errors that may occur, whatever their source may be.

In an error free algorithm, the successive values of the $NJ$-dimensional weight vector $\mathbf{w}$ all exactly satisfy the constraint equation (4.162) and therefore all lie on a constraint plane $\Lambda$ defined by

$$\Lambda = \{\mathbf{w} : \mathbf{C}^T\mathbf{w} = \mathbf{f}\} \tag{4.195}$$

This constraint plane [which is $(NJ-J)$-dimensional] may be indicated diagramatically as shown in Figure 4.15.

Any vectors that point in a direction that is normal to the constraint plane are all linear combinations of the constraint matrix column vectors, and therefore all have the form $\mathbf{Ca}$, where $\mathbf{a}$ is a constant vector whose components determine the linear combination. Consequently the initial weight vector in the algorithm (4.179), $\mathfrak{f} = \mathbf{C}(\mathbf{C}^T\mathbf{C})^{-1}\mathbf{f}$, points in a direction that is normal to the constraint plane. In addition to pointing in a normal (to the constraint plane) direction, the initial weight vector also terminates exactly on the constraint plane since $\mathbf{C}^T\mathfrak{f} = \mathbf{f}$, and therefore $\mathfrak{f}$ is the shortest vector that can terminate on the constraint plane, as illustrated in Figure 4.15.

By setting the constraint weight vector $\mathbf{f}$ equal to zero, the homogeneous form of the constraint equation,

$$\mathbf{C}^T\mathbf{w} = \mathbf{0} \tag{4.196}$$

defines a second plane [that is also $(NJ-J)$-dimensional] that passes through the coordinate space origin and is referred to as the "constraint subspace." This constraint subspace is also depicted diagramatically in Figure 4.15.

The constrained LMS algorithm (4.179) involves the premultiplication of a certain vector in the $\mathbf{W}$-space by the matrix $\mathbf{P}$, which can be regarded as a

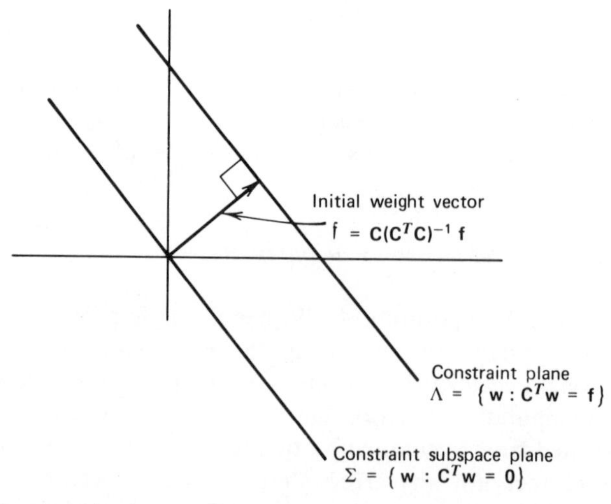


**Figure 4.15** Diagrammatic representation of the constraint plane, the constraint subspace plane, and initial weight vector in the w-space.

projection operator. Premultiplication of any weight vector by the matrix **P** results in the elimination of any vector components that are perpendicular to the plane $\Sigma$, thereby projecting the original weight vector onto the constraint subspace plane as illustrated in Figure 4.16.

The only factor in (4.179) remaining to be discussed is the vector $y(k)\mathbf{x}(k)$, which is an estimate of the unconstrained gradient of the performance measure. Recall from (4.163) that the unconstrained performance measure is $\frac{1}{2}\mathbf{w}^T\mathbf{R}_{xx}\mathbf{w}$, while from (4.164) the unconstrained gradient is given by $\mathbf{R}_{xx}\mathbf{w}$. Since the covariance matrix $\mathbf{R}_{xx}$ is unknown a priori, the estimate provided by $y(k)\mathbf{x}(k)$ is used in the algorithm.

The constrained optimization problem posed by (4.161) and (4.162) can be illustrated diagramatically in **w**-space as shown in Figure 4.17. The algorithm must succeed in moving from the initial weight vector $\mathfrak{f}$ to the optimum weight vector $\mathbf{w}_{\text{opt}}$ along the constraint plane $\Lambda$. The operation of the constrained LMS algorithm (4.179) in solving the above constrained optimization problem may now be considered.

In Figure 4.18 the current value of the weight vector is $\mathbf{w}(k)$, and this current value is to be modified by taking the unconstrained negative gradient estimate $-y(k)\mathbf{x}(k)$, scaling it by $\Delta_s$, and adding the result to $\mathbf{w}(k)$. In general, such a change results in a new vector that lies somewhere off the constraint plane. By premultiplying the vector $[\mathbf{w}(k)-\Delta_s y(k)\mathbf{x}(k)]$ by the matrix **P**, the projection onto the constraint subspace plane is obtained. Finally, by adding $\mathfrak{f}$ to constraint subspace plane projection a new weight vector that lies on the constraint plane is obtained. This new weight vector $\mathbf{w}(k+1)$ satisfies the constraint to within the accuracy of the arithmetic used in performing the above operations. This error correcting feature of the constrained LMS algorithm prevents any computational errors from accumulating.

The convergence properties of the constrained LMS algorithm are of course closely related to those for the unconstrained LMS algorithm and have

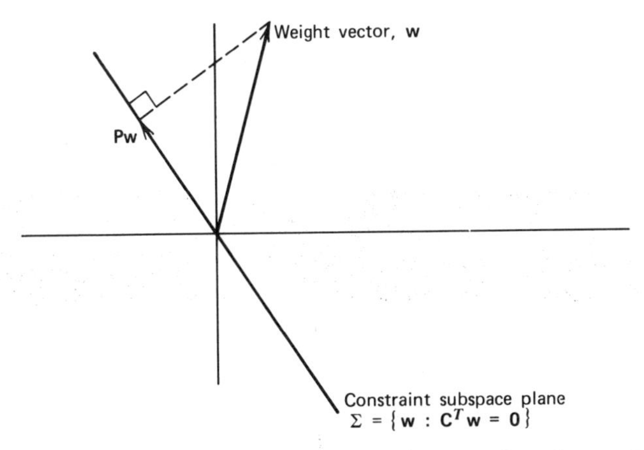


**Figure 4.16** Matrix **P** projects vectors onto the constraint subspace plane.

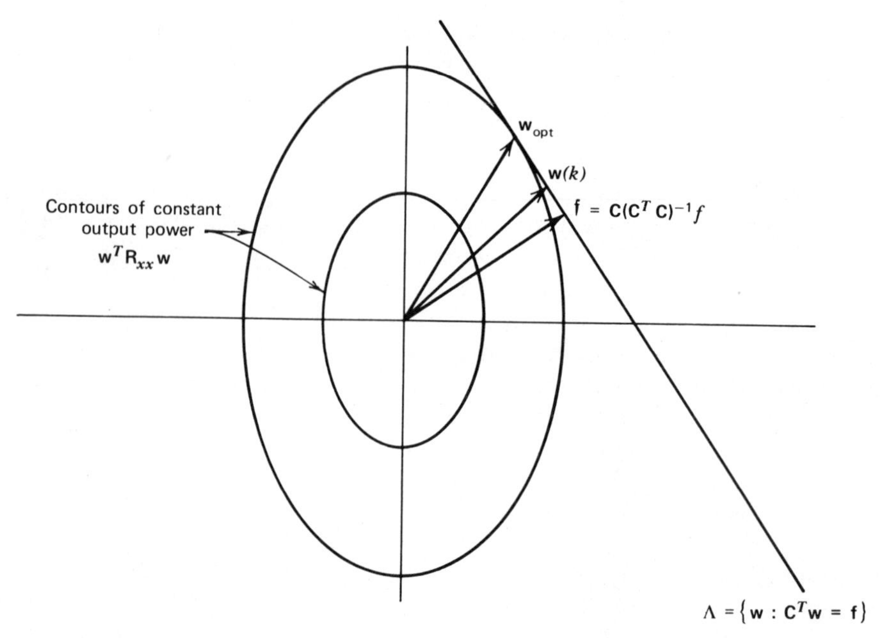


**Figure 4.17** Diagrammatic representation of the constrained optimization problem showing contours of constant output power, the constraint plane $\Lambda$, the initial weight vector $\mathfrak{f}$, and the optimum constrained weight vector $\mathbf{w}_{\text{opt}}$ that minimizes the output power.

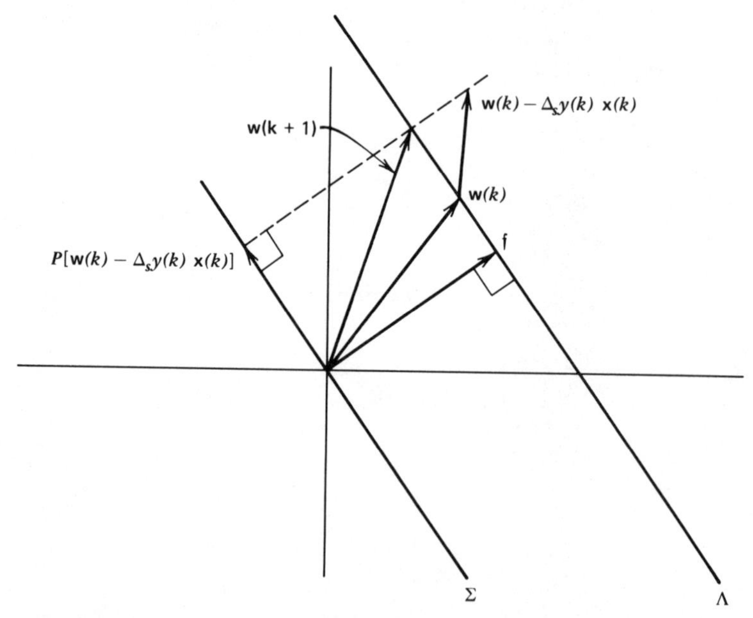


**Figure 4.18** Operation of the constrained LMS algorithm: $\mathbf{w}(k+1) = \mathbf{P}[\mathbf{w}(k) - \Delta_s y(k)\mathbf{x}(k)] + \mathfrak{f}$.

been discussed above. Likewise the same procedures that were employed to speed convergence for the LMS algorithm may also be applied to speed convergence of the constrained LMS algorithm.

## 4.6 SIMULATION RESULTS

It has been shown that the fundamental misadjustment versus speed of adaptation trade-off is less favorable for the DSD algorithm than for the LMS algorithm [15]. Consequently it remains to determine the improvement in this fundamental trade-off that can be realized using the Powell accelerated gradient (PAG) algorithm compared to the LMS algorithm where eigenvalue spread in the input signal covariance matrix is present.

To exercise the foregoing algorithms, an array geometry and a signal environment must be selected. The array geometry and signal environment to be used here are shown in Figure 4.19, which depicts a four-element $Y$ array having $d=0.787\lambda$ element spacing with the desired signal located at 0° and three distinct narrowband Gaussian jamming signals located at 15°, 90°, and 165°. The received signal covariance matrix is therefore given by

$$\frac{1}{n}\mathbf{R}_{xx}=\frac{s}{n}(\mathbf{u}\mathbf{u}^{\dagger})+\sum_{i=1}^{3}\frac{J_i}{n}(\mathbf{v}_i\mathbf{v}_i^{\dagger})+\mathbf{I} \tag{4.197}$$

where $n$ denotes the thermal noise power (taken to be unity), $s/n$ denotes the signal-to-thermal noise ratio, and $J_i/n$ denotes the jammer-to-thermal noise ratios for each of the three jammers ($i=1,2,3$). The elements of the signal steering vector $\mathbf{u}$ and the various jammer steering vectors $\mathbf{v}_i$ are easily defined

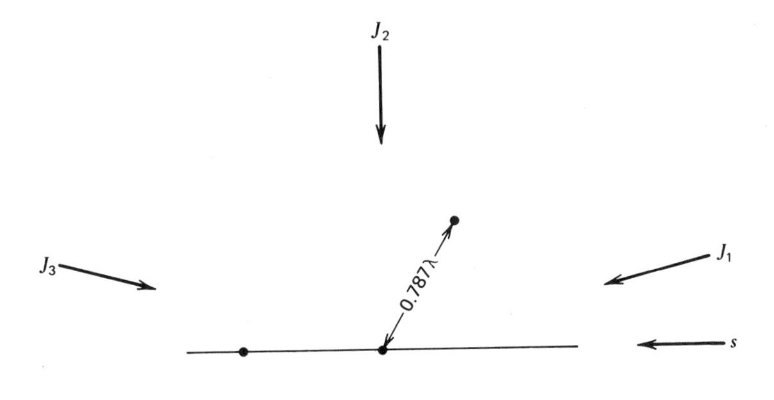


**Figure 4.19** Four-element $Y$-array geometry with signal and jammer locations for selected example.

from the array geometry and the various signal arrival angles. The desired signal is taken to be a biphase modulated signal having a phase angle of either 0° or 180° with equal probability at each sample.

Two different signal conditions were simulated corresponding to two values of eigenvalue spread in the received signal covariance matrix. The first condition represents a respectable eigenvalue spread of $\lambda_{max}/\lambda_{min}=2440$, while the second condition represents a more extreme eigenvalue spread of $\lambda_{max}/\lambda_{min}=16{,}700$. Choosing the jammer-to-thermal noise ratios to be $J_1/n=500$, $J_2/n=40$, and $J_3/n=200$ together with $s/n=10$ yields the corresponding eigenvalues $\lambda_1=2.44\times10^3$, $\lambda_2=4.94\times10^2$, $\lambda_3=25.62$, and $\lambda_4=1.0$ for which the optimum output SNR is $\text{SNR}_{opt}=15.0$ (11.7 dB). Likewise choosing the jammer-to-thermal noise ratios to be $J_1/n=4000$, $J_2/n=40$, and $J_3/n=400$ along with $s/n=10$ yields the eigenvalues $\lambda_1=1.67\times10^4$, $\lambda_2=10^3$, $\lambda_3=29$, and $\lambda_4=1.0$ for which the optimum output SNR is also $\text{SNR}_{opt}=15.0$. In all cases the initial weight vector setting was taken to be $\mathbf{w}^T(0)=[0.1,0,0,0]$. Figures 4.20 and 4.21 show the convergence results for the LMS and PAG algorithms, respectively, plotted as output SNR in decibels versus number of

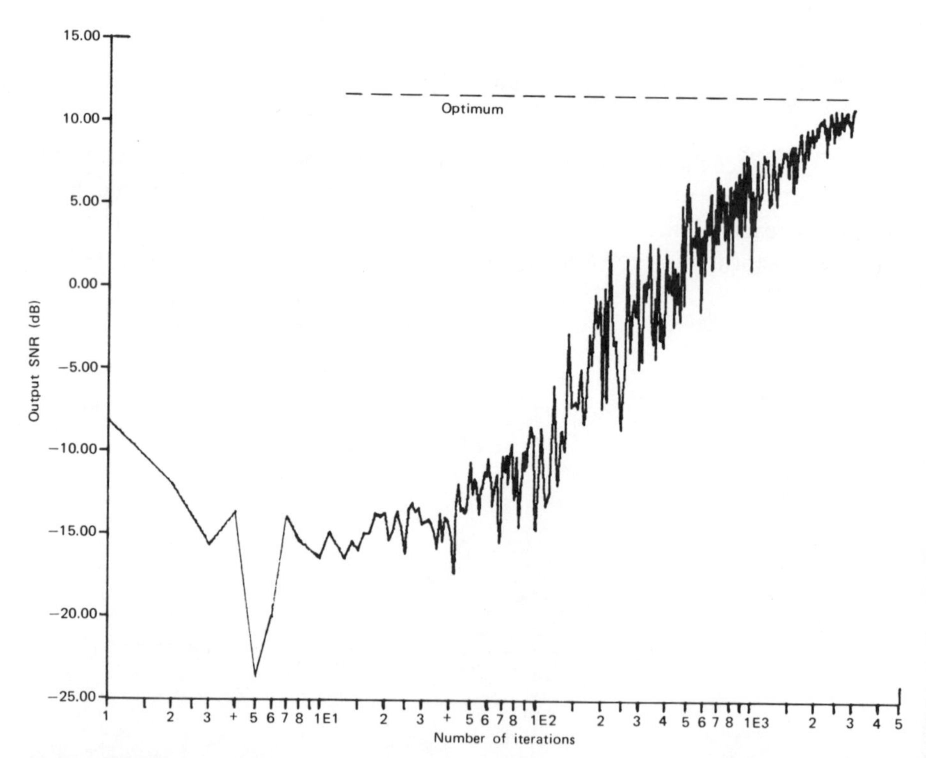


**Figure 4.20** Output SNR versus number of iterations for LMS algorithm with eigenvalue spread = 2440 and $\alpha_L=0.1$.

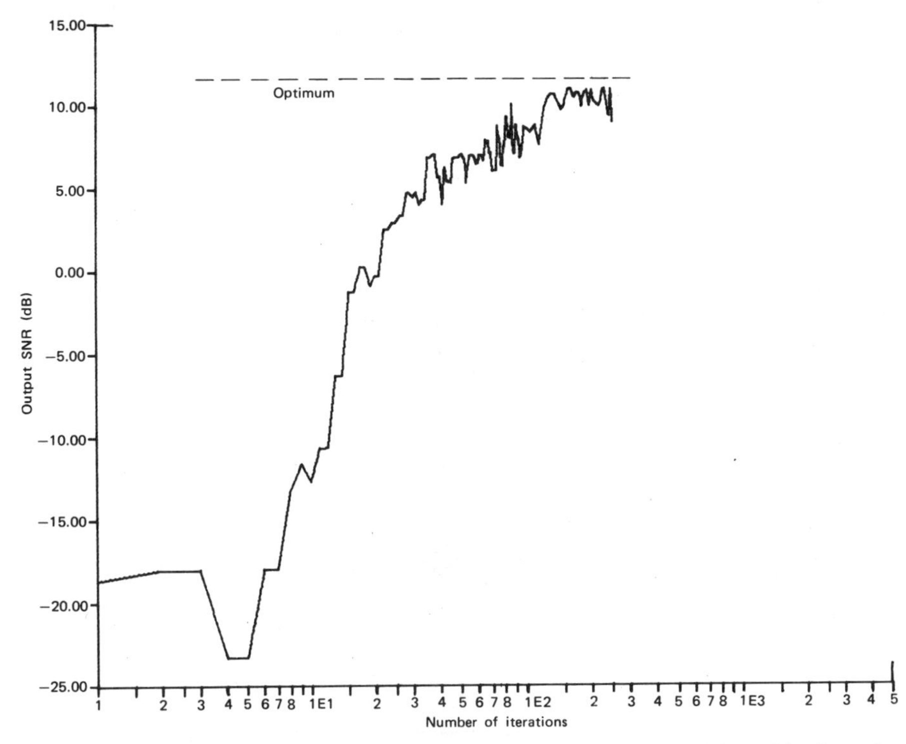


**Figure 4.21** Output SNR versus number of iterations for PAG algorithm with eigenvalue spread = 2440 and $K = 9$.

iterations for an eigenvalue spread of 2440 (here output SNR means output signal-to-jammer plus thermal noise ratio). The expected value of the gradient and of the quantity $\mathbf{v}^{\dagger}\mathbf{R}_{xx}\mathbf{v}$ required by the PAG algorithm was taken over $K = 9$ data samples, and one iteration of the PAG algorithm was therefore defined to occur every nine data samples, even though a weight update does not then occur on some iterations. The loop gain of the LMS loop was selected in accordance with (4.49) which requires that $\Delta_s \operatorname{tr}(\mathbf{R}_{xx}) < 1$ for stability. Letting $\Delta_s \operatorname{tr}(\mathbf{R}_{xx}) = \alpha_L$ and choosing $\alpha_L = 0.1$ therefore ensures stability while giving reasonably fast convergence with an acceptable degree of misadjustment error. As a consequence of the manner in which an iteration was defined for the PAG algorithm, the time scale for Figure 4.21 is nine times greater than the time scale for Figure 4.20. In Figure 4.21 the PAG algorithm is within 3 dB of the optimum after approximately 80 iterations (720 data samples), whereas in Figure 4.20 the LMS algorithm requires approximately 1500 data samples to reach the same point. Furthermore, it may be seen that the steady-state misadjustment for the two algorithms in these examples is very comparable so the PAG algorithm converges twice as fast as the LMS algorithm for a given level of misadjustment in this example.

Figures 4.22 and 4.23 likewise show the convergence of the LMS and PAG algorithms for the same algorithm parameters as in Figures 4.20 and 4.21 but with the eigenvalue spread = 16,700. In Figure 4.23 the PAG algorithm is within 3 dB of the optimum after approximately 200 iterations (1800 data samples), whereas the LMS algorithm in Figure 4.22 does not reach the same point even after 4500 data samples. The degree of convergence speed improvement that is attainable therefore increases as the degree of eigenvalue spread increases.

A word of caution is now required concerning the convergence that may be expected using the PAG algorithm. The simulation results given here were compiled for an array having only four elements; as the number of array elements increases, the number of consecutive steps to be made in orthogonal gradient directions also increases, thereby yielding significant direction errors in the later steps (since estimation errors accumulate over the consecutive step directions). Consequently for a given level of misadjustment the learning curve time constant does not increase linearly with $N$ (as with LMS adaptation), but rather increases more rapidly with the result that for $N$ greater than about 10, the PAG algorithm can actually converge more slowly than the LMS algorithm!

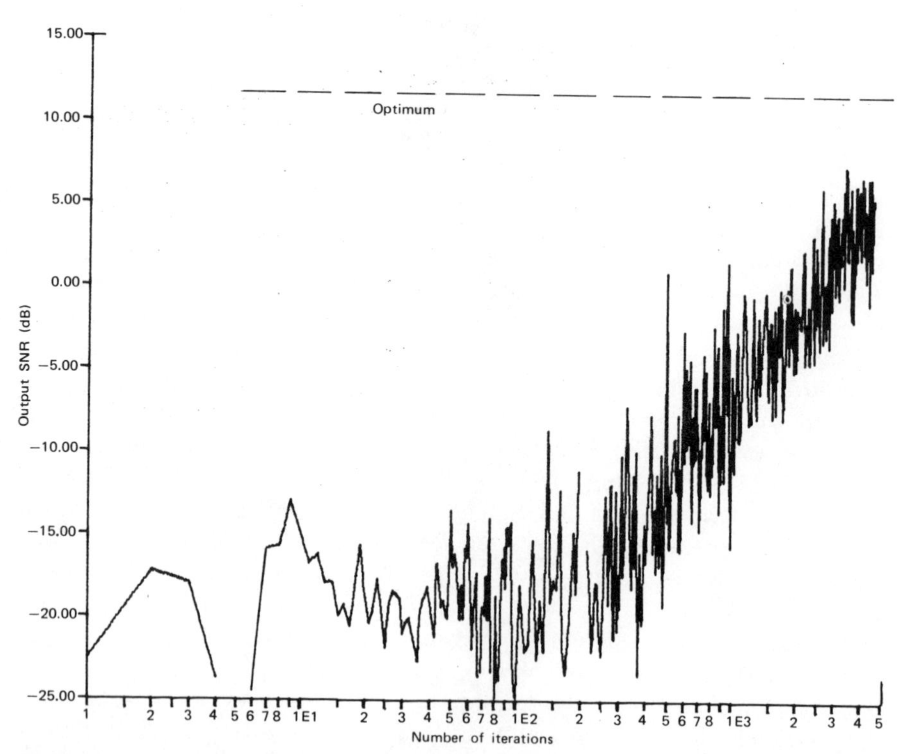


**Figure 4.22** Output SNR versus number of iterations for LMS algorithm with eigenvalue spread = 16,700 and $\alpha_L = 0.1$.

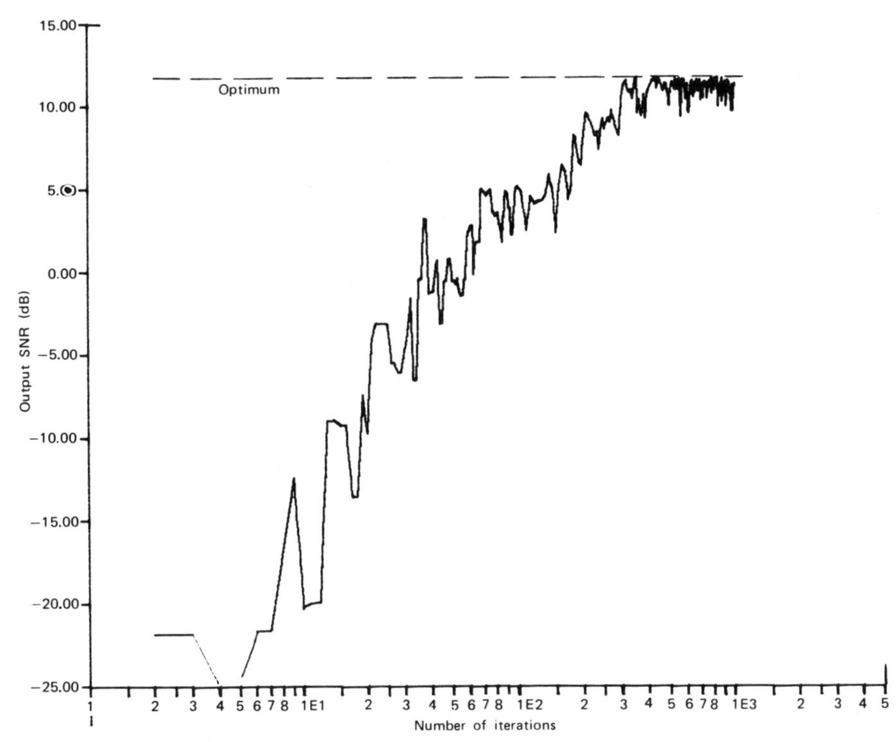


**Figure 4.23** Output SNR versus number of iterations for PAG algorithm with eigenvalue spread = 16,700 and $K=9$.

## 4.7 SUMMARY AND CONCLUSIONS

The LMS algorithm applies the method of steepest descent to the MSE performance measure to obtain a simple implementation that is particularly well suited to continuous signal communication systems. The LMS algorithm requires the generation of a reference signal that is compared with the array output to form an error signal. This technique is useful for adaptive arrays that are expected to distinguish between desired and undesired signals on the basis of differences in modulation characteristics. The heart of an LMS loop is the correlator (multiplier), which forms the product $e(t)x_i(t)$ that is required to obtain the estimated gradient. For an $N$-element array, $N$ correlators are therefore required to implement the LMS algorithm to control each array element.

For some practical applications it may be undesirable to require $N$ correlators as the LMS algorithm does. In such cases the alternative presented by the DSD algorithm which only requires direct performance index measurements (error power measurements in the case of the MSE criterion) may be attractive. The DSD algorithm does not have as favorable a convergence speed versus misadjustment trade-off as the LMS algorithm, and both the

DSD and LMS algorithms exhibit the same degree of convergence speed sensitivity to eigenvalue spread in the input signal covariance matrix.

One way of reducing the convergence speed sensitivity to eigenvalue spread is to employ an algorithm based on an accelerated gradient approach, provided the number of degrees of freedom of the array processor is not too high. An algorithm based on the Powell descent cycle was presented that illustrated the improvement in the speed of convergence that can be realized. Accelerated gradient approaches have certain implementation drawbacks, however, and other methods (discussed in later chapters) may be preferred to obtain the desired reduction in convergence speed sensitivity to eigenvalue spread.

In applications involving high energy, long-duration waveforms, it is often desirable to constrain the main beam of the array so that undesirable signal waveform distortion will not occur. A constrained LMS algorithm for this purpose was discussed, and additional constraint methods are introduced in the next chapter in connection with the Howells-Applebaum interference canceller loop.

## PROBLEMS

**1** [15] ***Misadjustment-Speed of Adaptation Trade-off for the LMS and DSD Algorithms.*** For the LMS algorithm the total misadjustment in the steady state is given by (4.83), whereas the total (minimum) misadjustment for the DSD algorithm is given by (4.133).

(a) Assuming all eigenvalues are equal so that $(T_{p_{\text{MSE}}})_{\text{av}} = T_{\text{MSE}}$ and that $M = 10\%$ for the LMS algorithm, plot $T_{\text{MSE}}$ versus $N$ for $N = 2, 4, 8, \ldots, 512$.

(b) Assuming all eigenvalues are equal so that $(T_{p_{\text{MSE}}})_{\text{av}} = T_{\text{MSE}}$ and that $(M_{\text{tot}})_{\text{min}} = 10\%$ for the DSD algorithm, plot $T_{\text{MSE}}$ versus $N$ for $N = 2, 4, 8, \ldots, 512$ and compare this result with the previous diagram obtained in part (a).

**2** [33] ***Reference Signal Generation for LMS Adaptation Using Polarization as a Desired Signal Discriminant.*** LMS adaptation requires a reference signal to be generated having properties sufficiently correlated either to the desired signal or the undesired signal to permit the adaptive system to preserve the desired signal in its output. Usually, the desired signal waveform properties (such as frequency, duration, type of modulation, signal format) are used to generate the reference signal, but if the signal and the interference can be distinguished by polarization, then polarization may be employed as a useful discriminant for reference signal generation.

Let $s$ denote a linearly polarized desired signal having the known polarization angle $\theta$, and let $n$ denote a linearly polarized interference signal having the polarization angle $\alpha$ (where it is only known that $\alpha \neq \theta$). Assume that the

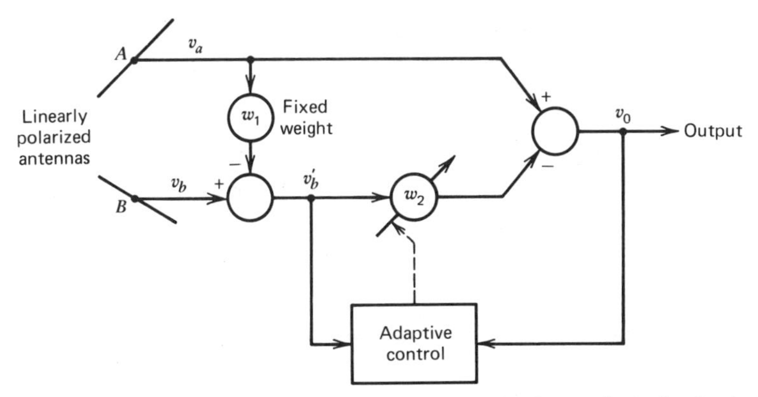


**Figure 4.24** Adaptive array configuration for interference rejection on the basis of polarization using LMS adaptation.

desired signal and interference impinge on two linearly polarized antennas ($A$ and $B$) as shown in Figure 4.24 where the antennas differ in orientation by the angle $\beta$. The two signals $v_a$ and $v_b$ may then be expressed as

$$v_a = s\cos\theta + n\cos\alpha$$

$$v_b = s\cos(\beta-\theta) + n\cos(\beta-\alpha)$$

(a) Show that by introducing the weight $w_1$ as illustrated in Figure 4.24, then the signal $v_b' = v_b - w_1 v_a$ can be made to be signal free (have zero desired signal content) by setting

$$w_1 = \frac{\cos(\beta-\theta)}{\cos\theta}$$

so that

$$v_b' = n\frac{\sin\beta}{\cos\theta}\sin(\alpha-\theta) = nf(\alpha,\beta,\theta)$$

(b) From the results of part (a), show that

$$v_0 = v_a - w_2 v_b' = s\cos\theta + n[\cos\alpha - w_2 f(\alpha,\beta,\theta)]$$

Since the output signal $v_0$ contains both desired signal and interference components, correlating it with the signal free voltage $v_b'$ yields a measure of the interference remaining in the output signal, and the adaptive weight $w_2$ can then be adjusted to reduce the interference content in the output.

(c) The error in the output signal $v_0$ is the interference signal component that is still present after $w_2 b_b'$ is subtracted from $v_a$. Assume that the interference and the desired signal are uncorrelated, then

$$E\{v_0^2\}=E\{s^2\}\cos^2\theta+E\{n^2\}[\cos\alpha-w_2 f(\alpha,\beta,\theta)]^2$$

If the rate of change of $w_2$ is proportional to $\partial E\{v_0^2\}/\partial w_2$, show that the final value of the weight occurs when $\partial E\{v_0^2\}/\partial w_2=0$ so that

$$w_2=\frac{\cos\alpha}{f(\alpha,\beta,\theta)}$$

(d) With $w_2$ set to the final value determined in part (c), show that the steady-state system output is given by

$$v_0=s\cos\theta$$

thereby showing that the system output is free of interference under steady-state conditions. The result above assumes that: (1) Knowledge of $\theta$ and the setting of $w_1$ are error free. (2) The circuitry is noiseless. (3) The number of input signals equals the number of antennas available. These ideal conditions are not met in practice, and reference [34] analyzes the system behavior under nonideal operating conditions.

3 [28] ***Relative Sensitivity of the Constrained Look-Direction Response Processor to Perturbed Wavefronts.*** The solution to the problem of minimizing the expected output power of an array $\eta=E\{\mathbf{w}^\dagger\mathbf{x}\mathbf{x}^\dagger\mathbf{w}\}$ subject to $\mathbf{x}_0^\dagger\mathbf{w}=f$ (or equivalently, $\eta=f^2$) is given by (4.169). Since the look-direction response is constrained by $\mathbf{x}_0^\dagger\mathbf{w}=f$ where $\mathbf{x}_0$ denotes a plane wave signal arriving from the angle $\theta_0$, the rationale behind this constraint is to regard the processor as a filter that will pass plane waves from the angle $\theta_0$ but attenuate plane waves from all other directions.

Let a perturbed plane wave be represented by $\mathbf{x}$, having components

$$\mathbf{x}_k=A_{k_0}(1+\alpha_k)\exp[j(\phi_{k_0}+\xi_k)]$$

where $\alpha_k$ represents amplitude deviations and $\xi_k$ represents phase deviations from the nominal plane wave signal $\mathbf{x}_0$. Assume that $\alpha_k,\xi_k$ are all uncorrelated zero-mean Gaussian random variables with variances $\sigma_\alpha^2,\sigma_\xi^2$ at each sensor of the array.

(a) Using $\eta=\mathbf{w}^\dagger E\{\mathbf{x}\mathbf{x}^\dagger\}\mathbf{w}$ and the fact that

$$E\{x_i x_j^*\}=x_{i_0}x_{j_0}^*\exp(-\sigma_\xi^2)\qquad \text{for } i\neq j$$

and

$$E\{x_i x_j^*\}=|x_{i_0}|^2(1+\sigma_\alpha^2)\qquad \text{for } i=j$$

show that

$$\eta=\exp(-\sigma_\xi^2)\mathbf{w}^\dagger\mathbf{x}_0\mathbf{x}_0^\dagger\mathbf{w}+\left[1-\exp(-\sigma_\xi^2)+\sigma_\alpha^2\right]\mathbf{w}^\dagger\mathbf{w}$$

or $\eta \cong f^2+(\sigma_\xi^2+\sigma_\alpha^2)\mathbf{w}^\dagger\mathbf{w}$ for small values of $\sigma_\xi^2, \sigma_\alpha^2$ assuming that $|x_{i_0}|^2=1$ (which is the case for a planar wave).

(b) The result in part (a) can be rewritten as

$$\eta=f^2\left[1+\mathfrak{S}\left(\sigma_\xi^2+\sigma_\alpha^2\right)\right]$$

where

$$\mathfrak{S} \triangleq \frac{\mathbf{w}^\dagger\mathbf{w}}{f^2}$$

Consequently the ratio $\mathfrak{S}$ can be regarded as the relative sensitivity of the processor to the perturbations whose variances are $\sigma_\xi^2, \sigma_\alpha^2$. Using the weights given by (4.169), show that

$$\mathfrak{S}=\frac{\mathbf{x}_0^\dagger\mathbf{R}_{xx}^{-2}\mathbf{x}_0}{\left(\mathbf{x}_0^\dagger R_{xx}^{-1}\mathbf{x}_0\right)^2}$$

The above relative sensitivity can become large if the eigenvalues of $\mathbf{R}_{xx}$ have a large spread, but if the eigenvalues of $\mathbf{R}_{xx}$ have a small spread, then $\mathfrak{S}$ cannot become large.

## REFERENCES

[1] B. Widrow, "Adaptive Filters," in *Aspects of Network and System Theory*, edited by R. E. Kalman and N. DeClaris, Holt, Rinehart and Winston, New York, 1970, pp. 563–587.

[2] F. R. Ragazzini and G. F. Franklin, *Sampled-Data Control Systems*, McGraw-Hill, New York, 1958.

[3] E. I. Jury, *Sampled-Data Control Systems,*, Wiley, New York, 1958.

[4] J. T. Tou, *Digital and Sampled-Data Control Systems*, McGraw-Hill, New York, 1959.

[5] B. C. Kuo, *Discrete-Data Control Systems*, Prentice-Hall, Englewood Cliffs, NJ, 1970.

[6] B. Widrow, "Adaptive Filters I: Fundamentals," Stanford Electronics Laboratories, Stanford, CA, Rept. SEL-66-126 (Tech. Rept. 6764-6), December 1966.

[7] J. S. Koford and G. F. Groner, "The Use of an Adaptive Threshold Element to Design a Linear Optimal Pattern Classifier," *IEEE Trans. Inf. Theory*, Vol. IT-12, January 1966, pp. 42–50.

[8] K. Steinbuch and B. Widrow, "A Critical Comparison of Two Kinds of Adaptive Classification Networks," *IEEE Trans. Electron. Comput.* (*Short Notes*), Vol. EC-14, October 1965, pp. 737–740.

[9] F. W. Smith, "Design of Quasi-Optimal Minimum-Time Controllers," *IEEE Trans. Autom. Control*, Vol. AC-11, January 1966, pp. 71–77.

[10] B. Widrow, P. E. Mantey, L. J. Griffiths, and B. B. Goode, "Adaptive Antenna Systems," *Proc. IEEE*, Vol. 55, No. 12, December 1967, pp. 2143–2159.

[11] L. J. Griffiths, "Signal Extraction Using Real-Time Adaptation of a Linear Multichannel Filter," Ph.D. Disseration, Stanford University, December 1967.

[12] K. D. Senne and L. L. Horowitz, "New Results on Convergence of the Discrete-Time LMS Algorithm Applied to Narrowband Adaptive Arrays," Proceedings of the 17th IEEE Conference on Decision and Control, San Diego, CA., 10–12 January, 1979, pp. 1166–1167.

[13] R. T. Compton, Jr., "An Adaptive Array in a Spread-Spectrum Communication System," *Proc. IEEE*, Vol. 66, No. 3, March 1978, pp. 289–298.

[14] D. M. DiCarlo and R. T. Compton, Jr., "Reference Loop Phase Shift in Adaptive Arrays," *IEEE Trans. Aerosp. Electron. Syst.*, Vol. AES-14, No. 4, July 1978, pp. 599–607.

[15] B. Widrow and J. M. McCool, "A Comparison of Adaptive Algorithms Based on the Methods of Steepest Descent and Random Search," *IEEE Trans. Antennas Propag.*, Vol. AP-24, No. 5, September 1976, pp. 615–638.

[16] B. W. Lindgren, *Statistical Theory*, Macmillan, New York, 1960, Ch. 5.

[17] M. R. Hestenes and E. Stiefel, "Method of Conjugate Gradients for Solving Linear Systems," *J. Res. Natl. Bur. Stand.*, Vol. 29, 1952, p. 409.

[18] J. D. Powell, "An Iterative Method for Finding Stationary Values of a Function of Several Variables," *Comput. J. (Br.)*, Vol. 5, No. 2, July 1962, pp. 147–151.

[19] R. Fletcher and M. J. D. Powell, "A Rapidly Convergent Descent Method for Minimization," *Comput. J. (Br.)*, Vol. 6, No. 2, July 1963, pp. 163–168.

[20] R. Fletcher and C. M. Reeves, "Functional Minimization by Conjugate Gradients," *Comput. J. (Br.)*, Vol. 7, No. 2, July 1964, pp. 149–154.

[21] D. G. Luenberger, *Optimization by Vector Space Methods*, Wiley, New York, 1969, Ch. 10.

[22] L. Hasdorff, *Gradient Optimization and Nonlinear Control*, Wiley, New York, 1976.

[23] L. S. Lasdon, S. K. Mitter, and A. D. Waren, "The Method of Conjugate Gradients for Optimal Control Problems," *IEEE Trans. Autom. Control*, Vol. AC-12, No. 2, April 1967, pp. 132–138.

[24] O. L. Frost, III, "An Algorithm for Linearly Constrained Adaptive Array Processing," *Proc. IEEE*, Vol. 60, No. 8, August 1972, pp. 926–935.

[25] A. E. Bryson, Jr. and Y. C. Ho, *Applied Optimal Control*, Blaisdell, Waltham, MA 1969, Ch. 1.

[26] E. J. Kelly, and M. J. Levin, "Signal Parameter Estimation for Seismometer Arrays," Massachusetts Institute of Technology, Lincoln Laboratories Technical Rept. 339, January, 1964.

[27] A. H. Nuttall, and D. W. Hyde, "A Unified Approach to Optimum and Suboptimum Processing for Arrays," U.S. Navy Underwater Sound Laboratory, New London, CT, USL Rep. 992, April 1969.

[28] R. N. McDonald, "Degrading Performance of Nonlinear Array Processors in the Presence of Data Modeling Errors," *J. Acoust. Soc. Am.*, Vol. 51, No. 4, April 1972, pp. 1186–1193.

[29] D. Q. Mayne, "On the Calculation of Pseudoinverses," *IEEE Trans. Autom. Control*, Vol. AC-14, No. 2, April 1969, pp. 204–205.

[30] K. Takao, M. Fujita, and T. Nishi, "An Adaptive Antenna Array Under Directional Constraint," *IEEE Trans. Antennas Propag.*, Vol. AP-24, No. 5, September 1976, pp. 662–669.

[31] O. L. Frost, III, "Adaptive Least Squares Optimization Subject to Linear Equality Constraints," Stanford Electronics Laboratories, Stanford, CA, DOC. SEL-70-053, Tech. Rep. TR6796-2, August 1970.

[32] J. L. Moschner, "Adaptive Fiitering with Clipped Input Data," Stanford Electronics Laboratories, Stanford, CA, Dec. SEL-70-053, Tech. Rep. TR6796-1, June 1970.

[33] H. S. Lu, "Polarization Separation by an Adaptive Filter," *IEEE Trans. Aerosp. Electron. Sys.*, Vol. AES-9, No. 6, November 1973, pp. 954–956.

# Chapter 5. The Howells-Applebaum Adaptive Processor

The key capability of adaptive interference nulling was developed for an intermediate frequency (IF) radar sidelobe canceller as represented by the patent of Howells [1]. An analysis of this approach by Applebaum [2] established the control-law theory governing the operation of an adaptive control loop for each array element. The Applebaum algorithm maximizes a generalized SNR with the assumptions that the desired signal is absent most of the time (as in a pulsed radar or sonar system) and the direction of arrival of the desired signal is known. Because the Howells-Applebaum processor is practical to implement, it has been applied extensively to the problem of clutter and interference rejection in radar systems [3]–[6]. Unless otherwise noted, the analysis of the maximum SNR processor and illustrative examples given here follow the treatment of this subject given by Gabriel [7].

## 5.1 INTRODUCTORY CONCEPTS

In introducing the Howells-Applebaum maximum SNR adaptive processor, it will be useful to discuss first the prerequisites of phase conjugacy, cross-correlation interferometers, and an *RC* integrator filter. With these introductory concepts in hand, the analysis of a simple two-element array having a single analog adaptive loop may then proceed. By studying this simple single-loop system behavior, a good perspective on adaptive system performance can be gained, since it is relatively easy to keep track of the various parameters affecting performance. For tutorial purposes analog circuits are utilized in the discussion, but digital processing is often preferred over analog processing in sophisticated systems.

### *5.1.1 Phase Conjugacy*

Spatial filtering is accomplished in an adaptive array by automatically sensing the direction of arrival of an interference source and forming a (retrodirective) beam in that direction to subtract from the original (unadapted) beam pattern. The idea of subtracting a retrodirective beam (having an amplitude

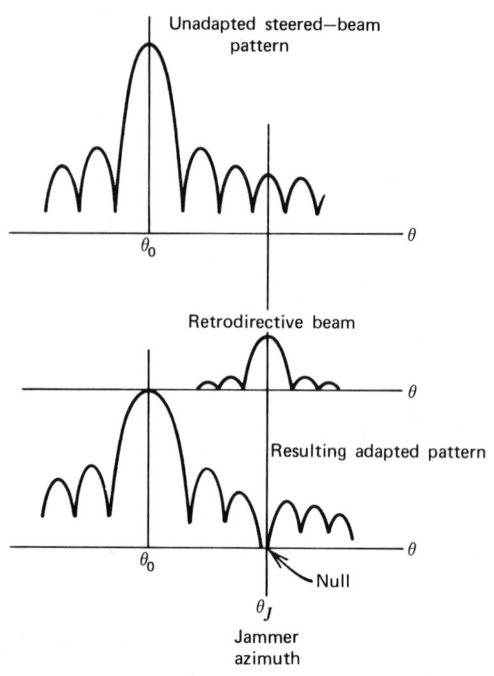


**Figure 5.1** Retrodirective beam principle illustrating subtraction of retrodirective beam from unadapted pattern to obtain adapted pattern with one interference source. From Gabriel, *Proc. IEEE*, February 1976.

and phase that matches the unadapted beam) from the unadapted beam response to obtain an overall beam null in the direction of an interference source is illustrated in Figure 5.1. The term "retrodirective beam" indicates that an auxiliary receive beam is automatically formed in the direction of the single interference source. To point the auxiliary receive beam in the direction of the interference source (i.e., to achieve "retrodirectivity"), the phase of each array element must be delayed (with respect to a selected phase reference) by the same amount that the received waveform was advanced. At any particular frequency, time delay is equivalent to a phase shift (to within $\pm 2\pi$), and consequently the retrodirective phase has a conjugate relationship to the phase of the original received signal at any element (when compared to the common reference element).

It is well known [8] that the desired phase conjugacy can easily be obtained by using a mixer whose reference sinusoid is assigned a frequency either higher than or equal to the received signal frequency, and then selecting the difference frequency as the output.

### 5.1.2 Cross-Correlation Interferometer

An adaptive array using a Howells-Applebaum adaptive processor derives the weight elements for obtaining phase conjugacy (thereby forming the retrodirective receive beam) by cross-correlating a received reference signal with each of the received element signals. The reference signal may be taken either as the output of a separate antenna or as the output of the entire array in which the element being controlled is located. In either case, the principle of operation is the same as that for the cross-correlation interferometer [9], [10], depicted in Figure 5.2. The information required for determining the conjugate phase for an incoming signal can be obtained from the cross-correlation interferometer in the following manner.

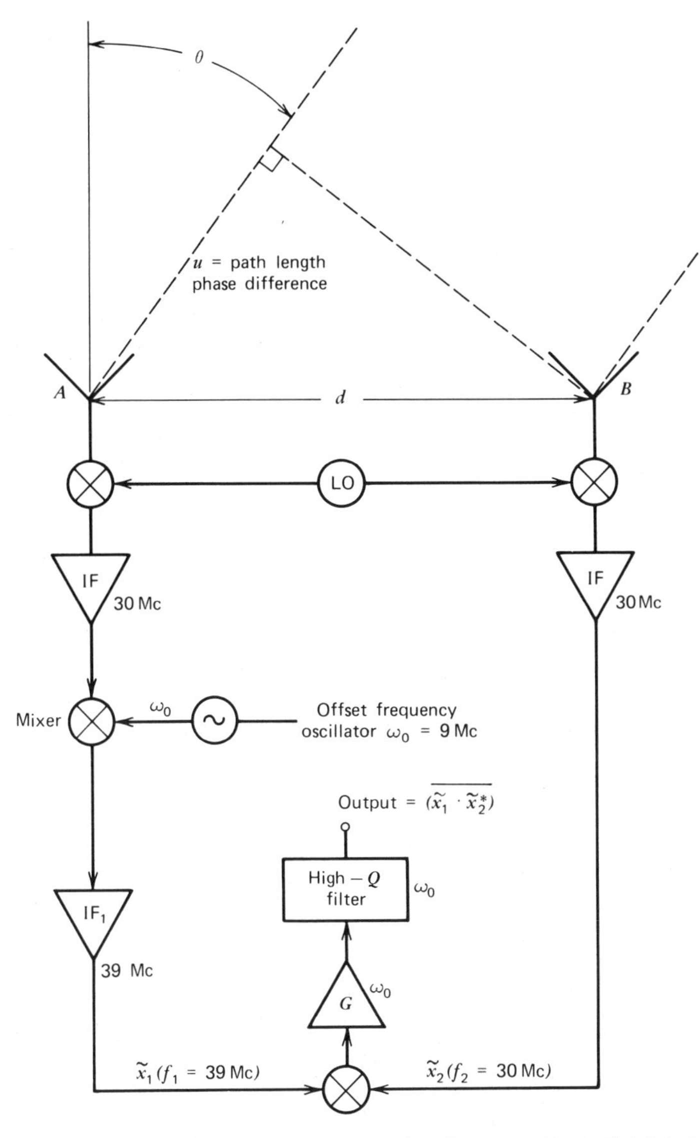


**Figure 5.2** Schematic diagram of cross-correlation interferometer. From Gabriel, *Proc. IEEE*, February 1976.

A single point source located at the angle $\theta$ off broadside generates a signal wavefront that arrives at the two sensors $A$ and $B$ of Figure 5.2 with a path-length phase difference of $u$. The path-length phase difference is given by

$$u = \frac{2\pi d}{\lambda}\sin\theta \tag{5.1}$$

where $d$ is the distance between the antenna phase centers and $\lambda$ is the wavelength of the incident wavefront. The RF signals at sensors $A$ and $B$ are shifted to a convenient IF band by two mixers having a common local oscillator so the RF phases and amplitudes are preserved in the shifted IF signals. If the complex IF signal at sensor $B$ is denoted by $\tilde{x}_2$, then the real part of this signal can be written as

$$x_2 = b\cos(\omega t - u) \tag{5.2}$$

where $b$ is the received signal amplitude at $B$, $\omega$ is the IF carrier frequency, and $u$ is the phase advance at $B$ with respect to $A$. The complex IF signal from sensor $A$ is offset to a higher frequency by mixing it with a constant-reference offset frequency $\omega_0$ in a second mixing operation. This offset frequency shift is introduced to avoid the problems associated with low-level dc detection, amplifier balancing, and flicker noise. The real part of the complex signal $\tilde{x}_1$ can therefore be expressed as

$$x_1 = a\cos[(\omega+\omega_0)t + \phi_0] \tag{5.3}$$

where $a$ is the amplitude of the signal received at $A$, $(\omega+\omega_0)$ is the shifted IF frequency, and $\phi_0$ is the phase constant of the reference offset oscillator.

The two IF signals $x_1$ and $x_2$ are combined in a final mixer as shown in Figure 5.2, where the difference-frequency component is selected as the output signal. Consequently the output signal of the final mixer can be regarded as the real part of the product of $\tilde{x}_1$ and the complex conjugate of $\tilde{x}_2$:

$$\text{mixer output} = \text{Re}\{\tilde{x}_1\tilde{x}_2^*\} = \frac{ab}{2}\cos(\omega_0 t + \phi_0 - u) \tag{5.4}$$

The final mixer therefore provides a cross-correlation of the signals received at sensors $A$ and $B$, and the output has a carrier frequency equal to the reference offset frequency $\omega_0$. The mixer output amplitude is proportional to the product of the two signal amplitudes ($a$ and $b$), and the phase is equal (except for the arbitrary phase constant $\phi_0$) to the path-length phase difference $u$ between the signals received at the two sensors.

Finally, the cross-correlation mixer signal is amplified in a high-gain amplifier (having gain $G$) with a passband centered at the offset frequency $\omega_0$

and integrated in a narrowband high-$Q$ filter to improve the output SNR. The high-$Q$ filter reduces the output noise both by reducing the output noise bandwidth and by integrating (or averaging) the correlation mixer output signal envelope as follows:

$$\text{filter output} = \int_{t}^{t+T} \text{Re}\{\tilde{x}_1 \tilde{x}_2^*\}\, dt \tag{5.5}$$

The cross-correlation interferometer system of Figure 5.2 therefore results in a constant complex output signal having a convenient reference offset frequency. The amplitude of this constant complex signal is proportional to the product of the amplitudes of the signals received at the two sensors, and the phase is equal to the path-length phase difference $u$ (which is also the conjugate phase of sensor $B$ with respect to sensor $A$). Consequently, the output of the cross-correlation interferometer contains the information needed (the exact conjugate phase angle) to form an adaptive element weight that results in a retrodirective lobe pointed toward the signal source.

### 5.1.3 *Integrating Filter*

For adaptive array use the cross-correlator high-$Q$ filter of Figure 5.2 must have outstanding phase stability and also permit easy control of the filter time constant. These factors favor an implementation in which the IF signal is down-converted to in-phase and quadrature ($I$ and $Q$) dc baseband channels at the output of the high-gain amplifier so that simple $RC$ filters can be employed as illustrated in Figure 5.3. The $I$ and $Q$ bipolar video signals at the output of the $RC$ filters can then be used to remodulate the offset reference signal (in the bipolar multipliers) and thereby establish the complex weight element $\tilde{w}$ at the offset reference frequency.

Assuming that the integrating filter is of the $RC$ type, we can base the transient analysis on the simple $RC$ circuit of Figure 5.4. The differential equation for output voltage from this $RC$ circuit may be written as

$$C\frac{dw}{dt} + \frac{w}{R} = \frac{v}{R} \tag{5.6}$$

or equivalently,

$$\tau_0 \frac{dw}{dt} + w = v \tag{5.7}$$

where $\tau_0 = RC$ is the filter time constant, and $v$ is an input step function voltage, that is,

$$v(t) = \begin{cases} 0 & \text{for } t < 0 \\ v_0 & \text{for } t \geq 0 \end{cases} \tag{5.8}$$

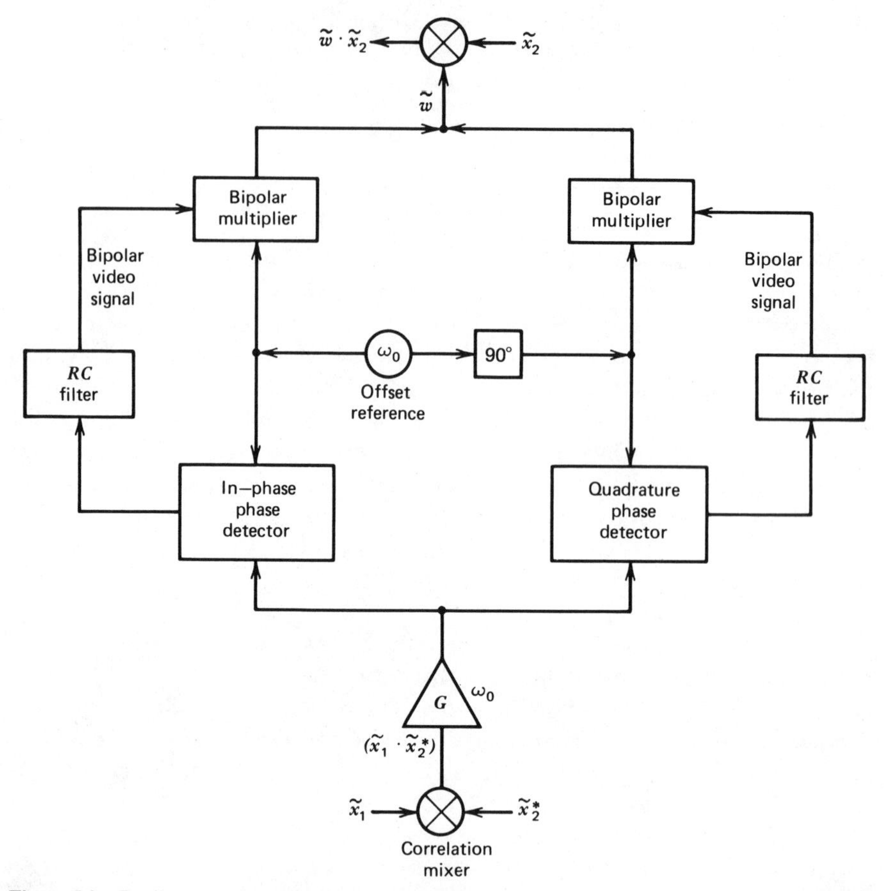


**Figure 5.3** Realization of high-$Q$ filter using simple $RC$ integrator filters in the quadrature and in-phase channels. From Gabriel, *Proc. IEEE*, February 1976.

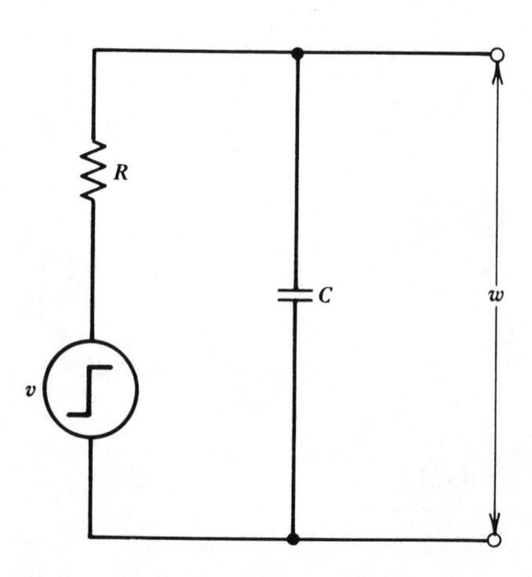


**Figure 5.4** Simple $RC$ integrator filter. From Gabriel, *Proc. IEEE*, February 1976.

Taking the Laplace transform of (5.7), we obtain the results

$$\tau_0 s w(s) - \tau_0 w(0+) + w(s) = \frac{v_0}{s} \tag{5.9}$$

so that

$$w(s) = \frac{w(0+)}{(s+\alpha_0)} + \frac{\alpha_0 v_0}{s(s+\alpha_0)} \tag{5.10}$$

where $\alpha_0 = \frac{1}{\tau_0}$ and $w(0+)$ is the initial value of the output voltage $w$ at time $t=0+$. Taking the inverse Laplace transform of (5.10) then yields

$$w(t) = w(0+)e^{-\alpha_0 t} + v_0(1-e^{-\alpha_0 t}) \tag{5.11}$$

or

$$w(t) = \left[w(0+) - v_0\right]e^{-\alpha_0 t} + v_0 \tag{5.12}$$

Since the *RC* filtering operation of Figure 5.4 can be done for both the *I* and *Q* channels, (5.12) can be written in terms of the entire complex numbers $\tilde{w}$ and $\tilde{v}_0$, so that

$$\tilde{w}(t) = \left[\tilde{w}(0+) - \tilde{v}_0\right]e^{-\alpha_0 t} + \tilde{v}_0 \tag{5.13}$$

Equation (5.13) therefore gives a simple transient equation for the complex weight $\tilde{w}$ representing the *I* and *Q* components of the bipolar video output signal from the integrating *RC* filters. The "~" notation to denote complex quantities will now be dropped, and unless otherwise noted it will be assumed that all quantities are represented by their complex envelopes.

## 5.2 TWO-ELEMENT ARRAY WITH ONE ADAPTIVE LOOP

The simplest adaptive array configuration to consider consists of a two-element array with a single Howells-Applebaum adaptive loop as shown in Figure 5.5. The adaptive loop indicated in Figure 5.5 is similar to the single-loop sidelobe canceller described in reference [1] except that a beam-steering signal $b_2^*$ has been added as suggested in reference [2]. It will be noted that the adaptive loop of Figure 5.5 incorporates the cross-correlation interferometer arrangement given in Figure 5.2. At this point it is useful to compare the adaptive loop of Figure 5.5 with the analog realization of the LMS adaptive loop given in Figure 4.6. It is seen that the analog version of the LMS loop is nearly identical with the Howells-Applebaum adaptive control loop. A Howells-Applebaum loop is usually realized with a lowpass

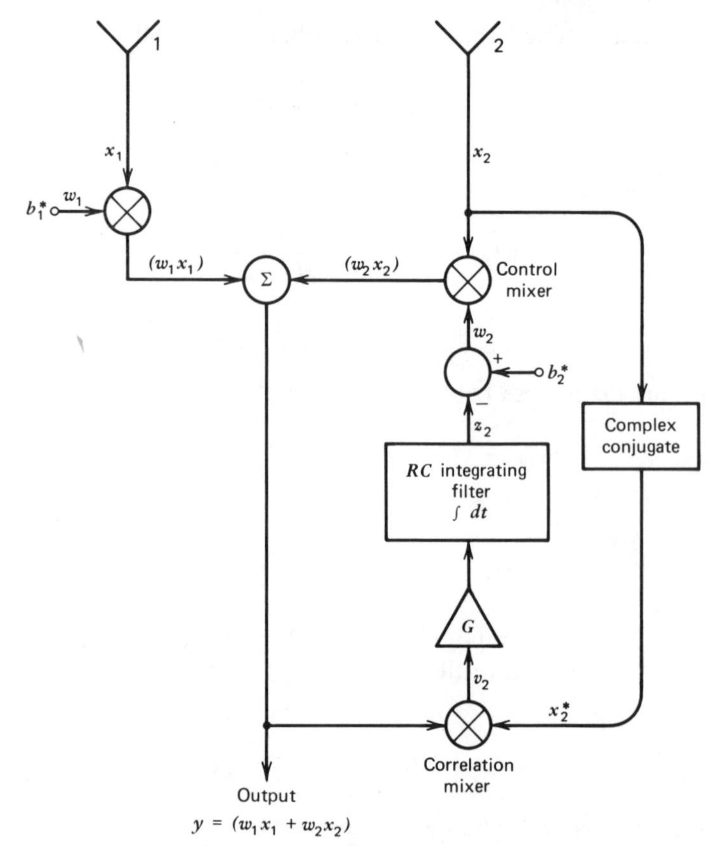


**Figure 5.5** Two-element array with a single Howells-Applebaum adaptive loop. From Gabriel, *Proc. IEEE*, February 1976.

filter replacing the integrator. Furthermore, the LMS loop employs a control signal (the "desired reference signal") introduced before the correlation mixer whereas the Howells-Applebaum loop employs a control signal (the "beam steering signal") introduced within the correlation mixer loop. Section 5.3.1 shows how these two control signals must be related in order for the LMS and maximum SNR algorithms to yield identical solutions for the adaptive weight vector. In view of the nearly identical loop configurations, it is hardly surprising that the characteristic closed-loop behavior of two adaptive loops is the same. It is nevertheless useful to carry out an analysis of the loop of Figure 5.5 under the assumed condition of desired signal absence since this represents an operating condition different from the previous analysis done for the LMS algorithm, and the loop represents an analog version (instead of the digital version treated before) of the LMS adaptive algorithm.

The element weight value $w_1$ is not adaptive and is simply set equal to its corresponding input beam-steering signal $b_1^*$. The element weight $w_2$ is, however, adaptive and is equal to the input beam steering signal $b_2^*$ minus the output voltage $z_2$ from the integrating $RC$ filter. The weighted element signals are summed to form the array output $(w_1x_1 + w_2x_2)$, and this output is in turn directed to the correlation mixer to be cross-correlated with $x_2^*$. The adaptive loop appearing in Figure 5.5 exhibits behavior that corresponds to a negative-feedback servo loop: the correspondence between a servo loop model and the indicated control loop can be established after considering in greater detail the signals and weight element values appearing within the control loop.

It should be noted that the diagram of Figure 5.5 is based on some simplifying assumptions that help focus attention on important control performance issues while eliminating nonessential details from consideration. It is assumed that all signals are represented by their complex envelopes so the RF or IF carrier modulated by that envelope never appears explicitly in any equation. Furthermore Figure 5.5 does not show the local oscillator, IF buffer amplifiers, and bandpass filters that are present in an actual system. The beam-steering signals $b_1^*$ and $b_2^*$ are both input with some reference offset frequency $\omega_0$, but the resulting $e^{j\omega_0 t}$ carrier frequency term is not included since it would merely be an additional nonessential quantity to be carried along in the signals of interest.

The purpose of the beam-steering signals $b_1^*$ and $b_2^*$ is to steer the receive beam in a desired azimuth direction $\theta_0$: this is equivalent to assuming that the direction of arrival of the desired signal is known. For quiescent conditions (when only receiver noise is present and the desired signal is absent), the adaptive weight $w_2$ settles to a quiescent steady-state value denoted by $w_q$. It is desired that $w_1$ and $w_q$ be exactly equal to the weight values needed to point the beam in the direction $\theta_0$. We select the reference phase point to be midway between the phase centers of the two sensor elements (which are spaced apart by the distance $d$), then from Figure 5.2 and (5.1) $w_1$ and $w_q$ may be selected as unit amplitude weights corresponding to the desired direction $\theta_0$ by choosing

$$w_1 = e^{ju_0} \tag{5.14}$$

and

$$w_q = w_1^* = e^{-ju_0} \tag{5.15}$$

where

$$u_0 = \frac{\pi d}{\lambda}\sin\theta_0 \tag{5.16}$$

The foregoing weights result in a quiescent beam pattern $G_q(\theta)$ that is given by

$$G_q(\theta) = \tfrac{1}{2}\left[e^{j(u-u_0)} + e^{-j(u-u_0)}\right] = \cos(u-u_0) \tag{5.17}$$

where

$$u=\frac{\pi d}{\lambda}\sin\theta \tag{5.18}$$

in which $\theta$ is the far-field azimuth angle variable. Now $b_1^*$ is simply set equal to $w_1$, but $b_2^*$ is related to $w_q$ through a gain constant $c_2$ (to be evaluated later), so that

$$b_2^*=c_2 w_q=c_2 e^{-ju_0} \tag{5.19}$$

Let the sensor element signals $x_1$ and $x_2$ consist of the receiver channel noise voltages $n_1$ and $n_2$ plus a statistically independent narrowband noisy signal $J_i$ caused by a single interference source located at the angle $\theta_i$. Furthermore assume that the interference is abruptly turned on at time $t=0$ in a step function manner. The element signals may then be described by

$$\left.\begin{array}{l} x_1=n_1 \\ x_2=n_2 \end{array}\right\} \quad \text{for } t<0 \text{ (}\textit{quiescent}\text{)} \tag{5.20}$$

and

$$\left.\begin{array}{l} x_1=n_1+J_i e^{-ju_i} \\ x_2=n_2+J_i e^{ju_i} \end{array}\right\} \quad \text{for } t>0 \tag{5.21}$$

where

$$u_i=\frac{\pi d}{\lambda}\sin\theta_i \tag{5.22}$$

The adaptive weight $w_2$ is given by

$$w_2=b_2^*-z_2 \tag{5.23}$$

The cross-correlator consists of the correlation mixer, amplifier, and integrating (or smoothing) filter. The transient behavior of the cross-correlator may be determined with the same *RC* filter approach used in Section 5.1.3, provided the input voltage can actually be represented by a step function (since a step function input was assumed in the analysis). Writing (5.7) in terms of the output signal $z_2$ and the input signal $Gv_2$, there results

$$\tau_0\frac{dz_2}{dt}+z_2=Gv_2 \tag{5.24}$$

where $G$ is the amplifier gain and $v_2$ is the output signal from the correlation mixer. The correlation mixer output signal is given by

$$v_2=k^2(w_1x_1+w_2x_2)x_2^* \tag{5.25}$$

in which $k$ is a mixer conversion constant. Since $v_2$ is a direct function of the element signals $x_1$ and $x_2$, any rapid RMS fluctuations occurring in $x_1$ and $x_2$ will be reflected in rapid RMS fluctuations in $v_2$ (to the extent permitted by the receiver channel bandwidth). Consequently for the $x_1$ and $x_2$ envelope step function defined by (5.20) and (5.21), there is a corresponding fast rise time in $v_2$ that is approximately equal to the reciprocal of the receive channel bandwidth. Thus, if $v_2$ is to satisfy the assumption of an input step function to the *RC* filter, the *RC* filter bandwidth must not exceed approximately 10% of the receiver channel bandwidth. This bandwidth relation between the *RC* filter and the receiver channel is considered further in the servo-loop model discussion.

Substituting (5.25) and (5.23) into (5.24), we then can write the filter output voltage as

$$\tau_0 \frac{dz_2}{dt} + z_2\left[1 + k^2 G |\overline{x}_2|^2\right] = k^2 G |\overline{x}_2|^2 \cdot \left[ b_2^* + \frac{w_1(\overline{x_1 x_2^*})}{|\overline{x}_2|^2} \right] \tag{5.26}$$

The differential equation represented by (5.26) has a solution similar to that of (5.12):

$$z_2(t) = \left[z_2(0+) - z_2(\infty)\right] e^{-\alpha t} + z_2(\infty) \tag{5.27}$$

where

$$z_2(\infty) = \frac{k^2 G |\overline{x}_2|^2 \{ b_2^* + [w_1(\overline{x_1 x_2^*})/|\overline{x}_2|^2] \}}{1 + k^2 G |\overline{x}_2|^2} \tag{5.28}$$

and

$$\alpha = \frac{1 + k^2 G |\overline{x}_2|^2}{\tau_0} \tag{5.29}$$

The overbars in (5.26)–(5.29) indicate that integration (or averaging) has taken place in accordance with the filter closed-loop bandwidth. The quantity $z_2(0+)$ is the initial value of the filter output voltage $z_2(t)$ at $t = 0+$, and $z_2(\infty)$ is the steady-state value after the transient response has decayed. This solution for $z_2(t)$ may be substituted into (5.23) to obtain a solution for the adaptive weight $w_2(t)$.

Define the quantity $w_{2_{\text{opt}}}$ as the optimum value of the weight $w_2$; namely, it is that value of $w_2$ that minimizes the array output noise power. The array output noise power is the sum of the quiescent receiver noise plus the external interference noise, weighted by the array weights $w_1$ and $w_2$. If $y_n$ is the array output noise voltage, then the mean square of $y_n$ (which is to be minimized) is given by

$$\overline{|y_n|^2} = \overline{|(w_1 x_1) + (w_2 x_2)|^2} \tag{5.30}$$

Setting the partial derivative of (5.30) with respect the weight $w_2$ equal to zero, we see that the optimum value of $w_2$ required to minimize (5.30) is given by

$$w_{2_{\text{opt}}} = -\frac{\overline{(w_1 x_1) x_2^*}}{\overline{|x_2|^2}} = -\frac{w_1 \overline{(x_1 x_2^*)}}{\overline{|x_2|^2}} \tag{5.31}$$

Recalling the discussion of Section 5.1.1, we see that $w_{2_{\text{opt}}}$ is the (normalized) retrodirective weight that results in placing an adapted spatial pattern null in the direction of the external interference source.

Referring to (5.28), we see that the expression for $z_2(\infty)$ contains the relationship (5.31) for $w_{2_{\text{opt}}}$ (assuming the same averaging effect is present), so the steady-state correlator output may be rewritten as

$$z_2(\infty) = \frac{k^2 G |\bar{x}_2|^2 \left(b_2^* - w_{2_{\text{opt}}}\right)}{1 + k^2 G |\bar{x}_2|^2} \tag{5.32}$$

### 5.2.1 *Servo-Loop Model of the Adaptive Loop*

It turns out that the adaptive loop of Figure 5.5 exhibits behavior that is similar to that of the Type-0 follower servo depicted in Figure 5.6a. The correspondence between a Type-0 servo loop and the Howells-Applebaum adaptive loop can be established by considering the equations describing the servo-loop behavior. First note that a Type-0 unit feedback circuit is inherently stable. By adoption of the same approach used to perform the transient analysis for the $RC$ integrating filter, the servo follower differential equation relating the output voltage $v_0$ and the input voltage $\mu\epsilon = \mu(v_i - v_0)$ can be written as

$$\tau_0 \frac{dv_0}{dt} + v_0 = \mu\epsilon = \mu(v_i - v_0) \tag{5.33}$$

or

$$\tau_0 \frac{dv_0}{dt} + (1 + \mu) v_0 = \mu v_i \tag{5.34}$$

Assuming that $v_i$ is an input step function, then the solution for $v_0$ can be written as

$$v_0 = \left[ v_0(0+) - \left( \frac{\mu v_i}{1 + \mu} \right) \right] e^{-\alpha t} + \frac{\mu v_i}{1 + \mu} \tag{5.35}$$

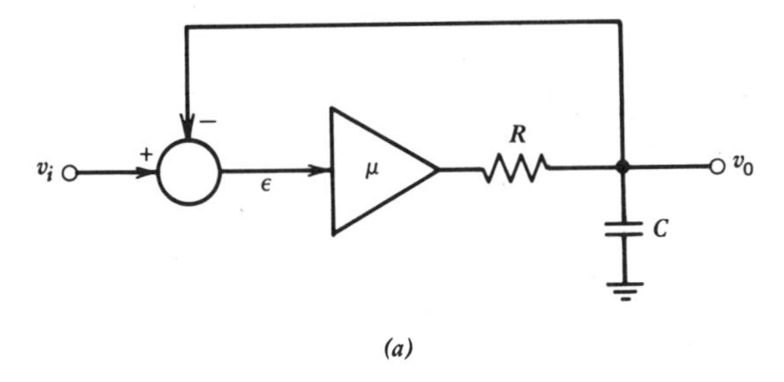


(a)

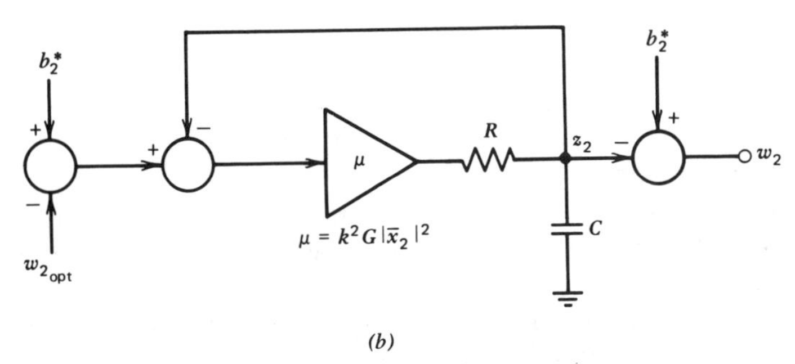


(b)

**Figure 5.6** Schematic diagrams for Type-O servo loop and equivalent servo-loop model of single adaptive loop. (*a*) Type-O follower servo. (*b*) Equivalent servo-loop model of Howells-Applebaum Adaptive Loop. From Gabriel, *Proc. IEEE*, February 1976.

where

$$\alpha = \frac{1+\mu}{\tau_0} \tag{5.36}$$

Compare (5.34), (5.35), and (5.36) with (5.26), (5.27), and (5.29), and it follows immediately that direct correspondences exist between:

$v_0$ and $z_2$

$\mu$ and $k^2 G|\bar{v}_2|^2$

$v_i$ and $(b_2^* - w_{2_{opt}})$

Because of the above correspondences, the adaptive loop of Figure 5.5 can be modeled by the equivalent servo loop shown in Figure 5.6b, so the equation for $z_2(\infty)$ becomes

$$z_2(\infty) = \frac{\mu}{1+\mu}\left(b_2^* - w_{2_{opt}}\right) \tag{5.37}$$

where

$$\mu = k^2 G |\bar{x}_2|^2 \tag{5.38}$$

The equivalent servo-loop concept is very revealing because it illuminates the relationship between the optimum weight $w_{2_{\text{opt}}}$, the servo gain factor $\mu$, and the degree to which the actual output $w_2(t)$ approaches the optimum. For values of $\mu \gg 1$, the steady-state value of $w_2$ is equal to $w_{2_{\text{opt}}}$. The equivalent servo gain factor defined by (5.38) is, however, proportional to the average power level from sensor element 2. This distinctive power-sensitive gain behavior has two important consequences as follows:

**1** It is necessary to establish a minimum value for $\mu$ based on the quiescent receiver noise power.

**2** It is necessary to establish a maximum value for $\mu$ based on the maximum interference power to be received at the sensor elements.

Since the minimum signal level of $x_2$ is just equal to the quiescent receiver noise voltage $n_2$ at the mixers, then

$$\text{minimum } \mu = \mu_{\text{min}} = k^2 G |\bar{n}_2|^2 \tag{5.39}$$

Therefore by choosing an appropriate value for the amplification gain $G$, the value of $\mu_{\text{min}}$ can be set to any convenient level.

Recalling that $k$ is the correlation mixer conversion constant, we see that the quantity $k^2|\bar{n}_2|^2$ appearing in (5.39) represents the output voltage from the correlation mixer due to $n_2$ alone. This output voltage is proportional to the noise power, so it is important to use enough preamplification to ensure the dominance of $|\bar{n}_2|^2$ over the thermal noise voltage generated internally within the correlation mixer.

The quiescent noise level of $n_2$ also determines the magnitude of the beam-steering signal $b_2^*$. This fact may be seen by first substituting (5.37) into (5.23) to yield

$$w_2(\infty) = b_2^* - \frac{\mu}{1+\mu}\left(b_2^* - w_{2_{\text{opt}}}\right) \tag{5.40}$$

In the quiescent signal condition $x_1 = n_1$ and $x_2 = n_2$. Since these independent noise voltages have zero average cross-correlation, that is,

$$E\{n_1 n_2^*\} = 0 \tag{5.41}$$

it follows from (5.31) that $w_{2_{\text{opt}}}$ is zero. Furthermore the quiescent value of $\mu$ is $\mu_{\min}$, so the quiescent value of $w_2(\infty)$ becomes:

$$\text{quiescent } w_2(\infty) = \frac{1}{1+\mu_{\min}} b_2^* = \frac{c_2}{1+\mu_{\min}} w_q \tag{5.42}$$

Recalling that the quiescent steady-state value of $w_2$ is by definition $w_q$, we see that (5.42) yields the result

$$c_2 = 1 + \mu_{\min} \tag{5.43}$$

Since the quiescent noise level of $n_2$ determines $\mu_{\min}$, it therefore also determines $c_2$ and hence the magnitude of the beam-steering signal $b_2^*$.

Now consider the effect on $\mu$ of increasing the interference power level at the sensor elements in order to determine a maximum value for $\mu$. From (5.36) and (5.38) it immediately follows that as the interference power increases, the ratio defined by

$$\frac{\text{response time}}{\text{integration time}} = \frac{1}{1+\mu} \tag{5.44}$$

decreases, with the result that less and less averaging of the input signals occurs until the conditions assumed for the simple transient solution of the differential equation (5.26) are no longer valid.

Once the ratio (response time/integration time) decreases and the input signal averaging effect of the integrating filter is degraded, then both $\mu$ and $w_2$ tend to follow the fast fluctuations in the envelopes of the element signals thereby causing the weight $w_2$ to become "noisy." In Section 5.3.2 the theory for control-loop noise is presented and expressions are given for the variance of the array element weights and for the additional noise appearing in the array output as a consequence of this element weight fluctuation. It is appropriate at this point to consider an upper-bound condition for avoiding the "noisy loop" problem.

To ensure that the parameters $\mu$ and $w_2$ do not follow fast fluctuations of the signal envelope, it is necessary that the closed-loop two-sided integrating filter bandwidth not exceed approximately one-tenth the bandwidth of the element signal channels. This bandwidth restriction ensures that enough integration time is available to average out any rapid fluctuations in $\mu$ thereby permitting $w_2$ to be reasonably (statistically) independent from any instantaneous fluctuations of the signal envelopes. For the simple servo loop of Figure 5.6, the closed-loop (filter) bandwidth is $\alpha$. If the sensor element signal channel bandwidth is denoted by $B_c$, then the upper-bound filter bandwidth restriction may be expressed as

$$2\alpha_{\max} \leqslant \frac{1}{10} 2\pi B_c \tag{5.45}$$

or [using (5.36)],

$$\mu_{\max} \leqslant \frac{\pi B_c \tau_0}{10} - 1 \tag{5.46}$$

where $\mu_{\max}$ is the maximum allowable value for the servo gain factor.

It is both convenient and informative to express the gain factor $\mu$ in terms of $\mu_{\min}$ and a power ratio; to do this note that when an interference signal is present, then $|\bar{x}_2|^2$ will be the sum of the squares of the magnitudes of $n_2$ and the interference signal $J_i$, so that

$$|\bar{x}_2|^2 = |\bar{n}_2|^2 + |\bar{J}_i|^2 \tag{5.47}$$

Now since $\mu$ is given by (5.38), it follows that

$$\frac{\mu}{\mu_{\min}} = \frac{k^2 G |\bar{x}_2|^2}{k^2 G |\bar{n}_2|^2} = \frac{|\bar{n}_2|^2 + |\bar{J}_i|^2}{|\bar{n}_2|^2} = 1 + \frac{|\bar{J}_i|^2}{|\bar{n}_2|^2} \tag{5.48}$$

The ratio of the squares of the voltage magnitudes in (5.48) is, of course, just the ratio of the interference signal power to the receiver noise power. If we denote this power ratio by $PR_i$, then from (5.48) $\mu$ can be expressed as

$$\mu = \mu_{\min}(1 + PR_i) \tag{5.49}$$

Combining the results of (5.49) and (5.46) then yields

$$\mu_{\max} = \mu_{\min}(1 + PR_{i_{\max}}) \leqslant \frac{\pi B_c \tau_0}{10} - 1 \tag{5.50}$$

or

$$\tau_0 \geqslant \frac{10}{\pi B_c}\left[1 + \mu_{\min}(1 + PR_{i_{\max}})\right] \tag{5.51}$$

Equation (5.51) relates the integrating filter time constant $\tau_0$ to the maximum interference power to be handled, since the channel bandwidth $B_c$ is generally fixed and cannot easily be changed. To illustrate the above result, select a maximum interference signal power that is 40 dB above the receiver noise level and a 50 MHz channel bandwidth. Assume $\mu_{\min} = 1$, then from (5.51) it follows that $\tau_0 \geqslant 6.37 \times 10^{-3}$ sec. The filter bandwidth corresponding to $\tau_0$ is $(1/2\pi\tau_0)$ and is therefore about 25 Hz.

The relationship between the servo-loop gain and the servo-loop bandwidth is most easily illustrated by a Bode plot [11] as in Figure 5.7, where the servo-loop gain is plotted versus frequency. In a Bode plot, the breakpoint for the 20-dB-per-decade slope line occurs at the *RC* filter bandwidth point $\omega = 1/\tau_0$, and an indication of the overall 3-dB servo-loop bandwidth is given

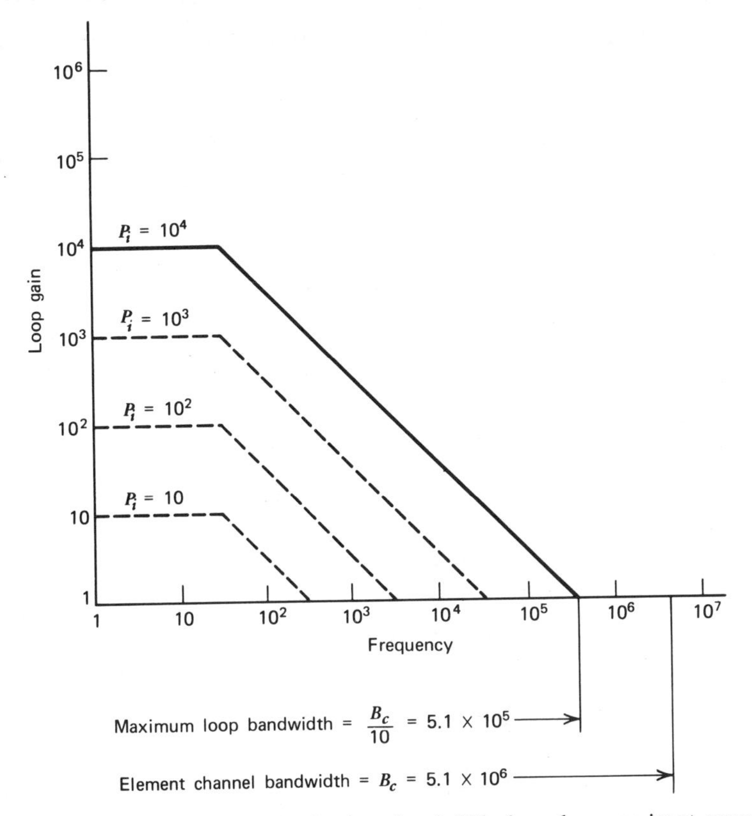


**Figure 5.7** Bode plot illustrating adaptive loop bandwidth dependence on input power level. From Gabriel, *Proc. IEEE*, February 1976.

by the intersection of the slope line with the unity gain axis which occurs approximately at $\omega = \mu/\tau_0$.

The servo-loop model of the adaptive control loop for weight adjustment discussed above has introduced the useful variables and constants $\mu$, $\mu_{\min}$, $PR_i$, and $PR_{i_{\max}}$, which facilitate the analysis of the adaptive loop performance. It is therefore desirable to rewrite the equations for $w_2(t)$ and $w_{2_{\text{opt}}}$ in terms of these new parameters. From (5.27) it follows that

$$w_2(t) = \left[ w_2(0+) - w_2(\infty) \right] e^{-\alpha t} + w_2(\infty) \tag{5.52}$$

From (5.23) it likewise follows that

$$w_2(\infty) = b_2^* - z_2(\infty) \tag{5.53}$$

Substituting (5.37) into (5.53) then gives

$$w_2(\infty) = \frac{1}{1+\mu} b_2^* + \frac{\mu}{1+\mu} w_{2_{\text{opt}}} \tag{5.54}$$

It will be noted from (5.54) that the steady-state value of $w_2$ consists of two distinct components:

$$\text{beam-steering component} = \frac{b_2^*}{1+\mu} \tag{5.55}$$

and

$$\text{retrodirective component} = \frac{\mu}{1+\mu} w_{2_{\text{opt}}} \tag{5.56}$$

Recalling that $b_2^* = (1+\mu_{\min})w_q = (1+\mu_{\min})w_1^*$, we may also write the beam-steering component as

$$\text{beam-steering component} = \frac{1+\mu_{\min}}{1+\mu} w_1^* \tag{5.57}$$

From (5.31)

$$w_{2_{\text{opt}}} = -\frac{w_1(\overline{x_1 x_2^*})}{|\bar{x}_2|^2} \tag{5.58}$$

and from (5.21) it follows that

$$\overline{x_1 x_2^*} = |\bar{J}_i|^2 e^{-j2u_i} \tag{5.59}$$

since $n_1$ and $n_2$ are zero mean statistically independent random processes. Recalling that $PR_i \triangleq |\bar{J}_i|^2/|\bar{n}_2|^2$ and $1+PR_i = \mu/\mu_{\min}$, $w_{2_{\text{opt}}}$ can then be rewritten as

$$w_{2_{\text{opt}}} = -w_1\left(\frac{\mu_{\min}}{\mu}\right) PR_i e^{-j2u_i} \tag{5.60}$$

where $u_i$ is given by (5.22). Furthermore,

$$\alpha = \frac{1+\mu}{\tau_0} \qquad \text{and } \mu = \mu_{\min}(1+PR_i) \tag{5.61}$$

The foregoing results clearly show that when $PR_i$ is nearly zero, then $w_{2_{\text{opt}}}$ is also close to zero, and the beam-steering component of $w_2(\infty)$ is dominant. Likewise as $PR_i$ increases, $\mu$ also increases, and the beam-steering component is attenuated while the retrodirective component magnitude increases. For large values of the ratio $PR_i$ where $\mu \gg 1$, the beam-steering component of $w_2(\infty)$ becomes negligible and the interference source retrodirective component dominates, thereby "capturing" $w_2$.

### 5.2.2 *Illustrative Example of Adaptive Loop Performance*

To illustrate the concepts introduced in the foregoing sections, it is useful to consider how the two-element array of Figure 5.5 responds to a specified set of signal conditions. The results obtained will help to emphasize important performance characteristics and their relation to selected constants and signal conditions.

***Assumptions and Initial Conditions.*** In conformity with (5.20) it is assumed that quiescent receiver noise prevails up to time $t=0$, when a single narrow-band interference signal source is switched on like a step function. The ratio $PR_i$ of interference signal power to receiver noise power is assumed to be limited to 40 dB. For a receiver channel bandwidth $B_c$ of 5 MHz and a minimum servo gain factor $\mu_{\min}=1$, the integrating filter time constant is $\tau_0=6.37$ millisec.

Assume that the two sensor elements are omnidirectional and are spaced apart by one-half wavelength so $d=\lambda/2$ and the resulting expressions for $u_0$, $u$, and $u_i$ in (5.16), (5.18), and (5.22) are thereby simplified. The initial value of $w_2(w_q)$, assuming the desired signal direction is $\theta_0=0°$, is then $1\angle 0°$.

***Transient Behavior of the Adaptive Weight $w_2$.*** The transient behavior of the adaptive weight $w_2$ can be calculated from (5.52) through (5.61), using the assumptions and initial conditions given above.

Once the selected constants have been chosen, the transient behavior of $w_2$ depends on two variable factors: the obvious power ratio $PR_i$ contained in both $\mu$ and $w_{2_{\text{opt}}}$, and the less obvious phase angle rotation that $w_2$ experiences in reaching the steady-state adapted value. Figure 5.8 illustrates the typical transient behavior of $w_2$ for an interference source with power ratio $PR_i=100$

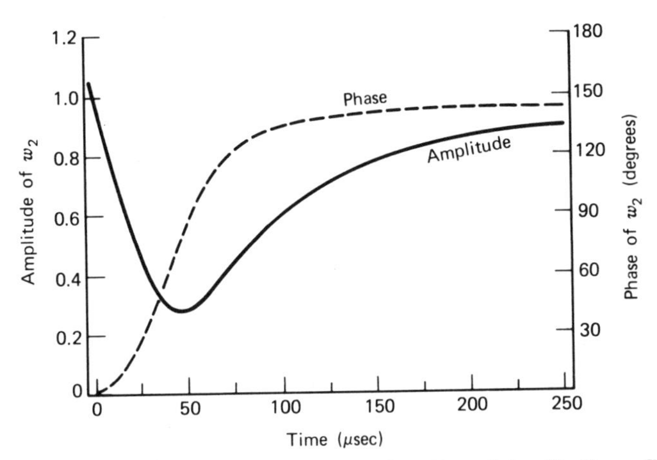


**Figure 5.8** Transient response of $w_2$ for $PR_i=100$, $\theta_i=10°$, and $\theta_0=0°$. From Gabriel, *Proc. IEEE*, February 1976.

located at $\theta_i = 10°$ off boresight so that $w_2$ must rotate through a phase angle of 148° before the angle appropriate for $w_2(\infty)$ is reached. If the interference source location were shifted from 10° over to the quiescent pattern null at $\theta = 90°$, then $w_2(0+)$ is already the correct value for $w_{2_{\text{opt}}}$, and no transient would occur. The only effect of changing the power ratio $PR_i$ is to change the time scale of the transient. For example, if the interference source for the example plotted in Figure 5.8 were increased in power level from $PR_i = 100$ to $PR_i = 1000$, the resulting transient curves would still appear as plotted in that figure with the time scale now divided by a factor of 10; the increase in interference source power level by a factor of 10 causes the adaptive loop to respond 10 times faster in adjusting $w_2$.

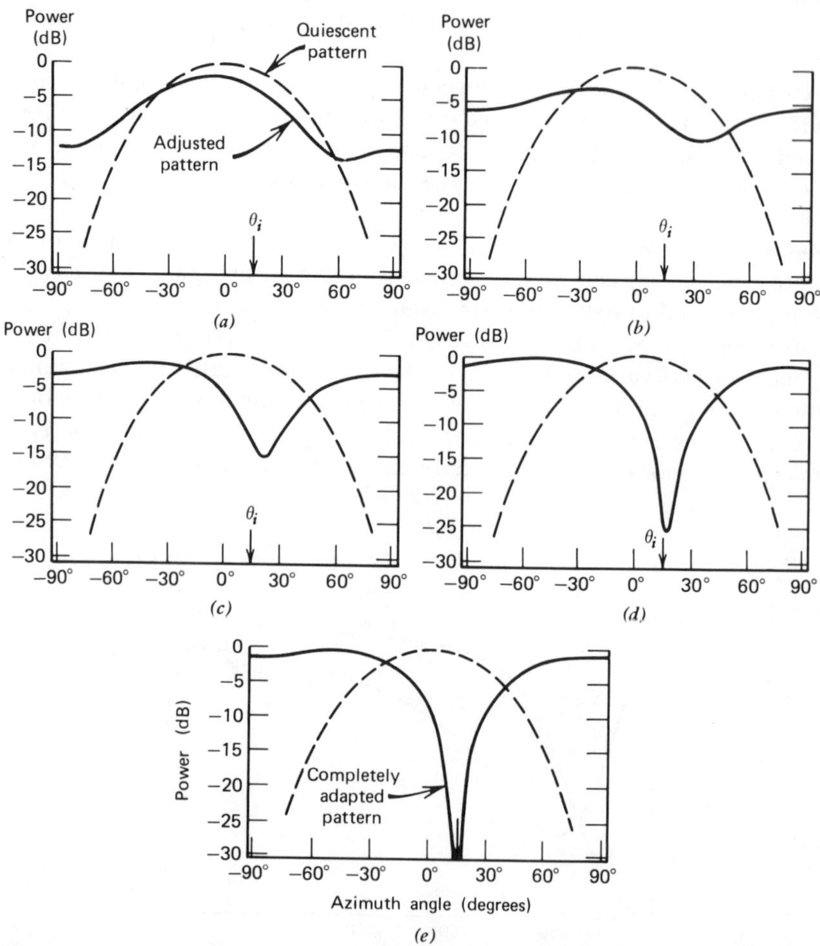


**Figure 5.9** Transient behavior of array beam pattern for $PR_i = 100$, $\theta_0 = 0°$, and $\theta_i = 15°$. (*a*) $t = 20$ μsec; (*b*) $t = 50$ μsec; (*c*) $t = 100$ μsec; (*d*) $t = 200$ μsec; (*e*) steady-state. From Gabriel, *Proc. IEEE*, February 1976.

***Transient Behavior of the Array Spatial Beam Pattern.*** The spatial beam pattern of the two-element array at different times during the transient adjustment phase can easily be obtained by fixing the weight $w_2$ at selected instants of time during its transient change and calculating the resulting beam pattern (represented by the array output voltage $y$) produced by a far-field test source located at the spatial angle $\theta$. The array beam pattern is directly proportional to $|y|$, and for convenience in plotting it is generally normalized to the maximum value of $y$. The normalized spatial-array beam pattern function $G(\theta,t)$ can be written as

$$G(\theta,t)=\tfrac{1}{2}\left[w_1 e^{-ju}+w_2(t)e^{ju}\right] \tag{5.62}$$

Figure 5.9 gives five plots of $G(\theta,t)$ at various different times ($t=20, 50, 100, 200$ $\mu$sec, and steady state) under the conditions $PR_i=100$, $\theta_0=0°$, and $\theta_i=15°$. The quiescent beam-steered pattern at $t=0$ is repeated in each plot, and the successive plots demonstrate the progressive development of a pattern null in the direction of the interference signal source at $\theta_i=15°$.

***Transient Behavior of the Output Noise Power.*** The performance measure that is of ultimate interest when using the Howells-Applebaum algorithm is the improvement in the output SNR as compared to a conventional array subjected to the same interference conditions. In obtaining the SNR, the signal component can be readily obtained from the change produced in the beam pattern $G(\theta,t)$. The output noise component (which forms the denominator of the SNR) is more fundamental to the improvement that can be obtained and is consequently often adopted as the performance measure itself.

The output noise power is the sum of the quiescent receiver noise and the interference noise, weighted by the array weights $w_1$ and $w_2$. From the expressions for the signals $x_1$ and $x_2$ given by (5.21), the output noise voltage $y_n$ can be written as

$$\begin{aligned} y_n &= w_1 x_1 + w_2 x_2 \\ &= w_1 n_1 + w_2 n_2 + J_i\left[w_1 e^{-ju_i}+w_2 e^{ju_i}\right] \end{aligned} \tag{5.63}$$

Since $n_1$ and $n_2$ are independent receiver noise sources, $w_1 n_1 + w_2 n_2$ can be written as $\sqrt{\overline{|w_1 n_1|^2+|w_2 n_2|^2}}$ . This square root term can be further simplified because the RMS amplitudes of $n_1$ and $n_2$ are assumed to be equal. The value of this square root term prior to $t=0$ represents the quiescent output receiver noise voltage, and since both $w_1$ and $w_2(0+)$ have unity amplitude,

$$|n_2|\sqrt{|w_1|^2+|w_2|^2}\,\Big|_{t=0}=\sqrt{2}\,|n_2| \tag{5.64}$$

The increase in the output noise power that results when an interference source is turned on can be expressed as a ratio of the square of the amplitude

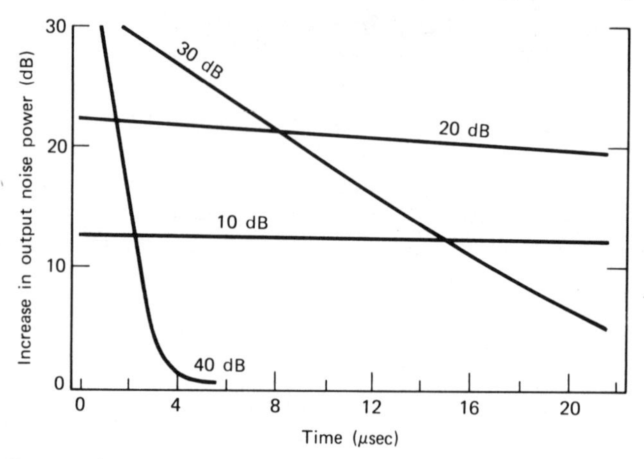


**Figure 5.10** Output noise power transient behavior of one adaptive loop when $\theta_0=0°$, $\theta_i=15°$, for $PR_i$ values of 10, 20, 30, and 40 dB. From Gabriel, *Proc. IEEE*, February 1976.

of $y_n$ to the square of the quiescent output receiver noise voltage:

$$\frac{|y_n|^2}{2|n_2|^2}=\frac{1}{2}\left[1+|w_2|^2+\frac{|J_i|^2}{|n_2|^2}\right.$$

$$\left.\cdot\left|w_1e^{-ju_i}+w_2e^{ju_i}\right|^2\right]$$

$$=\frac{1}{2}\left[1+|w_2|^2+PR_i\left|\left(w_1e^{-ju_i}+w_2e^{ju_i}\right)\right|^2\right] \tag{5.65}$$

The plot of (5.65) as a function of time appearing in Figure 5.10 illustrates the decibel increase in output noise power when the interference source is switched on at $t=0$ for $\theta_0=0°$ and $\theta_i=15°$, for power ratios of 10, 20, 30, and 40 dB. These curves clearly show the direct dependence of the speed of the transient response on the power ratio $PR_i$.

***Interference Signal Bandwidth Effects.*** The bandwidth of the interference signal affects the adaptive array performance because the array response is frequency sensitive. At any instant of time the adaptive weight $w_2$ can assume only one amplitude value and one phase value. The delay (or advance) of an incoming signal at one element with respect to the other involves an actual time-delay distance, as illustrated in Figure 5.11, where the time-delay distance $u_i$ is referenced to the geometric phase center of the array. For a given wavelength $\lambda$, $u_i$ is defined by (5.22). An element spacing of $\lambda/2$ has been assumed which will now be taken to mean $\lambda_0/2$ corresponding to the center frequency $f_0$ of the RF bandwidth of the interference signal. One may

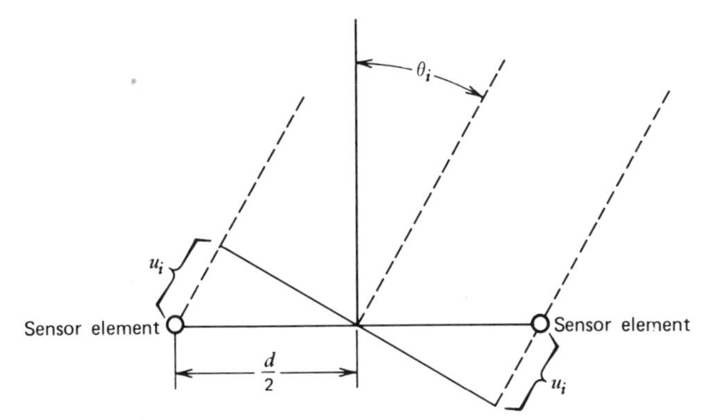


**Figure 5.11** Diagram of actual time-delay distance $u_i$ for two-element array. From Gabriel, *Proc. IEEE*, February 1976.

therefore write

$$d = \frac{\lambda_0}{2} \tag{5.66}$$

so that for a signal having instantaneous wavelength $\lambda$,

$$u_i = \frac{\lambda_0}{\lambda} \cdot \frac{\pi}{2} \sin\theta_i = \frac{f}{f_0} \cdot \frac{\pi}{2} \sin\theta_i \tag{5.67}$$

The instantaneous frequency $f$ can be further defined as equal to the center frequency $f_0$ plus an offset $\Delta f$ so that $u_i$ finally becomes

$$u_i = \left(1 + \frac{\Delta f}{f_0}\right) \frac{\pi}{2} \sin\theta_i \tag{5.68}$$

The foregoing frequency-dependent relationship for $u_i$ provides the means for handling a broadband interference source by dividing the interference power into a number of equally spaced discrete spectral lines. The $m$th discrete spectral line then has its associated offset frequency $\Delta f_m$, voltage amplitude at the mixers $J_m$, and power ratio $PR_m$. It is particularly convenient to assume a uniform amplitude spectrum of equally spaced lines as shown in Figure 5.12. Furthermore assume that the spectral lines do not cross-correlate with one another. With the above specifications and assuming a total of $M$ spectral lines, we may rewrite the element signals $x_1$ and $x_2$ in (5.21) as:

$$\left.\begin{aligned} x_1 &= n_1 + \sum_{m=1}^{M} J_m e^{-ju_m} \\ x_2 &= n_2 + \sum_{m=1}^{M} J_m e^{ju_m} \end{aligned}\right\} \quad \text{for } t \geqslant 0 \tag{5.69}$$

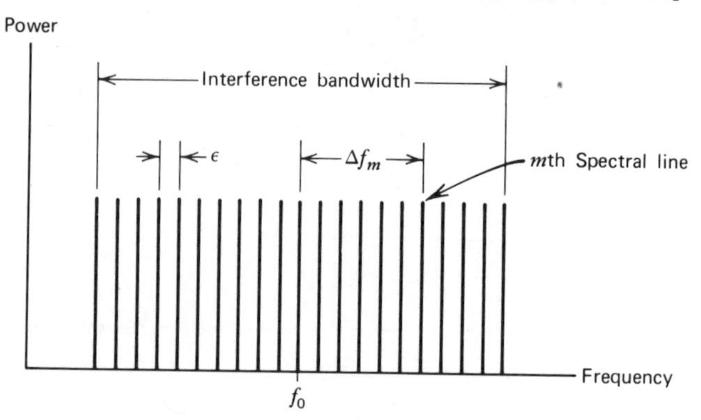


**Figure 5.12** Interference source power spectrum having uniform amplitude spectral lines spaced $\varepsilon$ apart. From Gabriel, *Proc. IEEE*, February 1976.

From the cross-correlation properties that exist between the various signal components in (5.69), it follows that

$$(\overline{x_1 x_2^*}) = \sum_{m=1}^{M} |\bar{J}_m|^2 e^{-j2u_m} \tag{5.70}$$

Therefore from (5.58) and (5.60) there results

$$w_{2_{\text{opt}}} = -\frac{w_1(\overline{x_1 x_2^*})}{|\bar{x}_2|^2} = -w_1 \frac{\mu_{\min}}{\mu} \sum_{m=1}^{M} PR_m e^{-j2u_m} \tag{5.71}$$

The expression for $\mu$ remains the same as in (5.61), where $PR_i$ for this case is interpreted as the sum of all the spectral-line power ratios $PR_i = \Sigma_m PR_m$.

Although $w_{2_{\text{opt}}}$ of (5.71) is optimum, it no longer represents a perfect solution for $w_2$ because there is now a different phase angle $2u_m$ associated with each individual spectral-line component, so there obtains a single resultant vector representing the sum of all the individual vector contributions. Therefore $w_{2_{\text{opt}}}$ represents an adaptation to the *power centroid* of the interference spectrum, rather than an adaptation in response to each individual spectral-line component.

The increase in output noise power resulting when a broadband interference source is turned on can be expressed by a ratio similar to (5.65) as follows:

$$\frac{|y_n|^2}{2|n_2|^2} = \frac{1}{2}\left[1 + |w_2|^2 + \sum_{m=1}^{M} PR_m |(w_1 e^{-ju_m} + w_2 e^{ju_m})|^2\right] \tag{5.72}$$

The sum term in (5.72) involves the summation of the individual power residues corresponding to the various spectral lines of the broadband interference source.

A plot of (5.72) versus interference bandwidth is given in Figure 5.13 for $PR_i = 2000$, $\theta_i = 45°$, and $\theta_0 = 0°$. The phase of the steady-state $w_2$ is also

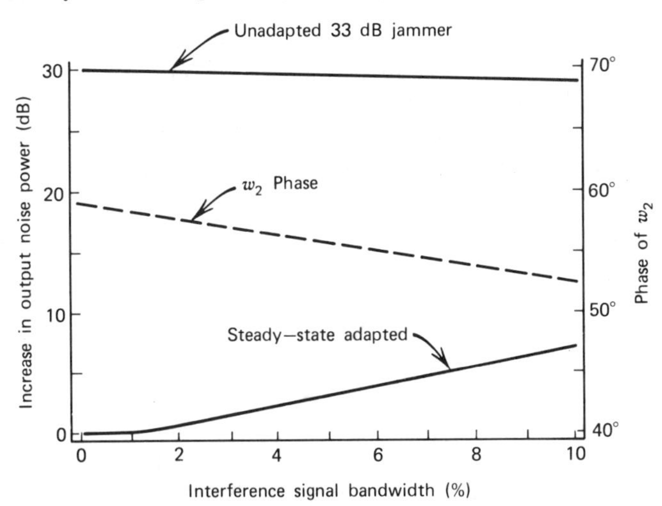


**Figure 5.13** Output noise power versus interference signal bandwidth for $PR_i = 2000$, $\theta_i = 45°$, and $\theta_0 = 0°$. From Gabriel, *Proc. IEEE*, February 1976.

plotted along with the output noise that would result from the quiescent beam pattern alone. The abscissa scale is percentage bandwidth, and each successive point represents an increase of one spectral line, which increases the overall interference signal bandwidth by 0.2%. For example, starting at the origin there is only a single interference spectral line where the entire power ratio of 2000 is concentrated, and the steady-state adapted weight $w_2(\infty)$ nulls out this interference source to a negligible residue. The next abscissa point adds a spectral line at a frequency 0.2% higher than the first line, giving an overall "bandwidth" of 0.2% for the two lines and yielding a power ratio of 1000 for each line. Likewise, the next point adds a third spectral line 0.2% higher in frequency than the second, resulting in an overall "bandwidth" of 0.4% for the three lines with a power ratio of 666 for each line.

Continuing in this manner, a total of 51 spectral lines are accumulated, resulting in a total overall bandwidth of 10% and a power ratio of 39.2 per line. The results shown in Figure 5.13 clearly show the progressive deterioration (increase) in output noise power as the constant total power of the interference source is spread over an increasingly wider bandwidth. It is worth noting that in order to be meaningful, the bandwidth of the interference source must always be entirely contained within the fixed channel bandwidth $B_c$.

### *5.2.3 Hard-Limiter Modification*

The adaptive loop configuration of Figure 5.5 has two drawbacks that derive from the fact that the output voltage $v_2$ from the correlation mixer is proportional to the power received at the sensor elements.

1 The dynamic range of the voltage $v_2$ is the square of the element signal dynamic range. Consequently, to handle a 40 dB range of interference power requires branch components having a linear dynamic range of 80 dB, which is very close to practical component limits. Therefore a severe restriction is placed on the system input signal dynamic range.

2 Since the speed of the transient response is proportional to interference signal power (as shown in Figure 5.10), this results in very sluggish response for weak sources of interference.

The need to limit the required dynamic range and the desire to increase the speed of the transient response to low-level interference sources leads to the suggestion of changing the output voltage dependence on power received by introducing a hard limiter; the incorporation of a hard limiter in the complex conjugate signal branch of the correlation mixer is the best solution so far discovered for alleviating these drawbacks [6]. This hard-limiter modification is introduced into the complex conjugate branch as shown in Figure 5.14. The constant amplitude level of the hard limiter output is denoted by $h$.

The hard-limiter modification changes the equations developed in the preceding sections because the correlation mixer product is now given by

$$v_2' = k^2(w_1x_1 + w_2x_2)\frac{hx_2^*}{|x_2|} \tag{5.73}$$

The amplitude variation in the conjugate input to the correlation mixer has been removed, and only the phase variation is retained (since a hard limiter preserves phase information). Whereas the loop gain used to be given by (5.38), it now becomes

$$\mu' = k^2G'x_2 \cdot \frac{hx_2^*}{|x_2|} = hk^2G'|\bar{x}_2| \tag{5.74}$$

so the loop gain is now proportional to voltage level rather than to power level. The amplifier gain $G'$ is again selected according to (5.39).

As before, it is convenient to express $\mu'$ in terms of $\mu'_{\min}$ and a power ratio as in (5.48)

$$\frac{\mu'}{\mu'_{\min}} = \frac{hk^2G'|\bar{v}_2|}{hk^2G'|\bar{n}_2|} = \sqrt{\frac{|\bar{n}_2|^2 + |\bar{J}_i|^2}{|\bar{n}_2|^2}} = \sqrt{1 + PR_i} \tag{5.75}$$

Therefore the ratio $\mu'/\mu'_{\min}$ is simply the square root of the previous expression obtained without the limiter.

The relationship between the maximum interference power and the $RC$ filter time constant $\tau_0'$ is given in the same manner as (5.50)

$$\mu'_{\max} = \mu'_{\min}\sqrt{1 + PR_{i_{\max}}} \leqslant \frac{\pi B_c \tau_0'}{10} - 1 \tag{5.76}$$

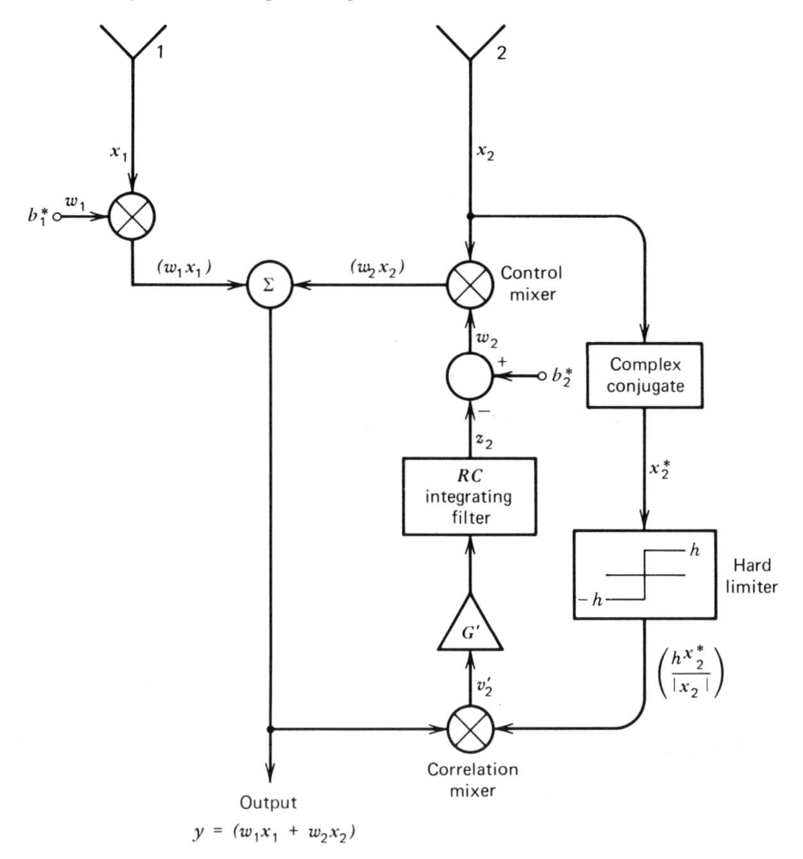


**Figure 5.14** Hard-limiter modification of two-element array with single adaptive control loop. From Gabriel, *Proc. IEEE*, February 1976.

If it is again assumed that the maximum interference power ratio is 40 dB and the channel bandwidth is 5MHz, then for $\mu'_{\text{min}} = 1$ the value of $\tau'_0$ is 64.3 $\mu$sec, and the corresponding filter bandwidth is 2475 Hz. It is also worth noting that since $\mu'_{\text{max}}$ involves the square root of $PR_{i_{\text{max}}}$ instead of $PR_{i_{\text{max}}}$ directly, the loop transient response is now faster by a factor of 100.

The optimum weight $w_{2_{\text{opt}}}$ given by (5.31) remains the same, but when expressed in terms of the new servo factor $\mu'$ it is then given by

$$w_{2_{\text{opt}}} = -\frac{w_1\left(\overline{x_1x_2^*}\right)}{|\bar{x}_2|^2} = -w_1\left(\frac{\mu'_{\text{min}}}{\mu'}\right)^2 PR_i e^{-j2u_i} \tag{5.77}$$

The appropriate equations for $w'_2(t)$ are the same as (5.52) and (5.56) with $\mu'$ replacing $\mu$ in those expressions.

Because of the presence of the square root in $\mu'$ (5.75), considerable changes have occurred in both $\alpha'$ and $w'_2(\infty)$. The nature of these changes is

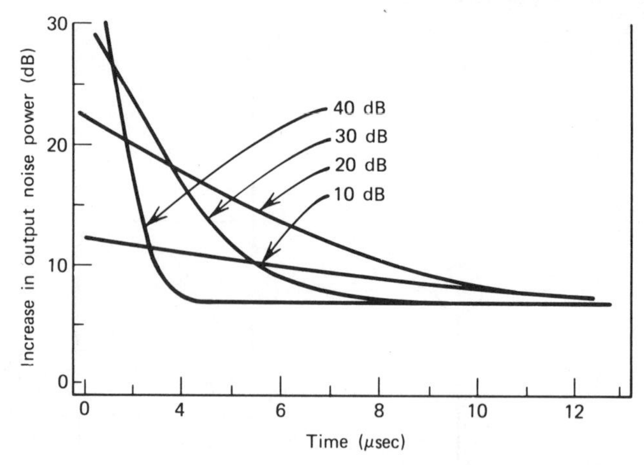


**Figure 5.15** Output noise power transient behavior with hard-limiter modification for $\theta_i=15°$, $\theta_0=0°$, and $PR_i=10$, 20, 30, and 40 dB. From Gabriel, *Proc. IEEE*, February 1976.

best illustrated by considering the transient behavior of the increase in output noise power represented by (5.65) for the same conditions under which Figure 5.10 was produced. The new transient curves resulting for the hard-limiter modification are shown in Figure 5.15, where $\theta_0=0°$, $\theta_i=15°$, and $PR_i=10$, 20, 30, and 40 dB. From Figure 5.15 it is seen that the transient response for $PR_i=40$ dB is identical with the corresponding response in Figure 5.10 because the maximum power condition for both cases was 40 dB, thereby yielding $\alpha'_m=\alpha=1.58\times 10^6(1/\text{sec})$. For $PR_i<40$ dB, however, the transient response decay is much faster with the limiter modification. It is tempting to conclude that the addition of a hard limiter will invariably lead to improved transient response, but when more than a single adaptive control loop is involved in the adaptive processor, it will be found that improved transient response does not always necessarily result.

Another very important difference that is manifest in Figure 5.15 is the rather high steady-state residue (approximately 7 dB compared to less than 1 dB in Figure 5.10) remaining after the transient behavior has decayed. This high level steady-state residue comes about because the new servo-loop gain $\mu'$ now involves the square root of $PR_i$ (instead of $PR_i$ directly) and consequently is much smaller than the previous value of $\mu$ for the same power-ratio values.

To reduce the large output noise residue to acceptable levels, it is only necessary to increase the quiescent loop gain $\mu'_{\text{min}}$ above the value previously chosen without the hard-limiter modification. Although from (5.76) it is seen that increasing $\mu'_{\text{min}}$ also (nearly directly) increases $\tau'_0$, the net effect on $\alpha'$ (and hence on the loop speed of response) is only minor as can be seen from (5.61). Therefore it is possible to improve the retrodirective component of $w'_2(\infty)$ by increasing $\mu'_{\text{min}}$ without paying any penalty in terms of increased response time. The steady-state increase in output noise power residue as a function of

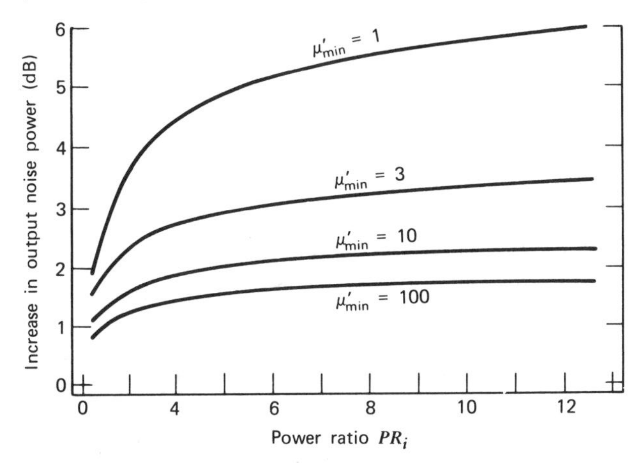


**Figure 5.16** Steady-state increase in output noise power with hard-limiter modification versus $PR_i$ under worst condition $\theta_i=\theta_0$ for quiescent loop gain $\mu'_{min}=1$, 3, 10, and 100. From Gabriel, *Proc. IEEE*, February 1976.

$PR_i$ for $\mu'_{min}=1$, 3, 10, and 100 is plotted in Figure 5.16. The results of Figure 5.16 indicate that it is desirable to maintain $\mu'_{min}\geqslant 10$ in order to hold the steady-state residue to an acceptably small value.

## 5.3 *N*-ELEMENT ARRAY WITH *N* ADAPTIVE LOOPS

Having considered the basic principles of operation and the performance characteristics of a single Howells-Applebaum adaptive loop, it is now appropriate to discuss the multiple-loop case in which each element of an *N*-element linear array has an associated adaptive loop for weight adjustment purposes. Begin by considering a multiple adaptive loop configuration in which each adaptive loop appears in the same manner as the single loop of Figure 5.5. An example of such a multiple-loop configuration is shown in Figure 5.17 for a six-element linear array.

Define an element signal vector $\mathbf{x}$ in which the $k$th component $x_k$ is similar to (5.21) and consists of the quiescent receiver channel noise voltage $n_k$ and a summation of voltage terms associated with $I$ external, narrowband interference sources:

$$\mathbf{x}^T=\left[x_1,x_2,\ldots,x_N\right] \tag{5.78}$$

where

$$x_k=n_k+\sum_{i=1}^{I} J_i e^{ju_i(2k-N-1)} \tag{5.79}$$

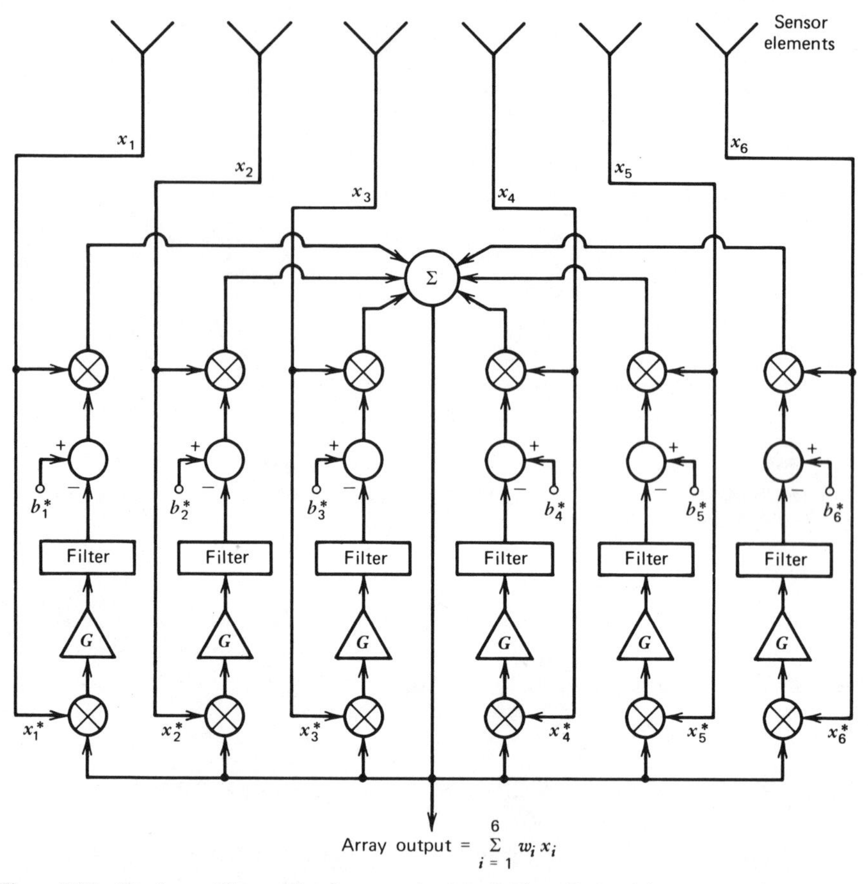


**Figure 5.17** Six-element linear adaptive array having six Howells-Applebaum control loops with beam-steering signals. From Gabriel, *Proc. IEEE*, February 1976.

and

$$u_i = \left(\frac{\pi d}{\lambda}\right)\sin\theta_i \tag{5.80}$$

The interference sources are assumed to be statistically independent where $J_i$ is the element channel voltage amplitude associated with the $i$th source and $\theta_i$ is its azimuth angle direction measured from the array boresight. It is furthermore assumed that a given interference source will produce equal voltage amplitudes at all of the array sensor elements. Each sensor element phase is referenced to the geometric center of the array.

As before, the beam-steering signals are selected to set up a shaped receive beam that is directed in a desired azimuth direction $\theta_0$. For quiescent conditions (when only receiver noise is present), the adaptive weights settle to steady-state values denoted by the quiescent weight vector $\mathbf{w}_q$ where

$$\mathbf{w}_q^T = [w_{q_1}, w_{q_2}, \ldots, w_{q_N}] \tag{5.81}$$

and

$$w_{q_k} = a_k e^{-ju_0(2k-N-1)} \tag{5.82}$$

where

$$u_0 = \left(\frac{\pi d}{\lambda}\right) \sin\theta_0 \tag{5.83}$$

The element coefficients $a_k$ have values selected to achieve the desired beam shape and desired sidelobe levels. The resulting quiescent array beam pattern can be expressed as

$$G_q(\theta) = (\mathbf{s}^T \mathbf{w}_q) = \sum_{k=1}^{N} a_k \exp[j(u-u_0)(2k-N-1)] \tag{5.84}$$

where $\mathbf{s}$ is a vector representing element signals of unit amplitude so that

$$\mathbf{s}^T = [s_1, s_2, \ldots, s_N] \tag{5.85}$$

where

$$s_k = \exp[ju(2k-N-1)] \tag{5.86}$$

and the phase factor $u$ is associated with the far-field angle variable $\theta$

$$u = \frac{\pi d}{\lambda} \sin\theta \tag{5.87}$$

The components of the input beam-steering vector $\mathbf{b}^*$,

$$\mathbf{b}^{*T} = [b_1^*, b_2^*, \ldots, b_N^*] \tag{5.88}$$

are directly related to the components of $\mathbf{w}_q$ by the relation

$$b_k^* = c_k w_{q_k} \tag{5.89}$$

where the constants $c_k$ are evaluated in the section immediately following.

### *5.3.1 Adaptive Weight Equations*

Each of the adaptive weight adjustment loops shown in Figure 5.17 is the same as the single loop of Figure 5.5, so the formation of the adaptive weight equations for the multiple-loop case can proceed in an exactly analogous manner. The weight $w_k$ associated with the $k$th sensor element is given by

$$w_k = b_k^* - z_k \tag{5.90}$$

Each correlation mixer voltage is once again given by the product of the signal $v_k^*$ with the summed array output:

$$v_k = k^2\left( x_k^* \sum_{i=1}^{N} w_i x_i \right) \tag{5.91}$$

where the voltage $v_k$ is governed by the same $RC$ filter differential equation (5.24), so that

$$\tau_0 \frac{dz_k}{dt} + z_k = \gamma\left( x_k^* \sum_{i=1}^{N} w_i x_i \right) \tag{5.92}$$

where

$$\gamma = k^2 G \tag{5.93}$$

The constant $\gamma$ represents a conversion-factor gain constant that is assumed to be the same for all the loops. Now whereas (5.26) contained only one unknown $z_2$ and could be solved immediately, (5.92) by contrast contains $N$ unknowns $z_1, z_2, \ldots, z_N$ and must be treated as a member of a set of $N$ simultaneous linear differential equations, one for each adaptive loop. It is convenient to use (5.90) to convert from $z_k$ to $w_k$, so that (5.92) now becomes

$$\tau_0 \frac{dw_k}{dt} + w_k = b_k^* - \gamma\left[ x_k^* \sum_{i=1}^{N} w_i x_i \right] \tag{5.94}$$

Using matrix notation, we may write the complete set of $N$ differential equations corresponding to (5.94) as

$$\tau_0 \frac{d\mathbf{w}}{dt} + \mathbf{w} = \mathbf{b}^* - \gamma[\mathbf{x}^* \mathbf{w}^T \mathbf{x}] \tag{5.95}$$

Now since $(\mathbf{w}^T\mathbf{x}) = (\mathbf{x}^T\mathbf{w}) = \sum_{i=1}^{N} w_i x_i$, it follows that the bracketed term in (5.95) can be rewritten as

$$[\mathbf{x}^* \mathbf{w}^T \mathbf{x}] = [\mathbf{x}^* \mathbf{x}^T]\mathbf{w} \tag{5.96}$$

The expected (averaged) value of $\mathbf{x}^*\mathbf{x}^T$ yields the input signal correlation matrix:

$$\mathbf{R}_{xx} \triangleq E\{\mathbf{x}^* \mathbf{x}^T\} \tag{5.97}$$

The averaged values of the correlation components forming the elements of $\mathbf{R}_{xx}$ are given by

$$\overline{x_k^* x_l} = \sum_{i=1}^{I} |\bar{J}_i|^2 \exp[j2u_i(l-k)] \qquad \text{for } l \neq k \tag{5.98}$$

and

$$\overline{x_k^* x_l} = |\overline{x_k}|^2 = |\bar{n}_k|^2 + \sum_{i=1}^{I} |\bar{J}_i|^2 \qquad \text{for } l = k \tag{5.99}$$

where (5.99) corresponds to the diagonal elements of $\mathbf{R}_{xx}$. Since the correlation matrix can be considered (in the absence of the desired signal) as the sum of the quiescent receiver noise matrix $\mathbf{R}_{nn_q}$ and the various individual interference source matrixes $\mathbf{R}_{nn_i}$, it follows that

$$\mathbf{R}_{nn} = \mathbf{R}_{nn_q} + \sum_{i=1}^{I} \mathbf{R}_{nn_i} \tag{5.100}$$

where $\mathbf{R}_{nn_q}$ can be expressed as

$$\mathbf{R}_{nn_q} = \begin{bmatrix} |\bar{n}_1|^2 & 0 & 0 & \cdots \\ 0 & |\bar{n}_2|^2 & 0 & \cdots \\ & & \ddots & \\ 0 & \cdots & \cdots & |\bar{n}_N|^2 \end{bmatrix} \tag{5.101}$$

and

$$\mathbf{R}_{nn_i} = |\bar{J}_i|^2 \begin{bmatrix} 1 & e^{j2u_i} & e^{j4u_i} & \cdots & \\ e^{-j2u_i} & 1 & e^{j2u_i} & \cdots & \\ e^{-j4u_i} & e^{-j2u_i} & 1 & \cdots & \\ & & & \ddots & \\ & & & & 1 \end{bmatrix} \tag{5.102}$$

Substituting $\mathbf{R}_{nn}$ of (5.100) into (5.95) and rearranging terms, we find the final expression for the adaptive weight matrix differential equation becomes:

$$\tau_0 \frac{d\mathbf{w}}{dt} + [\mathbf{I} + \gamma \mathbf{R}_{nn}]\mathbf{w} = \mathbf{b}^* \tag{5.103}$$

where $\mathbf{I}$ is the identity matrix.

The solution of (5.103) is very simple provided that the matrix $\mathbf{R}_{nn}$ is diagonal. In general, $\mathbf{R}_{nn}$ is not diagonal but is positive definite Hermitian, so that a simple transformation of coordinates can always be found such that $\mathbf{R}_{nn}$ can be diagonalized. The matrix that accomplishes the desired diagonalization is a nonsingular orthonormal modal matrix which will be denoted here by $Q$. The resulting diagonalized matrix has diagonal elements that are the eigenvalues of the matrix $\mathbf{R}_{nn}$. The eigenvalues of $\mathbf{R}_{nn}$ are given by the solutions of the equation

$$|\mathbf{R}_{nn}-\lambda_i\mathbf{I}|=0, \qquad i=1,2,\ldots,N \tag{5.104}$$

where the $\lambda_i$ are the eigenvalues of $\mathbf{R}_{nn}$, and $\mathbf{I}$ is the identity matrix. Corresponding to each eigenvalue there is an associated eigenvector $\mathbf{e}_i$ that satisfies

$$\mathbf{R}_{nn}\mathbf{e}_i=\lambda_i\mathbf{e}_i \tag{5.105}$$

These eigenvectors (which are normalized to unit length and are orthogonal to one another) make up the rows of the transformation matrix $\mathbf{Q}$, that is,

$$\mathbf{Q}=\begin{bmatrix} e_{11} & e_{12} & e_{13} & \cdots \\ e_{21} & e_{22} & e_{23} & \cdots \\ e_{31} & e_{32} & e_{33} & \cdots \\ \vdots & & & \\ e_{N1} & e_{N2} & e_{N3} & \cdots \end{bmatrix}, \qquad \text{where } \mathbf{e}_i=\begin{bmatrix} e_{i1} \\ e_{i2} \\ \vdots \\ e_{iN} \end{bmatrix} \tag{5.106}$$

Once $\mathbf{R}_{nn}$ is diagonalized by the $\mathbf{Q}$-matrix transformation, there results

$$[\mathbf{Q}^*\mathbf{R}_{nn}\mathbf{Q}^T]=\begin{bmatrix} \lambda_1 & 0 & & 0 \\ 0 & \lambda_2 & & 0 \\ 0 & 0 & \lambda_3 & 0 \\ \vdots & & \ddots & \\ \cdots & \cdots & \cdot & \lambda_N \end{bmatrix} \tag{5.107}$$

Now since $\mathbf{R}_{nn}=E\{\mathbf{x}^*\mathbf{x}^T\}$, it follows that (5.107) may be written as

$$[\mathbf{Q}^*\mathbf{R}_{nn}\mathbf{Q}^T]=[\overline{\mathbf{Q}^*\mathbf{x}^*\mathbf{x}^T\mathbf{Q}^T}]=\overline{[\mathbf{x}'^*\mathbf{x}'^T]}=\Lambda \tag{5.108}$$

where

$$\mathbf{x}'=\mathbf{Q}\mathbf{x} \tag{5.109}$$

The $\mathbf{Q}$ matrix therefore transforms the real signal vector $\mathbf{x}$ into the orthonormal signal vector $\mathbf{x}'$. Furthermore, the components of $\mathbf{x}'$ are determined by the eigenvectors of $\mathbf{R}_{nn}$, that is,

$$x'_k = (\mathbf{e}_k^T \mathbf{x}) \tag{5.110}$$

Note that the orthonormal signal vector components $x'_k$ have two special characteristics:

**1** They are uncorrelated so

$$E\{x'^*_k x'_l\} = 0 \qquad \text{for } l \neq k \tag{5.111}$$

**2** Their amplitudes are given by the square root of the corresponding eigenvalue so that

$$E\{x'^*_k x'_k\} = \lambda_k \tag{5.112}$$

The transformation matrix **Q** may therefore be regarded as yielding the same signal components that would obtain with the operation of an appropriately selected orthogonal beamforming network.

Just as the signal vector **x** was transformed into **x**′ by (5.109), the beam-steering vector **b*** may likewise be transformed to define a new beam-steering vector **b**′* as

$$\mathbf{b}' = \mathbf{Q}\mathbf{b} \tag{5.113}$$

where the *k*th component of **b**′ is determined by the *k*th eigenvector appearing in **Q**.

The *Q*-coordinate transformation operating on both **x** and **b*** suggests an equivalent circuit representation for the system that is illustrated in Figure 5.18b, where an equivalent "orthonormal adaptive array" system is shown alongside a simplified representation of the real system in Figure 5.18a. There are a set of weights forming the weight vector **w**′ in the orthonormal system, and the adaptive weigth matrix equation for the equivalent system is easily shown to be

$$\tau_0 \frac{d\mathbf{w}'}{dt} + [\mathbf{I} + \gamma \mathbf{R}'_{nn}]\mathbf{w}' = b'^* \tag{5.114}$$

where

$$\mathbf{R}'_{nn} = E\{\mathbf{x}'^* \mathbf{x}'^T\} = \Lambda \tag{5.115}$$

As a consequence of the diagonalization that results in the orthonormal system, a set of independent linear differential equations is obtained, each of which has a solution if the eigenvalues can be determined. Each of the

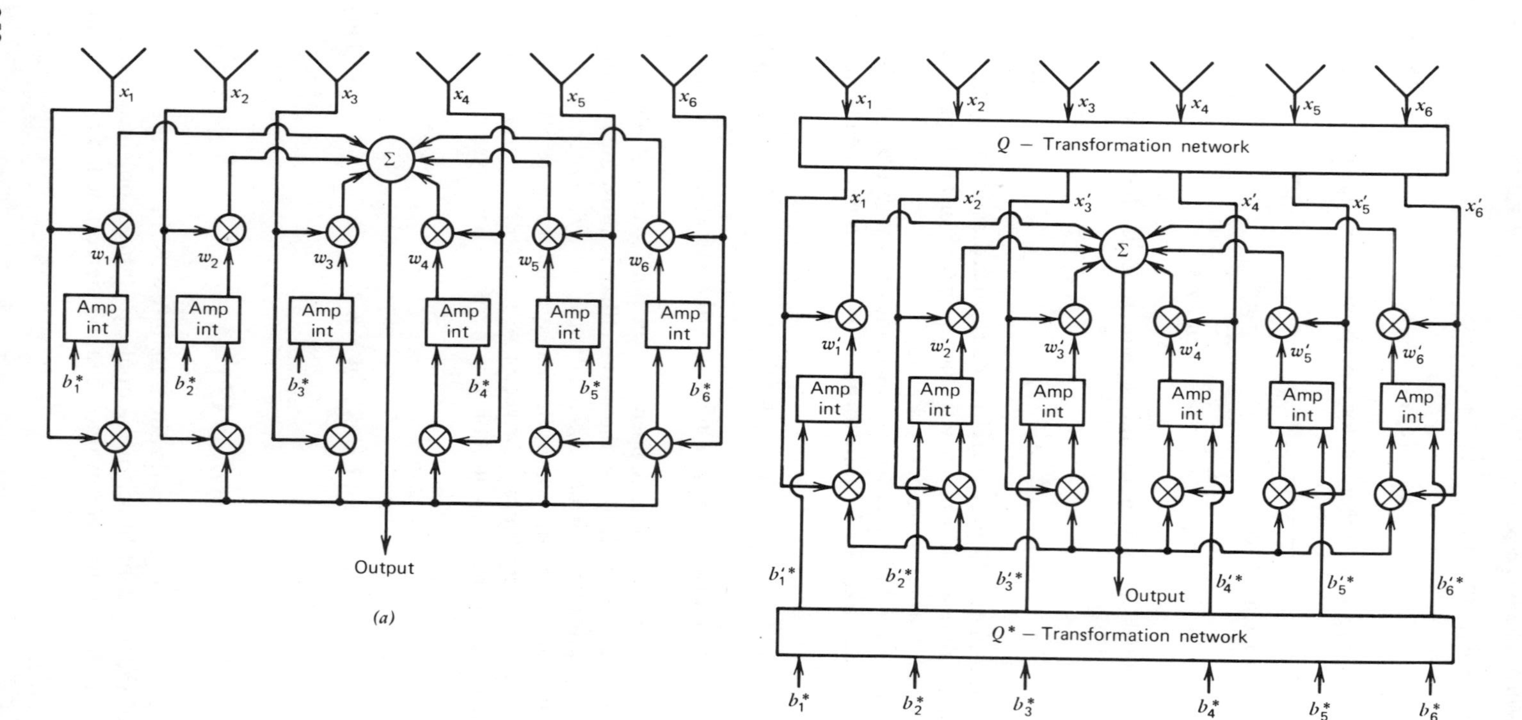


**Figure 5.18** Equivalent circuit representations for a six-element adaptive array system. (*a*) Real adaptive array system. (*b*) Equivalent orthonormal adaptive array system. From Gabriel, *Proc. IEEE*, February 1976.

orthonormal servo loops in the equivalent system responds independently of the other loops, because the $x_k'$ input signals are orthogonalized and are therefore completely uncorrelated with one another. The weight equation for the $k$th orthonormal servo loop can therefore be written as

$$\tau_0 \frac{dw_k'}{dt} + (1 + \gamma\lambda_k) w_k' = b_k'^* \tag{5.116}$$

Note that the equivalent servo gain factor can be defined from (5.116) as

$$\mu_k = \gamma\lambda_k \tag{5.117}$$

so the equivalent servo gain factors for the various orthonormal loops are now determined by the eigenvalues of the input signal covariance matrix. On comparing (5.117) with (5.38), it is seen that the positive, real eigenvalues $\lambda_k$ correspond to the square of a signal voltage amplitude, and any given eigenvalue is in fact proportional to the power appearing at the orthonormal network output port.

For the input beam-steering vector $\mathbf{b}^*$, the output desired signal power is given by

$$P_s = |\mathbf{w}^T \mathbf{b}|^2 \tag{5.118}$$

Likewise the array output noise power can be written as

$$P_n = \overline{|\mathbf{w}^T \mathbf{x}|^2} \tag{5.119}$$

where the signal vector $\mathbf{x}$ is assumed to be comprised only of quiescent receiver channel noise plus the directional noise signal components due to external sources of interference. The signal-to-noise performance measure is therefore just a ratio of the above two quadratic forms:

$$\left(\frac{s}{n}\right) = \frac{|\mathbf{w}^T \mathbf{b}|^2}{\overline{|\mathbf{w}^T \mathbf{x}|^2}} = \frac{\mathbf{w}^\dagger [\mathbf{b}^* \mathbf{b}^T] \mathbf{w}}{\mathbf{w}^\dagger \mathbf{R}_{nn} \mathbf{w}} \tag{5.120}$$

where $\mathbf{R}_{nn}$ is the covariance matrix of the noise signal environment. From the results of Chapter 3, it follows that the optimum weight vector that yields the maximum SNR for (5.120) is given by

$$\mathbf{w}_{\text{opt}} = \frac{1}{(\text{constant})} \mathbf{R}_{nn}^{-1} \mathbf{b}^* \tag{5.121}$$

On comparing (5.121) with (4.45), it is seen that both the LMS and maximum SNR algorithms yield precisely the same weight vector solution (to within a multiplicative constant when the desired signal is absent) provided that $\mathbf{r}_{xd} = \mathbf{b}^*$ since these two vectors play exactly the same role in determining the optimum weight vector solution. Consequently adopting a specific vector $\mathbf{r}_{xd}$ for the LMS algorithm is equivalent to selecting $\mathbf{b}^*$ for the maximum SNR algorithm, which represents direction of arrival information—this provides the relation between a reference signal and a beam-steering signal that must obtain for the LMS and maximum SNR algorithms to yield equivalent solutions.

From the foregoing discussion, it follows that the optimum orthonormal weight is given (to within a scalar multiple) by

$$w'_{k_{\text{opt}}} = \left(\frac{1}{\mu_k}\right) b_k'^* \tag{5.122}$$

If we substitute (5.117) and (5.122) into (5.116), there immediately results

$$\tau_0 \frac{dw'_k}{dt} + (1+\mu_k) w'_k = \mu_k w'_{k_{\text{opt}}} \tag{5.123}$$

The above differential equation is of exactly the same form as (5.34), so for a step-function change in the input signal the solution may be written as follows:

$$w'_k(t) = \left[ w'_k(0) - w'_k(\infty) \right] e^{-\alpha_k t} + w'_k(\infty) \tag{5.124}$$

where

$$w'_k(\infty) = \left(\frac{\mu_k}{1+\mu_k}\right) w'_{k_{\text{opt}}} \tag{5.125}$$

and

$$\alpha_k = \left(\frac{1+\mu_k}{\tau_0}\right) \tag{5.126}$$

In the foregoing equations $w'_k(\infty)$ represents the steady-state weight, $w'_k(0)$ is the initial weight value, and $\alpha_k$ is the transient decay factor. The adaptive weight transient responses can now be determined by the eigenvalues. The $k$th orthonormal servo loop may be represented by the simple type-0 position servo illustrated in Figure 5.19.

To relate the orthonormal system weights $w_k'$ to the actual weights $w_k$ note that the two systems shown in Figure 5.18 must be exactly equivalent so that

$$\sum_{k=1}^{N} w_k x_k = \sum_{k=1}^{N} w'_k x'_k \tag{5.127}$$

or

$$\mathbf{w}^T \mathbf{x} = \mathbf{w}'^T \mathbf{x}' = \mathbf{w}'^T \mathbf{Q} \mathbf{x} \tag{5.128}$$

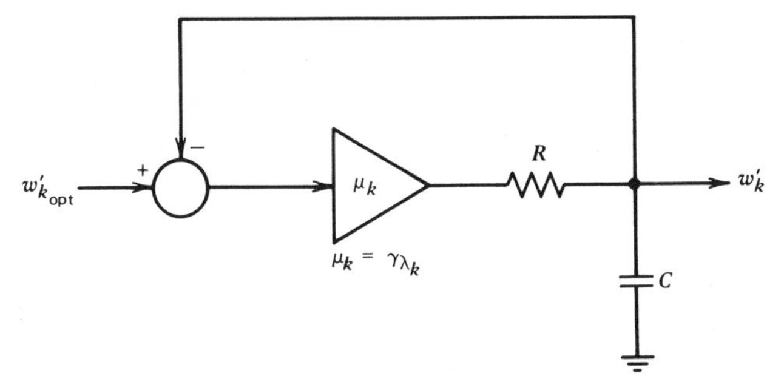


**Figure 5.19** Type-O servo model for $k$th orthonormal adaptive control loop. From Gabriel, *Proc. IEEE*, February 1976.

Consequently,

$$\mathbf{w} = \mathbf{Q}^T\mathbf{w}' \tag{5.129}$$

From (5.129) it follows that the solution for the $k$th actual weight can be written as

$$w_k = \left(e_{1_k}w'_1 + e_{2_k}w'_2 + \cdots + e_{N_k}w'_N\right) \tag{5.130}$$

where $e_{1_k}$ is the $k$th element of the first eigenvector,
$e_{2_k}$ is the $k$th element of the second eigenvector,
⋮
$e_{N_k}$ is the $k$th element of the $N$th eigenvector.

To evaluate the $c_k$ coefficients that relate $\mathbf{b}^*$ to $\mathbf{w}_q$ as in (5.89), once again assume that only receiver noise is present in any channel for quiescent conditions so that various channel signals are uncorrelated, the $\mathbf{Q}$ transformation matrix is an identity matrix, and therefore there is no difference between $\mathbf{w}'$ and $\mathbf{w}$. With $\mathbf{Q}_q = \mathbf{I}$, the quiescent noise covariance matrix $\mathbf{R}_{nn_q}$ is already diagonalized, and if it is further assumed that the receiver noise power in all channels is equal (and denoted by $|\bar{n}_0|^2$), then from (5.107) it follows that

$$\mathbf{Q}_q^*\mathbf{R}_{nn_q}\mathbf{Q}_q^T = \left[\lambda_0\delta_{ij}\right] \tag{5.131}$$

where

$$\lambda_0 = |\bar{n}_0|^2 \tag{5.132}$$

so the smallest eigenvalue is simply equal to the receiver channel noise power. This smallest eigenvalue then defines the minimum servo gain factor $\mu_{\min}$ as

$$\mu_{\min} = \gamma\lambda_0 \tag{5.133}$$

Since the quiescent steady-state weight $\mathbf{w}(\infty)$ must by definition be equal to

$\mathbf{w}_q$, (5.125), (5.122), and (5.89) can be applied to yield

$$w_{q_k} = \frac{1}{1+\mu_{\min}} b_k^* = \left(\frac{c_k}{1+\mu_{\min}}\right) w_{q_k}$$

or

$$c_k = (1+\mu_{\min}) \tag{5.134}$$

From (5.124)–(5.126) and (5.117), it follows that the effective time constant with which the $k$th component of $\mathbf{w}'$ converges to its optimum value is $\tau_0/(1+\gamma\lambda_k)$. In effect, therefore, $\lambda_{\min}$ determines how rapidly the adaptive array can follow changes in the noise environment. Equation (5.130) shows that each actual weight can be expressed as a weighted sum of exponentials, and the component that converges most slowly is the $\lambda_{\min}$ component.

When an adaptive array like that of Figure 5.17 operates with a distributed external noise field, the loop convergence will be very slow for some angular noise distributions [6]. Furthermore, if $\gamma$ is increased or $\tau_0$ is decreased to speed the weight convergence, the loop becomes "noisy." Slow weight convergence in fact occurs whenever trace $(\mathbf{R}_{nn})/\lambda_{\min}$ is large, and in these cases there is no choice of $\gamma$ and $\tau_0$ that will yield rapid convergence without excessive loop noise. These facts suggest that it is appropriate to consider the effects of noise on the solutions represented by (5.122)–(5.126).

### 5.3.2 *Loop Noise Considerations*

The adaptive weight elements in a Howells-Applebaum control loop are not fixed in the steady-state but have components that vary in a random manner. This random behavior can be modeled and the variance of the weights computed for the steady-state condition. This element weight "noise" results in an additional noise component in the array output signal. In this section expressions are given for the variance of the element weights and for the resulting additional noise in the array output. A complete derivation of these results is given by Brennan, Pugh, and Reed [6].

Let $\xi$ denote the random (or noise) component of the adaptive weight vector $\mathbf{w}$ and let $\mathfrak{N}$ denote the random component of $\mathbf{R}_{nn}$ so that

$$\mathbf{w} = \overline{\mathbf{w}} + \xi \tag{5.135}$$

and

$$\mathbf{R}_{nn} = \overline{\mathbf{R}}_{nn} + \mathfrak{N} \tag{5.136}$$

where now $\overline{\mathbf{w}}$ and $\overline{\mathbf{R}}_{nn}$ denote average values. Recalling that the adaptive weights must satisfy the differential equation

$$\tau_0 \frac{d\mathbf{w}}{dt} + (\mathbf{I} + \gamma \mathbf{R}_{nn})\mathbf{w} = \mathbf{b}^* \tag{5.137}$$

we may substitute the values $\overline{\mathbf{w}}$ and $\overline{\mathbf{R}}_{nn}$ into (5.137) and subtract the result from the equation resulting with (5.135) and (5.136) substituted into (5.137) to give

$$\tau_0 \frac{d\xi}{dt} + (\mathbf{I} + \gamma \overline{\mathbf{R}}_{nn})\xi = -\gamma \mathfrak{R}\mathbf{w} \tag{5.138}$$

Premultiplying (5.138) by the transformation matrix $\mathbf{Q}^*$ and using the fact that $\mathbf{Q}^*\mathbf{Q}^T = \mathbf{I}$ then yields

$$\frac{d\zeta}{dt} + \frac{1}{\tau_0}(\mathbf{I} + \gamma \Lambda)\zeta = -\beta \mathbf{Q}^* \mathfrak{R}\mathbf{w} = \mathbf{u} \tag{5.139}$$

where

$$\zeta = \mathbf{Q}^* \xi \tag{5.140}$$

and

$$\beta = \frac{\gamma}{\tau_0} \tag{5.141}$$

Equation (5.139) represents a system of $N$ independent linear differential equations of which the $n$th component can be written as

$$d\zeta_n + \sigma_n \zeta_n \, dt = u_n \, dt \tag{5.142}$$

where

$$\sigma_n = \frac{1 + \gamma \lambda_n}{\tau_0} \tag{5.143}$$

and

$$u_n = u_n(\tau, \overline{\mathbf{w}}, \zeta) = (-\beta \mathbf{Q}^* \mathfrak{R}\mathbf{w})_n \tag{5.144}$$

represents the $n$th component of the column vector $\mathbf{u}$.

Multiplying (5.142) by the factor $e^{\sigma_n t}$ and integrating each term from $t_0$ to $t$ then yields

$$\zeta_n(t) = \zeta_n(t_0) \exp[-\sigma_n(t - t_0)] + \int_{t_0}^{t} e^{-\sigma_n(t-\tau)} \cdot u_n(\tau, \overline{\mathbf{w}}, \zeta) \, d\tau \tag{5.145}$$

If only the steady-state case is considered, then the weights will be near their mean steady-state values. The steady-state solution for variations in the element weights can be obtained from (5.145) by setting $t_0 = -\infty$ and ignoring any effect of the initial value $\zeta_n(t_0)$ to give

$$\zeta_n(t)=\int_0^{\infty} e^{-\sigma_n \tau} u_n(t-\tau)\,d\tau \tag{5.146}$$

One important measure of the noise present in the adaptive loops is the variance of the weight vector denoted by var(**w**):

$$\operatorname{var}(\mathbf{w})=E\left\{\sum_{n=1}^{N}|\mathbf{w}_n-\overline{\mathbf{w}}_n|^2\right\}=E\{\boldsymbol{\xi}^{\dagger}\boldsymbol{\xi}\} \tag{5.147}$$

where $N$ is the dimension of the weight vector (or the number of degrees of freedom in the adaptive array system).

Now since $\boldsymbol{\zeta}=\mathbf{Q}^*\boldsymbol{\xi}$, (5.147) becomes

$$\operatorname{var}(\mathbf{w})=E\{\boldsymbol{\zeta}^{\dagger}\boldsymbol{\zeta}\} \tag{5.148}$$

The elements of the covariance matrix of $\boldsymbol{\zeta}(t)$ in (5.148) can be obtained from (5.146) and the definition of $u_n$

$$\begin{aligned}
E\{\zeta_j^*\zeta_k\}=\beta^2\int_0^{\infty}d\tau_1\int_0^{\infty}&E\{[\mathbf{Q}^*\mathfrak{N}(t-\tau_1)\overline{\mathbf{w}}]_j^*\\
&\cdot\exp(-\sigma_j\tau_1-\sigma_k\tau_2)[Q^*\mathfrak{N}(t-\tau_2)\overline{w}]_k\}\,d\tau_2\\
&+\beta^2\int_0^{\infty}d\tau_1\int_0^{\infty}E\{[\mathbf{Q}^*\mathfrak{N}(t-\tau_1)\boldsymbol{\xi}(t-\tau_1)]_j^*\\
&\cdot\exp(-\sigma_j\tau_1-\sigma_k\tau_2)[\mathbf{Q}^*\mathfrak{N}(t-\tau_2)\boldsymbol{\xi}(t-\tau_2)]_k\}\,d\tau_2
\end{aligned} \tag{5.149}$$

where the cross-product terms do not appear since $E\{\boldsymbol{\xi}(t)\}=0$ and $\mathfrak{N}(t)$ and $\boldsymbol{\xi}(t)$ are independent noise processes.

The evaluation of the double integrals in (5.149) involves several approximations and is rather laborious, but a useful lower bound results which is given by reference [6].

$$\operatorname{var}(\mathbf{w})\geqslant\frac{\beta^2\Delta}{2}\sum_{n=1}^{N}\frac{1}{\sigma_n}E\{|(\mathbf{Q}^*\mathfrak{N}\overline{\mathbf{w}})_n|^2\} \tag{5.150}$$

where $\Delta$ represents the time interval between successive independent samples of the input signal vector. For a pulse radar, $\Delta$ is approximately the same as the pulse width. For a communications system, $\Delta$ is approximately $1/B$, where $B$ is the signal bandwidth.

The bound expressed by (5.150) is useful in selecting parameter values for the Howells-Applebaum servo loops, for if this bound is not small, then the noise fluctuations at the output of the adaptive loops will be correspondingly large. For cases of practical interest [when var(**w**) is small compared to $\overline{\mathbf{w}}^{\dagger}\overline{\mathbf{w}}$], the right hand side of (5.150) will be an accurate estimate of var(**w**). Equation (5.150) can in turn be simplified (after considerable effort) to yield the expression

$$\mathrm{var}(\mathbf{w}) \geqslant \left[ \frac{\Re \beta \Delta}{2} - \frac{\beta \Delta}{2\gamma} \sum_{n=1}^{N} \frac{1}{\lambda_n + 1/\gamma} \right] \bar{\mathbf{w}}^{\dagger} \bar{\mathbf{R}}_{nn} \bar{\mathbf{w}} \tag{5.151}$$

where $\lambda_n$ represents the $n$th eigenvalue of $\bar{\mathbf{R}}_{nn}$,

$$\Re = \frac{N}{\beta} - \frac{1}{\gamma} \mathrm{trace}(\boldsymbol{\mu}) \tag{5.152}$$

$$\boldsymbol{\mu} = (\beta \mathbf{H})^{-1} \tag{5.153}$$

and

$$\mathbf{H} = \bar{\mathbf{R}}_{nn} \frac{\mathbf{I}}{\gamma} \tag{5.154}$$

The random component of the adaptive weights gives rise to additional noise in the output signal, which is of interest as one measure of how great a penalty is incurred by the adaptive processor's inability to maintain precisely optimum weight settings. Since the total output noise power is the noise power without noisy weights $\bar{\mathbf{w}}^{\dagger} \bar{\mathbf{R}}_{nn} \bar{\mathbf{w}}$ plus the additional noise due to the random weight components, it can be shown that the total output noise power is given by

$$\begin{aligned} E\left\{ |\mathbf{w}^T \mathbf{x}|^2 \right\} &= \bar{\mathbf{w}}^{\dagger} \bar{\mathbf{R}}_{nn} \mathbf{w} + E\left\{ \boldsymbol{\xi}^{\dagger} \bar{\mathbf{R}}_{nn} \boldsymbol{\xi} \right\} \\ &\cong \bar{\mathbf{w}}^{\dagger} \bar{\mathbf{R}}_{nn} \bar{\mathbf{w}} \left[ 1 + \frac{\beta \Delta}{2} \sum_{n=1}^{N} \lambda_n \right] \end{aligned} \tag{5.155}$$

when $\gamma \lambda_n \gg 1$ for $n = 1, 2, \ldots, N$. The quantity $\beta \Delta$ occurs both in (5.155) and in (5.151) is the ratio $\gamma \Delta / \tau_0$, which is the gain divided by the loop time constant where the time constant is measured in intervals of the independent-sample rate of the system.

When loop noise is present in the system, the total noise power output increases by the factor $(1 + K_n)$ where from (5.155),

$$K_n \geqslant \frac{\gamma \Delta}{2\tau_0} \sum_{n=1}^{N} \lambda_n = \frac{\gamma}{4 B \tau_0} \mathrm{trace}\left( \bar{\mathbf{R}}_{nn} \right) \tag{5.156}$$

where $\Delta = 1/2B$ (i.e., $B$ is the bandwidth of the input signal process) so that $K_n$ is a direct measure of algorithm misadjustment due to noise in the weight vector.

Recalling the solution to (5.123), we see that the effective time constant of the normal weight component $w'_{k'}$ having the slowest convergence rate is

$$\tau_{\mathrm{eff}} = \frac{\tau_0}{1 + \gamma \lambda_{\min}} \cong \frac{\tau_0}{\gamma \lambda_{\min}} \tag{5.157}$$

where $\gamma\lambda_{\min} \geqslant 1$ to avoid a steady-state bias error in the solution. On combining (5.156) and (5.157) there results

$$\frac{\tau_{\text{eff}}}{\Delta} \geqslant \frac{1}{2K_n\lambda_{\min}} \sum_{n=1}^{N} \lambda_n = \frac{\operatorname{trace}(\overline{\mathbf{R}}_{nn})}{2K_n\lambda_{\min}} \tag{5.158}$$

Equation (5.158) shows that when the smallest eigenvalue $\lambda_{\min}$ is small compared to trace($\overline{\mathbf{R}}_{nn}$), many independent samples of the input signal are required before the adaptive array can settle to a near-optimum set of weights without excessive loop noise; no set of loop parameters can yield both low loop noise and rapid convergence of the adaptive control loops in this case. Reference [13] gives an analysis of steady-state weight jitter in Howells-Applebaum control loops for the case where no assumption of statistical independence between the input signal and weight processes is made. This analysis shows that steady-state weight jitter is closely related to the occurrence of statistical dependence between the weight and signal processes.

### 5.3.3 *Adaptive Array Behavior in Terms of Eigenvector Beams*

The **Q**-matrix transformation defined by (5.106) is composed of normalized and mutually orthogonal eigenvectors. The components of these eigenvectors may be interpreted as array element weights, giving rise to a set of normalized orthogonal eigenvector beams. The eigenvector beam interpretation provides good insight into the operation of the adaptive array. The $k$th eigenvector beam may be expressed as

$$g_k(\theta) = (\mathbf{s}^T\mathbf{e}_k) = \sum_{i=1}^{N} e_{ki}s_i \tag{5.159}$$

where $\mathbf{s}$ and its components $s_i$ for a linear $N$-element array are defined by

$$\mathbf{s}^T = [s_2, s_2, \ldots, s_N] \tag{5.160}$$

where

$$s_i = e^{ju(2i-N-1)} \tag{5.161}$$

and

$$u = \frac{\pi d}{\lambda}\sin\theta \tag{5.162}$$

By defining the variable $z$ related to the spatial angle $\theta$ as

$$z \triangleq e^{j2u} \tag{5.163}$$

then the eigenvector beam may be conveniently rewritten as

$$g_k(\theta)=\left(\frac{1}{\sqrt{z}}\right)^{N-1}\left[e_{k1}+e_{k2}z+e_{k3}z^2+\cdots+e_{kN}z^{N-1}\right] \tag{5.164}$$

Equation (5.165) expresses the eigenvector beam in an array polynomial form in which the eigenvector components are the coefficients of the array polynomial. The array polynomial can also be expressed in the factored form,

$$g_k(\theta)=\left(\frac{1}{\sqrt{z}}\right)^{N-1}\left[a_{N-1}(z-z_1)(z-z_2)\cdots(z-z_{N-1})\right] \tag{5.165}$$

where the roots $z_1, z_2, \ldots, z_{N-1}$ are the zeros or null points of the eigenvector beam pattern. Knowing the null points, we can find the corresponding array polynomial coefficients, or knowing the coefficients, we can determine the corresponding null points. All the roots $z_1, z_2, \ldots, z_{N-1}$ are located on the $z$-plane unit circle. The foregoing representation of adaptive array beam patterns in terms of eigenvector beam components is very useful for the following reasons:

1. Adaptive arrays form nulls in the direction of interference sources.
2. An array consisting of $N$ elements has $(N-1)$ degrees of freedom and is therefore capable of forming $(N-1)$ distinct nulls in the array beam pattern. Adaptive array behavior therefore concerns how beam pattern nulls are formed in response to the interference environment.
3. Controlled null placement may be incorporated directly into the beam-steering array weights. For a practical example illustrating the use of this concept see reference [14].
4. Constrained null positions are usually associated with corresponding eigenvector beams.

If the interference environment consists of a single narrowband source of interference located at the off-boresight angle $\theta_1$, then the $\mathbf{R}_{nn}$ covariance matrix contains one unique eigenvalue, and the corresponding unique eigenvector produces a retrodirective eigenvector beam centered on the source at $\theta_1$ as illustrated in Figure 5.1. The $\mathbf{R}_{nn}$ matrix in this case also contains nonunique eigenvalues having arbitrary nonunique eigenvectors; these arbitrary nonunique eigenvector beams are not essential to array operation, and array pattern performance can be characterized solely in terms of the unique retrodirective eigenvector beams.

The overall array beam pattern is most easily derived by considering the output of the orthonormal system represented in Figure 5.18($b$) for the input signal vector $\mathbf{s}$, defined in (5.161). Since the output for the real system and the orthonormal system must be identical, it follows that

$$G(\theta,t)=\sum_{i=1}^{N} w_i s_i=\sum_{i=1}^{N} w_i' s_i' \tag{5.166}$$

or equivalently,

$$G(\theta,t)=\mathbf{w}'^T\mathbf{s}' \tag{5.167}$$

where

$$\mathbf{s}'=\mathbf{Qs} \tag{5.168}$$

Now the $i$th component of $\mathbf{s}'$ is given by

$$s_i'=(\mathbf{e}_i^T\mathbf{s})=\sum_{k=1}^{N} e_{ik}s_k \tag{5.169}$$

but the above summation just defines the $i$th eigenvector beam [as can be seen from (5.159)], so that

$$s_i'=(\mathbf{e}_i^T\mathbf{s})=g_i(\theta) \tag{5.170}$$

Consequently the overall array beam pattern can be expressed as

$$G(\theta,t)=\sum_{i=1}^{N} w_i' g_i(\theta) \tag{5.171}$$

which shows that the output beam pattern is the summation of the $N$ eigenvector beams weighted by the orthonormal system adaptive weights.

Since the $k$th component of the quiescent orthonormal weight vector is given by

$$w_{q_k}'=(\mathbf{e}_k^{\dagger}\mathbf{w}_q) \tag{5.172}$$

the steady-state solution for the $k$th component of the orthonormal weight vector given by (5.124) can be rewritten using (5.125), (5.89), and (5.134) to yield

$$w_k'(\infty)=\left(\frac{1+\mu_{\min}}{1+\mu_k}\right)w_{q_k}' \tag{5.173}$$

Assume as before that quiescent signal conditions up to time $t=0$ consist only of receiver noise and that the external interference sources are switched on at $t=0$, then

$$w_k'(0)=w_{q_k}' \tag{5.174}$$

and the solution for $w_k$ expressed by (5.124) can be rewritten in the more convenient form

$$w_k' = w_{q_k}' - (1 - e^{-\alpha_k t})\left[\frac{\mu_k - \mu_{\min}}{1 + \mu_k}\right] w_{q_k}' \tag{5.175}$$

It is immediately apparent that at time $t=0$ (5.171) results in

$$G(\theta, 0) = \sum_{i=1}^{N} w_{q_i}' g_i(\theta) = (\mathbf{w}_q'^T \mathbf{s}') = (\mathbf{w}_q'^T \mathbf{Q}\mathbf{s}) \tag{5.176}$$

From (5.129) it is seen that $\mathbf{w}_q^T = \mathbf{w}_q'^T \mathbf{Q}$ so that

$$G(\theta, 0) = (\mathbf{w}_q^T \mathbf{s}) = G_q(\theta) \tag{5.177}$$

where the quiescent pattern $G_q(\theta)$ was previously defined by (5.84).

Finally by substituting (5.177) and (5.175) into (5.171), there results

$$G(\theta, t) = G_q(\theta) - \sum_{i=1}^{N} (1 - e^{-\alpha_i t})\left[\frac{\mu_i - \mu_{\min}}{1 + \mu_i}\right] w_{q_i}' g_i(\theta) \tag{5.178}$$

where it will be recalled that

$$\alpha_i = \frac{1 + \mu_i}{\tau_0}$$

$$\mu_i = \gamma \lambda_i$$

*and*

$$\mu_{\min} = \gamma \lambda_{\min}$$

The foregoing result emphasizes that the output beam pattern function of the adaptive array consists of two parts:

1. The quiescent beam pattern $G_q(\theta)$.
2. The summation of weighted orthogonal eigenvector beams that is subtracted from $G_q(\theta)$.

Note also from (5.178) that the weighting associated with any eigenvector beams corresponding to eigenvalues equal to $\lambda_0$ (the quiescent eigenvalue) is zero since the numerator $(\mu_i - \mu_{\min})$ will be zero for such eigenvalues. Consequently any eigenvector beams associated with $\lambda_0$ can be disregarded, leaving only unique eigenvector beams (which are also retrodirective) to influence the resulting pattern. The transient response time of (5.178) is determined by the value of $\alpha_i$, which in turn is proportional to the eigenvalue. Therefore a large eigenvalue yields a fast transient response for its associated eigenvector beam, while a small eigenvalue results in a slow transient response.

The foregoing eigenvector beam interpretation of adaptive array behavior may be easily illustrated by considering an eight-element linear array having $\lambda/4$ element spacing and two narrowband interference sources having nearly equal power ratios of $PR_1 = 1250$ and $PR_2 = 1200$ located at $\theta_1 = 18°$ and $\theta_2 = 22°$, respectively. Forming the covariance matrix using (5.100) for this case and solving for the eigenvalues yields two unique solutions: $\lambda_1 = 18{,}544.4$ and $\lambda_2 = 1057.58$. These widely different eigenvalues result despite the nearly equal jammer powers because the interference sources are located close together compared to the array quiescent beamwidth.

Solving for the two (normalized) eigenvectors associated with the unique eigenvalues then permits the two retrodirective eigenvector beam patterns to be found $g_1'(\theta)$ and $g_2'(\theta)$, which are both illustrated in Figure 5.20. Beam $g_1'(\theta)$ covers both interference sources in the same manner as a centered beam pattern, and its total output power is equal to the first eigenvalue

$$\left(\frac{\lambda_1}{\lambda_0}\right) = 1 + P_1 g_1'^2(\theta_1) + P_2 g_2'^2(\theta_2) = 18{,}544 \tag{5.179}$$

The second eigenvector beam $g_2'(\theta)$ splits the interference sources in the manner of a difference beam, and its total output power is equal to the second eigenvalue

$$\left(\frac{\lambda_2}{\lambda_0}\right) = 1 + P_1 g_2'^2(\theta_1) + P_2 g_2'^2(\theta_2) = 1057 \tag{5.180}$$

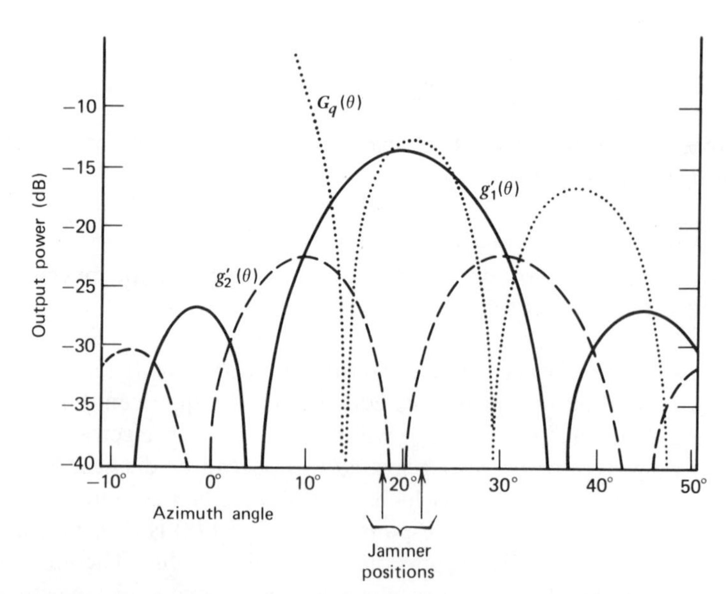


**Figure 5.20** Two retrodirective eigenvector beam patterns $g_1'(\theta)$ and $g_2'(\theta)$ for two-jammer example. From Gabriel, *Proc. IEEE*, February 1976.

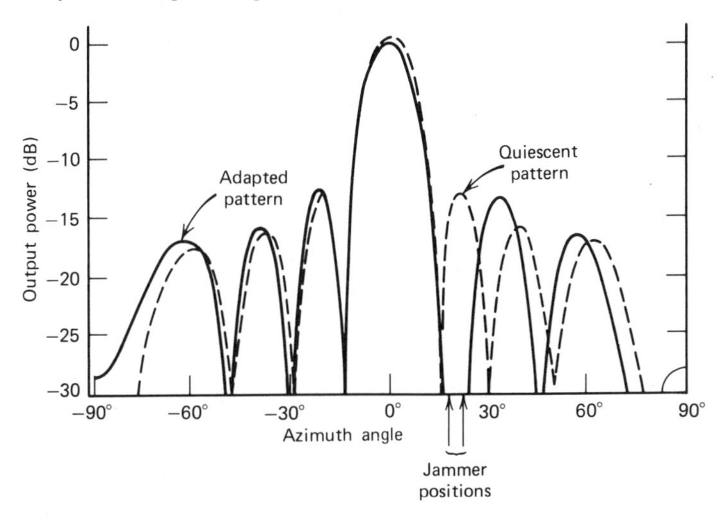


**Figure 5.21** Steady-state adapted array pattern and quiescent array pattern for two-jammer example. From Gabriel, *Proc. IEEE*, February 1976.

Although both eigenvector beams contain power from both of the sources, their respective output signals remain decorrelated. The cross-correlation product of the two eigenvector beam outputs is given by

$$E\{x_1'^* x_2'\} = E\{|J_1|^2\} g_1'(\theta_1) g_2'(\theta_1) + E\{|J_2|^2\} g_1'(\theta_2) g_2'(\theta_2) \quad (5.181)$$

This cross-correlation product can be zero if the product $[g_1'(\theta)g_2'(\theta)]$ is positive when $\theta=\theta_1$ and negative when $\theta=\theta_2$ thereby resulting in decorrelation between the two eigenvector beam signals. Figure 5.21 shows the overall

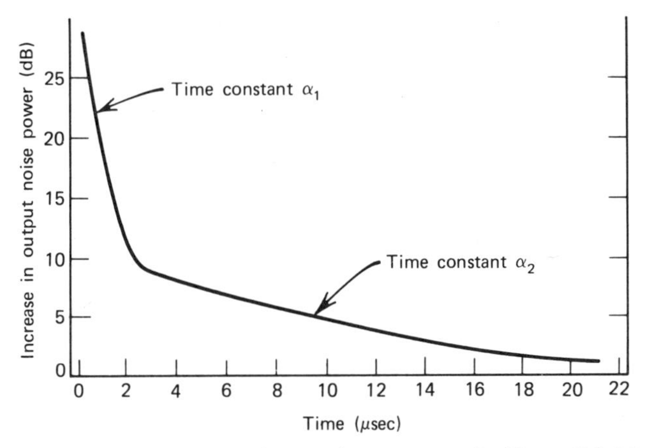


**Figure 5.22** Array transient response for two-jammer example. From Gabriel, *Proc. IEEE*, February 1976.

quiescent beam pattern and the resulting steady-state adapted pattern for this two-source example.

Figure 5.22 illustrates the transient response (in terms of increase in output noise power) of the adaptive array for this two-interference source example where it is seen that the response has two distinct slopes associated with the two distinct (and widely different) eigenvalues.

### *5.3.4 Illustrative Example of N-element Linear Adaptive Array Performance*

To illustrate the concepts and equations for the $N$-element linear array developed in the foregoing sections, it is useful to consider how an eight-element linear array responds to various different sets of specified signal conditions. The results obtained will prove useful in illustrating how important performance characteristics discussed above are related to selected array constants and signal conditions.

***Assumptions and Initial Conditions.*** Only the quiescent receiver noise is assumed to be present in each channel up to time $t=0$, when all the external interference sources are switched on in a single step function. The quiescent RMS noise voltage is equal in all channels and denoted by $n_0$, such that the quiescent eigenvalue $\lambda_0$ is given by $|n_0|^2$. This defines the quiescent servo gain factor $\mu_{\min}$ by way of (5.133). For the configuration of Figure 5.17, it is convenient to choose $\mu_{\min}=1$ so the amplifier gains $G$ are set accordingly.

Once $\mu_{\min}$ is selected, it is convenient to express $\mu_i$ (from 5.117) as a ratio of eigenvalues so that

$$\frac{\mu_i}{\mu_{\min}} = \frac{\lambda_i}{\lambda_0} \tag{5.182}$$

The quiescent steered-beam pattern $G_q(\theta)$ and its associated quiescent weight vector $\mathbf{w}_q$ must be specified as defined by (5.81)–(5.84). For illustrative purposes $G_q(\theta)$ will be chosen to correspond to the simple uniform-illumination beam resulting for an eight-element linear array with element spacing $\lambda/2$, $\mu=\pi/2\sin\theta$, and $a_k=1$ for all elements. As a result of this selection of array constants

$$w_{q_k} = e^{-ju_0(2k-9)} \tag{5.183}$$

and

$$G_q(\theta) = \frac{\sin 8(u-u_0)}{\sin(u-u_0)} \tag{5.184}$$

The coefficients of the input beam-steering vector $\mathbf{b}^*$ may then be found from

(5.134) and (5.82) to be

$$c_k = (1 + \mu_{\min}) = 2 \tag{5.185}$$

and

$$b_k^* = c_k w_{q_k} = 2e^{-ju_0(2k-9)} \tag{5.186}$$

The maximum power condition for each of the orthonormal loops of Figure 5.18(*b*) is exactly analogous to that for the single-loop case of (5.46) so the maximum servo gain factor $\mu_{\max}$ is given by

$$\mu_{\max} = \mu_{\min}\left(\frac{\lambda_{\max}}{\lambda_0}\right) = \left(\frac{\pi B_c \tau_0}{10}\right) - 1 \tag{5.187}$$

where $\lambda_{\max}$ represents the maximum eigenvalue of concern to the adaptive array. The channel bandwidth $B_c$ and filter time constant $\tau_0$ are assumed to be the same for all element channel servo loops. Solving for $\tau_0$ from (5.187) yields

$$\tau_0 = \left(\frac{10}{\pi B_c}\right)\left[1 + \mu_{\min}\left(\frac{\lambda_{\max}}{\lambda_0}\right)\right] \tag{5.188}$$

or

$$\tau_0 = \left(\frac{10}{\pi B_c}\right)\left[1 + \mu_{\min} + \mu_{\min} \sum_{r=1}^{R} P_r g_m^2(\theta_r)\right] \tag{5.189}$$

It is seen from (5.189) that the maximum power (or maximum eigenvalue) is much larger than the jammer-to-receiver-noise power ratios $P_r$ because the various $P_r$ are multiplied by the power gain of the retrodirective eigenvector beams.

***Output Noise Power and SNR Degradation.*** The output SNR of the adaptive array compared to the output SNR of a conventional array under the same interference conditions is the performance characteristic of ultimate interest. Instead of forming the actual SNR, it is usually sufficient to consider the output noise power by itself to illustrate the system transient behavior. Since the receiver noise and external interference sources are statistically independent, the total output noise power can be obtained by simply linearly adding these two separate output noise powers.

The receiver noise output power can be expressed as

$$|y_{0_n}(t)|^2 = \sum_{k=1}^{N} |w_k n_k|^2 = \sum_{i=1}^{N} |w_i' n_0|^2 \tag{5.190}$$

Substituting for $w_i'$ from (5.175) then yields

$$|y_{0_n}(t)|^2 = |\bar{n}_0|^2 \sum_{i=1}^{k} [1 - A_i(t)]^2 |w'_{q_i}|^2 \tag{5.191}$$

where

$$A_i(t) = (1 - e^{-\alpha_i t}) \left[ \frac{\mu_i - \mu_0}{1 + \mu_i} \right] \tag{5.192}$$

From (5.192) it is seen that $A_i(t)$ is zero for $t=0$ and for $\mu_i = \mu_0$ (for nonunique eigenvalues). Therefore for quiescent conditions at $t=0$, it follows that

$$|y_{0_n}(0)|^2 = |\bar{n}_0|^2 \sum_{i=1}^{N} |w'_{q_i}|^2 = |\bar{n}_0|^2 \sum_{k=1}^{N} |w_{q_k}|^2 \tag{5.193}$$

since the output noise power must be the same for either the real system or the equivalent orthonormal system. Consequently, (5.191) may be rewritten as

$$|y_{0_n}(t)|^2 = |\bar{n}_0|^2 \sum_{k=1}^{N} |w_{q_k}|^2 - \sum_{i=1}^{N} [2 - A_i(t)] A_i(t) |w'_{q_i}|^2 \tag{5.194}$$

Equation (5.194) is a particularly convenient form because the $w'_{q_i}$ associated with nonunique eigenvalues need not be evaluated since $A(t)=0$ for such eigenvalues.

The output noise power contributed by $R$ external interference sources is given by the sum of their output power pattern levels:

$$|y_{0_j}(t)|^2 = |\bar{n}_0|^2 \sum_{r=1}^{R} P_r G^2(\theta_r, t) \tag{5.195}$$

where $P_r$ is the power ratio of the $r$th source, $\theta_r$ is its angular location, and $G(\theta_r, t)$ is given by (5.178).

The total output noise power is now given by the sum of (5.194) and (5.195), and the increase in the output noise power (when the interference sources are turned on) is the ratio of this sum to the quiescent noise (5.193), so that

$$\frac{|y_0(t)|^2}{|y_{0_n}(0)|^2} = 1 + \left\{ \frac{\displaystyle\sum_{r=1}^{R} P_r G^2(\theta_r, t) - \sum_{i=1}^{N} [2 - A_i(t)] A_i(t) |w'_{q_i}|^2}{\displaystyle\sum_{k=1}^{N} |w_{q_k}|^2} \right\} \tag{5.196}$$

The increase in output noise power given by (5.196) is the quantity usually presented to illustrate the system transient behavior. An attractive feature of the increase in output noise power is the fact that it indicates the general magnitude of the adapted (steady-state) weights.

The degradation in the SNR, $D_{sn}$, enables one to normalize the effect of adapted-weight magnitude level. This degradation is simply the quiescent SNR divided by the adapted SNR and is given by

$$D_{sn}=\left(\frac{G_q^2(\theta_s)}{G^2(\theta_s,t)}\right)\left(\frac{|y_0(t)|^2}{|y_{0_n}(0)|^2}\right) \tag{5.197}$$

where the ratio in the second term is recognized as just (5.196), the increase in output noise power.

***Eigenvalues and Eigenvectors of the Noise Covariance Matrix.*** The eigenvalues of the Hermitian noise covariance matrix $\mathbf{R}_{nn}$ are given by solutions to

$$|\mathbf{R}_{nn}-\lambda_i\mathbf{I}|=0 \tag{5.198}$$

and the corresponding eigenvectors are given by solutions to

$$\mathbf{R}_{nn}\mathbf{e}_i=\lambda_i\mathbf{e}_i \tag{5.199}$$

where the covariance matrix is formed as indicated by (5.100)–(5.102). For computational convenience, receiver noise can be assigned a value of unity and all noise powers expressed as ratios to receiver noise power. Adopting this convention, the quiescent noise matrix $\mathbf{R}_{nn_q}$ becomes an identity matrix, and with $R$ narrowband interference sources the noise covariance matrix becomes

$$\mathbf{R}_{nn}=\mathbf{I}+\sum_{r=1}^{R}P_r\mathbf{M}_r \tag{5.200}$$

where $\mathbf{M}_r$ now represents the covariance matrix due to the $r$th interference source.

Wideband interference sources can be represented by dividing the jammer power spectrum into a series of discrete spectral lines (as described in Section 5.2.2). A uniform amplitude spectrum of uncorrelated lines spaced apart by a constant frequency increment $\epsilon$ is once again assumed as illustrated in Figure 5.12. If $P_r$ is the power ratio of the entire jammer power spectrum, then the power ratio of a single spectral line (assuming a total of $L_r$ spectral lines) is

$$P_{rl}=\left(\frac{P_r}{L_r}\right) \tag{5.201}$$

Furthermore if $B_r$ denotes the percent bandwidth of the jamming spectrum, then the frequency offset of the $l$th spectral line is

$$\frac{\Delta f_l}{f_0} = \left(\frac{B_r}{100}\right)\left[-\frac{1}{2} + \left(\frac{l-1}{L_r - 1}\right)\right] \tag{5.202}$$

The covariance matrix with $R$ broadband interference sources may therefore be written as

$$\mathbf{R}_{nn} = \mathbf{I} + \sum_{r=1}^{R} \sum_{l=1}^{L_r} P_{rl}\mathbf{M}_{rl} \tag{5.203}$$

The $mn$th component ($m$th row and $n$th column) of the matrix $\mathbf{M}_{rl}$ is in turn given by

$$(\mathbf{M}_{rl})_{mn} = e^{j2u_{rl}(n-m)} \tag{5.204}$$

where

$$u_{rl} = \left(\frac{f_l}{f_0}\right)\frac{\pi}{2}\sin\theta_r = \left(1 + \frac{\Delta f_l}{f_0}\right)\frac{\pi}{2}\sin\theta_r \tag{5.205}$$

Since the adaptive array cannot respond to signals outside the element channel receiver bandwidth $B_c$, assume that $B_r$ does not exceed $B_c$.

***Performance Characteristics for Various Signal Conditions.*** Using the representations introduced above for the signal covariance matrix, both narrow and broadband interference sources may be introduced at specified locations about a quiescent beam pattern and the resulting adaptive performance computed. It has already been found that a single narrowband source located in the sidelobe region yields a single unique eigenvalue, and the resulting adapted pattern has little distortion in the main beam. Furthermore, two narrowband interference sources located in the sidelobe region having nearly equal powers but closely spaced led to two widely different eigenvalues. The adapted performance for this two-source example was illustrated in Figures 5.21 and 5.22 where little distortion in the main beam was found to result.

Four narrowband sources located in the sidelobe region of the quiescent beam pattern yield four distinct eigenvalues and require four degrees of freedom to provide the retrodirective eigenvector beams required to place adapted beam pattern nulls at the jammer locations. Depending on how severe the adjustment required to place the desired null is, there may be appreciable main beam distortion in the overall adapted pattern.

The Howells-Applebaum adaptive loop places a single adaptive weight in each element channel of the adaptive array; this configuration is most appropriate for dealing with narrowband interference sources but can also be employed for broadband interference sources having percentage bandwidths up to about 20%. Gabriel [7] gives two examples as follows: a 2% bandwidth source in the sidelobe region for which two degrees of freedom (two pattern nulls) are required to provide proper cancellation, and a 15% bandwidth source in the sidelobe region for which three degrees of freedom are required. Very wide broadband interference sources require a transversal equalizer in each element channel (instead of a single adaptive weight) for proper compensation, with a Howells-Applebaum adaptive loop then required for every tap appearing in the tapped-delay line.

A single narrowband interference source located in the main beam of the quiescent pattern requires one degree of freedom to provide the pattern null for satisfactory jammer suppression, but the adapted pattern will now exhibit severe distortions compared with the quiescent pattern. For interference sources located in the main beam, the increase in output noise power is an unsatisfactory indication of array performance because there is a net SNR degradation due to the resulting main beam distortion in the adapted pattern. Main beam constraints for such cases can be introduced into a Howells-Applebaum adaptive loop in several ways (as discussed in Section 5.4).

### 5.3.5 *Hard Limiter Modification for N Adaptive Loops*

It was shown in Section 5.3.2 that the adaptive array performance depends on the external noise field, as well as on the parameters of the adaptive control loops. The power level and angular location of the external noise field determine the noise covariance matrix and therefore its eigenvalues. The eigenvalues in turn directly affect the array performance since both the transient response of the adaptive array and the control loop noise depend explicitly on these eigenvalues. For a nonstationary signal environment, wide variations in array performance may occur, ranging from excessive control loop noise (when the interference is strong) to very slow convergence (when the interference is weak). By introducing a hard limiter into the adaptive control loop, the effects of varying noise intensity can be greatly reduced, and the dynamic range of signals in the control loops can be reduced without degrading array performance [12].

A hard limiter is introduced in the conjugate signal branches in the same manner as discussed in Section 5.2.3 for the case of a single adaptive servo loop. The resulting modification of the adaptive array circuitry is illustrated in Figure 5.23 for a six-element linear array.

With the signal envelopes hard limited, the input to the correlation mixers (which were previously $x_k^*$) are now $u_k^* = x_k^*/|x_k|$ so that amplitude variations in the conjugate signals are removed, and only the phase variations are retained. The correlation mixer voltage $v_k$ [which was previously given by (5.91)] is now given by

$$v_k' = k^2\left(u_k^* \sum_{i=1}^{N} w_i x_i\right) \tag{5.206}$$

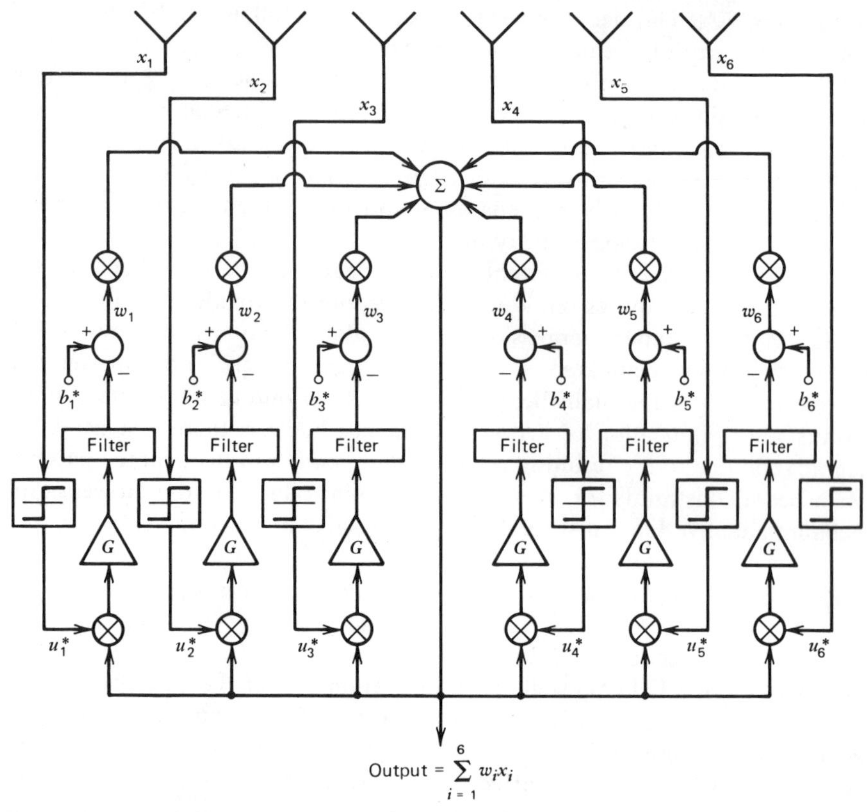


**Figure 5.23** Hard limiter modification of linear six-element adaptive array system. From Gabriel, *Proc. IEEE*, February 1976.

On comparing (5.206) with (5.91), it is seen that $u_k^*$ has simply replaced $x_k^*$, so the resulting adaptive weight matrix differential equation now becomes

$$\tau_0 \frac{d\mathbf{w}}{dt} + \left[\mathbf{I} + \gamma'\mathbf{M}\right]\mathbf{w} = \mathbf{b}^* \tag{5.207}$$

which is exactly analogous to (5.103) with $\mathbf{M}$ replacing $\mathbf{R}_{nn}$ and $\gamma' = k^2 G'$ where $\mathbf{M}$ is the modified noise covariance with envelope limiting having elements given by

$$M_{ml} = E\left\{ \frac{x_m^* x_l}{|x_m|} \right\} \tag{5.208}$$

Assuming the quadrature components of each signal $x_k$ are zero-mean Gaussian random variables having variance $\sigma^2$, we can then compute the elements of the covariance matrix $\mathbf{M}$ directly from the elements of $\mathbf{R}_{nn}$ by using the relation [12]

$$M_{ml}=\sqrt{\frac{\pi}{8}}\;\frac{1}{\sigma}(\mathbf{R}_{nn})_{ml} \tag{5.209}$$

It follows that the elements of $\mathbf{M}$ differ from the elements of $\mathbf{R}_{nn}$ only by a common factor $(1/\sigma)\sqrt{(\pi/8)}$ . Consequently the effective time constants that determine the rate of convergence and control loop noise are changed by this same factor, thereby reducing the dependence of array performance on the strength of the external noise field.

It is worthwhile to note that limiting does not change the relative values of the elements of the signal covariance matrix or the relative magnitudes of the eigenvalues presuming identical channels. Consequently, for cases involving widely different eigenvalues, limiting will not succeed in reducing the eigenvalue spread to provide both rapid transient response and low control loop noise. Nevertheless, limiting always reduces the dynamic range of signals in the control loops and thereby simplifies the loop implementation.

## 5.4 INTRODUCTION OF MAIN BEAM CONSTRAINTS

For some interference source distributions (especially when an interference source is located in the array main beam), while the array may succeed in suppressing the interference signal by placing a beam pattern null at the jammer location using the maximum SNR algorithm, the main beam may be so distorted as a result that a net degradation in SNR is experienced. Furthermore, desired signals may sometimes be of sufficient intensity to elicit undesirable reactions from the adaptive processor, resulting in desired signal waveform distortion or cancellation. To prevent such array performance degradation from occuring, it is possible to introduce constraints so the adaptive processor maintains desired mainlobe signals while realizing good cancellation of interference in the sidelobes. The constraint methods discussed here follow the development that is given by Applebaum and Chapman [15].

The primary methods that have been proposed for applying main beam constraints may be classified as one of the following types:

1. Time Domain: When the desired signal is not present in the main beam the processor is allowed to adapt to achieve interference suppression. The adaptive weights obtained during this period are then held constant until the next adaptation or sampling period. While this approach ensures that the adaptive processor remains unaffected by mainlobe desired signals, it does not protect against mainlobe distortion resulting from main beam jamming and is also vulnerable to blinking jammers.

2. Frequency Domain: When the interference sources all have substantially wider bandwidths than the desired signal, the adaptive processor can be constrained from responding to desired signals by permitting adaptation to occur only in response to signal energy outside the signal bandwidth. This approach somewhat degrades the cancellation capability, and the main beam gain pattern may still be seriously distorted by the jammer configuration.
3. Angle Domain: There are three angle domain techniques that provide main beam constraints in the steady state as noted below:
   (a) Pilot signals.
   (b) Preadaptation spatial filters.
   (c) Control loop spatial filters.

These techniques are also helpful for constraining the array response to short duration signals since they slow down the transient response to main beam signals.

The design and performance characteristics of both the time and frequency domain methods are highly dependent on the interference environment and the radar system of concern and provide no protection against main beam distortion when main beam jamming is present. The angle domain techniques are therefore of primary interest for providing the capability of introducing main beam constraints into the adaptive processor response. Each of the three angle domain techniques are therefore discussed in turn.

### *5.4.1 Pilot Signals*

To illustrate the use of pilot signal techniques, consider the multiple sidelobe canceller (MSLC) adaptive array configuration shown in Figure 5.24, where it will be noted that an integrator with feedback structure is taken to represent the integrating filter in the Howells-Applebaum adaptive loop.

The "pilot signals" are used to shape the array beam and to maintain the gain at the peak of the main beam (thereby helping to avoid SNR degradation). The pilot signals consist of a CW tone signal injected into each element channel at a frequency that is easily filtered from the normal signal band. It is not necessary to use the beam-steering phase shifters shown in Figure 5.24, since if they are not present the pilot signals may be injected with the proper phase relationship corresponding to the desired main beam direction instead of in phase with each other as shown. The amplitudes and phases of the injected pilot signals $s_1,\ldots,s_4$ may be represented by the vector $\mu\mathbf{s}$, where $\mathbf{s}$ has unit length and $\mu$ is a scalar amplitude factor. The reference channel (or main beam) signal is represented by the injected pilot signal $s_0$.

For the adaptive control loops shown in Figure 5.24, it follows that the vector differential equation for the weight vector may be written as

$$\frac{d\mathbf{w}}{dt}=\mathbf{u}^*(t)\epsilon(t)-\mathbf{w}(t) \tag{5.210}$$

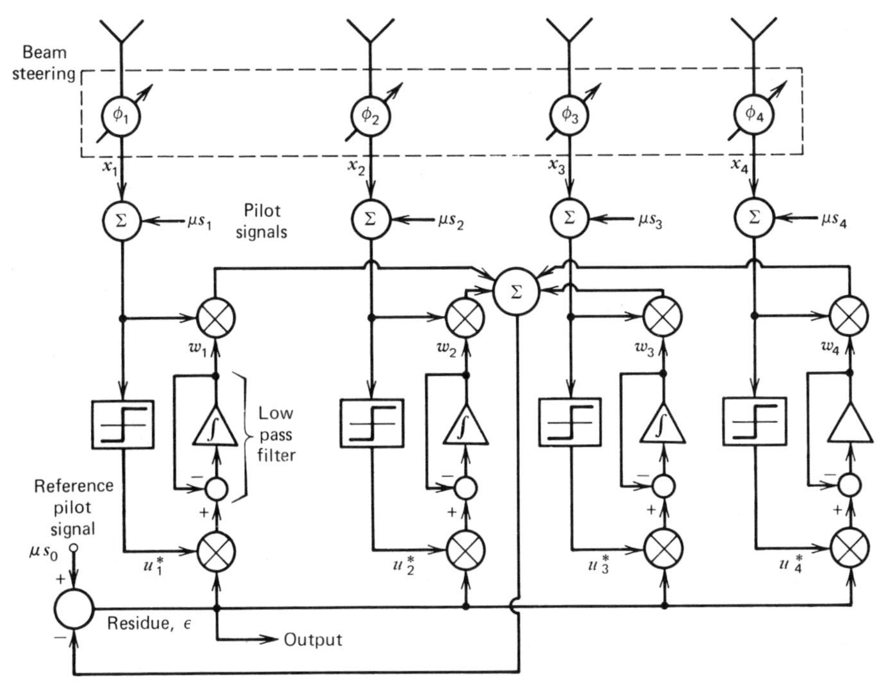


**Figure 5.24** Multiple sidelobe canceller (MSLC) adaptive array configuration with beam-steering pilot signals and main beam control.

Since $\epsilon = \mu s_0 - \mathbf{x}^T\mathbf{w}$, it follows by taking expected values of (5.210) and the results of Section 5.3.5, that

$$\frac{d\mathbf{w}}{dt} = g\mu\mathbf{r}_{xs_0} - \left[\mathbf{I} + g\mathbf{R}_{xx}\right]\mathbf{w} \tag{5.211}$$

where $g$ is a gain factor representing the correlation mixer gain and the effect of the limiter. The steady-state solution of (5.211) is given by

$$\mathbf{w}_{ss} = \left[\mathbf{I} + g\mathbf{R}_{xx}\right]^{-1} g\mu\mathbf{r}_{xs_0} \tag{5.212}$$

In the absence of any desired signal, then

$$\mathbf{x} = \mathbf{n} + \mu\mathbf{s} \tag{5.213}$$

where $\mathbf{n}$ is the noise signal vector and $\mu\mathbf{s}$ is the injected pilot signal vector. Consequently

$$\mathbf{R}_{xx} \triangleq E\{\mathbf{x}^*\mathbf{x}^T\} = \mathbf{R}_{nn} + \mu\mathbf{s}^*\mathbf{s}^T \tag{5.214}$$

and

$$\mathbf{r}_{xs_0} \triangleq E\{\mathbf{x}^* s_0\} = \mu\mathbf{s}^* s_0 \tag{5.215}$$

On substituting (5.214) and (5.215) into (5.212) it can be shown that

$$\mathbf{w}_{ss} = \frac{\mathbf{K}^{-1}\mathbf{s}^* g\mu^2 s_0}{1 + g\mu^2 \mathbf{s}^T \mathbf{K}^{-1} \mathbf{s}^*} \tag{5.216}$$

where $\mathbf{K} = \mathbf{I} + g\mathbf{R}_{nn}$. Substituting the expression for $\mathbf{K}^{-1}$ in (5.216) then yields

$$\mathbf{w}_{ss} = \frac{(\mathbf{I} + g\mathbf{R}_{nn})^{-1}\mathbf{s}^* g\mu^2 s_0}{1 + g\mu^2 \mathbf{s}^T (\mathbf{I} + g\mathbf{R}_{nn})^{-1} \mathbf{s}^*} \tag{5.217}$$

For large pilot signals $\mu^2 \to \infty$, and (5.217) becomes

$$\mathbf{w}_{ss} \cong \left[ \frac{(\mathbf{I} + g\mathbf{R}_{nn})^{-1}\mathbf{s}^*}{\mathbf{s}^T (\mathbf{I} + g\mathbf{R}_{nn})^{-1} \mathbf{s}^*} \right] s_0 \tag{5.218}$$

From (5.218) it follows that if $\mathbf{s}$ has equal amplitude components (corresponding to the sensor element voltage responses for a main beam signal), then the array response in the direction of the main beam will be

$$\mathbf{s}^T \mathbf{w}_{ss} \cong s_0 \tag{5.219}$$

and is therefore a constant, independent of the noise covariance matrix $\mathbf{R}_{nn}$ (and hence independent of any received waveforms).

The array configuration of Figure 5.24 employs one set of pilot signals to realize a single main beam constraint. Multiple constraints can also be realized by using multiple sets of pilot signals, with each set at a different frequency. Pilot signals have the advantage that they can be inserted close to the input of each element channel and thereby compensate for any amplitude and phase errors. Strong pilot signals, however, require the channel elements to have a large dynamic range, and they must be filtered out of the array output if interference with the desired signal is not to occur.

### *5.4.2 Preadaption Spatial Filters*

An entirely different approach to applying constraints to an adaptive array involves the use of the "preadaption spatial filtering," in which two beams are formed following the beam-steering phase shifters. The general structure of preadaption spatial filtering is illustrated in Figure 5.25, in which the main beam is formed with fixed weights $\mathbf{s}^*$ (forming a unit length weight vector) to form the desired quiescent pattern. The second beam (termed a "cancellation beam") is formed adaptively by an MSLC whose input channels are obtained from spatial filtering represented by the matrix transformation $\mathbf{A}$. The number of output channels from $\mathbf{A}$ is one less than the number of sensor elements, and the transformation $\mathbf{A}$ is selected to maintain a constant response in the main beam direction so that $\mathbf{A}\mathbf{s} = 0$.

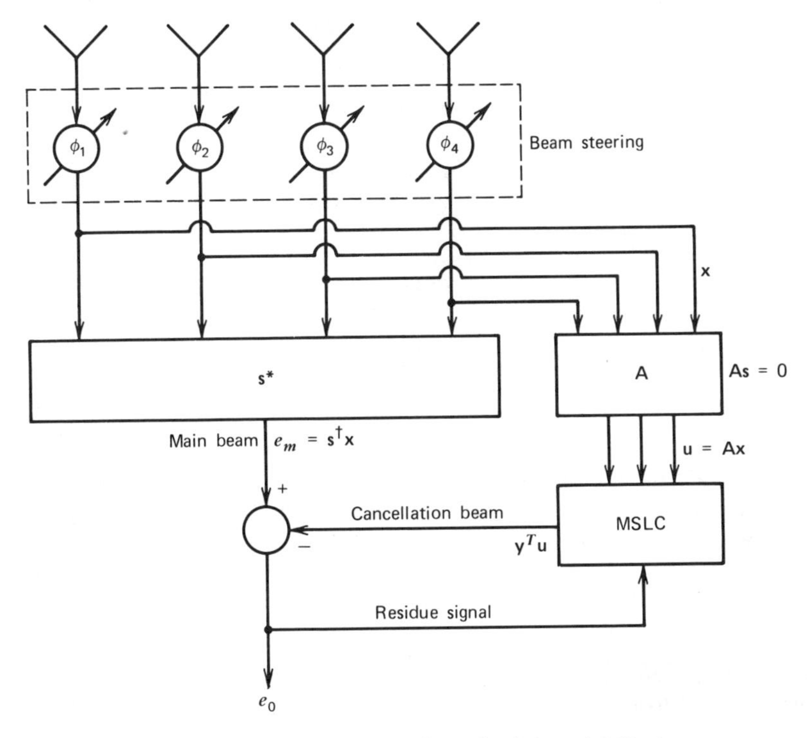


**Figure 5.25** General structure of preadaption spatial filtering.

From the signal vector definitions of Figure 5.25 it follows that:

$$e_m = \mathbf{s}^\dagger \mathbf{x} = \mathbf{x}^T \mathbf{s}^* \tag{5.220}$$

$$e_0 = e_m - \mathbf{y}^T \mathbf{u} \tag{5.221}$$

and

$$\mathbf{u} = \mathbf{A}\mathbf{x} \tag{5.222}$$

Therefore

$$e_0 = \mathbf{s}^\dagger \mathbf{x} - \mathbf{y}^T \mathbf{A}\mathbf{x} = (\mathbf{s}^* - \mathbf{A}^T \mathbf{y})^T \mathbf{x} = \mathbf{w}^T \mathbf{x} \tag{5.223}$$

The composite weight vector for the entire system can therefore be written as

$$\mathbf{w} = \mathbf{s}^* - \mathbf{A}^T \mathbf{y} \tag{5.224}$$

Consequently

$$\mathbf{s}^T \mathbf{w} = \mathbf{s}^T (\mathbf{s}^* - \mathbf{A}^T \mathbf{y}) = \|\mathbf{s}\|^2 - \mathbf{s}^T \mathbf{A}^T \mathbf{y} \tag{5.225}$$

Since $\mathbf{A}$ was selected so that $\mathbf{As} = 0$, it follows that

$$\mathbf{s}^T \mathbf{w} = \|\mathbf{s}\|^2 = 1 \tag{5.226}$$

Denote the covariance matrix associated with $\mathbf{u}$ by

$$\mathbf{R}_{uu} \triangleq E\{\mathbf{u}^*\mathbf{u}^T\} = E\{\mathbf{A}^*\mathbf{x}^*\mathbf{x}^T\mathbf{A}^T\} = \mathbf{A}^*\mathbf{R}_{xx}\mathbf{A}^T \tag{5.227}$$

The MSLC unit generates a weight vector $\mathbf{y}$ that satisfies the matrix differential equation [corresponding to (5.210)],

$$\frac{d\mathbf{y}}{dt} = g\mathbf{u}^* e_0 - \mathbf{y} \tag{5.228}$$

In the steady state [where $(d\mathbf{y}/dt) = 0$] it follows that

$$[\mathbf{I} + g\mathbf{R}_{uu}]\mathbf{y} = gE\{\mathbf{u}^* e_m\} \tag{5.229}$$

where $g$ is a gain factor and the right side of (5.229) represents the cross-correlation vector of $e_m$ with each component of $\mathbf{u}$. Using (5.220), (5.222), and (5.227) in (5.229), we find that

$$(\mathbf{I} + g\mathbf{A}^*\mathbf{R}_{xx}\mathbf{A}^T)\mathbf{y} = g\mathbf{A}^*\mathbf{R}_{xx}\mathbf{s}^* \tag{5.230}$$

Premultiply (5.230) by $\mathbf{A}^T$ and use (5.224); it then follows that the composite weight applied to the input signal vector satisfies the steady-state relation:

$$(\mathbf{I} + g\mathbf{A}^T\mathbf{A}^*\mathbf{R}_{xx})\mathbf{w}_{ss} = \mathbf{s}^* \tag{5.231}$$

when $\mathbf{x}$ does not contain a desired signal component, then $\mathbf{R}_{xx}$ may be replaced by $\mathbf{R}_{nn}$.

Now allow $g$ to become very large so that (5.230) yields

$$\mathbf{A}^*\mathbf{R}_{xx}\mathbf{A}^T\mathbf{y} = \mathbf{A}^*\mathbf{R}_{xx}\mathbf{s}^* \tag{5.232}$$

or

$$\mathbf{A}^*\mathbf{R}_{xx}(\mathbf{s}^* - \mathbf{A}^T\mathbf{y}) = \mathbf{A}^*\mathbf{R}_{xx}\mathbf{w}_{ss} = 0 \tag{5.233}$$

Now since $\mathbf{As} = 0$ and the rank of the transformation matrix $\mathbf{A}$ is $N-1$, (5.233) implies that $\mathbf{R}_{xx}\mathbf{w}$ is proportional to $\mathbf{s}^*$ so that

$$\mathbf{R}_{xx}\mathbf{w}_{ss} = \mu\mathbf{s}^* \tag{5.234}$$

or

$$\mathbf{w}_{ss} = \mu\mathbf{R}_{xx}^{-1}\mathbf{s}^* \tag{5.235}$$

where $\mu$ is a proportionality constant that may be evaluated using (5.226).

Substituting $\mu = (\mathbf{s}^T \mathbf{R}_{xx}^{-1} \mathbf{s}^*)^{-1}$ in (5.235) then yields

$$\mathbf{w}_{ss} = \frac{\mathbf{R}_{xx}^{-1} \mathbf{s}^*}{\mathbf{s}^T \mathbf{R}_{xx}^{-1} \mathbf{s}^*} \tag{5.236}$$

as the solution that the composite weight vector approaches when $g$ becomes very large.

While preadaption spatial filtering does not result in dynamic range problems, it does require the implementation of multiple beams, which can be expensive. It may also be noted that the accuracy of the beam-steering phase shifters also limits the effectiveness of the constraints, but this limit is true of all three methods considered here.

Applebaum and Chapman [15] consider two specific realizations of preadaption spatial filtering represented by Figure 5.25: (1) the use of a Butler matrix to obtain orthogonal beams, one of which is regarded as the "main" beam, and (2) the use of an **A** matrix transformation obtained by fixed element-to-element subtraction. The manner in which these specific realizations operate to introduce constraints nevertheless is the same as that already explained for the general structure of Figure 5.25.

### *5.4.3 Control Loop Spatial Filters*

The Howells-Applebaum adaptive control loop with constraints applied directly in the loop by means of a spatial matrix filter is illustrated by the configuration of Figure 5.26. The purpose of the spatial matrix filter is to remove any components of the signal vector **v** pointing in the direction of the unit length beam-steering vector $\mathbf{b}^*$ by means of a projection operator. The successful removal of such signal components will then constrain the array response in the direction of **b**.

For the configuration of Figure 5.26, it is seen that the amplified output of the correlation mixer is given by

$$\mathbf{v}(t) = \mathbf{u}^*(t)\mathbf{x}^T(t)\mathbf{w}(t) \tag{5.237}$$

Taking expected values of all quantities in (5.237), we see that it follows that expectations of the steady-state values may be written as

$$\mathbf{v} = g\mathbf{R}_{xx}\mathbf{w}_{ss} \tag{5.238}$$

where $g$ is a gain factor. Now

$$\mathbf{z} = \hat{\mathbf{P}}\mathbf{v} = g\hat{\mathbf{P}}\mathbf{R}_{xx}\mathbf{w}_{ss} \tag{5.239}$$

Since $\mathbf{w} = \mathbf{b}^* - \mathbf{z}$, however, it follows that (5.239) can be rewritten as a relation that the steady-state weight values must satisfy:

$$(\mathbf{I} + g\hat{\mathbf{P}}\mathbf{R}_{xx})\mathbf{w}_{ss} = \mathbf{b}^* \tag{5.240}$$

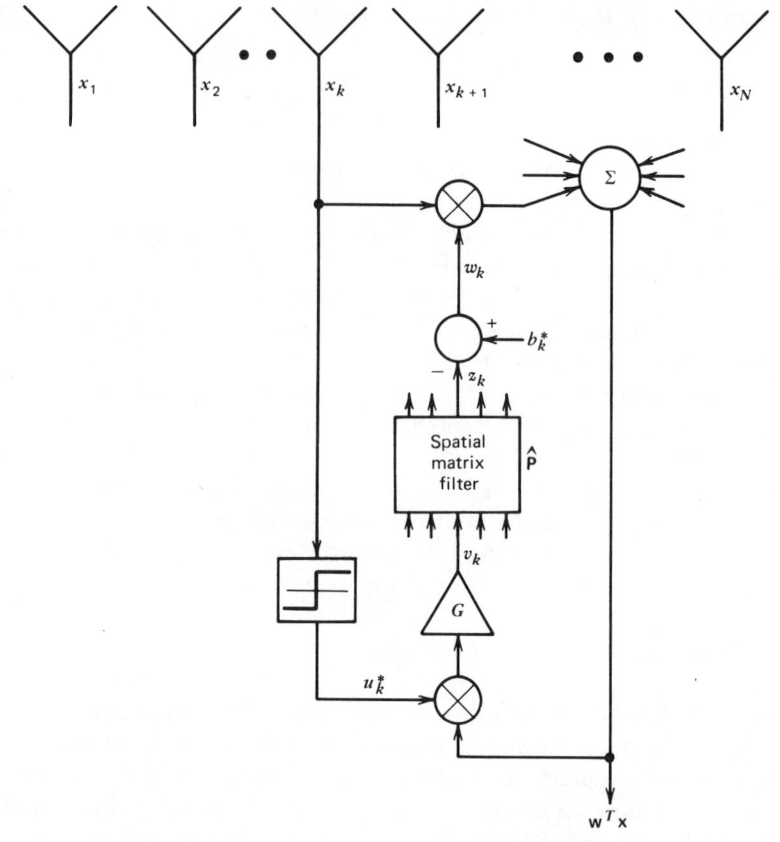


**Figure 5.26** Adaptive processor with control loop spatial filtering.

When the beam-steering vector is uniformly weighted, the projection performed by the spatial filter to remove signal components in the direction of **b** is

$$\hat{\mathbf{P}} = \mathbf{I} - \mathbf{b}^*\mathbf{b}^T \tag{5.241}$$

Substituting (5.241) into (5.240) results in

$$(\mathbf{I} + g\mathbf{R}_{xx} - g\mathbf{b}^*\mathbf{b}^T\mathbf{R}_{xx})\mathbf{w}_{ss} = \mathbf{b}^* \tag{5.242}$$

or

$$\left[(\mathbf{I} + g\mathbf{R}_{xx})\mathbf{R}_{xx}^{-1} - g\mathbf{b}^*\mathbf{b}^T\right]\mathbf{R}_{xx}\mathbf{w}_{ss} = \mathbf{b}^* \tag{5.243}$$

Rewrite (5.243) in the form

$$\mathbf{R}_{xx}\mathbf{w}_{ss} = \left[\mathbf{Q} - g\mathbf{b}^*\mathbf{b}^T\right]^{-1}\mathbf{b}^* \tag{5.244}$$

where $\mathbf{Q} \triangleq (\mathbf{I}+g\mathbf{R}_{xx})\mathbf{R}_{xx}^{-1}$, and apply a matrix inversion identity to obtain the result

$$[\mathbf{Q}-g\mathbf{b}^*\mathbf{b}^T]^{-1}\mathbf{b}^* = \frac{\mathbf{R}_{xx}(\mathbf{I}+g\mathbf{R}_{xx})^{-1}\mathbf{b}^*}{1-g\mathbf{b}^T\mathbf{R}_{xx}(\mathbf{I}+g\mathbf{R}_{xx})^{-1}\mathbf{b}^*} \tag{5.245}$$

The denominator of (5.240) may be simplified as follows:

$$\begin{aligned} g\mathbf{b}^T\mathbf{R}_{xx}(\mathbf{I}+g\mathbf{R}_{xx})^{-1}\mathbf{b}^* &= \mathbf{b}^T(\mathbf{I}+g\mathbf{R}_{xx}-\mathbf{I})(\mathbf{I}+g\mathbf{R}_{xx})^{-1}\mathbf{b}^* \\ &= \mathbf{b}^T\mathbf{b}^* - \mathbf{b}^T(\mathbf{I}+g\mathbf{R}_{xx})^{-1}\mathbf{b}^* \\ &= 1-\mathbf{b}^T(\mathbf{I}+g\mathbf{R}_{xx})^{-1}\mathbf{b}^* \end{aligned} \tag{5.246}$$

Substituting (5.245) and (5.246) into (5.244) then yields the following relationship that the steady-state weight vector must satisfy:

$$\mathbf{w}_{ss} = \frac{(\mathbf{I}+g\mathbf{R}_{xx})^{-1}\mathbf{b}^*}{\mathbf{b}^T(\mathbf{I}+g\mathbf{R}_{xx})^{-1}\mathbf{b}^*} \tag{5.247}$$

The introduction of $\hat{\mathbf{P}}=\mathbf{I}-\mathbf{b}^*\mathbf{b}^T$ only constrains the array response in the direction of $\mathbf{b}$. Additional constraints may be required should it be desired to constrain the response over a finite region of the mainlobe instead of only a single direction. Likewise when the steering vector is other than uniform weighting, an additional constraint is also required to obtain the desired quiescent pattern. It is furthermore desirable to transform the constraints to an orthogonal set thereby minimizing the accuracy requirements of the spatial matrix filter. To illustrate a projection filter constructed from an orthogonal set, consider the representation of a projection filter with $M+1$ orthogonal constraints:

$$\hat{\mathbf{P}} = \mathbf{I} - \sum_{m=0}^{M} \mathfrak{c}_m \mathfrak{c}_m^{\dagger} \tag{5.248}$$

where $\mathfrak{c}_m \mathfrak{c}_n = \delta_{mn}$, the Kronecker delta. The $\mathfrak{c}_m$ are the constraint vectors, of which some interesting examples are discussed below.

The constraint that maintains the array response at the peak of the beam is usually referred to as the "zero-order" constraint. The weight vector solution obtained with a zero-order constraint will differ from the unconstrained solution ($\hat{\mathbf{P}}=\mathbf{I}$) only by a multiplicative scale factor.

Since a single point constraint will usually not provide sufficient influence on the beam response to be of much use, consider the use of multiple constraints to increase the zone of beam constraint by controlling the first few derivatives of the pattern function in the direction of interest. A constraint that controls the $m$th derivative is usually referred to as an "$m$th-order" constraint.

To synthesize a $\mathbf{c}_m$ constraint vector corresponding to the $m$th derivative of the pattern function, note that the pattern function of a linear array can be written as

$$G(\theta)=\sum_{k=1}^{N} w_k e^{jk\theta} \tag{5.249}$$

It immediately follows that the $m$th derivative of $G(\theta)$ is given by

$$G^m(\theta)=\sum_{k=1}^{N} (jk)^m w_k e^{jk\theta} \tag{5.250}$$

Consequently the elements of $\mathbf{c}_m$ (for $m=0, 1,$ and 2) are given by

$$c_{0i}=d_0 \tag{5.251}$$

$$c_{1i}=e_0+e_1 i \tag{5.252}$$

and

$$c_{2i}=f_0+f_1 i+f_2 i^2 \tag{5.253}$$

The constants defining the elements of $\mathbf{c}_m$ are adjusted so that all $\mathbf{c}_m$ are of unit length and are mutually orthogonal.

Now consider how to establish a beam having nonuniform weighting as well as zero-, first-, and second-order constraints on the beam shape at the center of the main beam. First expand $\mathbf{w}_q$ (the vector of desired quiescent weights) in terms of the constraint vectors $\mathbf{c}_m$ (for $m=0$, 1, and 2) and a remainder vector $\mathbf{c}_r$ as

$$\mathbf{w}_q=a_0\mathbf{c}_0+a_1\mathbf{c}_1+a_2\mathbf{c}_2+a_r\mathbf{c}_r \tag{5.254}$$

where

$$a_i=\mathbf{w}_q^T\mathbf{c}_i \qquad \text{for } i=0,1,2 \tag{5.255}$$

and

$$a_r\mathbf{c}_r=w_q-\sum_{i=0}^{2} a_i\mathbf{c}_i \tag{5.256}$$

Now construct the complementary projection matrix filter according to

$$\hat{\mathbf{P}}=\mathbf{I}-\mathbf{c}_0\mathbf{c}_0^{\dagger}-\mathbf{c}_1\mathbf{c}_1^{\dagger}-\mathbf{c}_2\mathbf{c}_2^{\dagger}-\cdots-\mathbf{c}_r\mathbf{c}_r^{\dagger} \tag{5.257}$$

The subspace spanned by the constraint vectors in the $N$-dimension space of the adaptive processor is preserved by the foregoing construction. The spatial matrix filter constructed according to (5.257) will result in a signal vector $\mathbf{z}$ containing no components in the direction of $\mathbf{w}_q$ or its first and second derivatives. The vector $\mathbf{w}_q$ is now added back in (at the point in Figure 5.26 where $\mathbf{b}^*$ is inserted) to form the final weight vector $\mathbf{w}$.

## 5.5 CONSTRAINT FOR THE CASE OF KNOWN DESIRED SIGNAL POWER LEVEL

In some applications there may be no a priori knowledge concerning either the desired signal structure or the direction of arrival. It may nevertheless be possible to exploit the power discrimination capabilities of adaptive arrays to cope with the problem of acquiring a weak desired signal in the presence of strong jamming or interference. This can be accomplished by placing a constraint on the adaptation algorithm to prevent suppression of all signals (including interference) that are below a specified input power level. By selecting a signal power threshold that is greater than the desired signal input power, the weak desired signal will not be suppressed while all interference signals above the threshold will be suppressed.

The most common method of obtaining a power discrimination capability was formulated by Compton [16] and is based on the use of proportional feedback control in a Howells-Applebaum adaptive loop. To accomplish the same result Zahm [17] has also proposed another technique that employs a combination of a steering command vector and a bias signal. To illustrate the proportional feedback control method for imposing a power threshold constraint, consider the adaptive null-steering array illustrated in Figure 5.27. The weight adjustment control loops are governed by the differential equation

$$\frac{d\mathbf{w}}{dt}=\alpha\{\mathbf{x}(t)[x_0(t)-\mathbf{x}^{\dagger}(t)\mathbf{w}]-a\mathbf{w}\} \tag{5.258}$$

where $a$ is a real scalar constant. The additional feedback path around the integrator provides the means for setting a power threshold.

For $\alpha$ sufficiently small, (5.258) can be approximated by expected values so that

$$\frac{d\mathbf{w}}{dt}=\alpha(\mathbf{r}_{xx_0}-[\mathbf{R}_{xx}+a\mathbf{I}]\mathbf{w}) \tag{5.259}$$

The steady-state weight vector is then given by

$$\mathbf{w}=[\mathbf{R}_{xx}+a\mathbf{I}]^{-1}\mathbf{r}_{xx_0} \tag{5.260}$$

If a desired signal is present, then we may write the output signal to

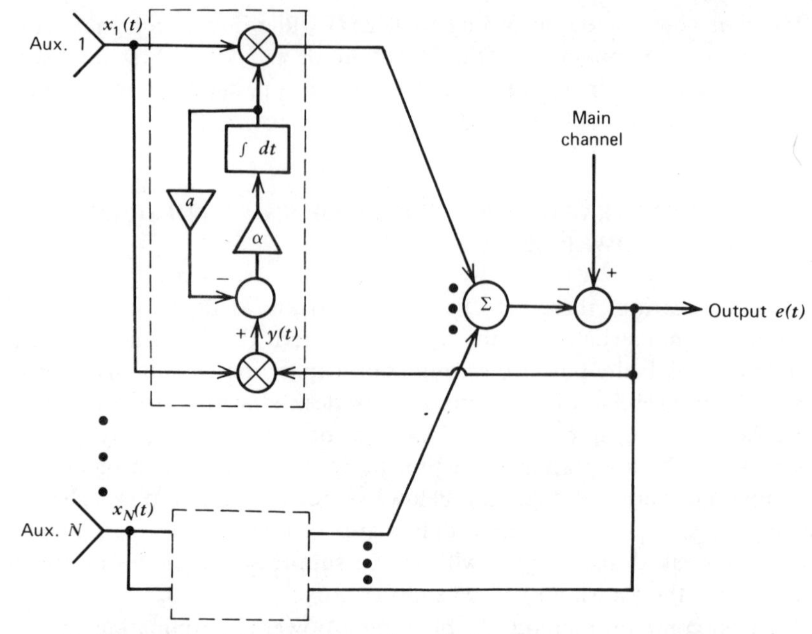


**Figure 5.27** Adaptive null-steering with $N$ auxiliary sensor elements.

interference plus noise ratio as

$$\mathrm{SNR} = \frac{1}{\left(P_e / \overline{|s_0 - \mathbf{w}^\dagger \mathbf{s}|^2}\right) - 1} \tag{5.261}$$

where $P_e$ represents the total output power, $\mathbf{s}$ is the desired signal direction vector, and $s_0$ is the main channel desired signal component. It will be convenient to define the parameter

$$SN' = \frac{\overline{|s_0 - \mathbf{w}^\dagger \mathbf{s}|^2}}{P_e} \tag{5.262}$$

so that

$$\mathrm{SNR} = \frac{1}{(1/SN') - 1} \tag{5.263}$$

It can be shown that

$$SN' = \frac{\left| s_0 - \sum_{i=1}^{N} \frac{(\mathbf{Qr}_{xx_0})_i^* (\mathbf{Qs})_i}{\lambda_i + a} \right|^2}{a^2 \sum_{i=1}^{N} \frac{|(\mathbf{Qr}_{xx_0})_i|^2}{\lambda_i (\lambda_i + a)^2} + P_{e_0}} \tag{5.264}$$

where $\mathbf{Q}$ is the unitary transformation that diagonalizes $\mathbf{R}_{xx}$, $\lambda_i$ are the eigenvalues of $R_{xx}$, and $P_{e_0}$ represents the minimum output power of $e(t)$ when $a=0$.

It can be shown from the above results and (5.258) that $a$ may be selected to prevent cancellation of the desired signal while suppressing high-level jammers. The output signal to interference plus noise ratio is therefore higher than when a pure integrator is used in the feedback loop.

In order to maximize the output SNR, the parameter $a$ is selected to maximize $SN'$. To illustrate how $a$ is selected, consider the case of one interfering jammer so that

$$\mathbf{R}_{xx} = J\mathbf{v}_J\mathbf{v}_J + P_s\mathbf{v}_s\mathbf{v}_s \tag{5.265}$$

and

$$\mathbf{r}_{xx_0} = \sqrt{J_0 J}\;\mathbf{v}_J e^{j\phi_J} + \sqrt{P_{s_0}P_s}\;\mathbf{v}_s e^{j\phi_s} \tag{5.266}$$

where $J_0$ = main channel jammer power,
$J$ = auxiliary channel jammer power (assumed equal in all auxiliary channels),
$\mathbf{v}_J$ = jammer direction delay vector.

$P_{s_0}$, $P_s$, and $\mathbf{v}_s$ are similarly defined for the desired signal. $\phi_J$ and $\phi_s$ represent the relative phase between the main and auxiliary channel signals for the jamming and desired signals, respectively.

If the desired signal and the interference signal angles of arrival are such that $\mathbf{v}_s$ and $\mathbf{v}_J$ are orthogonal (which simplifies the discussion for tutorial purposes), then (5.264) reduces to

$$SN' = \frac{P_{s_0}\left[\dfrac{\sigma^2 + a}{\sigma^2 + a + NP_s}\right]^2}{a^2 N\left[\dfrac{P_{s_0}P_s}{(NP_s + \sigma^2 + a)^2(NP_s + \sigma^2)} + \dfrac{J_0 J}{(NJ + \sigma^2 + a)^2(NJ + \sigma^2)}\right] + P_{e_0}} \tag{5.267}$$

where $\sigma^2$ = auxiliary channel thermal noise power.

For $a=0$ (which corresponds to the conventional LMS null-steering algorithm) $SN'$ becomes

$$SN' = \frac{P_{s_0}\left[\dfrac{\sigma^2}{NP_s + \sigma^2}\right]^2}{P_{e_0}}; \qquad a=0 \tag{5.268}$$

The result shows that $SN'$ decreases as the input desired signal power in the auxiliary channels $NP_s$ increases above the thermal noise level $\sigma^2$. When $NP_s \gg \sigma^2$ in (5.268), $SN'$ is inversely proportional to the input desired signal power; this is the power inversion characteristic of the minimum MSE performance criterion. When $a \gg NP_s + \sigma^2$ and $a \gg NJ + \sigma^2$, (5.267) reduces to

$$SN' = \frac{P_{s_0}}{\dfrac{NP_s}{NP_s+\sigma^2}P_{s_0} + \dfrac{NJ}{NJ+\sigma^2}J_0 + P_{e_0}} \tag{5.269}$$

This result shows that suppression of the main channel signal can be prevented by selecting $a$ to be sufficiently large. However, $a$ is too large in this example because jammer suppression has also been prevented, as indicated by the presence of the term $NJJ_0/(NJ+\sigma^2)$ in the denominator of (5.269).

Next, assume that the main channel jammer power $J_0$ is nominally equal to the auxiliary channel jammer power, and choose $a = NP_s$. Then $SN'$ becomes

$$SN' = \frac{0.25P_{s_0}}{\dfrac{P_{s_0}(NP_s-\sigma^2)^2}{4NP_s(NP_s+\sigma^2)} + \dfrac{(NP_s-\sigma^2)^2}{N[1+(P_s/J)]^2(NJ+\sigma^2)} + P_{e_0}} \tag{5.270}$$

For $J \gg P_s$, $P_s \gg \sigma^2$,

$$SN' \approx \frac{0.25P_{s_0}}{0.25P_{s_0} + P_{e_0}}$$

so that the expression for the output signal to interference plus noise ratio in (5.263) is approximated by

$$\text{SNR} \cong \frac{1}{4}\frac{P_{s_0}}{P_{e_0}} \tag{5.271}$$

Thus the output signal to interference plus noise ratio is now proportional to the main channel signal power divided by the output residue power $P_{e_0}$ (recall that $P_{e_0}$ is the minimum output residue power obtained when $a=0$). Equation (5.271) shows that when $J \gg P_s$ and $P_s \gg \sigma^2$, the output signal to interference plus noise ratio can be significantly improved by selecting the weight feedback gain as

$$a \approx NP_s \tag{5.272}$$

This value of $a$ (when $J \gg P_s$ and $P_s \gg \sigma^2$) then prevents suppression of the relatively weak desired signal while strongly suppressing higher power level jamming signals.

## 5.6 SUMMARY AND CONCLUSIONS

The Howells-Applebaum adaptive processor that maximizes the output SNR is introduced, and it is found to be very closely related to the LMS algorithm discussed in Chapter 4. The Howells-Applebaum adaptive processor is generally employed in situations where the desired signal is usually absent (in contrast to the LMS algorithm which requires that the desired signal be present) and makes use of a beam-steering vector instead of a reference signal. We find that the Howells-Applebaum processor behavior is characterized by a quiescent mode (when the desired signal is absent) and an adapted mode (when the desired signal is present). The transient behavior of the algorithm is most easily described in terms of eigenvector beams, which can be analyzed by introduction of a transformation to a normal coordinate system that diagonalizes the signal covariance matrix. The processor exhibits the characteristic of sensitivity to eigenvalue spread, so that strong interference sources are cancelled rapidly while weak interference sources are suppressed slowly. The dynamic range requirements of the circuitry used to implement the Howells-Applebaum processor can be reduced by introduction into the control loop of a hard limiter that modifies the effective signal covariance matrix but does not affect the eigenvalue spread.

Different methods for constraining the maximum SNR algorithm to maintain a favorable desired signal response are discussed which include the following:

1. The use of pilot signals.
2. Preadaptation spatial filters.
3. Control loop spatial filters.
4. Discrimination constraint for known desired signal power level.

The close relationship that exists between the Howells-Applebaum maximum SNR processor and the LMS algorithm makes the similar transient behavior characteristics of these two algorithms hardly surprising. The susceptibility of the algorithm performance to eigenvalue spread in the signal covariance matrix leads to a consideration of ways in which this susceptability can be reduced. One way of reducing this susceptability and maintaining fast transient response for all eigenvalue conditions is to employ a direct matrix inversion (DMI) algorithm, which is introduced in the next chapter.

## PROBLEMS

**1** [15] Show that (5.216) results from (5.212) by the following:

(a) Substitute (5.214) and (5.215) into (5.212).

(b) Let $\mathbf{K}=\mathbf{I}+g\mathbf{R}_{nn}$.

(c) Apply the matrix inversion lemma [(D.10) of Appendix D] to the resulting expression.

**2** [15] Show that (5.231) follows from the steady-state relationship given by (5.229).

**3** [15] Apply the matrix inversion identity

$$[\mathbf{Q}+\mathbf{e}\mathbf{f}^T]^{-1}\mathbf{e}=\frac{\mathbf{Q}^{-1}\mathbf{e}}{1+\mathbf{f}^T\mathbf{Q}^{-1}\mathbf{e}}$$

where $\mathbf{Q}$ is a nonsingular $N \times N$ matrix and $\mathbf{e}$ and $\mathbf{f}$ are $\mathbf{N}\times 1$ vectors to (5.244), and show that (5.245) results.

**4** [15] By substituting the relationships expressed by (5.245) and (5.246) into (5.244), show that the steady-state weight vector relationship given by (5.247) results.

**5** To show that (5.264) can be developed from (5.262), define the ratio

$$SN' \triangleq \frac{\mathbf{w}'^{\dagger}\mathbf{s}'\mathbf{s}'^{\dagger}\mathbf{w}'}{\mathbf{w}'^{\dagger}\mathbf{R}'_{xx}\mathbf{w}'}$$

where

$$\mathbf{w}'=\begin{bmatrix}-\mathbf{w}\\ \hline 1\end{bmatrix}, \qquad \mathbf{s}'=\begin{bmatrix}\mathbf{s}\\ \hline s_0\end{bmatrix}$$

and

$$\mathbf{R}'_{xx}=\left[\begin{array}{c|c}\mathbf{R}_{xx} & \mathbf{r}_{xx_0}\\ \hline \mathbf{r}^{\dagger}_{xx_0} & P_0\end{array}\right]$$

(a) Show that $|s_0-\mathbf{w}^{\dagger}\mathbf{s}|^2=\mathbf{w}'^{\dagger}\mathbf{s}'\mathbf{s}'^{\dagger}\mathbf{w}'$.

(b) Show that $P_e \triangleq \mathbf{w}'^{\dagger}\mathbf{R}'_{xx}\mathbf{w}'=\mathbf{w}^{\dagger}\mathbf{R}_{xx}\mathbf{w}-\mathbf{w}^{\dagger}\mathbf{r}_{xx_0}-\mathbf{r}^{\dagger}_{xx_0}w+P_0$.

(c) Since $\mathbf{w}=[\mathbf{R}_{xx}+a\mathbf{I}]^{-1}\mathbf{r}_{xx_0}$ from (5.260) show that

$$\mathbf{w}=\mathbf{w}_{\text{opt}}+\Delta\mathbf{w}$$

where $\mathbf{w}_{\text{opt}}=\mathbf{R}_{xx}^{-1}\mathbf{r}_{xx_0}$ (the Wiener solution), and $\Delta\mathbf{w}=-a\mathbf{R}_{xx}^{-1}\mathbf{w}$.

(d) Substitute $\mathbf{w}=\mathbf{w}_{\text{opt}}+\Delta\mathbf{w}$ into $P_e$ from part (b) and show that

$$P_e=P_{e_0}+\Delta\mathbf{w}^{\dagger}\mathbf{R}_{xx}\Delta\mathbf{w}$$

where

$$P_{e_0} = P_0 - \mathbf{w}_{\text{opt}}^{\dagger}\mathbf{r}_{xx_0} - \mathbf{r}_{xx_0}^{\dagger}\mathbf{w}_{\text{opt}} + \mathbf{w}_{\text{opt}}^{\dagger}\mathbf{R}_{xx}\mathbf{w}_{\text{opt}}$$
$$= P_0 - \mathbf{r}_{xx_0}^{\dagger}\mathbf{R}_{xx}^{-1}\mathbf{r}_{xx_0}$$
$$= P_0 - \mathbf{w}_{\text{opt}}^{\dagger}\mathbf{R}_{xx}\mathbf{w}_{\text{opt}}$$

*Hint.* Note that

$$\Delta\mathbf{w}^{\dagger}\mathbf{R}_{xx}\mathbf{w}_{\text{opt}} + \mathbf{w}_{\text{opt}}^{\dagger}\mathbf{R}_{xx}\Delta\mathbf{w} - \Delta\mathbf{w}^{\dagger}\mathbf{r}_{xx_0} - \mathbf{r}_{xx_0}^{\dagger}\Delta\mathbf{w} = 0$$

because

$$\Delta\mathbf{w}^{\dagger}\mathbf{R}_{xx}\mathbf{w}_{\text{opt}} = \Delta\mathbf{w}^{\dagger}\mathbf{R}_{xx}\mathbf{R}_{xx}^{-1}\mathbf{r}_{xx_0}$$
$$= \Delta\mathbf{w}^{\dagger}\mathbf{r}_{xx_0}$$

(e) Show that

$$\Delta\mathbf{w}^{\dagger}\mathbf{R}_{xx}\Delta\mathbf{w} = a^2 \sum_{i=1}^{N} \frac{|(\mathbf{Q}\mathbf{r}_{xx_0})_i|^2}{\lambda_i(\lambda_i + a)^2}$$

by using $\Delta\mathbf{w} = -a\mathbf{R}_{xx}^{-1}\mathbf{w}$.
*Hint.* Note that

$$\mathbf{r}_{xx_0}^{\dagger}\mathbf{Q}^{-1}\mathbf{Q}[\mathbf{R}_{xx} + a\mathbf{I}]^{-1}\mathbf{Q}\mathbf{Q}^{-1}\mathbf{R}_{xx}^{-1}\mathbf{Q}\mathbf{Q}^{-1} \cdot [\mathbf{R}_{xx} + a\mathbf{I}]^{-1}\mathbf{Q}\mathbf{Q}^{-1}\mathbf{r}_{xx_0}$$

is comprised entirely of diagonalized matrices since

$$\mathbf{Q}\mathbf{R}_{xx}^{-1}\mathbf{Q}^{-1} = \mathbf{\Lambda} \text{ and } \mathbf{Q}\mathbf{Q}^{-1} = \mathbf{I}$$

**6** [18] ***Performance Degradation Due to Errors in the Assumed Direction of Signal Incidence.*** The received signal vector can be represented by

$$\mathbf{x}(t) = \mathbf{s}(t) + \sum_{i=2}^{m} \mathbf{g}_i(t) + \mathbf{n}(t)$$

where

$$\mathbf{s}(t) = \text{desired signal vector} = s(t)\mathbf{v}_1$$
$$\mathbf{g}_i(t) = \text{directional noise sources} = g_i(t)\mathbf{v}_i$$

and

$\mathbf{n}(t)=$ thermal noise vector comprised of narrowband Gaussian noise components independent from one sensor element to the next.

The vectors $\mathbf{v}_i$, $i=1,\ldots,m$ can be regarded as steering vectors where

$$\mathbf{v}_i^T=\left[\exp(-j\omega_c\tau_{i1}),\exp(-j\omega_c\tau_{i2}),\ldots,\exp(-j\omega_c\tau_{iN})\right]$$

and $\tau_{ik}$ represents the delay of the $i$th directional signal at the $k$th sensor relative to the geometric center of the array; and $\omega_c$ is the carrier signal frequency.

The optimum weight vector should satisfy
$\mathbf{w}_{\text{opt}}=\mathbf{R}_{xx}^{-1}\mathbf{r}_{xd}$ where $\mathbf{R}_{xx}$ is the received signal covariance matrix and $\mathbf{r}_{xd}$ is the cross-correlation vector between the desired signal $s$ and the received signal vector $\mathbf{x}$. Direction of arrival information is contained in $\mathbf{r}_{xd}$, and if the direction of incidence is assumed known, then $\mathbf{r}_{xd}$ can be specified and only $\mathbf{R}_{xx}^{-1}$ need be determined to find $\mathbf{w}_{\text{opt}}$. If the assumed direction of incidence is in error, however, then $\mathbf{w}=\mathbf{R}_{xx}^{-1}\tilde{\mathbf{r}}_{xd}$ where $\tilde{\mathbf{r}}_{xd}$ represents the cross-correlation vector computed using the errored signal steering vector $\tilde{\mathbf{v}}_1$.

(a) For the foregoing signal model, the optimum weight vector can be written as $\mathbf{w}_{\text{opt}}=[S\mathbf{v}_1\mathbf{v}_1^{\dagger}+\mathbf{R}_{nn}]^{-1}\cdot(S\mathbf{v}_1)$ where $\mathbf{R}_{nn}$ denotes the noise covariance matrix and $S$ denotes the desired signal power per sensor. If $\mathbf{v}_1$ is in error, then $\tilde{\mathbf{r}}_{xd}=(S\tilde{\mathbf{v}}_1)$. Show that the resulting weight vector computed using $\tilde{\mathbf{r}}_{xd}$ is given by

$$\mathbf{w}=\frac{S}{1+S\mathbf{v}_1^{\dagger}\mathbf{R}_{nn}^{-1}\mathbf{v}_1}\left[(1+S\mathbf{v}_1^{\dagger}\mathbf{R}_{nn}^{-1}\mathbf{v}_1)\mathbf{R}_{nn}^{-1}\tilde{\mathbf{v}}_1-S\mathbf{v}_1^{\dagger}\mathbf{R}_{nn}^{-1}\tilde{\mathbf{v}}_1\mathbf{R}_{nn}^{-1}\mathbf{v}_1\right]$$

(b) Using the result obtained in part (a), show that the output signal-to-noise power ratio (when only the desired signal and thermal noise are present) from the array is given by

$$\left(\frac{S}{N}\right)_{\text{out}}=\frac{\mathbf{w}^{\dagger}E\{\mathbf{s}\mathbf{s}^{\dagger}\}\mathbf{w}}{\mathbf{w}^{\dagger}\mathbf{R}_{nn}\mathbf{w}}=\frac{S\mathbf{w}^{\dagger}(\mathbf{v}_1\mathbf{v}_1^{\dagger})\mathbf{w}}{\mathbf{w}^{\dagger}\mathbf{R}_{nn}\mathbf{w}}$$

$$=\frac{S|\mathbf{v}_1^{\dagger}\mathbf{R}_{nn}^{-1}\tilde{\mathbf{v}}_1|^2}{\tilde{\mathbf{v}}_1^{\dagger}\mathbf{R}_{nn}^{-1}\tilde{\mathbf{v}}_1-2S|\mathbf{v}_1^{\dagger}\mathbf{R}_{nn}^{-1}\tilde{\mathbf{v}}_1|^2+\mathbf{v}_1^{\dagger}\mathbf{R}_{nn}^{-1}\mathbf{v}_1\left[S^2\{(\mathbf{v}_1^{\dagger}\mathbf{R}_{nn}^{-1}\mathbf{v}_1)^*\times(\tilde{\mathbf{v}}_1^{\dagger}\mathbf{R}_{nn}^{-1}\tilde{\mathbf{v}}_1)-|\mathbf{v}_1^{\dagger}\mathbf{R}_{nn}^{-1}\tilde{\mathbf{v}}_1|^2\}+2S\tilde{\mathbf{v}}_1^{\dagger}\mathbf{R}_{nn}^{-1}\tilde{\mathbf{v}}_1\right]}$$

(c) Use the fact that $\mathbf{R}_{nn} = \sigma^2\mathbf{I}$ and the result of part (b) to show that

$$\left(\frac{S}{N}\right)_{\text{out}} = \frac{S\left(\dfrac{N}{\sigma^2}\right)\dfrac{|\mathbf{v}_1^\dagger\tilde{\mathbf{v}}_1|^2}{N^2}}{\left(1+\dfrac{NS}{\sigma^2}\right)^2\left[1-\dfrac{|\mathbf{v}_1^\dagger\tilde{\mathbf{v}}_1|^2}{N^2}\right]+2\left(\dfrac{NS}{\sigma^2}\right)\left[1-\dfrac{|\mathbf{v}_1^\dagger\tilde{\mathbf{v}}_1|^2}{N^2}\right]}$$

(d) Show for a uniform linear array that

$$|\mathbf{v}_1^\dagger\tilde{\mathbf{v}}_1| = \frac{|\sin\left[(N\pi d/\lambda_c)\sin\tilde{\theta}\right]|^2}{|\sin\left[(\pi d/\lambda_c)\sin\tilde{\theta}\right]|^2}$$

where $d$ represents the separation between sensors and $\tilde{\theta}$ represents the angular uncertainty from boresight.

## REFERENCES


[1] P. W. Howells, "Intermediate Frequency Side-Lobe Canceller," U.S. Patent 3202990, August 24, 1965.

[2] S. P. Applebaum, "Adaptive Arrays," Syracuse University Research Corp., Report SPL TR 66-1, August 1966.

[3] L. E. Brennan and I. S. Reed, "Theory of Adaptive Radar," *IEEE Trans. Aerosp. Electron. Syst.*, Vol. AES-9, No. 2, March 1973, pp. 237–252.

[4] L. E. Brennan and I. S. Reed, "Adaptive Space-Time Processing in Airborne Radars," Technology Service Corporation, Santa Monica, CA, Report TSC-PD-061-2, February 24, 1971.

[5] A. L. McGuffin, "Adaptive Antenna Compatibility with Radar Signal Processing," in Proceedings of the Array Antenna Conference, February 1972, Naval Electronics Laboratory, San Diego, CA.

[6] L. E. Brennan, E. L. Pugh, and I. S. Reed, "Control Loop Noise in Adaptive Array Antennas," *IEEE Trans. Aerosp. Electron. Syst.*, Vol. AES-7, No. 2, March 1971, pp. 254–262.

[7] W. F. Gabriel, "Adaptive Arrays—An Introduction," *Proc. IEEE*, Vol. 64, No. 2, February 1976, pp. 239–272.

[8] D. L. Margerum, "Self-Phased Arrays," in *Microwave Scanning Antennas*, Vol. 3, edited by R. C. Hansen, Academic Press, New York, 1966, Ch. 5.

[9] A. E. Covington and N. W. Broten, "An Interferometer for Radio Astronomy with a Single-Lobed Radiation Pattern," *IRE Trans. Antennas Propag.*, Vol. AP-5, July 1957, pp. 247–255.

[10] J. L. Pawsey and R. N. Bracewell, *Radio Astronomy, International Monographs on Radio*, Clarendon Press, Oxford, England, 1955.

[11] H. Chestnut and R. W. Mayer, *Servomechanisms and Regulating System Design*, Vol. 1, 2nd ed., Wiley, New York, 1959.

[12] L. E. Brennan and I. S. Reed, "Effect of Envelope Limiting in Adaptive Array Control Loops," *IEEE Trans. Aerosp. Electron. Syst.*, Vol. AES-7, No. 4, July 1971, pp. 698–700.

[13] A. J. Berni, "Weight Jitter Phenomena in Adaptive Control Loops," *IEEE Trans. Aerosp. Electron. Syst.*, Vol. AES-14, No. 4, July 1977, pp. 355–361.

[14] D. E. N. Davies, "Independent Angular Steering of Each Zero of the Directional Pattern for a Linear Array," *IEEE Trans. Antennas Propag.*, Vol. AP-15, March 1967, pp. 296–298.

[15] S. P. Applebaum and D. J. Chapman, "Adaptive Arrays with Main Beam Constraints," *IEEE Trans. Antennas Propag.*, Vol. AP-24, No. 5, September 1976, pp. 650–662.

[16] R. T. Compton, Jr., "Adaptive Arrays: On Power Equalization with Proportional Control," Ohio State University, Columbus, Quart. Rept. 3234-1, Contract N0019-71-C-0219, December 1971.

[17] C. L. Zahm, "Application of Adaptive Arrays to Suppress Strong Jammers in the Presence of Weak Signals," *IEEE Trans. Aerosp. Electron. Syst.*, Vol. AES-9, No. 2, March 1973, pp. 260–271.

[18] C. L. Zahm, "Effects of Errors in the Direction of Incidence on the Performance of an Adaptive Array," *Proc. IEEE*, Vol. 60, No. 8, August 1972, pp. 1008–1009.

# Chapter 6. Direct Inversion of the Sample Covariance Matrix

In many applications the practical usefulness of an adaptive array critically depends on the convergence rate that can be achieved. For example, when adaptive radars require simultaneous rejection of jamming and clutter and provide automatic platform motion compensation, then rapid convergence to steady-state solutions is essential. Adaptive control of sensor arrays with the popular maximum SNR or LMS algorithms may well result in slow adaptive weight vector convergence (depending on the eigenvalues of the noise covariance matrix). When the covariance matrix eigenvalues differ by orders of magnitude, then the algorithm convergence time can be exceedingly long, and in any case it is highly example dependent. One way to speed convergence and circumvent the convergence rate dependence on eigenvalue distribution is to employ a direct method of adaptive weight computation, based on the sample covariance matrix of the signal environment [1]–[3].

## 6.1 THE DIRECT MATRIX INVERSION (DMI) APPROACH

The signals impinging on the receiving elements of an $N$-element adaptive array are represented by the $N$-dimensional signal vector $\mathbf{x}$, whose associated covariance matrix is given by

$$\mathbf{R}_{xx} = E\{\mathbf{x}\mathbf{x}^{\dagger}\} \tag{6.1}$$

When the desired signal is absent, then only noise and interference are present and

$$\mathbf{R}_{xx} = \mathbf{R}_{nn} \tag{6.2}$$

When the desired signal is present, then from Chapter 3 the optimal weight vector solution is given by

$$\mathbf{w}_{\text{opt}} = \mathbf{R}_{xx}^{-1}\mathbf{r}_{xd} \tag{6.3}$$

where $\mathbf{r}_{xd}$ is the cross-correlation between the random vector $\mathbf{x}(t)$ and the reference signal $d(t)$. When the desired signal is absent, then the optimal weight vector solution is given by

$$\mathbf{w}_{\text{opt}} = \mathbf{R}_{nn}^{-1}\mathbf{r}_{xd} = \mathbf{R}_{nn}^{-1}\mathbf{b}^* \tag{6.4}$$

where now $\mathbf{b}^*$ is the vector of beam-steering signals matched to the target doppler frequency and angle of incidence. Note that specifying $\mathbf{r}_{xd}$ is fully equivalent to specifying $\mathbf{b}^*$.

If the signal, clutter, and interference situation were known *a priori*, then the covariance matrix could be evaluated and the optimal solution for the adaptive weights computed directly using either (6.3) or (6.4), both of which require matrix inversion. In practice the signal, clutter, and interference situation are not known *a priori*, and furthermore the interference environment frequently changes due to the presence of moving near-field scatterers, antenna motion, interference, and jamming. Consequently the adaptive processor must continually update the weight vector to meet the new requirements posed by varying conditions. This need to update the weight vector in the absence of detailed *a priori* information leads to the expedient of obtaining estimates of $\mathbf{R}_{xx}$ or $\mathbf{R}_{nn}$, and $\mathbf{r}_{xd}$ in a finite observation interval, and employing these estimates in (6.3) or (6.4) to obtain the desired weight vector estimate. This method for implementing the adaptive processor is referred to as the direct matrix inversion (DMI) technique. The estimates of $\mathbf{R}_{xx}$, $\mathbf{R}_{nn}$, and $\mathbf{r}_{xd}$ are based on the ML principle, which yields unbiased estimates having minimum variance [4].

Although an algorithm based on the DMI approach can be shown, in theory, to converge more rapidly than the LMS or maximum SNR algorithms, the practical difficulties of realizing such a pleasing result for any application should be kept firmly in mind. It is necessary to implement a considerable amount of circuitry (much more than the LMS or maximum SNR algorithms) to obtain estimates of the elements of $\mathbf{R}_{xx}$ (or $\mathbf{R}_{nn}$). Formation of the sample covariance matrix requires $KN(N+1)/2$ complex multiplications, where $K$ is the number of independent samples used, and $N$ is the total number of weights. To invert the resulting Hermitian matrix then requires $(N^3/2+N^2)$ complex multiplications, and computing the weights requires another $N^2$ multiplications [2]. Signal vector dimensionality (which is determined by the number of degrees of freedom of the array processor) is therefore crucial in determining whether this approach is practical for a specific application. Matrix inversion may present further computational problems when the matrix to be inverted is ill-conditioned, so the degree of eigenvalue spread in the covariance matrix to be inverted also affects the practicality of this approach. Furthermore, finite circuit operation speeds may preclude the possibility of achieving the theoretical convergence rate except when the input signals of concern are narrowband and the array size is small [3]. It is worth noting that when the covariance matrix to be inverted has the form of a Toeplitz matrix (a situation that arises when using tapped-delay line channel processing) then the Toeplitz matrix inversion algorithm of W. F. Trench [5] should be exploited to facilitate the computation. The convergence rates presented here assume that all computations are done with sufficient accuracy to overcome the effects of any ill-conditioned matrices and therefore represent an upper limit on how well any DMI approach can be expected to perform.

### 6.1.1 *Use of the Sample Covariance Matrix*

Suppose that the cross-correlation vector $\mathbf{r}_{xd}$ (or equivalently that the beam-steering vector $\mathbf{b}^*$) is known. The optimal weight vector estimate $\hat{\mathbf{w}}$ may then be determined by using

$$\hat{\mathbf{w}}_1 = \hat{\mathbf{R}}_{xx}^{-1}\mathbf{r}_{xd} \tag{6.5}$$

and assuming $\mathbf{x}(t)$ contains the desired signal where $\hat{\mathbf{R}}_{xx}$ is the sample covariance estimate of $\mathbf{R}_{xx}$, or by using

$$\hat{\mathbf{w}}_2 = \hat{\mathbf{R}}_{nn}^{-1}\mathbf{r}_{xd} = \hat{\mathbf{R}}_{nn}^{-1}\mathbf{b}^* \tag{6.6}$$

if we assume $\mathbf{x}(t)$ does not contain the desired signal where $\hat{\mathbf{R}}_{nn}$ is the sample covariance estimate of $\mathbf{R}_{nn}$.

The array output SNR using $\hat{\mathbf{w}}_1$ or $\hat{\mathbf{w}}_2$ can be written as

$$\left(\frac{s}{n}\right)_1 = \frac{\hat{\mathbf{w}}_1^\dagger \mathbf{s}\mathbf{s}^\dagger \hat{\mathbf{w}}_1}{\hat{\mathbf{w}}_1^\dagger \mathbf{R}_{nn}\hat{\mathbf{w}}_1} = \frac{\mathbf{r}_{xd}^\dagger \hat{\mathbf{R}}_{xx}^{-1}\mathbf{s}\mathbf{s}^\dagger \hat{\mathbf{R}}_{xx}^{-1}\mathbf{r}_{xd}}{\mathbf{r}_{xd}^\dagger \hat{\mathbf{R}}_{xx}^{-1}\mathbf{R}_{nn}\hat{\mathbf{R}}_{xx}^{-1}\mathbf{r}_{xd}} \tag{6.7}$$

and

$$\left(\frac{s}{n}\right)_2 = \frac{\mathbf{r}_{xd}^\dagger \hat{\mathbf{R}}_{nn}^{-1}\mathbf{s}\mathbf{s}^\dagger \hat{\mathbf{R}}_{nn}^{-1}\mathbf{r}_{xd}}{\mathbf{r}_{xd}^\dagger \hat{\mathbf{R}}_{nn}^{-1}\mathbf{R}_{nn}\hat{\mathbf{R}}_{nn}^{-1}\mathbf{r}_{xd}} \tag{6.8}$$

where $\mathbf{s}$ denotes the desired signal vector component of $\mathbf{x}$ [it will be recalled from Chapter 3 that $\mathbf{s}(t) = s(t)\mathbf{v}$]. The SNR $(s/n)_2$ obviously has meaning only during those time intervals when a desired signal is actually present; the weight adjustment in this case is assumed to take place when the desired signal is absent. The "rate of convergence" of the two algorithms (6.5) and (6.6) is to be based on the output SNR normalized to the optimum output SNR, $SN_o$, compared with the number of independent signal samples $K$ used to obtain the required sample covariance matrices.

Assuming that all signals present at the array input can be modeled as sample functions from zero-mean Gaussian processes, then a ML estimate of $\mathbf{R}_{xx}$ (or $\mathbf{R}_{nn}$ when the desired signal is not present) can be formed using the sample covariance matrix given by

$$\hat{\mathbf{R}}_{xx} = \frac{1}{K}\sum_{j=1}^{K} \mathbf{x}(j)\mathbf{x}^{\dagger}(j) \tag{6.9}$$

where $\mathbf{x}(j)$ denotes the $j$th time sample of the signal vector $\mathbf{x}(t)$. Note that the assumption of independent zero-mean samples implies $E[\mathbf{x}(i)\mathbf{x}^{\dagger}(j)]=0$ for $i \neq j$.

Since each element of the matrix $\hat{\mathbf{R}}_{xx}$ is a random variable, the output SNR is also a random variable. It is instructive to compare the actual SNR obtained using $\hat{\mathbf{w}}_1$ and $\hat{\mathbf{w}}_2$ of (6.5) and (6.6) with the optimum SNR obtained using (6.3) and (6.4) ($SN_o = \mathbf{s}^{\dagger}\mathbf{R}_{nn}^{-1}\mathbf{s}$), by forming the normalized SNR as follows:

$$\rho_1 = \frac{(s/n)_1}{SN_o} \tag{6.10}$$

and

$$\rho_2 = \frac{(s/n)_2}{SN_o} \tag{6.11}$$

It can be shown [2] that the probability distribution of $\rho_2$ is described by the incomplete Beta distribution given by

$$Pr(\rho_2 \leqslant y) = \frac{K!}{(N-2)!(K+1-N)!}\int_0^y (1-u)^{N-2}u^{K+1-N}\,du \tag{6.12}$$

where

$K=$ total number of independent time samples used in obtaining $\hat{\mathbf{R}}_{nn}$

and

$N=$ number of adaptive degrees of freedom

The probability distribution function of (6.12) contains important information concerning the convergence of the DMI algorithm that is easily seen by considering the mean and the variance of $\rho_2$. From (6.12) it follows that the average value of $\rho_2$ is given by

$$E\{\rho_2\} = \bar{\rho}_2 = \frac{K+2-N}{K+1} \tag{6.13}$$

The variance of $\rho_2$ is given by

$$\text{var}(\rho_2) = \frac{(K+2-N)(N-1)}{(K+1)^2(K+2)} \tag{6.14}$$

Equation (6.14) initially appears to be a surprising result since for fixed $K$ and $N$ it suggests that $\text{var}(\rho_2)$ is independent of both the amount of noise the system must contend with and the eigenvalue spread of the noise covariance matrix. Recalling that $\rho$ is a normalized SNR, however, we see that both the actual SNR $(s/n)$ and the optimum SNR $SN_o$ will be affected in the same way by any noise power increase, so the normalized ratio remains the same, and the variance of the normalized ratio likewise remains unchanged. Eigenvalue spread has no effect on (6.14) since this expression was obtained with the assumption that the sample matrix inversion is computed exactly. Consequently the only errors reflected in (6.14) are sample covariance matrix estimation errors. The effect of eigenvalue spread on the matrix inversion computation is addressed in a later section.

Equation (6.13) is plotted in Figure 6.1 (we assume that $N$ is significantly larger than 2) and shows that so long as $K \geqslant 2N$, the loss in $\bar{\rho}_2$ due to nonoptimum weights is less than 3 dB. This result leads to the convenient rule of thumb that the number of time samples required to obtain a useful sample covariance matrix (when the desired signal is absent) is just twice the number of adaptive degrees of freedom.

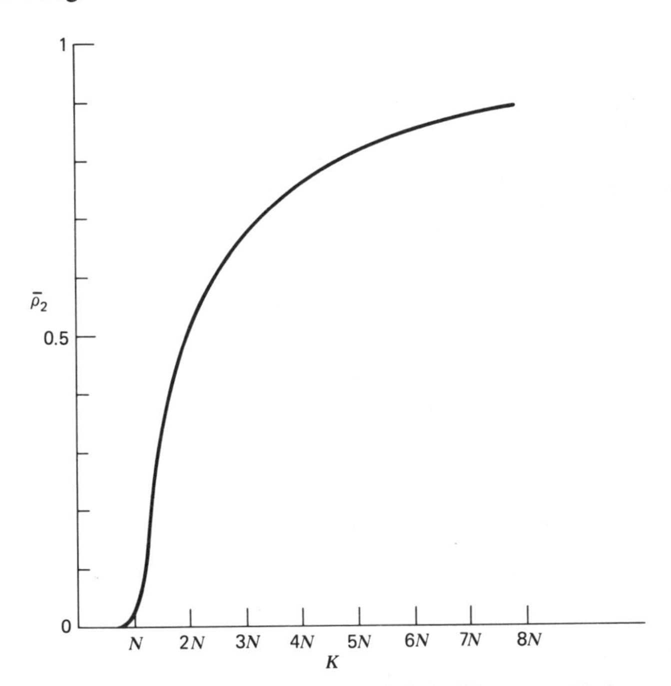


**Figure 6.1** Plot of $\bar{\rho}_2$ versus $K$ of (6.13). We assume $N \gg 2$.

Next consider the convergence behavior when the signal is present while estimating $\mathbf{w}$ as with (6.5). Rather than attempt to derive the probability distribution function of $\rho_1$ directly, it is more convenient to exploit the results obtained for $\rho_2$ by defining the random variable

$$\rho_1' = \frac{\mathbf{r}_{xd}^{\dagger}\hat{\mathbf{R}}_{xx}^{-1}\mathbf{s}\mathbf{s}^{\dagger}\hat{\mathbf{R}}_{xx}^{-1}\mathbf{r}_{xd}}{\mathbf{r}_{xd}^{\dagger}\hat{\mathbf{R}}_{xx}^{-1}\mathbf{R}_{xx}\hat{\mathbf{R}}_{xx}^{-1}\mathbf{r}_{xd}\mathbf{s}^{\dagger}\mathbf{R}_{xx}^{-1}\mathbf{s}} \tag{6.15}$$

which has the same probability distribution function as that of $\rho_2$. Knowing the statistical properties of $\rho_1'$ and the relationship between $\rho_1'$ and $\rho_1$ then enables the desired information about $\rho_1$ to be easily obtained. It can be shown that the relationship between $\rho_1'$ and $\rho_1$ is given by [3]

$$\rho_1 = \frac{\rho_1'}{SN_o(1-\rho_1')+1} \tag{6.16}$$

Since $\rho_1'$ has the same probability distribution function as $\rho_2$, it immediately follows that:

$$E\{\rho_1\} < E\{\rho_2\} \tag{6.17}$$

and

$$\lim_{SN_o \to 0} E\{\rho_1\} = E\{\rho_2\} \tag{6.18}$$

The inequality expressed by (6.17) implies that, on the average, the output SNR achieved by using $\hat{\mathbf{w}}_1 = \hat{\mathbf{R}}_{xx}^{-1}\mathbf{r}_{xd}$ is less than the output SNR achieved using $\hat{\mathbf{w}}_2 = \hat{\mathbf{R}}_{nn}^{-1}\mathbf{r}_{xd}$ (except in the limit as $K \to \infty$, in which case both estimates are equally accurate). This behavior derives from the fact that the presence of the desired signal increases the time required (or number of samples required) to obtain accurate estimates of $\mathbf{R}_{xx}$ from the sample covariance matrix compared to the time required to obtain accurate estimates of $\mathbf{R}_{nn}$ when the desired signal is absent. The limit expressed by (6.18) indicates that for $SN_o < 1$, the difference in SNR performance obtained using $\hat{\mathbf{w}}_1$ or $\hat{\mathbf{w}}_2$ is negligible.

By use of (6.16) it can be shown that the mean of $\rho_1$ can be expressed as the following infinite series [3]:

$$E\{\rho_1\} = \frac{a}{a+b}\left\{1 + \sum_{i=1}^{\infty}(-SN_o)^i\left(\frac{b}{a+b+1}\right)\left(\frac{b+1}{a+b+2}\right)\cdots \cdot\left(\frac{i+b-1}{a+b+i}\right)\right\} \tag{6.19}$$

where $a = K-N+2$ and $b = N-1$. The manner in which $E\{\rho_1\}$ depends on

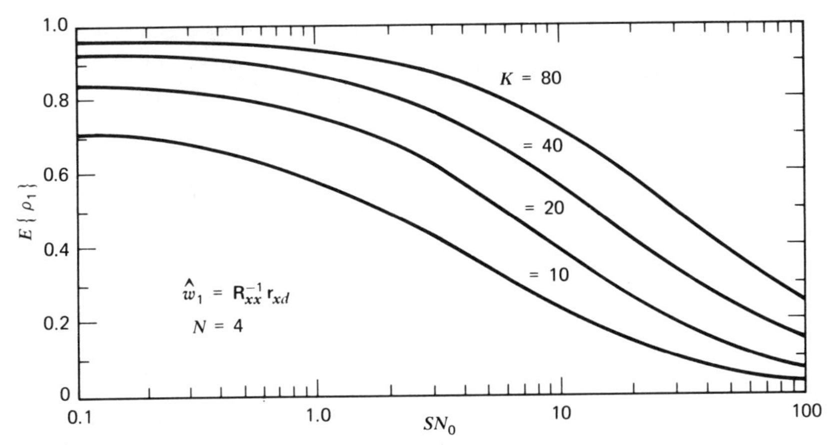


**Figure 6.2** The average normalized output SNR versus optimum output SNR for several sample sizes.

$SN_o$ is illustrated in Figure 6.2 for $N=4$ with $K$ as an independent parameter.

Since $\hat{\mathbf{w}}_2$ significantly outperforms $\hat{\mathbf{w}}_1$ when $SN_o \gg 1$, it is advantageous to remove the signal components from $\mathbf{x}(t)$ before forming $\hat{\mathbf{R}}_{xx}$. When the desired signal vector component $\mathbf{s}(t)$ is completely known, then subtracting $\mathbf{s}(t)$ from the received signal vector $\mathbf{x}(t)$ enables a desired signal-free noise vector to be formed which can be used to generate the sample noise covariance matrix $\hat{\mathbf{R}}_{nn}$. An improper procedure that is occasionally suggested for eliminating the desired signal component is to form the sample covariance matrix $\hat{\mathbf{D}}$ where

$$\hat{\mathbf{D}} = \hat{\mathbf{R}}_{xx} - \mathbf{s}\mathbf{s}^\dagger \tag{6.20}$$

so that

$$E\{\hat{\mathbf{D}}\} = \mathbf{R}_{nn} \tag{6.21}$$

The procedure suggested by (6.20) is unsatisfactory for obtaining the fast convergence associated with $\hat{\mathbf{w}}_2$ because even though $E\{\hat{\mathbf{D}}\} = \mathbf{R}_{nn}$, the weight vector estimate obtained using $\hat{\mathbf{w}}_3 = \hat{\mathbf{D}}^{-1}\mathbf{r}_{xd}$ only results in an estimate that is a scalar multiple of $\hat{\mathbf{w}}_1 = \hat{\mathbf{R}}_{xx}^{-1}\mathbf{r}_{xd}$ and therefore has the associated convergence properties of $\hat{\mathbf{w}}_1$. This fact may easily be seen by forming

$$\begin{aligned}\hat{\mathbf{w}}_3 &= \hat{\mathbf{D}}^{-1}\mathbf{r}_{xd}\\ &= \left[\hat{\mathbf{R}}_{xx} - \mathbf{s}\mathbf{s}^\dagger\right]^{-1}\mathbf{r}_{xd}\\ &= \frac{1}{1+\mathbf{s}^\dagger\hat{\mathbf{R}}_{xx}^{-1}\mathbf{s}}\hat{\mathbf{R}}_{xx}^{-1}\mathbf{r}_{xd}\end{aligned} \tag{6.22}$$

The coefficient of $\hat{\mathbf{R}}_{xx}^{-1}\mathbf{r}_{xd}$ in (6.22) is a scalar, and therefore the output SNR resulting from the use of $\hat{\mathbf{w}}_3$ is identical to that obtained using $\hat{\mathbf{w}}_1$, so no transient performance improvement can be achieved in this manner. If any transient response improvement were to be possible, it would be necessary for $\hat{\mathbf{w}}_3$ of (6.22) to be a scalar multiple of $\hat{\mathbf{R}}_{nn}^{-1}\mathbf{r}_{xd}$. Since $\hat{\mathbf{w}}_3 = \alpha\hat{\mathbf{R}}_{xx}^{-1}\mathbf{r}_{xd}$, however, the transient response of $\hat{\mathbf{w}}_3$ is the same as that of $\mathbf{w}_1$, and a priori knowledge of s as employed in (6.20) does not improve the DMI algorithm response.

### 6.1.2 *Use of the Sample Covariance Matrix and the Sample Cross-Correlation Vector*

In many practical radar and communications systems it is unrealistic to expect that the cross-correlation vector $\mathbf{r}_{xd}$ (or the beam-steering vector $\mathbf{b}^*$) would be known a priori. An alternative approach is to determine the optimal weight vector by using

$$\hat{\mathbf{w}}_4 = \hat{\mathbf{R}}_{xx}^{-1}\hat{\mathbf{r}}_{xd} \tag{6.23}$$

where $\hat{\mathbf{r}}_{xd}$ is the sample cross-correlation vector given by

$$\hat{\mathbf{r}}_{xd} = \frac{1}{K}\sum_{j=1}^{K} \mathbf{x}(j)\,d^*(j) \tag{6.24}$$

The transient behavior of the DMI algorithm represented by $\hat{\mathbf{w}}_4$ of (6.23) will be different from that found for $\hat{\mathbf{w}}_1$ and $\hat{\mathbf{w}}_2$ of (6.5) and (6.6), respectively.

The transient response characteristics of $\hat{\mathbf{w}}_4$ can be determined by considering the least-squares estimate of $\mathbf{w}_{\text{opt}}$ based on $K$ independent samples of the input vector $\mathbf{x}$. It is convenient in posing this problem to assume the adaptive array configuration shown in Figure 6.3. This configuration can represent two important adaptive array structures in the following manner: If $x_0(t) = d(t)$, the reference signal representation of the desired signal, then the minimum mean square error (MMSE) and the maximum output SNR are both given by the Wiener solution $\mathbf{w}_{\text{opt}}$. Likewise if $x_0(t)$ represents the output of a reference antenna (usually a high-gain antenna pointed in the direction of the desired signal source), then the configuration represents a coherent sidelobe canceller (CSLC) system, for which obtaining the MMSE weight vector solution minimizes the output error (or residue) power and hence minimizes the interference power component of the array output.

The transient response of $\hat{\mathbf{w}}_4$ given by (6.23) may be characterized in terms of the output SNR versus the number of data samples $K$ used to form the estimates $\hat{\mathbf{R}}_{xx}$ and $\hat{\mathbf{r}}_{xd}$. An alternate way of characterizing performance that is appropriate for CSLC applications is to consider the output residue power versus the number of data samples. System performance in terms of the output residue power will be considered. The output residue power (or MSE) is given by

$$\xi(\hat{\mathbf{w}}) = E|e(t)|^2 = \sigma_0^2 - \hat{\mathbf{w}}^\dagger\mathbf{r}_{xd} - \mathbf{r}_{xd}^\dagger\hat{\mathbf{w}} + \hat{\mathbf{w}}^\dagger\mathbf{R}_{xx}\hat{\mathbf{w}}^\dagger \tag{6.25}$$

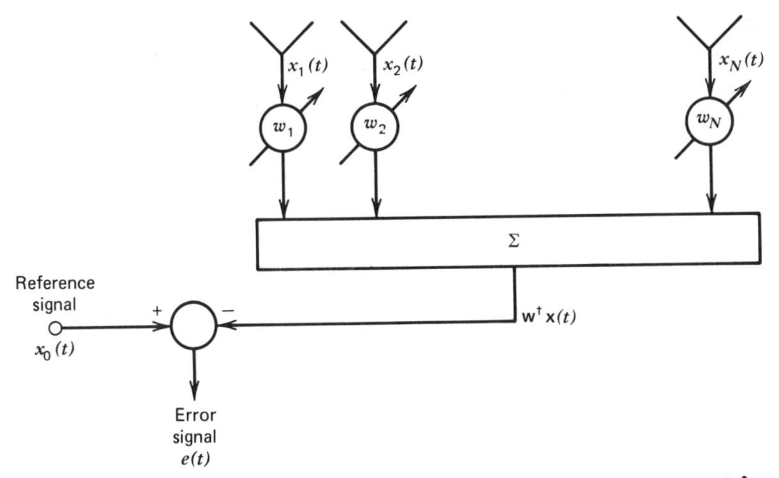


**Figure 6.3** Adaptive array configuration for considering transient behavior of $\hat{\mathbf{w}}_4$.

and the estimated MSE (or sample MSE) may be expressed as

$$\hat{\xi}(\hat{\mathbf{w}}) = \frac{1}{K}\sum_{i=1}^{K}|e(i)|^2 = \hat{\sigma}_0^2 - \hat{\mathbf{w}}^\dagger\hat{\mathbf{r}}_{xd} - \hat{\mathbf{r}}_{xd}^\dagger\hat{\mathbf{w}} + \hat{\mathbf{w}}^\dagger\hat{\mathbf{R}}_{xx}\hat{\mathbf{w}} \tag{6.26}$$

where $\sigma_0^2$ is the signal power in the reference antenna and

$$\hat{\sigma}_0^2 = \frac{1}{K}\sum_{i=1}^{K} x_0(i)x_0^*(i) \tag{6.27}$$

The transient behavior of the system can be characterized by evaluating the statistical properties of $\xi(\hat{\mathbf{w}})$ for a given number of data samples $K$. This can be accomplished by assuming that $\mathbf{x}(i)$ and $x_0(i)$ are sample functions from an $(N+1)$ variate, zero-mean, Gaussian random vector $\mathbf{x}'$ having the density function

$$p(\mathbf{x}') = \pi^{-(N+1)}|\mathbf{R}'_{xx}|^{-1}\exp\{-\mathbf{x}'^\dagger\mathbf{R}'^{-1}_{xx}\mathbf{x}'\} \tag{6.28}$$

where $\mathbf{R}'_{xx}$ is the $(N+1)\times(N+1)$ covariance matrix:

$$\mathbf{R}'_{xx} = E\left\{\begin{bmatrix}\mathbf{x}(t)\\ \hdashline x_0(t)\end{bmatrix}\left[\mathbf{x}^\dagger(t) \,\vdots\, x_0^*(t)\right]\right\}$$

$$= \begin{bmatrix}\mathbf{R}_{xx} & \mathbf{r}_{xx_0}\\ \mathbf{r}_{xx_0}^\dagger & \sigma_0^2\end{bmatrix} \tag{6.29}$$

where $\mathbf{r}_{xx_0}$ plays the role of $\mathbf{r}_{xd}$ so that $\hat{\mathbf{R}}'_{xx}$ is given by the estimates $\hat{\mathbf{R}}_{xx}$, $\hat{\mathbf{r}}_{xd}$, and $\hat{\sigma}_0^2$. The sample covariance matrix $\hat{\mathbf{R}}'_{xx}$ has the following important properties:

1. The elements of $\hat{\mathbf{R}}'_{xx}$ are jointly distributed according to the complex Wishart probability density function [6]

$$p(\mathbf{A})=\frac{|\mathbf{A}|^{K-N-1}\exp\left[-\mathrm{tr}\left(\mathbf{R}'^{-1}_{xx}\mathbf{A}\right)\right]}{\pi^{1/2(N+1)N}\Gamma(K)\Gamma(K-1)\cdots\Gamma(K-N)|\mathbf{R}'_{xx}|^{K}} \tag{6.30}$$

where

$$\mathbf{A}=K\hat{\mathbf{R}}'_{xx}$$

and

$$\Gamma(k)=(k-1)!$$

2. $\hat{\mathbf{R}}'_{xx}$ is the ML estimate of $\mathbf{R}'_{xx}$ [6]. Therefore $\hat{\mathbf{R}}_{xx}$, $\hat{\mathbf{r}}_{xd}$, and $\hat{\sigma}_0^2$ are the ML estimates of $\mathbf{R}_{xx}$, $\mathbf{r}_{xd}$, and $\sigma_0^2$, respectively.

By using a series of transformations on the partitioned matrix $\hat{\mathbf{R}}'_{xx}$, the following important results can also be obtained [7]:

1. The mean and variance of the sample MSE $\hat{\xi}$, realized using $\hat{\mathbf{w}}_4$ of (6.23) is given by

$$E\{\hat{\xi}\}=\left(1-\frac{N}{K}\right)\xi_{\min} \tag{6.31}$$

and

$$\mathrm{var}\{\hat{\xi}\}=\frac{1}{K}\left(1-\frac{N}{K}\right)\xi^2_{\min} \tag{6.32}$$

where

$$\xi_{\min}=\sigma_0^2-\mathbf{r}^{\dagger}_{xd}\mathbf{R}^{-1}_{xx}\mathbf{r}_{xd}$$

2. The difference between the output residue power $\xi(\hat{\mathbf{w}}_4)$ and the minimum output residue power $\xi_{\min}$ is a direct measure of the quality of the array performance relative to the optimum. It is convenient to define the normalized performance quality parameter (or "misadjustment") $M=r^2$ as

$$r^2 \triangleq \frac{\xi(\hat{\mathbf{w}}_4)-\xi_{\min}}{\xi_{\min}} \tag{6.33}$$

The parameter $r$ is a random variable having the density function

$$p(r)=2\frac{K!}{(K-N)!(N-1)!}\cdot\frac{r^{2N-1}}{(1+r^2)^{K+1}}, \qquad 0<r<\infty \tag{6.34}$$

The statistical moments of the misadjustment $r^2$ are easily obtained by recognizing that the variable $y$ defined by

$$y=\frac{1}{1+r^2} \tag{6.35}$$

is governed by an incomplete Beta function distribution with parameters $a=K-N+1$ and $b=N$. Consequently the mean and variance of $r^2$ are given by

$$E\{r^2\}=\frac{N}{K-N} \tag{6.36}$$

and

$$\text{var}\{r^2\}=\frac{NK}{(K-N)^2(K-N-1)} \tag{6.37}$$

3. For the case when $x_0(t)=d(t)$, define a normalized SNR as in (6.10),(6.11) for $\hat{\mathbf{w}}_4$ according to

$$\rho_3=\frac{(s/n)_3}{SN_o}=\frac{\hat{\mathbf{w}}_4^\dagger \mathbf{s}\mathbf{s}^\dagger \hat{\mathbf{w}}_4}{\hat{\mathbf{w}}_4^\dagger \mathbf{R}_{nn}\hat{\mathbf{w}}_4 \mathbf{s}^\dagger \mathbf{R}_{nn}^{-1}\mathbf{s}} \tag{6.38}$$

The probability density function of $\rho_3$ is difficult to evaluate in closed form, but the mean and variance of $\rho_3$ can be determined numerically using the relations given below.

$$\rho_3=\frac{1}{(1+SN_o)\left[\dfrac{C+\sin^2\phi_1}{C+\sin^2\phi_1\cos^2\phi_2}\right]-SN_o} \tag{6.39}$$

where $C=(1/r)\sqrt{(s/n)_3}$ and where the joint density function of $r$, $\phi_1$, and $\phi_2$ is given by

$$p(r,\phi_1,\phi_2)=\frac{2}{\pi}\frac{K!}{(K-N)!(N-2)!}\frac{r^{2N-1}}{(1+r^2)^{K+1}} \cdot(\sin\phi_1)^{2N-2}(\sin\phi_2)^{2N-3} \tag{6.40}$$

for $0<r<\infty, 0\leqslant\phi_1<\pi, 0\leqslant\phi_2<\pi$. Note that $r$, $\phi_1$, and $\phi_2$ are statistically independent and that $r$ has the same density as (6.34). With the foregoing expressions the numerical evaluation of $E\{\rho_3\}$ can be obtained from

$$E\{\rho_3\}=\int_R P(r)\int_{\Phi_1} P(\phi_1)\int_{\Phi_2}\rho_3 P(\phi_2)\,d\phi_2\,d\phi_1\,dr \tag{6.41}$$

$E\{\rho_3^2\}$ can likewise be obtained from (6.41) with $\rho_3^2$ replacing $\rho_3$. The variance of $\rho_3$ is then given by $\operatorname{var}\{\rho_3\}=E\{\rho_3^2\}-E^2\{\rho_3\}$.

4. The normalized MSE performance measure defined by

$$\hat{\xi}_N \triangleq \frac{2K\hat{\xi}}{\xi_{\min}} \tag{6.42}$$

is statistically independent of both $\hat{\mathbf{w}}$ and $\hat{\mathbf{R}}_{xx}$.

The results just summarized have important implications for the transient performance that can be achieved with a DMI algorithm. Let us first consider the results expressed by (6.36) and (6.37). From (6.36) it is seen that the output residue power is within 3 dB of the optimum value after only $2N$ distinct time samples, or within 1 dB after $5N$ samples; thereby indicating rapid convergence independent of the signal environment or the array configuration. We see this rapid convergence property, however, applies directly to the interference suppression of a sidelobe canceller (SLC) system *assuming no desired signal is present* when forming $\hat{\mathbf{R}}_{xx}$.

In communications systems the desired signal is usually present and the SNR performance measure is the primary quantity of interest rather than the MSE, $\xi$. Furthermore, in radar systems the SLC is often followed by a signal processor that rejects clutter returns so only that portion of the output residue power due to RF interference (rather than clutter) must be suppressed by the SLC system. Let us now show that the presence of either the desired signal or clutter returns in the main beam of an SLC system acts as a disturbance that tends to slow the rate of convergence of a DMI algorithm.

Consider the radar SLC configuration shown in Figure 6.4. The system consists of an SLC designed to cancel only interference followed by a signal processor to remove clutter. To simplify the discussion, the clutter power $\xi_c^2$ received in the main antenna is assumed much larger than clutter entering the low-gain auxiliary antennas so that clutter in the auxiliary channels can be neglected and $\xi_{\min}=\sigma_c^2+\xi_{N_0}$ where $\xi_{N_0}$ represents the minimum output interference plus thermal noise power. Furthermore assume that the clutter returns can be represented as a sample function from a stationary, zero-mean Gaussian process. It then follows from (6.25), (6.33), and (6.36) that

$$\frac{E[\xi(\hat{\mathbf{w}})-\xi_{\min}]}{\xi_{N_0}}=\frac{N}{K-N}\left[1+\frac{\sigma_c^2}{\xi_{N_0}}\right] \tag{6.43}$$

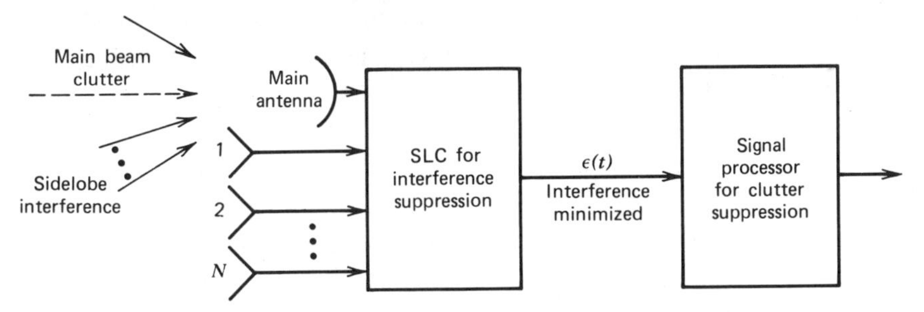


**Figure 6.4** SLC for interference suppression in a radar system.

From (6.43) it is evident that the presence of main beam clutter prolongs convergence of the DMI algorithm by an amount that is approximately proportional to the main antenna clutter power divided by the minimum output interference power. This slower convergence is due to the presence of clutter-jammer cross-terms in $\hat{\mathbf{r}}_{xd}$ which result in noisier estimates of the weights. For rapid convergence, adaptation should be confined to time intervals that are relatively clutter free; or a means for minimizing clutter terms present in the estimate of $\hat{\mathbf{r}}_{xd}$ must be found.

The foregoing result is also applicable to communications systems where the main antenna is pointed toward an active desired signal source. Let $\sigma_s^2$ represent the desired signal power received in the main channel and assume the desired signal entering the auxiliary channels can be neglected. Then the SLC output signal to interference plus noise ratio SNR can easily be shown to be given by

$$\mathrm{SNR}=\frac{\sigma_s^2}{\sigma_s^2 r^2+\xi_{N_0}(1+r^2)} \tag{6.44}$$

The maximum value of SNR is denoted by $SN_o$ where $SN_o=\sigma_s^2/\xi_{N_0}$ is obtained when $r^2=0$ or $\hat{\mathbf{w}}=\mathbf{w}_{\text{opt}}$. Normalizing (6.44) to $SN_o$ yields the expression for the normalized output signal to interference plus noise ratio:

$$\rho_4=\frac{\mathrm{SNR}}{SN_o}=\frac{1}{1+[1+(SN_o)]r^2} \tag{6.45}$$

Note that $0\leqslant\rho_4\leqslant 1$ so the optimum value of $\rho_4$ is unity. When the output signal to interference plus noise ratio is small so that $SN_o\ll 1$, then the probability distribution of $\rho_4$ is approximated by the following Beta probability density function:

$$P(\rho_4)=\frac{K!}{(N-1)!(K-N)!}(1-\rho_4)^{N-1}\rho_4^{K-N} \tag{6.46}$$

On comparing (6.46) with the probability density function contained in (6.12), it immediately follows that the two density functions are identical provided that $N$ in (6.12) is replaced by $N+1$. It then follows from (6.13) for the CSLC system of Figure 6.3 that

$$E\{\rho_4\} = \frac{K-N+1}{K+1}; \qquad SN_o \ll 1 \tag{6.47}$$

Hence, for small $SN_o$ only $K=2N-1$ independent samples are needed to converge within 3 dB of $SN_o$. For large SNR ($\gg K/N$), the expected SNR (unnormalized) is approximated by

$$\overline{(\text{SNR})} \cong \frac{K}{N} - 1; \qquad SN_o \gg \frac{K}{N}; \qquad K > N \tag{6.48}$$

The presence of a strong desired signal in the main channel therefore slows convergence to the optimum SNR, but does not affect the average output SNR after $K$ samples under the conditions of (6.48).

Finally, consider the results given in (6.39) and (6.40) for the case $x_0(t) = d(t)$ (continuous reference signal present). For large $SN_o$, the distribution of $\rho_3$ in (6.39) is approximated by the density function of (6.46) so that

$$E\{\rho_3\} \cong \frac{K-N+1}{K+1}; \qquad SN_o \gg 1 \tag{6.49}$$

However, the rate of convergence decreases as $SN_o$ decreases below zero decibels. This effect is illustrated by the plot of $E\{\rho_3\}$ versus $K$ in Figure 6.5.

The behavior described for a reference signal configuration is just the converse of that obtained for the SLC configuration with the desired signal present. The presence of the desired signal in the main channel of the SLC introduced a disturbance that reduced the accuracy of the weight estimate and slowed the convergence. For the reference signal configuration, however,

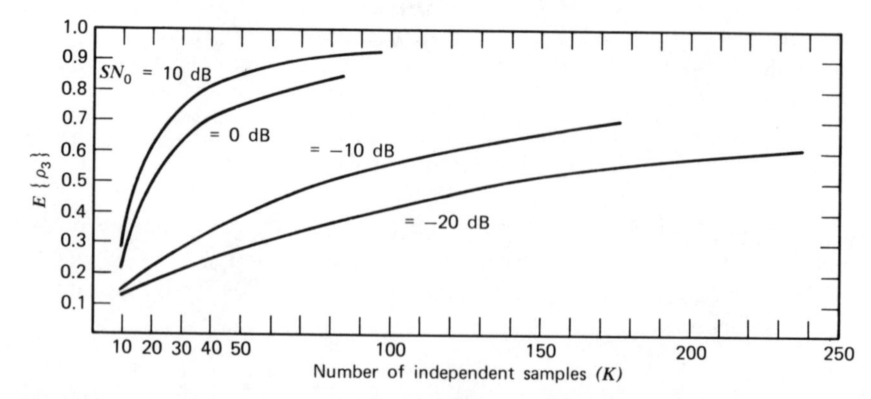


**Figure 6.5** $E\{\rho_3\}$ of (6.38) versus number of independent samples for $N=8$ and selected values of $SN_o$.

the estimates $\hat{\mathbf{r}}_{xd}$ and $\hat{\mathbf{R}}_{xx}$ are highly correlated under strong desired signal conditions, and the errors in each estimate tend to compensate each other thereby yielding an improved weight estimate and faster convergence.

## 6.2 TRANSIENT RESPONSE COMPARISONS

The transient response characteristics of the DMI algorithms corresponding to $\hat{\mathbf{w}}_1$ of (6.5) and $\hat{\mathbf{w}}_4$ of (6.23) may be obtained by examining

$$\bar{\rho} = \frac{E\{\text{output SNR}\}}{SN_o}$$

as a function of $K$ (the number of independent samples used in obtaining $\hat{\mathbf{R}}_{xx}$ and $\hat{\mathbf{r}}_{xd}$) for selected signal conditions. Assuming an input interference-to-signal ratio of 14 dB and an input signal-to-thermal noise ratio of 0 dB, $\bar{\rho}$ was obtained for $\hat{\mathbf{w}}_1$ and $\hat{\mathbf{w}}_4$ by averaging 50 independent responses for a four-element linear array for which $d/\lambda = \frac{1}{2}$. A single directional interference source was assumed that was separated from the desired signal location by 75°. The resulting transient response is given in Figure 6.6.

It is interesting to compare the transient response obtained using $\hat{\mathbf{w}}_1$ and $\hat{\mathbf{w}}_4$ with the transient response that would result using the LMS algorithm having an iteration period equal to the intersample collection period of the DMI algorithm. We do this by considering the behavior of $\bar{\rho}(K)$. For the DMI approach, $K$ represents the number of independent signal samples used to

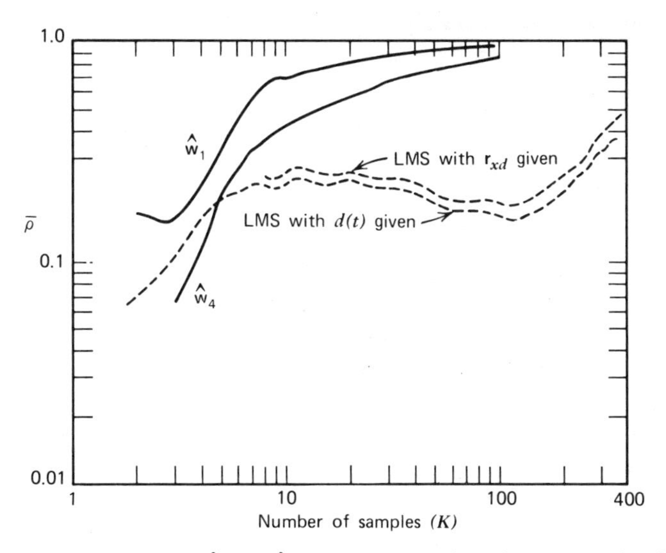


**Figure 6.6** Transient response of $\hat{\mathbf{w}}_1$ and $\hat{\mathbf{w}}_4$ for $SN_o = 3.8$ and one interference signal located 75° away from the desired signal.

compute the sample covariance matrix, while for the LMS algorithm $K$ represents the number of iterations completed. A comparison of the DMI and LMS algorithms in this manner is not entirely satisfactory for the following reasons:

1. The transient response of the LMS algorithm depends on the selection of the step size, which can be made arbitrarily small (thereby resulting in an arbitrarily long transient response time constant).
2. The transient response of the LMS algorithm depends on the starting point at which the initial weight vector guess is set. A good initial guess may result in excellent transient response.
3. Whereas the variance of $\rho(K)$ decreases as $K$ increases with the DMI approach, the variance of $\rho(K)$ remains constant (in the steady-state) as $K$ increases for the LMS algorithm since the steady-state variance is determined by the step size selection.

Nevertheless, a comparison of $\bar{\rho}(K)$ behavior does yield an indication of the output SNR response speed and is therefore of some value.

The LMS algorithm convergence condition (4.49) is violated if the step size $\Delta_s$ exceeds $1/P_{\text{IN}}$ where $P_{\text{IN}}$ = total array input power from all sources. Consequently by selecting $\Delta_s = 0.4/P_{\text{IN}}$, the step size is only 4 dB below the convergence condition limit, the LMS control loop is quite noisy, and the resulting speed of the transient response is close to the theoretical maximum. Since the maximum time constant for the LMS algorithm transient response is associated with the minimum covariance matrix eigenvalue,

$$\tau_{\max} \cong \frac{1}{k_s \Delta t \lambda_{\min}} \tag{6.50}$$

and $\lambda_{\min} \geqslant \sigma^2$, the thermal noise power variance,[1] it follows that

$$\tau_{\max} \cong \frac{P_{\text{IN}}}{0.4\sigma^2} \tag{6.51}$$

With a four-element array, $P_{\text{IN}} = 4(P_D + \sigma^2 + P_I)$ where

$P_D$ = desired signal power

$\sigma^2$ = thermal noise power

$P_I$ = interference signal power

Therefore if $P_I/P_D = 14$ dB and $P_D/\sigma^2 = 0$ dB then $P_{\text{IN}}/\sigma^2 = 108$ and $\tau_{\max} =$ 270 samples.

[1]It is assumed that each weighted channel contains thermal noise (with noise power $\sigma^2$) that is uncorrelated between channels.

The transient response time constant of the desired signal power is proportional to the eigenvalue $\lambda_1$ associated with the desired signal power. Since in the current example $\lambda_1 \cong 4P_D$, the time constant associated with the desired signal power is approximately

$$\tau_1 \cong \frac{1}{\Delta_s \lambda_1} = \frac{P_{\text{IN}}}{(0.4)4P_D} = 67 \text{ samples} \tag{6.52}$$

A convenient starting point for the weight vector with the LMS algorithm is $\mathbf{w}_0^T = [1,0,0,0]$, since this corresponds to an omnidirectional array sensitivity pattern. For the foregoing conditions, the behavior of $\bar{\rho}(K)$ resulting from the use of $\hat{\mathbf{w}}_3, \hat{\mathbf{w}}_4$ and the two versions of the LMS algorithm given by

$$\mathbf{w}(k+1) = \mathbf{w}(k) + k_s \Delta t \left[\mathbf{r}_{xd} - \mathbf{x}(k)\mathbf{x}^\dagger(k)\mathbf{w}(k)\right], \qquad \mathbf{r}_{xd} \text{ given} \tag{6.53}$$

and

$$\mathbf{w}(k+1) = \mathbf{w}(k) + k_s \Delta t \left[\mathbf{x}(k) d^*(k) - \mathbf{x}(k)\mathbf{x}^\dagger(k)\mathbf{w}(k)\right], \qquad d(k) \text{ given} \tag{6.54}$$

was determined by simulation. The results are illustrated in Figure 6.6 where for the specified conditions $SN_o = 3.8$.

The results of Figure 6.6 indicate that the response resulting from the use of the DMI derived weights $\hat{\mathbf{w}}_1$ and $\hat{\mathbf{w}}_4$ is superior to that obtained from the LMS derived weights. Whereas the initial response of the LMS derived weights indicated improved output SNR with increasing $K$, this trend reverses when the LMS algorithm begins to respond along the desired signal eigenvector, since any decrease in the desired signal response without a corresponding decrease in the thermal noise response causes the output signal to thermal noise ratio to decrease. Once the array begins to respond along the thermal noise eigenvectors, then $\bar{\rho}$ again begins to increase.

The undesirable transient behavior of the two LMS algorithms in Figure 6.6 can be avoided by selecting an initial starting weight vector different from that chosen for the foregoing comparison. For example, by selecting $\mathbf{w}(0) = \mathbf{r}_{xd}$ the initial LMS algorithm response can be greatly improved since this initial starting condition biases the array pattern toward the desired signal direction thereby providing an initially high output SNR. Furthermore by selecting $\mathbf{w}(0) = \alpha \mathbf{r}_{xd}$ where $\alpha$ is a scalar constant the initial starting condition can also result in a monotonically increasing $\bar{\rho}(K)$ as $K$ increases for the two LMS algorithms since by appropriately weighting $\mathbf{r}_{xd}$, the magnitude of the initial array response to the desired signal can be made small enough so that as adaptation proceeds, the resulting $\bar{\rho}(K)$ always increases.

Improvement of the transient response through judicious selection of the initial starting condition can also be introduced into the DMI derived weight vectors as well. For example, by selecting

$$\hat{\mathbf{w}}_1 = \left[ \frac{1}{K} \left( \sum_{j=1}^{K} \mathbf{x}(j)\mathbf{x}^{\dagger}(j) + \alpha \mathbf{I} \right) \right]^{-1} \mathbf{r}_{xd} \tag{6.55}$$

then even before an estimate of $\mathbf{R}_{xx}$ is formed, the weight vector can be biased toward the desired signal direction, and the transient responses corresponding to $\hat{\mathbf{w}}_1$ and $\hat{\mathbf{w}}_4$ in Figure 6.6 can be greatly improved. Nevertheless, even if improved transient response for an LMS algorithm is obtained by appropriately biasing the initial weight vector, the speed of convergence for DMI derived weights remains faster.

It is also interesting to obtain the transient response of $\bar{\rho}(K)$ with two interference signals present. Assume one interference-to-signal ratio of 30 dB and a second interference-to-signal ratio of 10 dB, where the stronger interference signal is located 30° away from the desired signal, the weaker interference signal is located 60° away from the desired signal, and with all other conditions the same as for Figure 6.6. The resulting transient response for the two DMI derived weight vectors and for the two LMS algorithms [with $d(t)$ given and with $\mathbf{r}_{xd}$ given] with initial starting weight vector = [1,0,0,0] is illustrated in Figure 6.7. The presence of two directional interference sources with widely different power levels results in a wide eigenvalue spread and a consequent slow convergence rate for the LMS algorithms. Since $P_{\text{IN}}/\lambda_1$ is now 40 times larger than for the conditions of Figure 6.6, the time constant $\tau_1$ is now 40 times greater than before. The DMI derived weight transient response, however, is virtually unaffected by this eigenvalue spread.

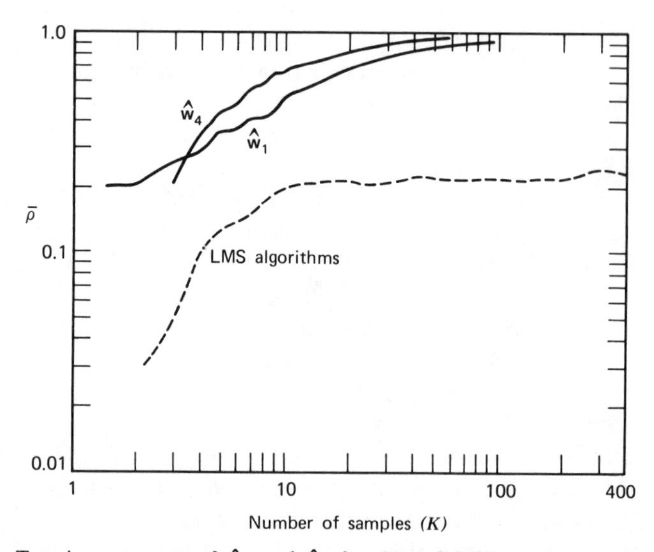


**Figure 6.7** Transient response of $\hat{\mathbf{w}}_1$ and $\hat{\mathbf{w}}_4$ for $SN_o = 0.4158$ and two interference signals located 30° and 60° away from the desired signal.

**Table 6.1 Comparison of DMI Algorithm Convergence Rates for Selected Array Configurations and Signal Conditions**

| Array Configuration | $\hat{\mathbf{w}}_{\text{opt}}$ | Signal Conditions | Performance Measure | Approximate Number of Independent Samples Required for Convergence (to within 3 dB of Optimum) |
|---|---|---|---|---|
| Sidelobe canceller | $\hat{\mathbf{R}}_{xx}^{-1}\hat{\mathbf{r}}_{xx_0}$ | Desired signal absent | Minimum MSE | $2N$, $N$ = number of array elements |
| | | Desired signal present in main beam only | Maximum SNR, $SN_o$ | $2N\left[1+\frac{SN_o}{2}\right]$, $SN_o \gg 1$<br>$2N$, $SN_o \ll 1$ |
| | | Clutter returns in main beam only | Minimum output (interference + noise) power, $\xi_{N_0}$ | $2N\left[1+\frac{\sigma_c^2}{2\xi_{N_0}}\right]$, $\sigma_c^2$ = main channel clutter power |
| Fully adaptive array | $\hat{\mathbf{R}}_{xx}^{-1}\hat{\mathbf{r}}_{xd}$ | Known desired signal present $d(t)=s(t)$ | $2N$,<br>Maximum SNR, $SN_o$ | $SN_o \gg 1$<br>$2N\left[1+\frac{1}{2SN_o}\right]$, $SN_o \ll 1$ |
| | $\hat{\mathbf{R}}_{xx}^{-1}\mathbf{r}_{xd}$ | Desired signal direction of arrival known but desired signal absent | Maximum SNR, $SN_o$ | $2N-3$ |
| | | Desired signal direction of arrival known and desired signal present | Maximum SNR, $SN_o$ | $2N$, $SN_o \ll 1$<br>$2N\left[1+\frac{SN_o}{2}\right]$, $SN_o \gg 1$ |

The principal convergence results for DMI algorithms under various array configurations and signal conditions can be conveniently summarized as shown in Table 6.1. The derivation of these results may be found in [2], [3].

## 6.3 SENSITIVITY TO EIGENVALUE SPREAD

Despite the fact that the convergence speed of a DMI algorithm is insensitive to eigenvalue spread in $\mathbf{R}_{xx}$, the accuracy of the steady-state solution will exhibit sensitivity to eigenvalue spread once that spread exceeds a certain critical amount. This steady-state accuracy sensitivity arises in the following manner.

Suppose that a sufficient number of independent time samples have been collected to ensure that the sample covariance matrix estimate is arbitrarily close to $\mathbf{R}_{xx}$ (or $\mathbf{R}_{nn}$). In that event, the only reason an exact solution for the adaptive weight vector would not result is that the sample matrix inversion cannot be accomplished with sufficient accuracy due to matrix ill-conditioning (as measured by the eigenvalue spread). As the number of bits available in the computer to perform the matrix inversion increases, the degree of matrix ill-conditioning that can be handled by the matrix inversion subroutine also increases. Nevertheless, the eigenvalue spread can always reach a

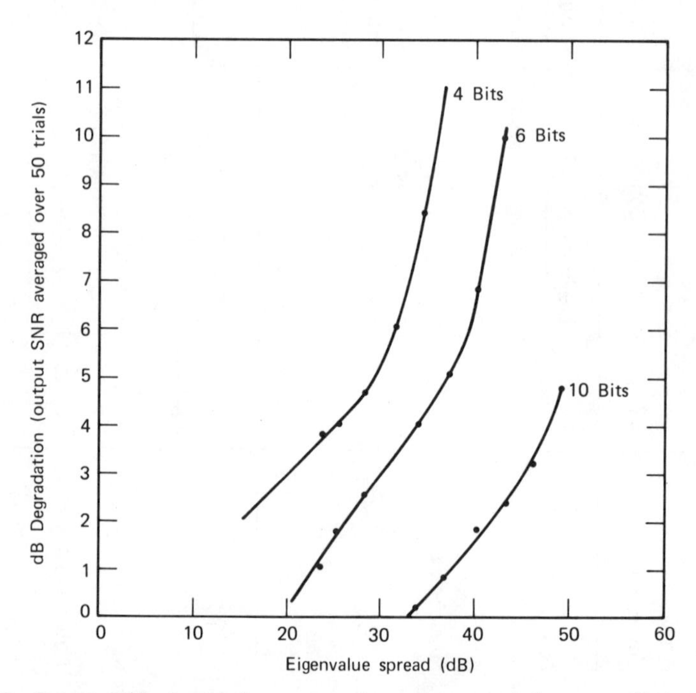


**Figure 6.8** Output SNR degradation versus eigenvalue spread for a specified number of available bits.

point where the number of bits available in the computer to perform the matrix inversion does not permit accurate inversion to be performed, and output SNR degradation will result.

From the foregoing discussion it follows that there is a fundamental trade-off between the number of bits available in the computer to accomplish the desired sample matrix inversion and the allowable eigenvalue spread in the input signal covariance matrix for any DMI algorithm. This trade-off was examined by simulating the desired matrix inversion for a prescribed number of available bits, and Figure 6.8 illustrates the nature of this trade-off for a four-element linear array in a two-jammer interference signal environment where the eigenvalue spread is defined by $\lambda_{max}/\lambda_{min}$, the ratio of maximum to minimum eigenvalues in the input signal covariance matrix. It is seen that so long as the sample covariance matrix has an eigenvalue spread less than a critical value (that depends on the number of bits available in the computer to accomplish the required matrix inversion), then a DMI algorithm is insensitive to eigenvalue spread. Once this critical value of eigenvalue spread is exceeded, however, very rapid degradation in the output SNR results with any additional increase in eigenvalue spread.

## 6.4 SUMMARY AND CONCLUSIONS

DMI algorithms yield convergence speeds that cannot be surpassed by the LMS algorithm. The weight vector obtained depends only on the signal covariance matrix (through the Wiener solution) and knowledge of the signal covariance matrix is gained directly by forming the sample covariance matrix. When the desired signal is absent from the signal environment, then use of a beam-steering vector $\mathbf{r}_{xd}$ yields the most desirable transient response characteristics. When the desired signal is strongly present, however, formation of the estimate $\hat{\mathbf{r}}_{xd}$ yields transient response characteristics that are superior to the use of a priori information represented by $\mathbf{r}_{xd}$ alone.

DMI algorithms are furthermore insensitive to eigenvalue spread until a certain critical level is exceeded that depends on the number of bits available in the computer to perform the matrix inversion computation. The practicality of the DMI approach is restricted by the number of degrees of freedom of the adaptive processor. When the feasibility of the DMI approach is not precluded however, the additional complexity introduced by directly obtaining the sample covariance matrix is rewarded by the rapid convergence and insensitivity to eigenvalue spread noted above.

## PROBLEMS

**1** The DMI algorithm requires a matrix inversion each time a new weight is to be calculated (as when the signal environment changes, for example). By applying the matrix inversion lemma

$$\left[A+\mathbf{u}^{\dagger}\mathbf{R}\mathbf{u}\right]^{-1}=A^{-1}-A^{-1}\mathbf{u}^{\dagger}\left[\mathbf{R}^{-1}+\mathbf{u}A^{-1}\mathbf{u}^{\dagger}\right]^{-1}\mathbf{u}A^{-1}$$

show that the weights can be updated at each sample using the recursive formulas

$$W(k)=W(k-1)+\frac{\mathbf{P}(k-1)\mathbf{x}(k)\epsilon^*(k)}{1+\mathbf{x}^{\dagger}(k)\mathbf{P}(k-1)\mathbf{x}(k)}$$

where

$$\epsilon^*(k)=-\mathbf{x}^{\dagger}(k)\mathbf{P}(k-1)\mathbf{b}^*$$

and

$$\mathbf{P}(k)=\mathbf{P}(k-1)-\frac{\mathbf{P}(k-1)\mathbf{x}(k)\mathbf{x}^{\dagger}(k)\mathbf{P}(k-1)}{1+\mathbf{x}^{\dagger}(k)\mathbf{P}(k-1)\mathbf{x}(k)}$$

*Hint*: Apply the lemma to the inversion of

$$\left[\hat{\mathbf{R}}_{xx}(k-1)+\mathbf{x}(k)\mathbf{x}^{\dagger}(k)\right]$$

**2** Consider an adaptive filter consisting of a tapped-delay line with a complex weight at each tap.

(a) Show that the DMI algorithm for minimizing the MSE between the filter output and the reference signal is of the same form as for an adaptive array if the tap outputs are taken as analogous to the array inputs.

(b) What restrictions on sampling are imposed if we desire independent samples of the inputs?

(c) Apply the analytical expressions derived for the adaptive array to determine the transient behavior of the DMI algorithm applied to adaptive filters. Assume independent samples.

**3** Starting with the probability density for $r^2=[\xi(\hat{\mathbf{w}})-\xi_{\min}]/\xi_{\min}$, determine the density of the weight vector $\Delta\mathbf{w}$ where $\Delta\mathbf{w}=\hat{\mathbf{w}}-\mathbf{w}_{\text{opt}}$. Show that the covariance of $\Delta\mathbf{w}$ depends on the covariance matrix $\mathbf{R}_{xx}$.

**4** Starting with the following expression for $\rho$ (for the reference signal case):

$$\rho=\frac{\hat{\mathbf{w}}^{\dagger}\mathbf{s}\mathbf{s}^{\dagger}\hat{\mathbf{w}}}{\hat{\mathbf{w}}^{\dagger}(\mathbf{R}_{xx}-\mathbf{s}\mathbf{s}^{\dagger})\hat{\mathbf{w}}}\frac{1}{SN_o}$$

show that

$$\rho=\frac{1}{1+(1+SN_o)\sum_{i=2}^{N}\dfrac{|\Delta q_i|^2}{|\sqrt{SN_o}+\Delta q_1|^2}}$$

Where

$$\Delta\mathbf{q} = \sqrt{SN_o}\ \hat{G} - \begin{bmatrix} \sqrt{SN_o} \\ 0 \\ 0 \\ \vdots \\ 0 \end{bmatrix}$$

$$\hat{G} = \sqrt{\frac{SN_o}{1+SN_o}}\ \mathbf{P}^{-1}\mathbf{R}_{xx}^{1/2}\hat{\mathbf{w}}$$

and where $\mathbf{P}$ is the unitary transformation that gives

$$\sqrt{\frac{SN_o}{1+SN_o}}\ \mathbf{R}_{xx}^{1/2}\mathbf{w}_0 = \mathbf{P}\begin{bmatrix} 1 \\ 0 \\ \vdots \\ 0 \end{bmatrix}$$

The variable $\Delta\mathbf{q}$ has the probability density function

$$p(\Delta\mathbf{q}) = \frac{k!}{\pi^N(k-N)!}\frac{1}{(1+\Delta\mathbf{q}^\dagger\Delta\mathbf{q})^{k+1}}$$

**5** Using the density for $\Delta\mathbf{q}$ in Problem 4 and the following transformation:

$$\begin{aligned}
\mathrm{Re}\{\Delta q_1\} &= r\cos\phi_1 \\
\mathrm{Im}\{\Delta q_1\} &= r\sin\phi_1\cos\phi_2 \\
\mathrm{Re}\{\Delta q_2\} &= r\sin\phi_1\sin\phi_2\cos\phi_3 \\
\mathrm{Im}\{\Delta q_2\} &= r\sin\phi_1\sin\phi_2\sin\phi_3\cos\phi_4 \\
&\ \ \vdots \\
\mathrm{Re}\{\Delta q_m\} &= r\sin\phi_1\cdots\sin\phi_{2N-2}\cos\phi_{2N-1} \\
\mathrm{Im}\{\Delta q_m\} &= r\sin\phi_1\cdots\sin\phi_{2N-2}\sin\phi_{2N-1}
\end{aligned}$$

where

$$\begin{aligned}
&0 \leqslant r < \infty; \qquad i+1,2,\cdots,2N-1 \\
&0 \leqslant \phi_i < \pi
\end{aligned}$$

derive the density $p(r,\phi_1,\phi_2)$ of (6.40) and the expression for $\rho(r,\phi_1,\phi_2)$.

*Hint.* The Jacobian of the transformation is

$$J = r^{2N-1}(\sin\phi_1)^{2N-2}(\sin\phi_2)^{2N-3}\cdots\sin\phi_{2N-2}$$

and

$$\int_0^{\pi}(\sin\phi)^n\,d\phi = \frac{\Gamma[(n+1)/2]}{\Gamma[(n+2)/2]}\sqrt{\pi}$$

**6** Derive the result in (6.44) using (6.33) and the fact that for a CSLC $\xi(\hat{\mathbf{w}}) = \sigma_s^2 + \sigma_N^2$ and $\xi_{\min} = \xi_{N_0} + \sigma_s^2$, and $\sigma_N^2$ is the output noise plus jammer power.

**7** Define the following transformation of $\mathbf{R}'_{xx}$ in (6.29):

$$x = K\left[\hat{\sigma}_0^2 - \hat{\mathbf{r}}_{xx_0}^{\dagger}\hat{\mathbf{R}}_{xx}^{-1}\hat{\mathbf{r}}_{xx_0}\right]$$
$$\mathbf{Y} = K\hat{\mathbf{R}}_{xx}$$
$$\hat{\mathbf{w}} = \hat{\mathbf{R}}_{xx}^{-1}\hat{\mathbf{r}}_{xd}$$

Then the joint density of $x$, $\mathbf{Y}$, and $\hat{\mathbf{w}}$ is given by

$$p(x,\mathbf{Y},\hat{\mathbf{w}}) = P(x)P(\hat{\mathbf{w}},\mathbf{Y})$$

where

$$p(\hat{w},\mathbf{Y}) = \frac{|Y|^{k-N+1}\exp\{-\mathrm{tr}[I + (1/\xi_0)\Delta\mathbf{w}\Delta\mathbf{w}^{\dagger}\mathbf{R}_{xx}]\mathbf{R}_{xx}^{-1}\mathbf{Y}\}}{\pi^N\pi^{1/2N(N-1)}\Gamma(k)\cdots\Gamma(k-N+1)|\mathbf{R}_{xx}|^k|\xi_0|^N}$$

and where $\Delta\mathbf{w} = \hat{\mathbf{w}} - \mathbf{w}_0$. Derive the density of $r^2 = (1/\xi_0)\Delta\mathbf{w}^{\dagger}\mathbf{R}_{xx}\Delta\mathbf{w}$ [see (6.34)] from the foregoing density function.

**8** The random variable $x$ defined in Problem 7 has the probability density function

$$p(x) = \frac{1}{(\xi_0)^{K-N}}\frac{1}{\Gamma(K-N)}|x|^{K-N-1}\exp\left[-\frac{x}{\xi_0}\right]$$

(a) Show that $2x/\xi_0$ is $\chi^2$ distributed with $2(K-N)$ degrees of freedom.
(b) Show that $\xi(\hat{\mathbf{w}}) = (1/K)x$ has mean and variance [see (6.31), (6.32)]

$$E\{\xi(\hat{\mathbf{w}})\} = \left(1 - \frac{N}{k}\right)\xi_0$$
$$\mathrm{var}[\xi(\hat{\mathbf{w}})] = \frac{K-N}{K^2}\xi_0^2$$

**9** [8], [9] ***Inversion of Complex Matrices Using Real Arithmetic.*** Techniques for inverting complex matrices using only real matrix operations are of considerable value since although complex inversion routines are generally superior (in terms of accuracy and computing time) to real matrix approaches, nevertheless mathematical packages for small computers generally include matrix inversion routines that only apply to real matrices.

(a) To invert the complex $n\times n$ matrix $M=A+jB$ to obtain $M^{-1}=E+jF$ is equivalent to solving the two simultaneous equations

$$AE-BF=I$$
$$AF+BE=0$$

for the unknown matrices $E$ and $F$. Premultiply the two above equations by $B$ and $A$, respectively, and by subtracting show that

$$[AB-BA]E+[A^2+B^2]F=-B$$

Similarly, premultiply the original equation pair by $A$ and $B$, respectively, and by adding show that

$$[BA-AB]F+[A^2+B^2]E=A$$

If $A$ and $B$ commute (so that $AB=BA$), show that $M^{-1}=[A^2+B^2]^{-1}[A-jB]$. The foregoing result involves the inverse of a real $n\times n$ matrix to obtain $M^{-1}$, but is restricted by the requirement that $A$ and $B$ commute.

(b) Let $C=A+B$ and $D=A-B$. Show that the original equation pair in part (a) reduce to

$$CE+DF=I$$
$$-DE+CF=-I$$

From the results expressed in the equation pair immediately preceding, show that either

$$M^{-1}=[C+DC^{-1}D]^{-1}[(DC^{-1}+I)+j(DC^{-1}-I)]$$

or

$$M^{-1}=[D+CD^{-1}C]^{-1}[(CD^{-1}+I)+j(-CD^{-1}+I)]$$

provided the indicated inverses exist. The foregoing equation pair represents alternate ways of obtaining $M^{-1}$ by inverting real $n\times n$ matrices without the restriction that $A$ and $B$ commute.

(c) An isomorphism exists between the field of complex numbers and a special set of $2 \times 2$ matrixes, that is,

$$a + jb \sim \begin{bmatrix} a & b \\ -b & a \end{bmatrix}$$

Consider the $2n \times 2n$ real matrixes defined by

$$\mathbf{G} \triangleq \begin{bmatrix} A & B \\ -B & A \end{bmatrix} \quad \text{and} \quad \mathbf{H} = \begin{bmatrix} E & F \\ -F & E \end{bmatrix}$$

Show that $H$ is the inverse of $G$ only if the original equation pair in part (a) is satisfied. Therefore one way of obtaining $M^{-1}$ is to compute $G^{-1}$ and identify the $n \times n$ submatrixes $E$ and $F$ appearing in $G^{-1}$. Then $M^{-1} = E + jF$. This approach does not involve the restrictions that beset the approaches of (a) and (b) above, but suffers from the fact that it requires the inversion of a $2n \times 2n$ matrix which drastically increases computer storage requirements, and therefore should only be used as a last resort.

## REFERENCES

[1] I. S. Reed, J. D. Mallett, and L. E. Brennan, "Sample Matrix Inversion Technique," Proceedings of the 1974 Adaptive Antenna Systems Workshop, March 11–13, Volume I, NRL Report 7803, Naval Research Laboratory, Washington, DC, pp. 219–222.

[2] I. S. Reed, J. D. Mallett, L. E. Brennan, "Rapid Convergence Rate in Adaptive Arrays," *IEEE Trans. Aerosp. Electron. Syst.*, Vol. AES-10, No. 6, pp. 853–863, November 1974.

[3] T. W. Miller, "The Transient Response of Adaptive Arrays in TDMA Systems," Ph.D. Dissertation, Department of Electrical Engineering, The Ohio State University, 1976.

[4] H. L. Van Trees, *Detection, Estimation, and Modulation Theory*, Part I, Wiley, New York, 1968, Ch. 1.

[5] S. Zohar, "Toeplitz Matrix Inversion: The Algorithm of W. F. Trench," *J. Assoc. Comput. Mach.*, Vol. 16, October 1969, pp. 592–601.

[6] N. R. Goodman, "Statistical Analysis Based on a Certain Multivariate Gaussian Distribution," *Ann. Math. Stat.*, Vol. 34, March 1963, pp. 152–177.

[7] A. B. Baggeroer, "Confidence Intervals for Regression (MEM) Spectral Estimates," *IEEE Trans. Inf. Theory*, Vol. IT-22, No. 5, September 1976, pp. 534–545.

[8] W. W. Smith, Jr. and S. Erdman, "A Note on Inversion of Complex Matrices," *IEEE Trans. Autom. Control*, Vol. AC-19, No. 1, February 1974, p. 64.

[9] M. E. El-Hawary, "Further Comments on 'A Note on the Inversion of Complex Matrices'," *IEEE Trans. Autom. Control*, Vol. AC-20, No. 2, April 1975, pp. 279–280.

# Chapter 7. Recursive Methods for Adaptive Array Processing

To avoid the computational problems associated with the direct calculation of a set of adaptive weights for an array processor, both the LMS and maximum SNR algorithms can be used. It is shown in Chapter 9 that random search algorithms also provide a means of circumventing computational problems. These algorithms all have the advantage that the required calculations are usually much simpler than the corresponding direct calculation, they are less susceptable to hardware inaccuracy, and they are continually updated to compensate for a time-varying signal environment.

Another class of processors based on recursive methods can also be used to circumvent computational problems [1]–[4]. The basic approach taken by these methods is to recursively perform the matrix inversion required by the direct calculation approach so that at no time is a direct matrix inversion computation required. The recursive algorithms should therefore exhibit the same kind of steady-state sensitivity to eigenvalue spread in the signal covariance matrix as that already found for DMI algorithms. Furthermore, since the principal difference between the recursive methods and the DMI algorithms lies in the manner in which the matrix inversion is computed, their rates of convergence are comparable. The recursive algorithms are primarily based on least-square estimation techniques and are closely related to Kalman filtering methods [5]. These methods assume that the sensor signals are available in sampled data form, and they define digital processors for updating the adaptive weights in the array processor. For stationary environments these recursive procedures compute the best possible selection of weights (based on a least-squares fit to the data received) at each sampling instant, while in contrast the LMS, maximum SNR, and random search methods are only asymptotically optimal.

## 7.1 THE WEIGHTED LEAST-SQUARES ERROR PROCESSOR

Consider the conventional $N$-element array of Figure 7.1 having a sequence of (real or complex) weights multiplying the received signals to form the adaptive processor. Assume that the received signals $x_i(t)$ contain a

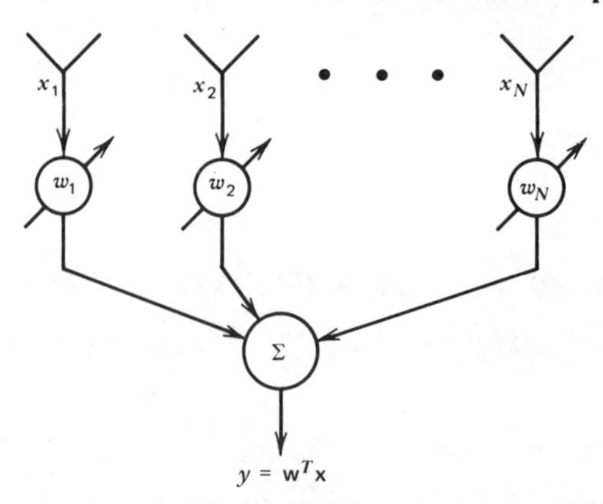


**Figure 7.1** Conventional $N$-element adaptive array processor.

directional desired signal component $s_i(t)$ and a purely random component $n_i(t)$, due to both directional and thermal noise so that $x_i(t)=s_i(t)+n_i(t)$. Collecting the received signals $x_i(t)$ and the multiplicative weights $w_i(t)$ as components in the $N$-dimensional vectors $\mathbf{x}(t)$ and $\mathbf{w}(t)$, we can write the adaptive processor output signal $y(t)$ as

$$y(t)=\mathbf{w}^T(t)\mathbf{x}(t) \tag{7.1}$$

The narrowband processor model of Figure 7.1 has been chosen instead of the more general tapped-delay line wideband processor in each element channel merely because the required mathematical manipulations are thereby simplified. The weighted least-squares error processor can easily be extended to the more general tapped-delay line form.

Consider the weighted least-squares performance measure based on $k$ data samples following Baird [6]

$$\mathfrak{P}(\mathbf{w})=\frac{1}{2}\sum_{i=1}^{k}\alpha_i\left[\mathbf{w}^T\mathbf{x}(i)-d(i)\right]^2=\frac{1}{2}\left[\mathbf{X}(k)\mathbf{w}-\mathbf{d}(k)\right]^T\mathbf{A}_k^{-1}\left[\mathbf{X}(k)\mathbf{w}-\mathbf{d}(k)\right] \tag{7.2}$$

where the elements of $\mathbf{X}(k)$ are received signal vector samples and the elements of $\mathbf{d}(k)$ are desired (or reference) signal samples as follows:

$$\mathbf{X}(k)\triangleq\begin{bmatrix}\mathbf{x}^T(1)\\ \mathbf{x}^T(2)\\ \vdots\\ \mathbf{x}^T(k)\end{bmatrix} \quad (7.3) \qquad \mathbf{d}(k)\triangleq\begin{bmatrix}d(1)\\ d(2)\\ \vdots\\ d(k)\end{bmatrix} \quad (7.4)$$

It will be noted that (7.4), and hence (7.2), presumes that the desired array output signal $d(t)$ is available (or known) so the discrete desired signal samples $d(1), d(2),\ldots, d(k)$ can be obtained. In the performance measure of (7.2), $\mathbf{A}_k$ is taken to be a diagonal weighting matrix whose role is to de-emphasize old data points and is of the form

$$\mathbf{A}_k = \begin{bmatrix} \alpha^{k-1} & 0 & \cdots & \cdot & 0 \\ 0 & \alpha^{k-2} & \cdots & \cdot & \cdot \\ & & & & \vdots \\ 0 & \cdots & & \alpha & 0 \\ 0 & 0 & & & 1 \end{bmatrix} \tag{7.5}$$

where $0<\alpha \leqslant 1$, so that older data is given increasingly less importance. If the signal environment is stationary so that all data samples are equally important, then we may select $\mathbf{A}_k = \mathbf{I}$, the identity matrix. The performance measure given by (7.2) is to be minimized by selecting the weight vector to yield the "best" (weighted least-squares) estimate of the desired signal vector $\mathbf{d}(k)$.

To minimize the weighted least squares performance measure (7.2), set the derivative of $\mathfrak{P}(\mathbf{w})$ with respect to $\mathbf{w}$ equal to zero thereby yielding the optimum weight setting as

$$\mathbf{w}_{ls}(k) = \left[\mathbf{X}^T(k)\mathbf{A}_k^{-1}\mathbf{X}(k)\right]^{-1}\mathbf{X}^T(k)\mathbf{A}_k^{-1}\mathbf{d}(k) \tag{7.6}$$

When an additional data sample is taken, the foregoing weight vector solution should be updated in the most efficient manner. The updated signals $\mathbf{X}(k+1)$ and $\mathbf{d}(k+1)$ as well as the updated matrix $\mathbf{A}_{k+1}$ can each be partitioned as follows:

$$\mathbf{X}(k+1) = \begin{bmatrix} \mathbf{X}(k) \\ \hdashline \mathbf{x}^T(k+1) \end{bmatrix} \tag{7.7}$$

$$\mathbf{d}(k+1) = \begin{bmatrix} \mathbf{d}(k) \\ \hdashline d(k+1) \end{bmatrix} \tag{7.8}$$

and

$$\mathbf{A}_{k+1} = \left[\begin{array}{c:c} \alpha \mathbf{A}_k & \begin{matrix} 0 \\ \vdots \\ 0 \end{matrix} \\ \hdashline 0 \cdots 0 & 1 \end{array}\right] \tag{7.9}$$

With this partitioning, the updated weight vector can be written as

$$\begin{aligned} \mathbf{w}_{ls}(k+1) &= \left[\mathbf{X}^T(k+1)\mathbf{A}_{k+1}^{-1}\mathbf{X}(k+1)\right]^{-1}\mathbf{X}^T(k+1)\mathbf{A}_{k+1}^{-1}\mathbf{d}(k+1) \\ &= \left[\mathbf{X}^T(k+1)\mathbf{A}_{k+1}^{-1}\mathbf{X}(k+1)\right]^{-1} \\ &\quad \cdot \left[\alpha \mathbf{X}^T(k)\mathbf{A}_k^{-1}\mathbf{d}(k) + \mathbf{x}(k+1)d(k+1)\right] \end{aligned} \tag{7.10}$$

From (7.10) it is seen that the updated weight vector solution requires the inversion of the matrix $[\mathbf{X}^T(k+1)\mathbf{A}_{k+1}^{-1}\mathbf{X}(k+1)]$, which can also be expanded by the partitioning given above as

$$\left[\mathbf{X}^T(k+1)\mathbf{A}_{k+1}^{-1}\mathbf{X}(k+1)\right]^{-1}$$

$$=\left\{\left[\mathbf{X}^T(k) \;\vdots\; \mathbf{x}(k+1)\right]\left[\begin{array}{c|c} \alpha\mathbf{A}_k^{-1} & \begin{matrix}0\\ \vdots\\ 0\end{matrix} \\ \hline 0\cdots 0 & 1\end{array}\right]\left[\begin{array}{c}\mathbf{X}(k)\\ \hline \mathbf{x}^T(k+1)\end{array}\right]\right\}^{-1}$$

$$=\left\{\alpha\left[\mathbf{X}^T(k)\mathbf{A}_k^{-1}\mathbf{X}^T(k)\right]+\mathbf{x}(k+1)\mathbf{x}^T(k+1)\right\}^{-1} \quad (7.11)$$

Now define

$$\mathbf{P}^{-1}(k) \triangleq \mathbf{X}^T(k)\mathbf{A}_k^{-1}\mathbf{X}(k) \quad (7.12)$$

Likewise define

$$\begin{aligned}\mathbf{P}^{-1}(k+1) &\triangleq \mathbf{X}^T(k+1)\mathbf{A}_{k+1}^{-1}\mathbf{X}(k+1)\\ &=\alpha\left[\mathbf{X}^T(k)\mathbf{A}_k^{-1}\mathbf{X}(k)\right]+\mathbf{x}(k+1)\mathbf{x}^T(k+1)\\ &=\alpha\left[\mathbf{P}^{-1}(k)+\frac{1}{\alpha}\mathbf{x}(k+1)\mathbf{x}^T(k+1)\right] \end{aligned} \quad (7.13)$$

Inverting both sides of (7.13) and applying the matrix inversion lemma [(D.4) of Appendix D] then yields

$$\mathbf{P}(k+1)=\frac{1}{\alpha}\left\{\mathbf{P}(k)-\frac{\mathbf{P}(k)\mathbf{x}(k+1)\mathbf{x}^T(k+1)\mathbf{P}(k)}{\alpha+\mathbf{x}^T(k+1)\mathbf{P}(k)\mathbf{x}(k+1)}\right\} \quad (7.14)$$

By our use of (7.14) in (7.10) and recognition that $\mathbf{w}_{ls}(k)=\mathbf{P}(k)\mathbf{X}(k)\mathbf{A}_k^{-1}\mathbf{d}(k)$, the updated solution for the weight vector now becomes

$$\begin{aligned}\mathbf{w}_{ls}(k+1)=\mathbf{w}_{ls}(k)+&\frac{\mathbf{P}(k)\mathbf{x}(k+1)}{\alpha+\mathbf{x}^T(k+1)\mathbf{P}(k)\mathbf{x}(k+1)}\\ &\cdot\left[d(k+1)-\mathbf{w}_{ls}^T(k)\mathbf{x}(k+1)\right]\end{aligned} \quad (7.15)$$

Equations (7.14) and (7.15) are the iterative relations of the recursive least squares algorithm. Equations (7.14) and (7.15) are started by adopting an initial guess for the weight vector $\mathbf{w}(o)$ and the initial Hermitian matrix $\mathbf{P}(o)$. It is common practice to select as an initial weight vector $\mathbf{w}(o)=[1\angle 0°,0,0,\ldots,0]$, thereby obtaining an omnidirectional array pattern (provided the sensor elements each have omnidirectional patterns) and to select $\mathbf{P}(o)$ as the identity matrix.

Equations (7.14) and (7.15) are seen to yield the updated weight vector in a computationally efficient manner that avoids calculating the matrix inverses present in (7.6) and (7.10). It is instructive to consider (7.6) in more detail for the additional insight to be gained into the mechanics of the processor. Since the trace of $\mathbf{A}_k$, $\mathrm{tr}[\mathbf{A}_k]$, is a scalar, (7.6) can be rewritten as

$$\mathbf{w}_{ls}(k) = \left\{ \frac{\mathbf{X}^T(k)\mathbf{A}_k^{-1}\mathbf{X}(k)}{\mathrm{tr}[\mathbf{A}_k]} \right\}^{-1} \left\{ \frac{\mathbf{X}^T(k)\mathbf{A}_k^{-1}\mathbf{d}(k)}{\mathrm{tr}[\mathbf{A}_k]} \right\} \tag{7.16}$$

The first bracketed term on the right-hand side of (7.16) is an estimate of the autocorrelation matrix $\hat{\mathbf{R}}_{xx}$ based on $k$ data samples, that is,

$$[\hat{\mathbf{R}}_{xx}]_{i,j}(k) = \sum_{n=1}^{k} \alpha^{k-n} x_i(k) x_j(k) \tag{7.17}$$

Equation (7.17) is an expression for forming exponentially de-weighted estimates of the matrix $\mathbf{R}_{xx}$ that was also used by Mantey and Griffiths [2]. Similarly, the second bracketed term on the right-hand side of (7.16) is an estimate for the cross-correlation vector $\hat{\mathbf{r}}_{xd}(k)$. Consequently the $\mathbf{P}$ matrix defined by (7.12) is directly related to the autocorrelation matrix since

$$\begin{aligned} \hat{\mathbf{R}}_{xx}^{-1}(k) &= \mathrm{tr}[\mathbf{A}_k]\mathbf{P}(k) = (1+\alpha+\alpha^2+\cdots+\alpha^{k-1})\mathbf{P}(k) \\ &= \left[\frac{1-\alpha^k}{1-\alpha}\right]\mathbf{P}(k) \end{aligned} \tag{7.18}$$

The form of the algorithm given by (7.14) and (7.15) requires that the desired signal be known at each sample point, which is an unrealistic assumption that obviates the need for a processor. For a practical implementation, it is sufficient to adopt an approximation or estimate of the desired signal. Consequently replacing $d(k+1)$ by $\hat{d}(k+1)$ in (7.15) yields a realization of the weighted least square error recursive processor that is useful for practical applications.

Equivalent forms of (7.15) that are useful for different signal conditions may be obtained by replacing certain instantaneous quantities by their known average values [3], [7]. To obtain these equivalent forms, rewrite (7.15) as

$$\begin{aligned} \mathbf{w}(k+1) = \mathbf{w}(k) &\frac{\mathbf{P}(k)}{\alpha + \mathbf{x}^T(k+1)\mathbf{P}(k)\mathbf{x}(k+1)} \\ &\cdot [\mathbf{x}(k+1)d(k+1) - \mathbf{x}(k+1)y(k+1)] \end{aligned} \tag{7.19}$$

where $y(k+1)$ is the array output. The product $\mathbf{x}(k+1)\cdot d(k+1)$ can now be replaced by its average value, which is an estimate of the cross-correlation vector $\hat{\mathbf{r}}_{xd}$. Since the estimate $\hat{\mathbf{r}}_{xd}$ does not follow instantaneous fluctuations of $\mathbf{x}(t)$ and $d(t)$, it may be expected that the convergence time would be greater using $\hat{\mathbf{r}}_{xd}$ than when using $\mathbf{x}(k)d(k)$ as shown in Chapter 6.

In the event that only the direction of arrival of the desired signal is known and the desired signal is absent, then the cross-correlation vector $\mathbf{r}_{xd}$ (which conveys direction of arrival information) is known, and the algorithm for updating the weight vector becomes

$$\mathbf{w}(k+1)=\mathbf{w}(k)+\frac{\mathbf{P}(k)}{\alpha+\mathbf{x}^T(k+1)\mathbf{P}(k)\mathbf{x}(k+1)} \cdot\left[\mathbf{r}_{xd}-\mathbf{x}(k+1)y(k+1)\right] \tag{7.20}$$

Equation (7.20) may of course also be used when the desired signal is present, but the rate of convergence is then slower than for (7.19) with the same desired signal present conditions.

For stationary signal environments $\alpha$ is selected equal to 1, but this choice can lead to a practical difficulty. As long as $0<\alpha<1$, (7.14) and (7.15) lead to stable numerical procedures. When $\alpha=1$, however, many iterations can result in the components of $\mathbf{P}(k)$ becoming so small that round-off errors have a significant effect on the result. To avoid this numerical sensitivity problem when $\alpha=1$, both sides of (7.14) can be multiplied by the factor $(k+1)$ to yield numerically stable equations, and (7.18) then becomes

$$\hat{\mathbf{R}}_{xx}^{-1}(k)=k\mathbf{P}(k) \tag{7.21}$$

## 7.2 UPDATED COVARIANCE MATRIX INVERSE

The weighted least squares error processor of the previous section was based on weighting the current received signal vector data compared with past data according to

$$\mathbf{P}^{-1}(k+1)=\alpha\mathbf{P}^{-1}(k)+\mathbf{x}^*(k+1)\mathbf{x}^T(k+1) \tag{7.22}$$

where $0\leqslant\alpha\leqslant 1$ and $\mathbf{P}^{-1}(k)\triangleq\mathbf{X}^T(k)\mathbf{A}_k^{-1}\mathbf{X}(k)$. A closely related alternative data weighting scheme is to use the sample covariance matrix $\hat{\mathbf{R}}_{xx}$ as the vehicle for summarizing the effect of old data so that

$$\hat{\mathbf{R}}_{xx}(k+1)=\alpha\hat{\mathbf{R}}_{xx}(k)+\mathbf{x}^*(k+1)\mathbf{x}^T(k+1) \tag{7.23}$$

where $0\leqslant\alpha\leqslant 1$. Inverting both sides of (7.24) yields

$$\hat{\mathbf{R}}_{xx}^{-1}(k+1)=\frac{1}{\alpha}\left\{\hat{\mathbf{R}}_{xx}(k)+\frac{1}{\alpha}\mathbf{x}^*(k+1)\mathbf{x}^T(k+1)\right\}^{-1} \tag{7.24}$$

Applying the matrix inversion lemma to (7.24) then results in

$$\hat{\mathbf{R}}_{xx}^{-1}(k+1)=\frac{1}{\alpha}\left\{\hat{\mathbf{R}}_{xx}^{-1}(k)-\frac{\hat{\mathbf{R}}_{xx}^{-1}(k)\mathbf{x}^*(k+1)\mathbf{x}^T(k+1)}{\alpha+\mathbf{x}^T(k+1)\hat{\mathbf{R}}_{xx}^{-1}(k)\mathbf{x}^*(k+1)}\hat{\mathbf{R}}_{xx}^{-1}(k)\right\} \tag{7.25}$$

In the absence of a desired signal,

$$\mathbf{w}_{\text{opt}}=\mathbf{R}_{xx}^{-1}\mathbf{b}^*$$

(or $\mathbf{R}_{xx}^{-1}\mathbf{r}_{xd}$) so that each side of (7.25) can be postmultiplied by $\mathbf{b}^*$ to give

$$\hat{\mathbf{w}}(k+1)=\frac{1}{\alpha}\left\{\hat{\mathbf{w}}(k)-\frac{\hat{\mathbf{R}}_{xx}^{-1}(k)\mathbf{x}^*(k+1)}{\alpha+\mathbf{x}^T(k+1)\hat{\mathbf{R}}_{xx}^{-1}(k)\mathbf{x}^*(k+1)}\cdot\mathbf{x}^T(k+1)\hat{\mathbf{w}}(k)\right\} \tag{7.26}$$

The data weighting represented by (7.26) implies that past data [represented by $\hat{\mathbf{R}}_{xx}(k)$] is never more important than current data [represented by $\mathbf{x}^*(k+1)\mathbf{x}^T(k+1)$]. An alternative data weighting scheme that permits past data to be regarded either as less important or more important than current data is to use

$$\hat{\mathbf{R}}_{xx}(k+1)=(1-\beta)\hat{\mathbf{R}}_{xx}(k)+\beta\mathbf{x}^*(k+1)\mathbf{x}^T(k+1), \qquad 0\leqslant\beta\leqslant 1 \tag{7.27}$$

The data weighting scheme represented by (7.27) has been successfully employed [using $\beta=1/(k+1)$ so that each sample is then equally weighted] to reject clutter, compensate for platform motion, and compensate for near-field scattering effects in an airborne moving target indication (AMTI) radar system [8]. Inverting both sides of (7.27) and applying the matrix inversion lemma results in [9]

$$\hat{\mathbf{R}}_{xx}^{-1}(k+1)=\frac{1}{(1-\beta)}\hat{\mathbf{R}}_{xx}^{-1}(k)-\frac{\beta}{(1-\beta)}\cdot\frac{\left[\hat{\mathbf{R}}_{xx}^{-1}(k)\mathbf{x}^*(k+1)\right]\left[\mathbf{x}^T(k+1)\hat{\mathbf{R}}_{xx}^{-1}(k)\right]}{(1-\beta)+\beta\left[\mathbf{x}^T(k+1)\hat{\mathbf{R}}_{xx}^{-1}(k)\mathbf{x}^*(k+1)\right]} \tag{7.28}$$

To obtain the updated weight vector $\hat{\mathbf{R}}_{xx}^{-1}(k+1)$ must be postmultiplied by the beam-steering vector $\mathbf{b}^*$. The beam-steering vector for AMTI radar systems can be matched to a moving target by including the expected relative Doppler and polarization phase factors thereby minimizing the effects of main beam clutter due to (for example) stationary targets [8], [10]. Carrying out the postmultiplication of (7.28) by $\mathbf{b}^*$ then yields

$$\hat{\mathbf{w}}(k+1) = \frac{1}{(1-\beta)}\left\{\hat{\mathbf{w}}(k) - \beta\frac{\hat{\mathbf{R}}_{xx}^{-1}(k)\mathbf{x}^*(k+1)}{(1-\beta)+\beta\left[\mathbf{x}^T(k+1)\hat{\mathbf{R}}_{xx}^{-1}(k)\mathbf{x}^*(k+1)\right]} \cdot \mathbf{x}^T(k+1)\hat{\mathbf{w}}(k)\right\} \tag{7.29}$$

The updating computation represented by (7.28) requires $N^2$ complex multiplications to form $\hat{\mathbf{R}}_{xx}^{-1}(k)\mathbf{x}^*(k+1)$ where $N$ is the number of degrees of freedom present in the adaptive processor. Furthermore, (7.28) also requires an additional $0.75N^2$ and $2.25N$ multiplications to complete the computation of $\hat{\mathbf{R}}_{xx}^{-1}(k+1)$. On comparing (7.29) and (7.26) with (7.20), it may appear that the direction-of-arrival information contained in $\mathbf{r}_{xd}$ (or $\mathbf{b}^*$) is missing in (7.26) and (7.29). We realize that the initial starting weight selected in either (7.26) or (7.29) reflects any direction-of-arrival information, however, so that (7.20), (7.26), and (7.29) merely represent different data weighting versions of the same basic weight update equation.

## 7.3 KALMAN FILTER METHODS FOR ADAPTIVE ARRAY PROCESSING

The adaptive control problems presented by small communications and data collection arrays are relatively simple and are often adequately handled by means of adaptive control algorithms based on gradient and random search methods. For more complex problems such as command and control of remote vehicles or rapid angular tracking in radar systems, however, more sophisticated processing is often required. For such demanding applications adaptive processing methods based on Kalman filtering [11], [12] have recently been proposed [5]. These methods require the inclusion of more *a priori* signal environment data than other methods and open the possibility of constructing processors that integrate adaptive array control with other system functions such as position location and navigation.

### *7.3.1 Development of a Kalman-Type Array Processor*

Consider again the simple $N$-element linear narrowband array model of Figure 7.1 for which the (sampled) signal vector is $\mathbf{x}(k)$, the adaptive processor weight vector (at corresponding sample times) is $\mathbf{w}(k)$, and the array output is $y(k) = \mathbf{w}^T(k)\mathbf{x}(k)$. To accommodate wider bandwidth signals, each channel of the array should contain a tapped-delay line filter, but the tap weights and delayed signals can still be represented as vectors $\mathbf{w}(k)$ and $\mathbf{x}(k)$ so the mathematical development for this case remains the same as for the narrowband case.

Let the dynamic behavior of the optimal array weights be represented by

$$\mathbf{w}_{\text{opt}}(k+1)=\mathbf{\Phi}(k+1,k)\mathbf{w}_{\text{opt}}(k),\ \mathbf{w}_{\text{opt}}(0)=\mathbf{w}_0 \tag{7.30}$$

where $\mathbf{\Phi}(k+1,k)$ is a transition matrix. If the array signal environment is stationary, then the optimal weights will be fixed and $\mathbf{\Phi}(k+1,k)$ is the identity matrix. For time-varying environments a more complex model for $\mathbf{\Phi}(k+1,k)$ must be developed that reflects how the optimal array weights must change in response to the changing environment. Let the system measurements be represented by a noise corrupted version of the optimal (not the actual) array output:

$$d(k)=\mathbf{x}^T(k)\mathbf{w}_{\text{opt}}(k)+\nu(k) \tag{7.31}$$

where $\nu(k)$ is a member of a white, Gaussian noise sequence having zero mean and variance given by

$$E\{\nu(k)\nu(j)\}=\sigma^2(k)\delta_{kj} \tag{7.32}$$

The selection of a value for $\sigma^2(k)$ is discussed later.

For the dynamical model (7.30) having state vector $\mathbf{w}_{\text{opt}}(k)$ and measurement from (7.31), a minimum MSE estimator for the array weights (rather than for the desired signal) can be obtained by straightforward application of Kalman filter theory from which it immediately follows that the optimum filtered estimate of the optimal array weight vector $\hat{\mathbf{w}}_{\text{opt}}(k/k)$ is given by the relationship [12]

$$\hat{\mathbf{w}}_{\text{opt}}(k/k)=\hat{\mathbf{w}}_{\text{opt}}(k/k-1)+\mathbf{K}(k)\left[d(k)-\mathbf{x}^T(k)\hat{\mathbf{w}}_{\text{opt}}(k/k-1)\right] \tag{7.33}$$

where $(k/k)$ denotes a filtered quantity at sample time $k$ based on measurements through (and including) $k$, $(k/k-1)$ denotes a predicted quantity at sample time $k$ based on measurements through $k-1$, and $\mathbf{K}(k)$ is the Kalman gain matrix. For complex quantities (7.33) can be rewritten as

$$\begin{aligned}\hat{\mathbf{w}}_{\text{opt}}(k/k)=&\,\hat{\mathbf{w}}_{\text{opt}}(k/k-1)\\&+\mathbf{K}(k)\left[d^*(k)-\mathbf{x}^\dagger(k)\hat{\mathbf{w}}_{\text{opt}}(k/k-1)\right]\end{aligned} \tag{7.34}$$

where * denotes complex conjugate and † denotes complex conjugate transpose.

Now

$$\hat{\mathbf{w}}_{\text{opt}}(k/k-1)=\mathbf{\Phi}(k,k-1)\hat{\mathbf{w}}_{\text{opt}}(k-1/k-1) \tag{7.35}$$

and the quantity in brackets of (7.34) is the difference between the optimal array output (or desired reference signal) and the actual array output. The

Kalman-type processor based on the foregoing equations is shown in Figure 7.2 where the Kalman gain vector is given by

$$\mathbf{K}(k)=\frac{\mathbf{P}(k/k-1)\mathbf{x}(k)}{\left[\mathbf{x}^T(k)\mathbf{P}(k/k-1)\mathbf{x}(k)+\sigma^2(k)\right]} \tag{7.36}$$

or for complex quantities,

$$\mathbf{K}(k)=\frac{\mathbf{P}(k/k-1)\mathbf{x}^*(k)}{\left[\mathbf{x}^T(k)\mathbf{P}(k/k-1)\mathbf{x}^*(k)+\sigma^2(k)\right]} \tag{7.37}$$

The predicted error covariance matrix is given by

$$\mathbf{P}(k/k-1)=\mathbf{\Phi}(k,k-1)\mathbf{P}(k-1/k-1)\mathbf{\Phi}^T(k,k-1) \tag{7.38}$$

The filtered error covariance matrix is defined by

$$\mathbf{P}(k/k) \triangleq E\left\{\left[\mathbf{w}_{\text{opt}}(k)-\hat{\mathbf{w}}_{\text{opt}}(k/k)\right]\left[\mathbf{w}_{\text{opt}}(k)-\hat{\mathbf{w}}_{\text{opt}}(k/k)\right]^T\right\} \tag{7.39}$$

Equation (7.39) can be expressed in the equivalent form

$$\mathbf{P}(k/k)=\mathbf{P}(k/k-1)-\mathbf{K}(k)\mathbf{x}^T(k)\mathbf{P}(k/k-1) \tag{7.40}$$

On substituting (7.36) into (7.40) there results

$$\mathbf{P}(k/k)=\mathbf{P}(k/k-1)-\frac{\mathbf{P}(k/k-1)\mathbf{x}(k)\mathbf{x}^T(k)\mathbf{P}(k/k-1)}{\left[\sigma^2(k)+\mathbf{x}^T(k)\mathbf{P}(k/k-1)\mathbf{x}(k)\right]} \tag{7.41}$$

By comparing (7.41) with (7.14), the great similarity between the recursive equations for $\mathbf{P}(k+1)$ of (7.14) and $\mathbf{P}(k/k)$ of (7.41) may be seen. Equation (7.41) can easily be rewritten by application of the matrix inversion lemma [(D.14) of Appendix D] as

$$\mathbf{P}^{-1}(k/k)=\mathbf{P}^{-1}(k/k-1)+\frac{1}{\sigma^2(k)}\mathbf{x}(k)\mathbf{x}^T(k) \tag{7.42}$$

On comparing (7.42) with (7.13), it is seen that the Kalman error covariance matrix with $\sigma^2(k)=1$ corresponds to $\mathbf{P}(k)$ for the weighted least square error processor with $\alpha=1$. In this case it follows from (7.21) that the error covariance matrix $\mathbf{P}(k/k)$ is related to the finite-time average estimate for the received signal autocorrelation matrix $\hat{\mathbf{R}}_{xx}(k)$ by the relationship

$$\hat{\mathbf{R}}_{xx}^{-1}(k)=k\mathbf{P}(k/k) \tag{7.43}$$

To begin the recursive equations (7.33)–(7.41), the initial values for $\hat{\mathbf{w}}(0/0)$

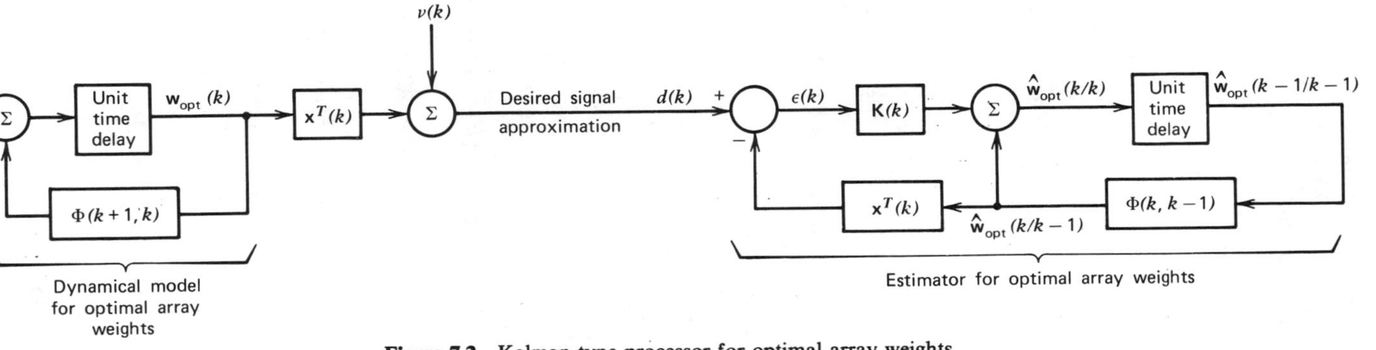


**Figure 7.2** Kalman-type processor for optimal array weights.

and $\mathbf{P}(0/0)$ must be specified. It is desirable to select $\hat{\mathbf{w}}_{\text{opt}}(0/0)=E\{\mathbf{w}_{\text{opt}}\}$ and $\mathbf{P}(0/0)=E\{\Delta\mathbf{w}(0)\Delta\mathbf{w}^T(0)\}$ [5] where $\Delta\mathbf{w}(0)=\hat{\mathbf{w}}(0/0)-\mathbf{w}_{\text{opt}}$ and from Chapter 3 $\mathbf{w}_{\text{opt}}$ is given by the Wiener solution,

$$\mathbf{w}_{\text{opt}}=\mathbf{R}_{xx}^{-1}(k)\mathbf{r}_{xd}(k) \tag{7.44}$$

In general, the signal statistics represented by the solution (7.44) are unknown, so a different procedure (discussed below) is employed to initialize the recursive equations. In the event that *a priori* environment data is available, then such information can be used to form a refined initial estimate of $\mathbf{w}_{\text{opt}}$ using (7.44), as well as to construct a dynamical system model by way of (7.30).

In situations where *a priori* information concerning the signal environment is not available, then $\hat{\mathbf{w}}_{\text{opt}}(0)$ is generally chosen to yield an omnidirectional array beam pattern, and $\mathbf{P}(0/0)$ can merely be set equal to the identity matrix. Furthermore, some means of selecting a value for the noise statistic $\sigma^2(k)$ must be given.

As can be seen from (7.33), the bracketed quantity represents the difference between the actual array output (using the estimated weights) and a noise corrupted version of the optimal array output derived from an identical array employing the optimal weight vector on the received signal set. This optimal output signal $d(t)$ can be interpreted as a reference signal approximation for the actual desired signal since the optimal array weights are designed to provide a MMSE estimate of the desired signal. Such an interpretation of the optimal array output signal in turn suggests a procedure for selecting a value for the noise statistic $\sigma^2(k)$. Since $d(t)$ is an approximation for the actual desired signal, $s(t)$, one may write [5]

$$d(k)=\mathbf{x}^T(k)\mathbf{w}_{\text{opt}}(k)+\nu(k)=s(k)+\eta(k) \tag{7.45}$$

where $\eta(k)$ now indicates the error in this approximation for the desired signal $s(k)$. Consequently,

$$\nu(k)=s(k)+\eta(k)-\mathbf{x}^T(k)\mathbf{w}_{\text{opt}}(k) \tag{7.46}$$

If $s(k)$, $\eta(k)$, and $\mathbf{x}(k)$ are all zero mean processes, then $E\{\nu(k)\}=0$. Furthermore if the noise sequence $\eta(k)$ is not correlated with $s(k)$ and the noise components of $\mathbf{x}(k)$, then

$$\begin{aligned}\sigma^2(k)=E\{\nu(k)\nu(k)\}&=\mathbf{w}_{\text{opt}}^T\mathbf{R}_{xx}\mathbf{w}_{\text{opt}}-2\mathbf{w}_{\text{opt}}^T\mathbf{r}_{xs}\\&\quad+E\{s^2\}+E\{\eta^2\}\\&=\text{MMSE}+E\{\eta^2\}\end{aligned} \tag{7.47}$$

It follows that if the "measurement" $d(k)$ is generated by using a realistic approximation for the desired signal (possibly obtained from a modem connected to the array output), then the value of $\sigma^2(k)$ is chosen based on the MMSE that can be achieved by an optimal array and the quality (as measured by the error variance) of the desired signal approximation.

If we rewrite (7.33) as

$$\hat{\mathbf{w}}(k/k)=\hat{\mathbf{w}}(k/k-1)+\frac{\mathbf{P}(k/k-1)}{\left[\mathbf{x}^T(k)\mathbf{P}(k/k-1)\mathbf{x}(k)+\sigma^2(k)\right]} \cdot\left[\mathbf{x}(k)d(k)-\mathbf{x}(k)y(k)\right] \tag{7.48}$$

where $y(k)=\hat{\mathbf{w}}^T(k/k-1)\mathbf{x}(k)$, it is apparent that the quantity $\mathbf{x}(k)d(k)$ can be replaced by its average value, in exactly the same manner as with the weighted least squares recursive algorithm of the preceding section. Consequently the Kalman filter weight update equation can also be modified to accommodate two additional forms involving the use of either $\hat{\mathbf{r}}_{xd}(k)$ (when either the desired signal or a reference signal model is available) or $\mathbf{r}_{xd}(k)$ (when only direction of arrival information is available).

Yet another alternative for handling a nonstationary environment with a Kalman-type processor exists when the time-varying nature of the signal statistics are unknown. Rather than attempt to construct an accurate dynamical model representation by way of (7.30), simply use

$$\mathbf{w}_{\text{opt}}(k+1)=\mathbf{w}_{\text{opt}}(k)+\boldsymbol{\xi}(k) \tag{7.49}$$

where $\boldsymbol{\xi}(k)$ is a zero mean white noise process with

$$\text{cov}\left[\boldsymbol{\xi}(k)\right]=\mathbf{Q}(k) \tag{7.50}$$

The elements of the matrix $\mathbf{Q}$ then represent the degree of uncertainty associated with adopting the stationary environment assumption represented by using the identity state transition matrix in (7.30). Equation (7.38) then becomes

$$\mathbf{P}(k/k-1)=\boldsymbol{\Phi}(k,k-1)\mathbf{P}(k-1/k-1)\boldsymbol{\Phi}^T(k,k-1)+\mathbf{Q} \tag{7.51}$$

The practical effect of the above modification is to prevent the Kalman gains in $\mathbf{K}(k)$ from decaying to values that are too small so when variations in the environment occur sufficient importance is attached to the most recent measurements. The estimate $\hat{\mathbf{w}}_{\text{opt}}$ then "follows" variations in the actual value of $\mathbf{w}_{\text{opt}}$, although the resulting optimal weight vector estimates are more "noisy" than when the $\mathbf{Q}$ matrix was absent.

### 7.3.2 *Speed of Convergence*

The fact that the value of $\sigma^2(k)$ is related to the MMSE and to the quality of the desired signal approximation makes it possible to obtain an expression from which the rate of convergence for the algorithm can be deduced under certain conditions [13]. Equation (7.36) can be rewritten as

$$\mathbf{K}(k)\mathbf{x}^T(k)\mathbf{P}(k/k-1)\mathbf{x}(k)+\mathbf{K}(k)\sigma^2(k)=\mathbf{P}(k/k-1)\mathbf{x}(k) \tag{7.52}$$

or

$$\mathbf{K}(k)\sigma^2(k)=\left[\mathbf{I}-\mathbf{K}(k)\mathbf{x}^T(k)\right]\mathbf{P}(k/k-1)\mathbf{x}(k) \tag{7.53}$$

Equation (7.53) may now be substituted into (7.40) to obtain

$$\mathbf{P}(k/k)\mathbf{x}(k)=\mathbf{K}(k)\sigma^2(k) \tag{7.54}$$

so that

$$\mathbf{K}(k)=\frac{\mathbf{P}(k/k)\mathbf{x}(k)}{\sigma^2(k)} \tag{7.55}$$

When we substitute the result expressed by (7.55) into (7.40) there results

$$\mathbf{P}(k/k)=\mathbf{P}(k/k-1)-\frac{\mathbf{P}(k/k)\mathbf{x}(k)}{\sigma^2(k)}\cdot\mathbf{x}^T(k)\mathbf{P}(k/k-1) \tag{7.56}$$

or

$$\mathbf{P}(k/k)=\mathbf{P}(k/k-1)-\frac{\left[\mathbf{P}(k/k)\mathbf{x}(k)\mathbf{x}^T(k)\mathbf{P}(k/k-1)\right]}{\sigma^2(k)} \tag{7.57}$$

Premultiplying both sides of (7.57) by $\mathbf{P}^{-1}(k/k)$ and postmultiplying both sides of (7.58) by $\mathbf{P}^{-1}(k/k-1)$, we see it follows that [also see (7.42)]

$$\mathbf{P}^{-1}(k/k)=\mathbf{P}^{-1}(k/k-1)+\frac{\mathbf{x}(k)\mathbf{x}^T(k)}{\sigma^2(k)} \tag{7.58}$$

Equation (7.58) can be rewritten as

$$\mathbf{P}^{-1}(k/k)=\frac{1}{\sigma^2(k)}\left[\sigma^2(k)\mathbf{P}^{-1}(k/k-1)+\mathbf{x}(k)\mathbf{x}^T(k)\right] \tag{7.59}$$

so that

$$\mathbf{P}(k/k)=\sigma^2(k)\left[\sigma^2(k)\mathbf{P}^{-1}(k/k-1)+\mathbf{x}(k)\mathbf{x}^T(k)\right]^{-1} \tag{7.60}$$

The recursive relationship expressed by (7.60) can be repeatedly applied beginning with $\mathbf{P}^{-1}(0/-1)$ to obtain

$$\mathbf{P}(k/k)=\sigma^2(k)\left[\sigma^2(k)\mathbf{P}^{-1}(0/-1)+\sum_{i=1}^{k}\mathbf{x}(i)\mathbf{x}^T(i)\right]^{-1} \tag{7.61}$$

For cases where the desired signal approximation is quite good $\hat{\sigma}^2(k)\approx$ MMSE, and the diagonal matrix $\hat{\sigma}^2(k)\mathbf{P}^{-1}(0/-1)$ can be neglected in comparison with $\sum_{i=1}^{k}\mathbf{x}(i)\mathbf{x}^T(i)$ so that

$$\mathbf{P}(k/k)\cong\hat{\sigma}^2(k)\left[\sum_{i=1}^{k}\mathbf{x}(i)\mathbf{x}^T(i)\right]^{-1} \tag{7.62}$$

When we use the result (7.55), it follows immediately that

$$\mathbf{K}(k)\cong\left[\sum_{i=1}^{k}\mathbf{x}(i)\mathbf{x}^T(i)\right]^{-1}\mathbf{x}(k) \tag{7.63}$$

which is independent of $\sigma^2(k)$.

The arithmetic average

$$\frac{1}{k}\sum_{i=1}^{k}\mathbf{x}(i)\mathbf{x}^T(i)\to\mathbf{R}_{xx}(k)\qquad\text{as } k\to\infty \tag{7.64}$$

The MSE at the $k$th sampling instant, $\xi^2(k)$, can be written as [13]

$$\xi^2(k)=\text{trace}\left[\mathbf{P}(k/k)\mathbf{R}_{xx}(k)\right]+\text{MMSE} \tag{7.65}$$

From (7.62) and (7.64) it follows that

$$\text{trace}\left[\mathbf{P}(k/k)\mathbf{R}_{xx}(k)\right]\cong\sigma^2(k)Nk^{-1} \tag{7.66}$$

where $N$ is the dimension of $\mathbf{x}(k)$ so the MSE at the $k$th sampling instant becomes

$$\xi^2(k)\cong\text{MMSE}\left[1+Nk^{-1}\right] \tag{7.67}$$

The result expressed by (7.67) means that convergence for this Kalman-type algorithm can theoretically be obtained within $2N$ iterations. This result is similar to the convergence results obtained for the DMI algorithm.

### 7.4 THE MINIMUM VARIANCE PROCESSOR

It is useful at this point to consider the application of the concepts used in obtaining a recursive algorithm to a broadband signal aligned array processor to illustrate the slight modifications involved in handling a multichannel tapped-delay line processor. If the direction of arrival of the desired signal is known, this information may be used to construct a signal aligned array as shown in Figure 7.3. The signal aligned array processor uses spatial correction filters (SCF) and knowledge of the desired signal's direction of arrival to align the desired signal components in each channel of the array by properly selecting the time delays $\tau_i$. That is [14],

$$\mathbf{x}(t)=\begin{bmatrix} z_1(t-\tau_1) \\ z_2(t-\tau_2) \\ \vdots \\ z_N(t-\tau_N) \end{bmatrix}=d(t)\mathbf{1}+\mathbf{n}(t) \tag{7.68}$$

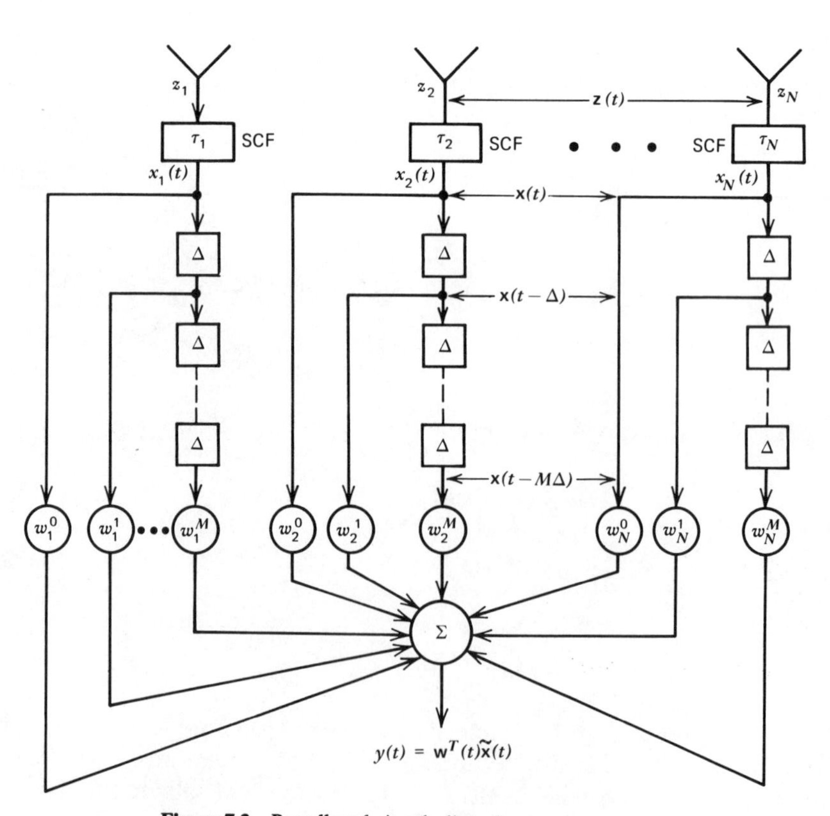


**Figure 7.3** Broadband signal aligned array processor.

where $\mathbf{1}=[1,1,\ldots,1]^T$, $d(t)$ is the the desired reference signal, and $\mathbf{n}(t)$ is the vector of the interference terms after the time delays. Collecting the steered received signal vector $\mathbf{x}(t)$ and its delayed components along the tapped-delay line into a single $(M+1)N\times 1$ dimensional vector gives

$$\mathbf{x}'(t)=\begin{bmatrix}\mathbf{x}(t)\\ \mathbf{x}(t-\Delta)\\ \vdots\\ \mathbf{x}(t-M\Delta)\end{bmatrix}=\begin{bmatrix}d(t)\mathbf{1}\\ d(t-\Delta)\mathbf{1}\\ \vdots\\ d(t-M\Delta)\mathbf{1}\end{bmatrix}+\mathbf{n}'(t) \tag{7.69}$$

In what follows it will be convenient to drop the prime notation and simply remember that all signals under consideration are the collection of terms in the tapped-delay line filter. Furthermore, collecting the $N$-dimensional weight vectors $\mathbf{w}^0(t),\mathbf{w}^1(t),\ldots,\mathbf{w}^M(t)$, as

$$\mathbf{w}(t)=\begin{bmatrix}\mathbf{w}^0(t)\\ \mathbf{w}^1(t)\\ \vdots\\ \mathbf{w}^M(t)\end{bmatrix} \tag{7.70}$$

the array output can be written in the form

$$\begin{aligned} y(t)&=\mathbf{w}^T\mathbf{x}(t)=d(t)\sum_{i=1}^{N} w_i^0\\ &=d(t)\sum_{i=1}^{N} w_i^0+d(t-\Delta)\sum_{i=1}^{N} w_i^1+\cdots+d(t-M\Delta)\sum_{i=1}^{N} w_i^M+\mathbf{w}^T\mathbf{n}(t) \end{aligned} \tag{7.71}$$

since $\mathbf{w}^{l^T}\mathbf{1}=\sum_{i=1}^{N} w_i^l$. If the adaptive weights are constrained according to

$$\sum_{i=1}^{N} w_i^l=\begin{cases}1, & l=0\\ 0, & l=1,2,\cdots,k\end{cases} \tag{7.72}$$

then the output signal can be written as

$$y(t)=d(t)+\mathbf{w}^T\mathbf{n}(t) \tag{7.73}$$

The output signal of (7.73) is unbiased since

$$E\{y(t)\}=E\{d(t)+\mathbf{w}^T\mathbf{n}(t)\}=d(t) \tag{7.74}$$

if we assume the noise vector has zero mean. The variance of the output signal is then given by

$$\operatorname{var}[y(t)] = E\{\mathbf{w}^T\mathbf{n}(t)\mathbf{n}^T(t)\mathbf{w}\} = \mathbf{w}^T\mathbf{R}_{nn}\mathbf{w} \tag{7.75}$$

Define the $N(M+1)\times(M+1)$ matrix,

$$\mathbf{I}_1 \triangleq \begin{bmatrix} \mathbf{1} & \mathbf{0} & \cdots & \mathbf{0} \\ \mathbf{0} & \mathbf{1} & \cdots & \mathbf{0} \\ \vdots & \vdots & & \\ \mathbf{0} & \mathbf{0} & \cdots & \mathbf{1} \end{bmatrix} \tag{7.76}$$

where $\mathbf{0}$ is an $N\times 1$ vector with all zero components.

If it is desired to minimize the output noise variance, then the noise variance performance measure can be defined by

$$\mathfrak{P}_{\text{mv}} = \mathbf{w}^T\mathbf{R}_{nn}\mathbf{w} \tag{7.77}$$

It is now desired to minimize (7.77) subject to the constraint (7.72), which can be rewritten as

$$\mathbf{I}_1^T\mathbf{w} = \begin{bmatrix} 1 \\ 0 \\ \vdots \\ 0 \end{bmatrix} = \mathbf{c} \tag{7.78}$$

The weight vector $\mathbf{w}$ that minimizes (7.77) subject to (7.78) can be chosen by using a vector Lagrange multiplier to form the modified performance measure:

$$\mathfrak{P}_{\text{mvm}} = \tfrac{1}{2}\mathbf{w}^T\mathbf{R}_{nn}\mathbf{w} + \boldsymbol{\lambda}[\mathbf{c} - \mathbf{I}_1^T\mathbf{w}] \tag{7.79}$$

Setting the derivative of $\mathfrak{P}_{\text{mvm}}$ with respect to $\mathbf{w}$ equal to zero to obtain $\mathbf{w}_{\text{mv}}$, requiring $\mathbf{w}_{\text{mv}}$ to satisfy (7.78) to evaluate $\boldsymbol{\lambda}$, and substituting the resulting value of $\boldsymbol{\lambda}$ into $\mathbf{w}_{\text{mv}}$ gives the minimum variance weight vector solution:

$$\mathbf{w}_{\text{mv}} = \mathbf{R}_{nn}^{-1}\mathbf{I}_1[\mathbf{I}_1^T\mathbf{R}_{nn}^{-1}\mathbf{I}_1]^{-1}\mathbf{c} \tag{7.80}$$

For a signal aligned array like that of Figure 7.3, it can also be established (as was done in Chapter 3 for a narrowband processor) that the minimum variance estimator resulting from the use of (7.80) is identical to the maximum likelihood estimator [15].

The weight vector computation defined by (7.80) requires the measurement and inversion of the noise autocorrelation matrix $\mathbf{R}_{nn}$. The noise autocorrelation matrix can be obtained from the data that also includes desired signal terms, and use of a recursive algorithm will circumvent the necessity of directly inverting $\mathbf{R}_{nn}$.

The difficulty of measuring the noise autocorrelation matrix when desired signal terms are present can be avoided by reformulating the optimization problem posed by (7.77) and (7.78). The minimization of (7.77) subject to the constraint (7.78) is completely equivalent to the following problem:

$$\text{Minimize } \mathfrak{P} = \mathbf{w}^T \mathbf{R}_{xx} \mathbf{w} \tag{7.81}$$

subject to the constraint

$$\mathbf{I}_1 \mathbf{w} = \mathbf{c} \tag{7.82}$$

The solution of (7.81),(7.82) may be found by once again using Lagrange multipliers with the result that

$$\mathbf{w}_{\text{opt}} = \mathbf{R}_{xx}^{-1} \mathbf{I}_1 \left[ \mathbf{I}_1^T \mathbf{R}_{xx}^{-1} \mathbf{I}_1 \right]^{-1} \mathbf{c} \tag{7.83}$$

To show the equivalence between the problem (7.81) and (7.82) and the original problem (7.77) and (7.78), expand the matrix $\mathbf{R}_{xx}$ as follows:

$$\mathbf{R}_{xx} = E\{\mathbf{x}(t)\mathbf{x}^T(t)\} = \mathbf{R}_{dd} + \mathbf{R}_{nn} \tag{7.84}$$

where

$$\mathbf{R}_{dd} = E\left\{ \begin{bmatrix} d(t)\mathbf{1} \\ d(t-\Delta)\mathbf{1} \\ \vdots \\ d(t-M\Delta)\mathbf{1} \end{bmatrix} \left[ d(t)\mathbf{1}, d(t-\Delta)\mathbf{1}, \ldots, d(t-M\Delta)\mathbf{1} \right] \right\} \tag{7.85}$$

Substituting (7.84) into (7.81), the problem now becomes

$$\text{Minimize } \mathfrak{P} = \mathbf{w}^T \mathbf{R}_{xx} \mathbf{w} = \mathbf{w}^T \mathbf{R}_{nn} \mathbf{w} + \mathbf{w}^T \mathbf{R}_{dd} \mathbf{w} \tag{7.86}$$

With the constraint (7.72), however, it follows that

$$\mathfrak{P} = \mathbf{w}^T \mathbf{R}_{nn} \mathbf{w} + E\{d^2(t)\} \tag{7.87}$$

Now the minimization of (7.87) subject to (7.82) must give exactly the same solution as the optimization problem of (7.77) and (7.78), since $E\{d^2(t)\}$ is not a function of $\mathbf{w}$ and the two problems are therefore completely equivalent.

The received signal correlation matrix at the $k$th sample time $\mathbf{R}_{xx}(k)$ can be measured using the exponentially de-weighted finite time average,

$$\hat{\mathbf{R}}_{xx}(k)=\frac{1}{\left(\sum_{n=1}^{k}\alpha^{k-n}\right)}\sum_{n=1}^{k}\alpha^{k-n}\mathbf{x}(n)\mathbf{x}^T(n) \tag{7.88}$$

where $0<\alpha\leqslant 1$. By defining the matrices

$$\mathbf{X}(k)\triangleq\begin{bmatrix}\mathbf{x}^T(1)\\ \mathbf{x}^T(2)\\ \vdots\\ \mathbf{x}^T(k)\end{bmatrix} \tag{7.89}$$

and

$$\mathbf{A}(k)\triangleq\begin{bmatrix}\alpha^{k-1} & 0 & \cdots & 0\\ 0 & \alpha^{k-2} & & \vdots\\ \vdots & & \alpha & 0\\ 0 & \cdots & 0 & 1\end{bmatrix} \tag{7.90}$$

then $\hat{\mathbf{R}}_{xx}(k)$ can be rewritten as

$$\hat{\mathbf{R}}_{xx}(k)=\frac{1}{\operatorname{tr}[\mathbf{A}(k)]}\mathbf{X}^T(k)\mathbf{A}(k)\mathbf{X}(k)=\frac{1}{\operatorname{tr}[\mathbf{A}(k)]}\mathbf{P}^{-1}(k) \tag{7.91}$$

where $\mathbf{P}^{-1}(k)\triangleq\mathbf{X}^T(k)\mathbf{A}(k)\mathbf{X}(k)$. Partitioning $\mathbf{x}(k+1)$ and $\mathbf{A}(k+1)$ as in (7.7) and (7.9) and applying the matrix inversion lemma immediately leads to (7.14). The desired updated weights may then be written in terms of $\mathbf{P}(k+1)$ from (7.83) as

$$\mathbf{w}_{\text{opt}}(k+1)=\mathbf{P}(k+1)\mathbf{I}_1\left[\mathbf{I}_1^T\mathbf{P}(k+1)\mathbf{I}_1\right]^{-1}\mathbf{c} \tag{7.92}$$

The inverse of $\mathbf{I}_1^T\mathbf{P}(k+1)\mathbf{I}_1$ required in (7.92) can be efficiently computed by application of the matrix inversion lemma to yield

$$\mathbf{w}_{\text{opt}}(k+1)=\left\{\mathbf{I}-\left[\frac{\mathbf{P}(k)}{\alpha+\mathbf{x}^T(k+1)\mathbf{P}(k)\mathbf{x}(k+1)}-\mathbf{P}(k+1)\right]\cdot\mathbf{x}(k+1)\mathbf{x}^T(k+1)\right\}\mathbf{I}_1^{-1}\mathbf{c} \tag{7.93}$$

The minimum variance recursive processor for the narrowband case takes a particularly simple form. It was found in Chapter 3 for this case that

$$\mathbf{w}_{\text{mv}}=\frac{\mathbf{R}_{nn}^{-1}\mathbf{1}}{\mathbf{1}^T\mathbf{R}_{nn}^{-1}\mathbf{1}} \tag{7.94}$$

The use of (7.94) presents an additional difficulty since the received signal vector $\mathbf{x}(t)$ generally contains signal as well as noise components. This difficulty can be circumvented as explained below.

The input signal covariance matrix is given by

$$\mathbf{R}_{xx}\triangleq E\{\mathbf{x}^*(t)\mathbf{x}^T(t)\}=E\{d^2(t)\}\mathbf{1}\mathbf{1}^T+\mathbf{R}_{nn} \tag{7.95}$$

where $E\{d^2(t)\}=\beta$, a scalar quantity. Inverting both sides of (7.95) and applying the matrix inversion lemma yields

$$\mathbf{R}_{xx}^{-1}=\left[\beta\mathbf{1}\mathbf{1}^T+\mathbf{R}_{nn}\right]^{-1}=\mathbf{R}_{nn}^{-1}-\frac{\beta\mathbf{R}_{nn}^{-1}\mathbf{1}\mathbf{1}^T\mathbf{R}_{nn}^{-1}}{1+\beta\mathbf{1}^T\mathbf{R}_{nn}^{-1}\mathbf{1}} \tag{7.96}$$

Substituting (7.96) into the ratio

$$\frac{\mathbf{R}_{xx}^{-1}\mathbf{1}}{\mathbf{1}^T\mathbf{R}_{xx}^{-1}\mathbf{1}}$$

the following matrix identity results:

$$\frac{\mathbf{R}_{xx}^{-1}\mathbf{1}}{\mathbf{1}^T\mathbf{R}_{xx}^{-1}\mathbf{1}}\equiv\frac{\mathbf{R}_{nn}^{-1}\mathbf{1}}{\mathbf{1}^T\mathbf{R}_{nn}^{-1}\mathbf{1}} \tag{7.97}$$

Exploit (7.21) for the case when $\alpha=1$; it then follows from (7.97) and (7.94) that

$$\mathbf{w}_{\text{mv}}(k+1)=\frac{(k+1)\mathbf{P}(k+1)\mathbf{1}}{(k+1)\mathbf{1}^T\mathbf{P}(k+1)\mathbf{1}}=\frac{\mathbf{P}(k+1)\mathbf{1}}{\mathbf{1}^T\mathbf{P}(k+1)\mathbf{1}} \tag{7.98}$$

where $\mathbf{P}(k+1)$ is given by (7.14). Note that when the desired signal is absent, $\mathbf{w}_{\text{mv}}$ of (7.98) converges more rapidly than when the desired signal is present, a result already found in Chapter 6.

## 7.5 SIMULATION RESULTS

The recursive processor defined by (7.34), (7.38), and (7.41) can represent both the Kalman and the weighted least squares error processor since the Kalman error covariance matrix with $\sigma^2(k)=1$ corresponds to $\mathbf{P}(k)$ for the weighted least square error processor with $\alpha=1$. The parameter $\sigma^2(k)$ for the Kalman processor should be selected to be equal to the MMSE in

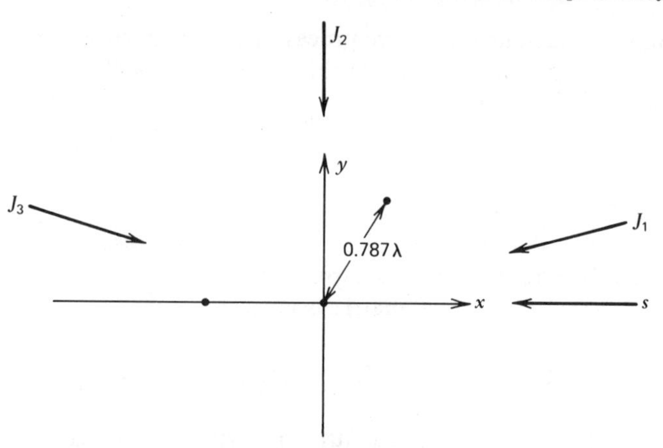


**Figure 7.4** Array geometry with signal and jammer locations for selected example.

accordance with (7.47). To simulate this recursive algorithm, it is necessary to define an array geometry and signal environment as in Figure 7.4 (which duplicates Figure 4.19 and is repeated here for convenience). This figure depicts a four element $Y$ array having $d=0.787\lambda$ element spacing with one desired signal at 0° and three distinct narrowband Gaussian jamming signals located at 15°, 90°, and 165°.

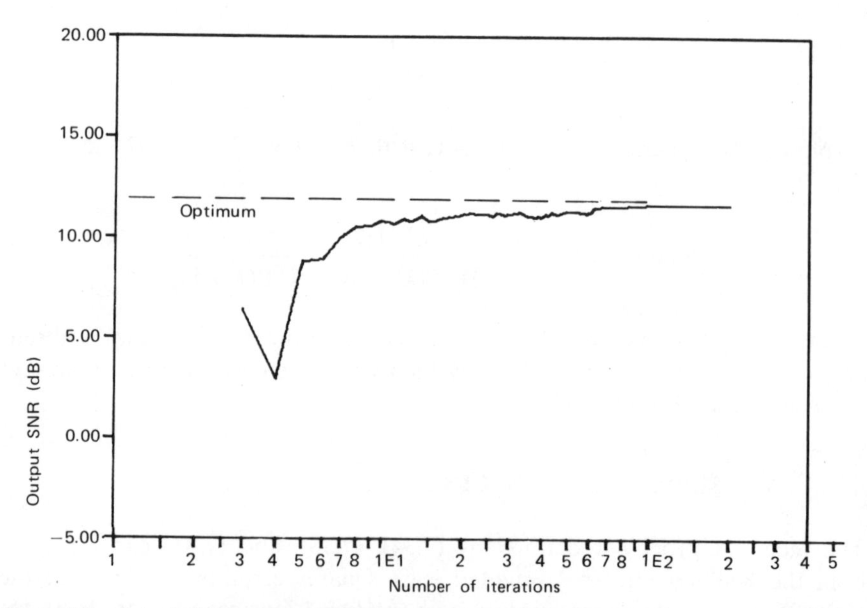


**Figure 7.5** Output SNR versus number of iterations for Kalman processor with eigenvalue spread of 16,700. Input $s/n=10$ for which the output $\text{SNR}_{\text{opt}}=15$ (11.76 dB) with algorithm parameters $\sigma^2=\text{MMSE}$, $\mathbf{w}(0)=\mathbf{0}$, and $\mathbf{P}(0)=\mathbf{I}$.

The desired signal in each case was taken to be a biphase modulated signal having a phase angle of either 0° or 180° with equal probability at each sample. Two signal environments are considered corresponding to eigenvalue spreads of 16,700 and 2440. Figures 7.5 to 7.8 give convergence results for an eigenvalue spread of 16,700, where the jammer-to-thermal noise ratios are $J_1/n=4000$, $J_2/n=400$, and $J_3/n=40$, for which the corresponding noise covariance matrix eigenvalues are given by $\lambda_1=1.67\times10^4$, $\lambda_2=1\times10^3$, $\lambda_3=29.0$, and $\lambda_4=1.0$. The input signal-to-thermal noise ratio is $s/n=10$ for Figures 7.5 and 7.6, $s/n=0.1$ for Figure 7.7, and $s/n=0.025$ for Figure 7.8. The performance of the algorithm in each case is recorded in terms of the output SNR versus number of iterations, where one weight iteration occurs with each new independent data sample.

Figures 7.9 and 7.10 give convergence results for an eigenvalue spread of 2440, where the jammer-to-thermal noise ratios are $J_1/n=500$, $J_2/n=200$, and $J_3/n=40$, for which the corresponding noise covariance matrix eigenvalues are given by $\lambda_1=2.44\times10^3$, $\lambda_2=4.94\times10^2$, $\lambda_3=25.62$, and $\lambda_4=$

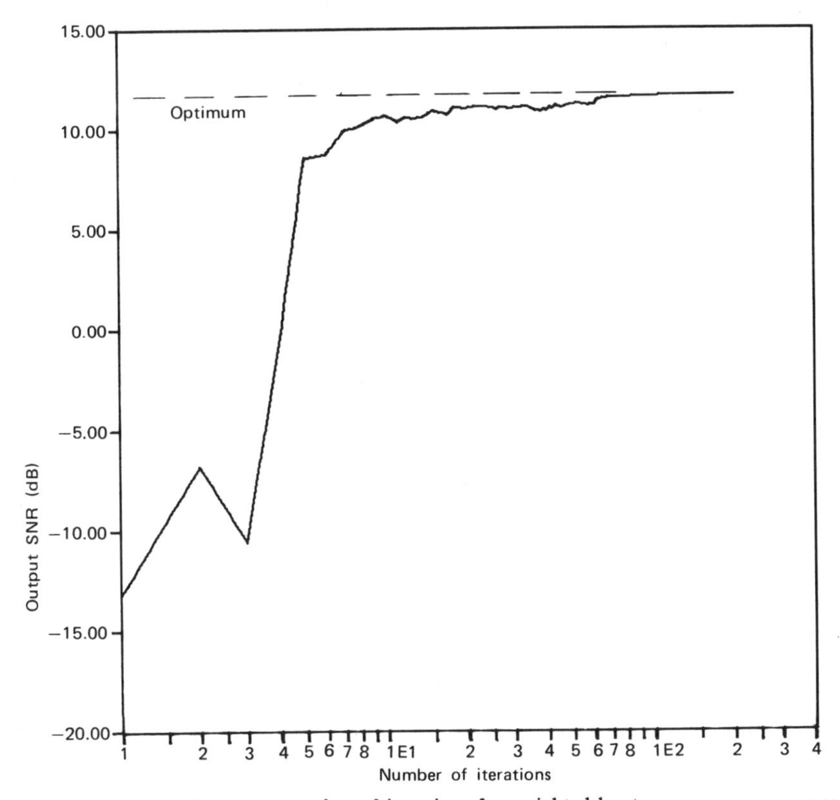


**Figure 7.6** Output SNR versus number of iterations for weighted least squares error processor with eigenvalue spread of 16,700. Input $s/n=10$ for which output $\text{SNR}_{\text{opt}}=15$ with algorithm parameters $\alpha=1$, $\mathbf{w}^T(0)=[1,0,0,0]$, and $\mathbf{P}(0)=\mathbf{I}$.

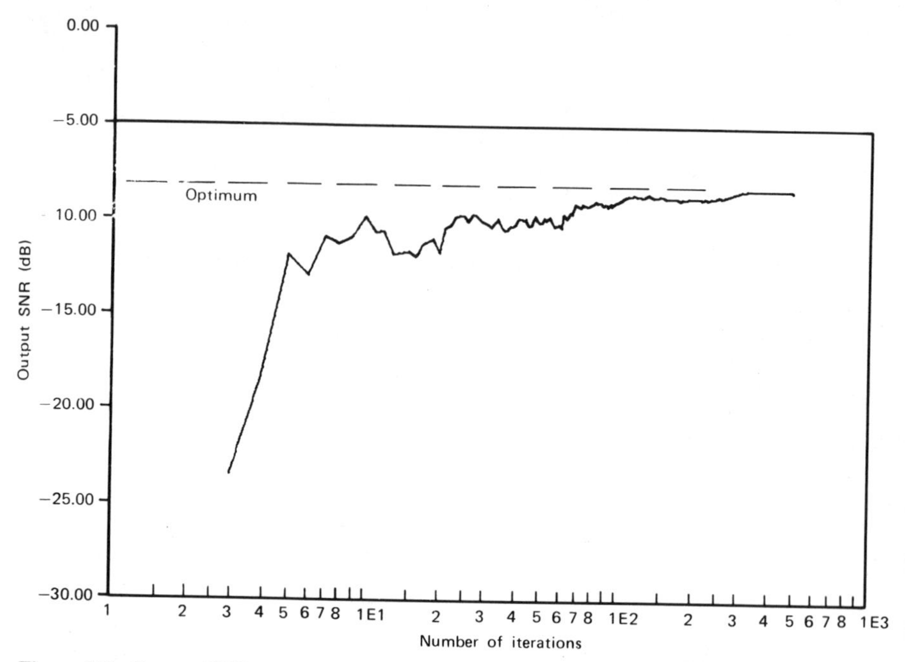


**Figure 7.7** Output SNR versus number of iterations for Kalman processor with eigenvalue spread of 16,700. Input $s/n=0.1$ for which output $\text{SNR}_{\text{opt}}=0.15$ ($-8.24$ dB) with algorithm parameters $\sigma^2=\text{MMSE}$, $\mathbf{w}(0)=\mathbf{0}$, and $\mathbf{P}(0)=\mathbf{I}$.

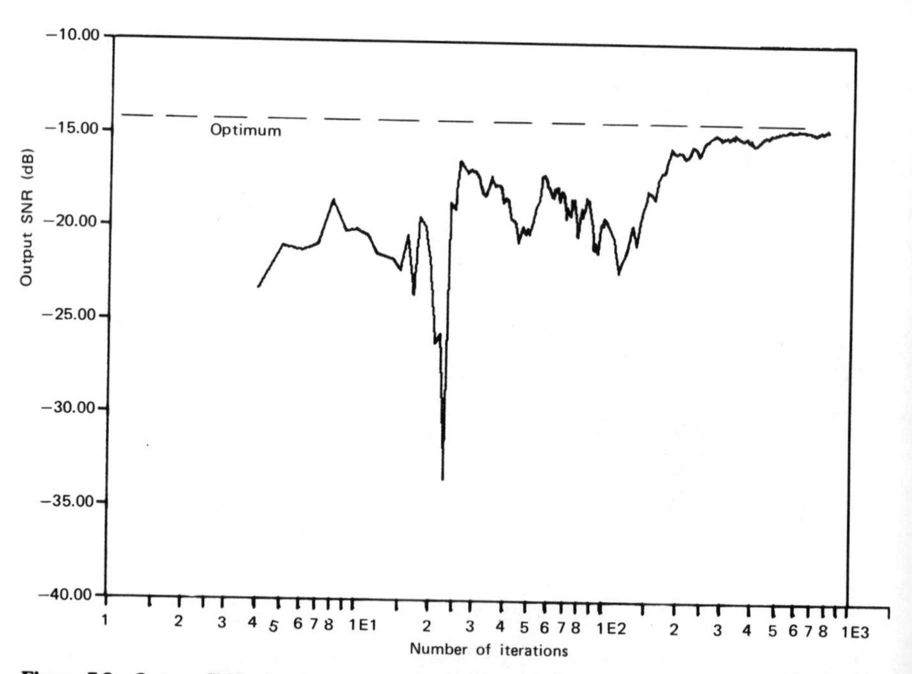


**Figure 7.8** Output SNR versus number of iterations for Kalman processor with eigenvalue spread of 16,700. Input $s/n=0.025$ for which output $\text{SNR}_{\text{opt}}=0.038$ ($-14.2$ dB) with algorithm parameters $\sigma^2=\text{MMSE}$, $\mathbf{w}(0)=\mathbf{0}$, and $\mathbf{P}(0)=\mathbf{I}$.

1.0. The input signal-to-thermal noise ratio is $s/n=0.1$ for Figure 7.9 and $s/n=10.0$ for Figure 7.10.

The simulation results shown in Figures 7.5 to 7.10 illustrate the following interesting and important properties that recursive algorithms exhibit:

1. Recursive algorithms exhibit fast convergence comparable to that of DMI algorithms, especially when the output $\text{SNR}_{\text{opt}}$ is large ($\sim$5 or 6 iterations when $\text{SNR}_{\text{opt}}=15.0$ for the examples shown). On comparing Figure 7.5 (for an eigenvalue spread of $1.67\times10^4$) with Figure 7.10 (for an eigenvalue spread of $2.44\times10^3$), it is seen that the algorithm convergence speed is insensitive to eigenvalue spread, which also reflects the similar property exhibited by DMI algorithms.
2. Algorithm convergence is relatively insensitive to the value selected for the parameter $\sigma^2(k)$ in (7.37). On comparing the results obtained in

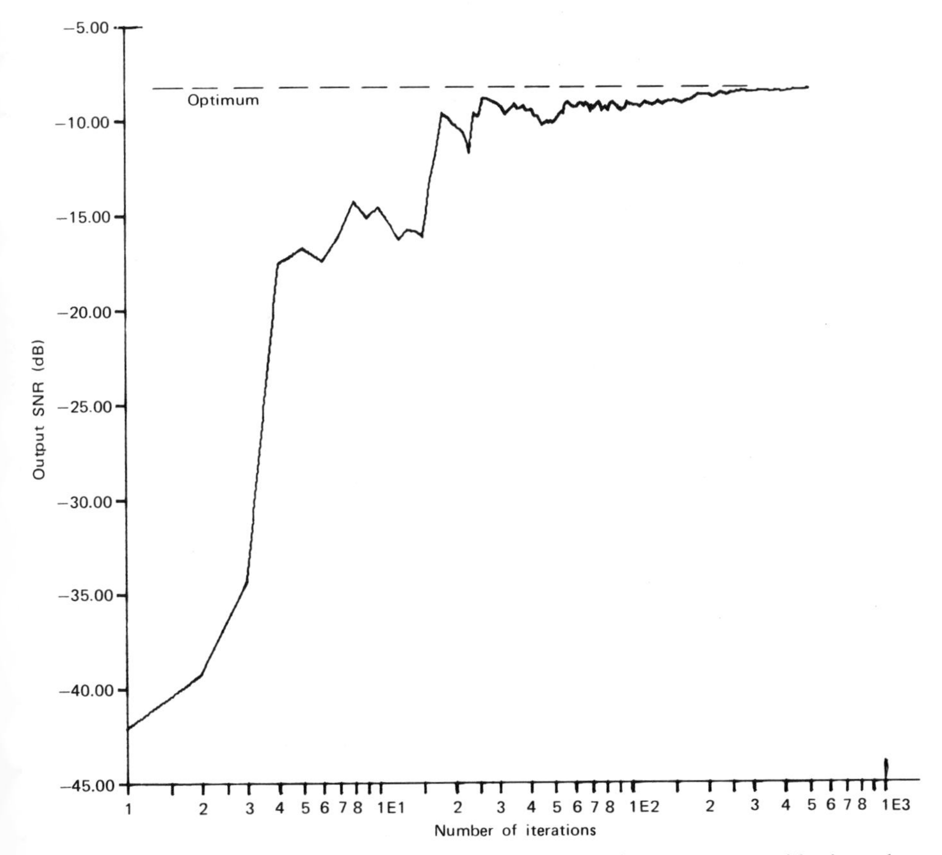


**Figure 7.9** Output SNR versus number of iterations for Kalman processor with eigenvalue spread of 2440. Input $s/n=0.1$ for which output $\text{SNR}_{\text{opt}}=0.15$ ($-8.24$ dB) with algorithm parameters $\sigma^2=\text{MMSE}$, $\mathbf{w}(0)=\mathbf{0}$, and $\mathbf{P}(0)=\mathbf{I}$.

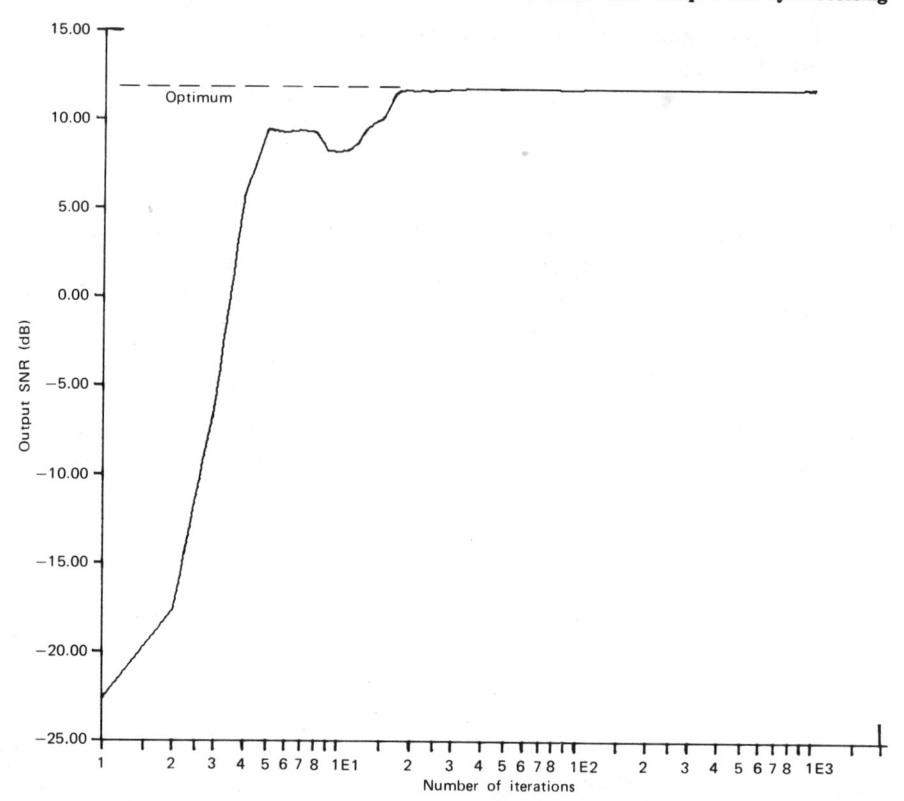


**Figure 7.10** Output SNR versus number of iterations for Kalman processor with eigenvalue spread of 2440. Input $s/n = 10$ for which output $\text{SNR}_{\text{opt}} = 15$ (11.76 dB) with algorithm parameters $\sigma^2 = \text{MMSE}$, $\mathbf{w}(0) = \mathbf{0}$, and $\mathbf{P}(0) = \mathbf{I}$.

Figure 7.5 where $\sigma^2 = \text{MMSE}$ with the results obtained in Figure 7.6 where $\sigma^2 = 1$, it is seen that virtually the same number of iterations are required to arrive within 3 dB of $\text{SNR}_{\text{opt}}$ even with different initial weight vectors.

3. The convergence speed of the recursive algorithm for the examples simulated here is slower for small values of $\text{SNR}_{\text{opt}}$ and faster for large values of $\text{SNR}_{\text{opt}}$, as seen in Figures 7.5, 7.7, and 7.8. This behavior is also exhibited by the DMI algorithm (and to some degree by all algorithms that do not assume direction-of-arrival information). When no direction-of-arrival information is assumed, such information can be "learned" from a strong desired signal component.

## 7.6 SUMMARY AND CONCLUSIONS

The specific form selected for a recursive processor should reflect the data weight scheme that is appropriate for the desired application. The various recursive algorithms may be developed by applying the matrix inversion lemma to the same basic weight update equation.

Since the recursive algorithms are different from a DMI algorithm primarily because the required matrix inversion is accomplished in a recursive manner, it is hardly surprising that many of the desirable properties that were found to apply to DMI algorithms also hold for recursive algorithms. Rapid convergence rates and insensitivity to eigenvalue spread are characteristics that make recursive processors attractive algorithm candidates provided sufficient computational power and accuracy are available to carry out the required calculations.

## PROBLEMS

**1** ***The Minimum Variance Weight Vector Solution***

(a) Show that setting the derivative of $\mathfrak{P}_{\text{mvm}}$ of (7.79) equal to zero yields

$$\mathbf{w}_{\text{mv}} = \mathbf{R}_{nn}^{-1}\mathbf{I}_1\lambda$$

(b) Show that requiring $\mathbf{w}_{\text{mv}}$ obtained in part (a) to satisfy (7.78) results in

$$\lambda = \left[\mathbf{I}_1^T\mathbf{R}_{nn}^{-1}\mathbf{I}_1\right]^{-1}\mathbf{c}$$

(c) Show that substituting the result obtained in part (b) into $\mathbf{w}_{\text{mv}}$ of part (a) results in (7.80).

**2** [15] ***Equivalence of the Maximum Likelihood and Minimum Variance Estimates*** In some signal reception applications the desired signal waveform is completely unknown and cannot be treated as a known waveform or even as a known function of some unknown parameters. Hence, no *a priori* assumptions regarding the signal waveform are made, and the waveform is regarded as an unknown time function that is to be estimated. One way of obtaining an undistorted estimate of an unknown time function with a signal-aligned array like that of Figure 7.3 is to employ a maximum likelihood estimator that assumes that the noise components of the received signal have a multidimensional Gaussian distribution. The likelihood function of the received signal at the SCF output can then be written as

$$\mathfrak{L} = \frac{1}{(2\pi)^{N/2(2k+1)}|\boldsymbol{\rho}|^{1/2}}\exp\left[-\frac{1}{2}\sum_{m,n=-k}^{k}(\mathbf{x}_m - \mathbf{s}_m)^T\boldsymbol{\rho}^{-1}(\mathbf{x}_n - \mathbf{s}_n)\right]$$

where $n, m$ denote distinct sample times and where $\boldsymbol{\rho}$ is the noise covariance matrix that is a matrix of $N \times N$ submatrices corresponding to the various tap points along the tapped-delay line. Differentiate the logarithm of the likelihood function with respect to $\mathbf{s}_n$ and equate the result to zero to obtain $\hat{\mathbf{s}}_m$, the maximum likelihood estimator for $\mathbf{s}_m$. Show that this result corresponds to the signal estimate obtained from the processor defined by the result in Problem 1.

**3** Show that the solution to the optimization problem posed by (7.81) and (7.82) is given by (7.83).

**4** Show that (7.92) leads to (7.93) by means of the following steps:

(a) Pre- and postmultiply (7.14) by $\mathbf{I}_1^T$ and $\mathbf{I}_1$, respectively, to obtain

$$\mathbf{I}_1^T\mathbf{P}(k+1)\mathbf{I}_1 = \frac{1}{\alpha}\left[\mathbf{I}_1^T\mathbf{P}(k)\mathbf{I}_1 - \frac{\mathbf{I}_1^T\mathbf{P}(k)\mathbf{x}(k+1)\mathbf{x}^T(k+1)\mathbf{P}(k)\mathbf{I}_1}{\alpha+\mathbf{x}^T(k+1)\mathbf{P}(k)\mathbf{x}(k+1)}\right]$$

(b) Apply the matrix identity (D.4) of Appendix D to the result obtained in part (a) to show that

$$\left[\mathbf{I}_1^T\mathbf{P}(k+1)\mathbf{I}_1\right]^{-1} = \alpha\left[\mathbf{I}_1^T\mathbf{P}(k)\mathbf{I}_1\right]^{-1} + \mathbf{I}_1^{-1}\mathbf{x}(k+1)\mathbf{x}^T(k+1)\mathbf{I}_1^{-T}$$

(c) Using the result of part (b) in (7.92), show that

$$\mathbf{w}_{\text{opt}}(k+1) = \mathbf{P}(k+1)\left[\alpha\mathbf{P}^{-1}(k) + \mathbf{x}(k+1)\mathbf{x}^T(k+1)\right]\mathbf{I}_1^{-T}\mathbf{c}$$

(d) Noting that

$$\mathbf{P}(k+1)\mathbf{P}^{-1}(k) = \frac{1}{\alpha}\left[\mathbf{I} - \frac{\mathbf{P}(k)\mathbf{x}(k+1)\mathbf{x}^T(k+1)}{\alpha+\mathbf{x}^T(k+1)\mathbf{P}(k)\mathbf{x}(k+1)}\right]$$

show that the result obtained in part (c) leads to (7.93).

**5** Show that the matrix identity (7.97) results from the substitution of (7.96) into the ratio:

$$\frac{\mathbf{R}_{xx}^{-1}\mathbf{1}}{\mathbf{1}^T\mathbf{R}_{xx}^{-1}\mathbf{1}}$$

## REFERENCES

[1] C. A. Baird, "Recursive Algorithms for Adaptive Arrays," Final Report, Contract No. F30602-72-C-0499, Rome Air Development Center, September 1973.

[2] P. E. Mantey and L. J. Griffiths, "Iterative Least-Squares Algorithms for Signal Extraction," Second Hawaii International Conference on System Sciences, January 1969, pp. 767–770.

[3] C. A. Baird, "Recursive Processing for Adaptive Arrays," Proceedings of the Adaptive Antenna Systems Workshop, March 11–13, 1974, Vol. I, NRL Report 7803, Naval Research Laboratory, Washington, DC, pp. 163–182.

[4] R. D. Gitlin and F. R. Magee, Jr., "Self-Orthogonalizing Adaptive Equalization Algorithms," *IEEE Trans. Commun.*, Vol. COM-25, No. 7, July 1977, pp. 666–672.

[5] C. A. Baird, "Kalman-Type Processing for Adaptive Antenna Arrays," IEEE International Conference on Communications, June 1974, Minneapolis, Minnesota, pp. 10G-1–10G-4.

[6] C. A. Baird, Jr. and J. T. Rickard, "Recursive Estimation in Array Processing," Fifth Asilomar Conference on Circuits and Systems, Pacific Grove, CA., November 1971, pp. 509–513.

[7] L. J. Griffiths, "A Simple Adaptive Algorithm for Real-Time Processing in Antenna Arrays," *Proc. IEEE*, Vol. 57, No. 10, October 1969, pp. 1695–1704.

[8] L. E. Brennan, J. D. Mallet, and I. S. Reed, "Adaptive Arrays in Airborne MTI Radar," *IEEE Trans. Antennas Propag.* Vol. AP-24, No. 5, September 1976, pp. 607–615.

[9] J. M. Shapard, D. Edelblute, and G. Kinnison, "Adaptive Matrix Inversion," Naval Undersea Research and Development Center, NUC-TN-528, May 1971.

[10] L. E. Brennan and I. S. Reed, "Theory of Adaptive Radar," *IEEE Trans. Aerosp. Electron. Syst.*, Vol. AES-9, No. 2, March 1973, pp. 237–252.

[11] R. E. Kalman, "A New Approach to Linear Filtering and Prediction Problems," *Trans. ASME, J. Basic Eng.*, Series D, Vol. 82, 1960, pp. 35–45.

[12] R. E. Kalman and R. S. Bucy, "New Results in Linear Filtering and Prediction Theory," *Trans. ASME, J. Basic Eng.* Series D, Vol. 83, March 1961, pp. 95–108.

[13] D. Godard, "Channel Equalization Using a Kalman Filter for Fast Data Transmission," *IBM J. Res. Dev.*, May 1974, pp. 267–273.

[14] C. A. Baird, Jr., "Recursive Minimum Variance Estimation for Adaptive Sensor Arrays," Proceedings of the IEEE 1972 International Conference on Cybernetics and Society, October 9–12, Washington, DC, pp. 412–414.

[15] J. Capon, R. J. Greenfield, and R. J. Kolker, "Multidimensional Maximum-Likelihood Processing of a Large Aperture Seismic Array," *Proc. IEEE*, Vol. 55, No. 2, February 1967, pp. 192–211.

# Chapter 8. Cascade Preprocessors

The problem of slow convergence using the LMS or maximum SNR algorithms arises whenever there is a wide spread in the eigenvalues of the input signal correlation matrix. The condition of wide eigenvalue spread occurs if the signal environment includes a very strong source of interference together with other weaker but nevertheless potent interference sources. This condition also obtains if two or more very strong interference sources arrive at the array from closely spaced but not identical directions.

It was shown in Chapter 4 that by appropriately selecting the step size and moving in suitably chosen directions an accelerated gradient procedure offers marked improvement in the convergence rate over that obtained with an algorithm that moves in directions determined by the gradient alone. Another approach for obtaining rapid convergence is to rescale the space in which the minimization is taking place by appropriately transforming the input signal coordinates so that the constant cost contours of the performance surface (which are represented by ellipses in Chapter 4) are approximately spherical and no eigenvalue spread is present in the rescaled space. If such a rescaling can be done, then in principle it would be possible to correct all the error components in a single step by choosing an appropriate step size.

With a method called scaled conjugate gradient descent (SCGD) [1], this philosophy is followed with a procedure that employs a CGD cycle of $N$ iterations and uses the information gained from this cycle to construct a scaling matrix that yields very rapid convergence on the next CGD cycle.

The philosophy behind the development of cascade preprocessors is similar to that of the SCGD method. The cascade preprocessor introduced by White [2], [3] attempts to overcome the problem of (sometimes) slow convergence by reducing the eigenvalue spread of the input signal correlation matrix through the introduction of an appropriate transformation (represented by the preprocessing network). Used in this manner, the function of the cascade network may be regarded as one of resolving the input signals into their eigenvector components. By then equalizing the resolved signals with automatic gain control (AGC) amplifiers, the eigenvalue spread may be considerably reduced, thereby simplifying the task of any gradient type algorithm.

It will be seen that by modifying the performance measure governing the control of the adaptive elements in a cascade preprocessor, a cascade network can perform the complete task of array pattern null steering without any need for a gradient type processor [4]. Using a cascade preprocessor in this manner reduces the complexity and cost of the overall processor and represents an attractive alternative to conventional gradient approaches.

Finally, let us introduce the use of a cascade preprocessor developed by Brennan et al. [5]–[7] to achieve adaptive null steering based on the Gram-Schmidt orthogonalization procedure. A Gram-Schmidt cascade preprocessor is both simpler than the other cascade networks discussed and possesses very fast convergence properties that make it a most appealing practical alternative. The discussion of eigenvector component preprocessing networks is presented here principally because it gives perspective to the development and use of cascade preprocessors.

## 8.1 NOLEN NETWORK PREPROCESSOR

Suppose that it were possible to construct a lossless transformation network to insert between the sensor elements and the adaptive weights that would resolve the input signals into their eigenvector components so the correlation matrix of the transformed signals was diagonal with the eigenvalues $\{\lambda_1, \lambda_2, \ldots, \lambda_N\}$ as the elements. The insertion of such a transformation network ahead of the adaptive weights to resolve the signals into orthogonal normalized eigenvector beams was suggested by Gabriel [8]. If a second transformation were introduced to equalize the various eigenvalues, then in principle it should be possible to select the step size so that complete convergence could be fully realized in a single step. If we assume the necessary *a priori* eigenvalue information were available to the designer, then the transformation matrix and equalizing network would appear in the adaptive array as shown in Figure 8.1 for a five-element array.

Since in general the eigenvalues of the input signal correlation matrix are unknown, an approximation of the eigenvector network must be constructed. If it were necessary for the eigenvector network to be exact, then this approach would suffer the disadvantage that when the correlation matrix is ill-conditioned (that is, the eigenvalues are widely diverse) then small errors made in estimating certain matrix components would become magnified and the attempt to resolve signals into eigenvector components would fail to achieve the desired effect. Fortunately, the method to be described appears to work quite well with only a rough approximation, and the conditions that are the most difficult for gradient type algorithms to handle are those with which the preprocessor approach succeeds most easily.

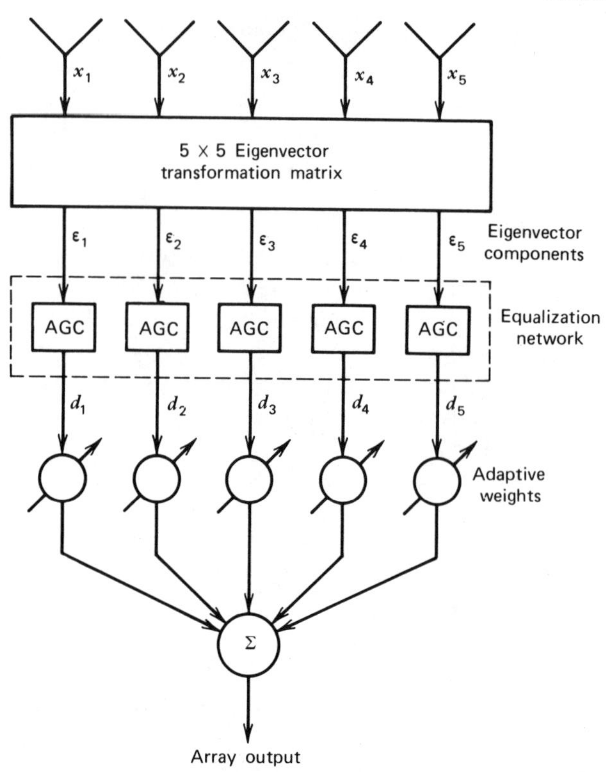


**Figure 8.1** Five-element adaptive array with eigenvector component transformation and eigenvalue equalization network.

### 8.1.1 *Single-Stage Nolen Network*

Consider the lossless, passive, reflectionless network shown in Figure 8.2. This network has $N$ input ports (one for each array sensor output) and $N$ output ports. The network elements consists of $N-1$ variable phase shifters and $N-1$ variable directional couplers that are connected in a series feed configuration. This network was first described by Nolen in an unpublished Bendix memorandum [9] concerned with the problem of synthesizing multiple beam antennas and is therefore termed a Nolen transformation network. Since the network is passive, lossless, and reflectionless, the total output power is the same as the total input power and the overall transformation matrix is unitary.

The signals in the transformation network are denoted by $v_n^k$, where $k$ indicates the level in the processor, and $n$ indicates the element channel to which the signal corresponds. The output signal in Figure 8.2 can be expressed as

$$v_1^2 = \sum_{n=1}^{N} a_n v_n^1 \tag{8.1}$$

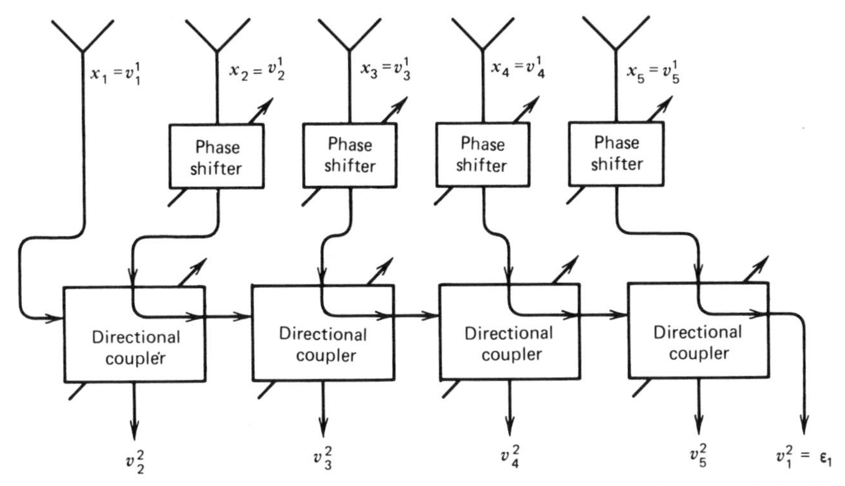


**Figure 8.2** Single-stage lossless, passive, reflectionless Nolen transformation network for $N=5$.

where the weighting factors are constrained by the unitary condition $\Sigma_n\ |a_n|^2=1$ and the condition that $a_1$ must be real.

Now suppose that it is desired to maximize the output power resulting from the single-level Nolen network. The output power can be expressed as

$$P_1 = E\{|v_1^2|^2\} = E\{v_1^2 v_1^{2*}\} = \sum_{n=1}^{N}\sum_{l=1}^{N} a_n E\{v_n^1 v_l^{1*}\} a_l^*$$

$$= \sum_{n=1}^{N}\sum_{l=1}^{N} a_n m_{nl}^1 a_l^* \tag{8.2}$$

where

$$m_{nl}^k = E\{v_n^k v_l^{k*}\} \tag{8.3}$$

is an element of the correlation matrix of the input signals. To introduce the unitary constraint $\Sigma_n|a_n|^2=1$ while maximizing $P_1$, employ the method of Lagrange multipliers by maximizing the quantity

$$Q = P_1 + \lambda\left[1 - \sum_{n=1}^{N}|a_n|^2\right] \tag{8.4}$$

where $\lambda$ denotes a Lagrange multiplier. Setting $a_n = u_n + jv_n$, the partial derivatives of $Q$ with respect to $u_L$ and $v_L$ can be found to be

$$\frac{\partial Q}{\partial u_L} = \sum_n m_{Ln}^1(u_n - jv_n) + \sum_l (u_l + iv_l) m_{lL}^1 - 2\lambda u_L$$

$$= \sum_n a_n m_{nL}^1 + \sum_l m_{Ll}^1 a_l^* - \lambda(a_L + a_L^*) \tag{8.5}$$

and

$$\frac{\partial Q}{\partial v_L} = j\sum_n m^1_{Ln}(u_n - jv_n) - j\sum_l (u_l + jv_l)m^1_{lL} - 2\lambda v_L$$

$$= -j\sum_n a_n m^1_{nL} + j\sum_l m^1_{Ll}a^*_l + j\lambda(a_L - a^*_L) \tag{8.6}$$

Setting both $\partial Q/\partial u_L$ and $\partial Q/\partial v_L = 0$, the solution for $a_L$ must satisfy the following conditions:

$$\left.\begin{aligned} \lambda(a_L + a^*_L) &= \sum_n a_n m^1_{nL} + \sum_l m^1_{Ll}a^*_l \\ \lambda(a_L - a^*_L) &= \sum_n a_n m^1_{nL} - \sum_l m^1_{Ll}a^*_l \end{aligned}\right\} \tag{8.7}$$

The conditions represented by (8.7) immediately simplify to

$$\lambda a_L = \sum_{n=1}^{N} a_n m^1_{nL} \tag{8.8}$$

which is classical eigenvector equation. Nontrivial solutions of (8.8) exist only when $\lambda$ has values corresponding to the eigenvalues of the matrix $\mathbf{M}^1$ (of which $m^1_{nL}$ is the $nL$th element).

Substituting (8.8) into (8.2) yields

$$P_1 = \lambda \sum_{n=1}^{N} a_n a^*_n \tag{8.9}$$

In view of the unitary constraint $\Sigma|a_n|^2 = 1$, (8.9) becomes

$$P_1 = \lambda \tag{8.10}$$

Consequently the largest value of $P_1$ (which is the maximum power available at the right-hand output port) is just the largest eigenvalue of the matrix $\mathbf{M}^1$.

When the phase shifters and directional couplers have been adjusted so that $P_1$ is maximized, then the signal $v^2_1$ in Figure 8.2 is orthogonal to all the other signals $v^2_k$. That is,

$$E\{v^2_1 v^{2*}_k\} = 0 \qquad \text{for } k = 2, 3, \ldots, N \tag{8.11}$$

This orthogonality condition reflects the fact that no other unitary combination of $v^2_1$ and any of the other $v^2_k$ signals can possibly deliver more power than is obtained with $v^2_1$ alone—such a result would contradict the fact that $P_1$ has been maximized.

The result of (8.11) shows that all the off-diagonal elements in the first row and the first column of the covariance matrix of the signals $v_k^2$ have been set equal to zero. Consequently, the transformation introduced by the single-stage network of Figure 8.2 has completed one step in the process of diagonalizing the covariance matrix of the input signals. To complete the diagonalization process, additional networks can be introduced as described in the next section.

### 8.1.2 *Cascaded Networks*

Now let the single-stage transformation network of Figure 8.2 be followed by a cascade of similar networks as shown in Figure 8.3. The second transformation network operates only on the signals $v_2^2$ through $v_5^2$, leaving $v_1^2=\varepsilon_1$ undisturbed. Because of the reflectionless character of the transformation networks, the parameters of the second network can be adjusted without affecting the prior maximization of $P_1$. Let the parameters of the second transformation network be adjusted to maximize $P_2$—the output power from the right-hand port of Level 2. The maximum power now available is equal to the largest eigenvalue of the covariance matrix of the input set $\{v_2^2, v_3^2, \ldots, v_N^2\}$. If $P_1$ was truly maximized in the first level, then the largest eigenvalue of the submatrix will be equal to the second largest eigenvalue of the complete covariance matrix.

The signals emerging from network 2 form the signal set $\{v_2^2, v_3^2, \ldots, v_N^2\}$, and adjusting the parameters of network 2 to maximize $P_2$ results in setting all the off-diagonal elements in the first row and first column of the submatrix for the signal set $\{v_2^2, v_3^2, \ldots, v_N^2\}$ to zero. As far as the complete covariance matrix is concerned, the first two transformation networks have diagonalized the first two rows and the first two columns. Each succeeding transformation network likewise diagonalizes one row and one column at a time. Once the output of the last transformation in the cascaded network system is reached, the entire matrix diagonalization is complete. Since the adjustment of parameters in each network leaves previous networks in the cascade undisturbed, there is no need to go back to the top and repeat the process unless the input signal statistics change.

In Figure 8.1 it is seen that the eigenvector transformation matrix is followed by an equalization network that consists of AGC amplifiers through which the signals are now passed. As a result the signal powers on the various output leads $\{d_1, d_2, \ldots, d_N\}$ are equalized. Since the covariance matrix corresponding to the eigenvector component signals $\{\varepsilon_1, \varepsilon_2, \ldots, \varepsilon_N\}$ is diagonalized, the eigenvalues are consequently equalized. Therefore the covariance matrix of the output signals from the equalization network is a scalar constant times the identity matrix.

The transformation resulting from the entire cascade of Figure 8.3 is unitary since each of the networks in the cascade is unitary. Consequently not

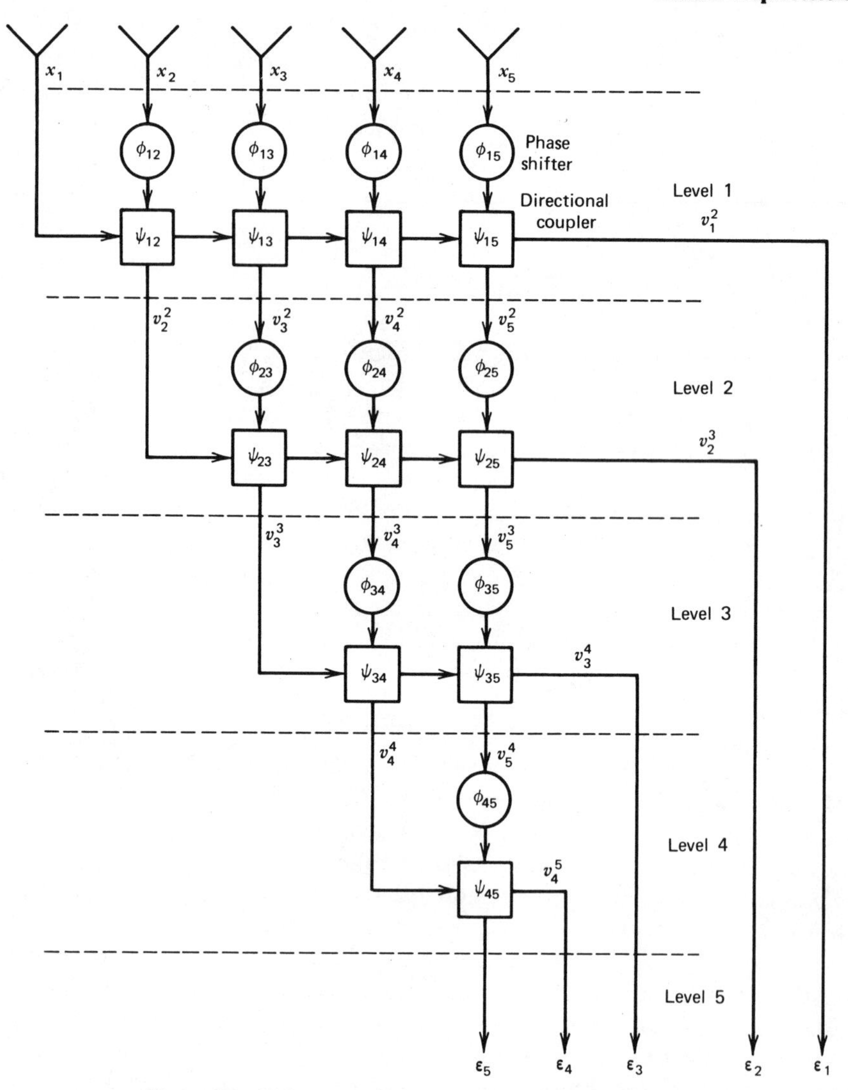


**Figure 8.3** Nolen cascade network for five-element array.

only is the total output power strictly equal to the total input power, but also the eigenvalues of the covariance matrix are unchanged. If the element $\mathbf{G}_J$ represents the overall transfer matrix of the first $J$ stages, then $\mathbf{G}_J$ can be represented as a product of factors in which each factor represents a single stage, that is,

$$\mathbf{G}_J = \mathbf{F}_J \cdot \mathbf{F}_{J-1} \cdots \mathbf{F}_2 \cdot \mathbf{F}_1 \tag{8.12}$$

where

$$\begin{bmatrix} \varepsilon_1 \\ \varepsilon_2 \\ \vdots \\ \varepsilon_J \\ v_{J+1}^{J+1} \\ \vdots \\ v_N^{J+1} \end{bmatrix} = \mathbf{G}_J \begin{bmatrix} x_1 \\ x_2 \\ \vdots \\ x_N \end{bmatrix} \tag{8.13}$$

Furthermore, the second and lower stages of the network have no effect on the signal $\varepsilon_1$. It follows that the first row of $\mathbf{G}_J$ must be the same as the first row of

$$\mathbf{G}_1 = \mathbf{F}_1 \tag{8.14}$$

Likewise, since the third and lower stages of the network have no effect on $F_2$, the second row of $\mathbf{G}_J$ must be identical with the second row of

$$\mathbf{G}_2 = \mathbf{F}_2 \cdot \mathbf{F}_1 \tag{8.15}$$

Continuing in the foregoing manner, the entire $\mathbf{G}_N$ matrix can be implemented in the Nolen form using a step-by-step process. Since the first row of $\mathbf{G}_J$ must be identical with the first row of $\mathbf{G}_1$, the second row of $\mathbf{G}_J$ must be identical with the second row of $\mathbf{G}_2, \ldots,$ and the $(J-1)$st row of $\mathbf{G}_J$ must be identical with the $(J-1)$st row of $\mathbf{G}_{J-1}$, it follows that the first $J-1$ rows and $J-1$ columns of $\mathbf{F}_J$ must be the same as those of an identity matrix. Furthermore, since there is no phase shifter between the leftmost input port at any stage and the right hand output port, the $J$th diagonal element of $\mathbf{F}_J$ must be real. These constraints in addition to the unitary constraint on the transfer matrix then define the bounds within which the element values (phase shift and directional coupling) of the $J$th row can be chosen.

If the output power maximization at each stage of the cascade transformation network is only approximate, the off-diagonal elements of the covariance matrix will not be completely nulled. However, with only a rough approximation the off-diagonal elements are at least reduced in amplitude, and, while the equalization network will no longer exactly equalize the eigenvalues, it will reduce the eigenvalue spread.

### *8.1.3 Control of the Eigenvector Transformation Networks*

Now consider the problem of adjusting the parameters in each stage of an eigenvector transformation network to realize the desired covariance matrix diagonalization. It turns out that those signal environment conditions that prove the most difficult for the LMS algorithm to handle are the same conditions that the preprocessor adapts to most easily [3].

For the single-stage network of Figure 8.2, the power at the right-hand output port must be maximized by appropriately adjusting each phase shifter and directional coupler in the network. Denote the signal flowing downward into the phase shifter directional coupler combination located in the $l$th row and the $k$th column of the cascade of Figure 8.3 by $v_k^l$. Likewise denote the signal flowing into this same coupler from the left by $y_k^l$. The signal coming out of the right-hand port (which is equivalent to $y_{N+1}^l$) is then $\varepsilon_l$.

Regard the directional coupler as equivalent to a goniometer having shaft angle $\psi_{lk}$; the signals into and out of one phase shifter directional coupler combination can be written as

$$y_{k+1}^l = y_k^l \cos\psi_{l,k} + y_k^l e^{j\phi_{l,k}} \sin\psi_{l,k} \tag{8.16}$$

$$v_k^{l+1} = y_k^l \sin\psi_{l,k} + v_k^l e^{j\phi_{l,k}} \cos\psi_{l,k} \tag{8.17}$$

Now $P_l = E\{|\varepsilon_l|^2\}$ is the power out of the $l$th stage where

$$\varepsilon_l = \sum_{k=l}^{N} a_k^l v_k^l \tag{8.18}$$

and

$$P_l = \sum_{i=l}^{N} \sum_{k=l}^{N} a_i^l m_{ik}^l a_k^{l*} \tag{8.19}$$

where $m_{ik}^l$ is given by (8.3) and

$$a_i^l = \begin{cases} \prod_{k=l+1}^{N} \cos\psi_{lk} & \text{for } i=1 \\ e^{j\phi_{li}} \sin\psi_{li} \prod_{k=i+1}^{N} \cos\psi_{lk} & \text{for } i>1 \end{cases} \tag{8.20}$$

Substituting the expression $\sum_{k=l}^{N-1} a_k^l v_k^l = \varepsilon_l - a_N v_N^l$ and (8.20) into (8.19) and taking partial derivatives with respect to $\phi_{lk}$ and $\psi_{lk}$ results in the following:

$$\frac{\partial P_l}{\partial \phi_{lk}} = c_1 \operatorname{Im}\left[ E\{\varepsilon_l v_k^{l*}\} e^{-j\phi_{lk}} \right] \tag{8.21}$$

$$\frac{\partial P_l}{\partial \psi_{lk}} = c_2 \operatorname{Re}\left[ E\{\varepsilon_l v_k^{l*}\} e^{j\phi_{lk}} \cos\psi_{lk} - E\{\varepsilon_l y_k^{l*}\} \sin\psi_{lk} \right] \tag{8.22}$$

where $c_1$ does not depend on $\phi_{lk}$ and $c_2$ does not depend on $\psi_{lk}$. Setting both of the above partial derivatives equal to zero then yields

$$\phi_{lk} = \arg\left[E\{\varepsilon_l v_k^{l*}\}\right] \tag{8.23}$$

and

$$\psi_{lk} = \tan^{-1}\left[\frac{\operatorname{Re}(E\{\varepsilon_l v_k^{l*}\}e^{-j\phi_{lk}})}{\operatorname{Re}(E\{\varepsilon_l y_k^{l*}\})}\right] \tag{8.24}$$

where the arctangent function is a multiple valued function. It turns out that by taking the value lying in the range $0 \rightarrow \pi$ radians, the output power is maximized. From (8.23) and (8.24) it is seen that measuring the correlations $E\{\varepsilon_l v_l^{k*}\}$ and $E\{\varepsilon_l y_k^{l*}\}$ yields an indication of the correct settings for $\phi_{lk}$ and $\psi_{lk}$. Starting with the first phase shifter coupler pair at the extreme left of any stage and proceeding with each successive pair until the last phase shift coupler pair is reached, $\phi_{lk}$ and $\psi_{lk}$ can be adjusted in accordance with (8.23) and (8.24). However, each time a new setting of these parameters is made while proceeding toward the output port, the value of $\varepsilon_l$ changes, and the values of the correlations change so new settings are required and the entire process must be applied recursively. There is presently no proof of convergence for this recursive adjustment process, but experience indicates that convergence in five or six complete row adjustments is commonplace [3]. When the eigenvalues of the input signal covariance matrix are widely separated, then convergence is even faster.

The eigenvector transformation network parameter adjustment time can be reduced dramatically by taking advantage of the fact that an exact eigenvector decomposition is not required in order to realize substantial eigenvalue spread reduction. This may be done by resorting to a piecemeal maximization technique in which each phase shift coupler combination is set to produce maximum power at its horizontal output port without regard to power at the entire stage output port, and considering only the signals present at the two inputs of each individual combination. With this piecemeal procedure, it is easy to show that the phase shift and directional coupler are now set according to

$$\phi_{lk} = \arg(E\{y_k^l v_k^{l*}\}) \tag{8.25}$$

and

$$\psi_{lk} = \frac{1}{2}\tan^{-1}\left[\frac{2|E\{y_k^l v_k^{l*}\}|}{|E\{[y_k^l]^2\}| - |E\{[v_k^l]^2\}|}\right] \tag{8.26}$$

where $0 \leqslant \tan^{-1}(\cdot) \leqslant \pi$.

The piecemeal adjustment process is a sequential one in which each phase shift coupler combination is set as we proceed from left to right along each row. When the end of a row is reached, drop down to the next stage and repeat the process. The adjustment process is then complete when one reaches the bottom of the cascade. The eigenvector beams produced by piecemeal adjustment procedure are shown by White [3] to be surprisingly close to those obtained by the complete recursive adjustment procedure. If the adjustment of each phase shifter directional coupler combination is regarded as one iteration, then a total of $N(N-1)/2$ iterations are required to complete the piecemeal adjustment procedure.

## 8.2 INTERFERENCE CANCELLATION WITH A NOLEN NETWORK PREPROCESSOR

Instead of using a Nolen eigenvector component preprocessor in cascade with a gradient type processor to alleviate the eigenvalue spread problem and speed convergence, the entire task of interference cancellation and desired signal preservation can be accomplished using a Nolen cascade network alone by selecting a different performance measure from that adopted in the preceding section [4]. Each stage in the Nolen network of Figure 8.3 has the capability of introducing one null in the radiation pattern corresponding to the bottom output port. Errors in parameter settings in the lower stages do not disturb the nulls set by the upper stages, but the inverse statement is not true, so a nonadaptive Nolen network is vulnerable to errors in the early stages. When adaptive control is applied to the parameter adjustment, however, a portion of the lower stage adjustment capability can be used to partially compensate for upper stage parameter setting errors. By virtue of the properties of the Nolen network, an appropriate set of parameter adjustments for interference suppression can be determined by means of a step-by-step process in which each step involves only a single stage.

### *8.2.1 Problem Formulation*

Consider the representation of a Nolen beamforming network shown in Figure 8.4 in which there is an array having $N$ elements followed by a general (passive, lossless, and matched) network having $N$ input ports and $N$ output ports. Under these conditions, the total output power will be exactly equal to the total input power, and the transfer function of the network may be regarded as a unitary matrix.

Let $\mathbf{x}$ be a complex vector representing the input signal envelopes at each sensor element, and let $\boldsymbol{\varepsilon}$ be a complex vector representing the output signal envelopes, that is,

$$\mathbf{x}=\begin{bmatrix} x_1 \\ x_2 \\ \vdots \\ x_N \end{bmatrix}, \qquad \boldsymbol{\varepsilon}=\begin{bmatrix} \varepsilon_1 \\ \varepsilon_2 \\ \vdots \\ \varepsilon_N \end{bmatrix} \tag{8.27}$$

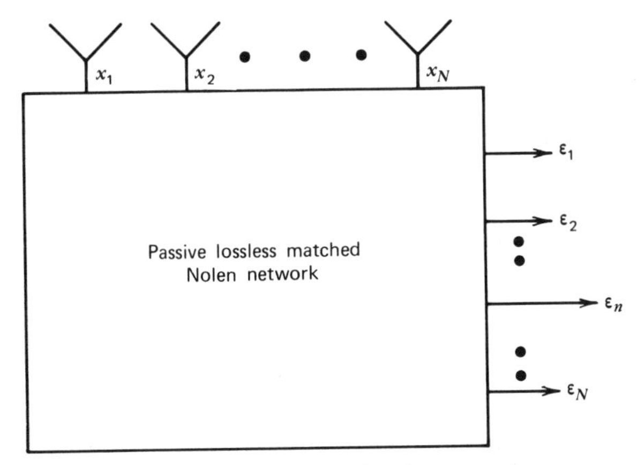


**Figure 8.4** Nolen beamforming network.

The vectors $\mathbf{x}$ and $\boldsymbol{\varepsilon}$ are related by

$$\boldsymbol{\varepsilon}=\mathbf{G}\mathbf{x} \tag{8.28}$$

where $\mathbf{G}$ is the transfer matrix of the Nolen network. Since $\mathbf{G}$ is unitary it follows that

$$\boldsymbol{\varepsilon}^{\dagger}\boldsymbol{\varepsilon}=\mathbf{x}^{\dagger}\mathbf{x} \tag{8.29}$$

and

$$\mathbf{G}^{\dagger}\mathbf{G}=\mathbf{I} \tag{8.30}$$

Equation (8.30) is referred to as the "unitary constraint" that $\mathbf{G}$ must satisfy. Define the covariance matrices of the input and the output signals by

$$\mathbf{R}_{xx}=E\{\mathbf{x}^*\mathbf{x}^T\} \tag{8.31}$$

and

$$\mathbf{R}_{\varepsilon\varepsilon}=E\{\boldsymbol{\varepsilon}^*\boldsymbol{\varepsilon}^T\}=\mathbf{G}^*\mathbf{R}_{xx}\mathbf{G}^T \tag{8.32}$$

The total output power appearing at the $n$th output port is just the $n$th diagonal element of $\mathbf{R}_{\varepsilon\varepsilon}$. Each output signal in general consists of a desired signal component and an interference signal component ("interference" here denotes the sum of external jamming and internal thermal noise). Assume that the desired signal is not correlated with any of the interference signal sources, then $\mathbf{R}_{\varepsilon\varepsilon}$ can be divided into signal and interference components:

$$\mathbf{R}_{\varepsilon\varepsilon}=\mathbf{G}^*[\mathbf{R}_{ss}+\mathbf{R}_{II}]\mathbf{G}^T \tag{8.33}$$

where

$$\mathbf{R}_{ss}=E\{\mathbf{x}_s^*\mathbf{x}_s^T\} \tag{8.34}$$

and

$$\mathbf{R}_{II} = E\{\mathbf{x}_I^* \mathbf{x}_I^T\} \tag{8.35}$$

The vectors $\mathbf{x}_s$ and $\mathbf{x}_I$ represent the signal and interference components of the input signal envelopes. At the $n$th output port denote the desired signal power by $\mathfrak{S}_n$ and the interference power by $\mathfrak{I}_n$ so that

$$\mathfrak{S}_n = [\mathbf{R}_{ss}]_{nn} \tag{8.36}$$

and

$$\mathfrak{I}_n = [\mathbf{R}_{II}]_{nn} \tag{8.37}$$

That is, the output power of interest is the $n$th diagonal element of the corresponding covariance matrix.

It is desired to maximize the signal power $\mathfrak{S}_n$ and minimize the interference power $\mathfrak{I}_n$ at the output port $n$. Since there is a conflict between these two objectives, a trade-off between them must be made. One approach to this trade-off problem is to select that unitary transfer matrix $\mathbf{G}$ which maximizes the ratio $\mathfrak{R}_n = \mathfrak{S}_n / \mathfrak{I}_n$. A more convenient way of attacking the problem is to adopt as the performance measure

$$\Omega_n = \mathfrak{S}_n - \Gamma \mathfrak{I}_n \tag{8.38}$$

where $\Gamma$ is a fixed scalar constant that reflects the relative importance on minimizing interference compared to maximizing the signal. If $\Gamma = 0$, then only maximizing the desired signal is of concern, while if $\Gamma \to \infty$, then only minimizing the interference is of concern.

There is a value $\Gamma = \mathfrak{R}_{\text{opt}}$ for which maximizing $\Omega_n$ produces exactly the same result as maximizing the ratio $\mathfrak{R}_n$. The value $\mathfrak{R}_{\text{opt}}$ is a function of the environment and is not ordinarily known in advance. By setting $\Gamma = \mathfrak{R}_{\text{min}}$ where $\mathfrak{R}_{\text{min}}$ is the minimum acceptable signal-to-interference ratio that provides acceptable performance, then maximizing $\Omega_n$ ensures that $\mathfrak{R}_n$ is maximized under the conditions when it is most needed. If the signal environment improves, then $\mathfrak{R}_n$ also improves, although not quite as much as if $\mathfrak{R}_n$ were maximized directly.

It is useful to define the matrix

$$\mathbf{Z} \triangleq \mathbf{G}^* \mathbf{R}_{ss} \mathbf{G}^T - \Gamma \mathbf{G}^* \mathbf{R}_{II} \mathbf{G}^T \tag{8.39}$$

The performance measure $\Omega_n$ is then the $n$th diagonal element of $\mathbf{Z}$. Furthermore,

$$\mathbf{Z} = \mathbf{G}^* \mathbf{M} \mathbf{G}^T \tag{8.40}$$

where

$$\mathbf{M} = \mathbf{R}_{ss} - \mathbf{R}_{II} \tag{8.41}$$

Note that as a consequence of (8.41), the matrix $\mathbf{M}$ (and hence $\mathbf{Z}$) will have both positive and negative eigenvalues. To maximize the element $\Omega_n$ one must select the transfer matrix $\mathbf{G}$ that maximizes the $n$th diagonal element of $\mathbf{Z}$ subject to (8.30). Using the method of Lagrangian multipliers to maximize the element $Z_{nn}$ with the unitary constraint on $\mathbf{G}$, and setting the resulting gradient equal to zero yields the relation

$$\sum_L M_{nL} G_{Lm} = \lambda G_{mn} \tag{8.42}$$

Equation (8.42) is precisely the form for an eigenvector equation. Consequently the $n$th column of $\mathbf{G}^T$ should be the eigenvector of the matrix $\mathbf{M}$ corresponding to the largest eigenvalue. When the $n$th column of $\mathbf{G}^T$ is so constructed, then the element $Z_{nn}$ will equal this largest eigenvalue, and all other elements of the $n$th row and the $n$th column of $\mathbf{Z}$ will vanish.

It is desired to obtain a unitary transfer matrix $\mathbf{G}$ for which the elements of the $n$th row are the elements of the eigenvector of the matrix $\mathbf{M}$ corresponding to the maximum eigenvalue (resulting in maximizing $\Omega_n$). If the first stage of the cascade Nolen network is adjusted to minimize $\Omega_1$, there is no conflict with maximizing $\Omega_n$, and the first diagonal element of $\mathbf{Z}$ will be set equal to the most negative eigenvalue of $\mathbf{M}$, the off-diagonal elements of the first row and first column of $\mathbf{Z}$ will disappear, and the first row of $\mathbf{G}$ will correspond to the appropriate eigenvector. Likewise proceed to adjust the second stage to minimize $\Omega_2$. This second adjustment results in the diagonalization of the second row and the second column of $\mathbf{Z}$, and the second row of $\mathbf{G}$ corresponds to the second eigenvector. Continue in this manner adjusting in turn to minimize the corresponding $\Omega$ until reaching stage $n$. At this point (the $n$th stage) it is desired to maximize $\Omega_n$ so the adjustment criterion must be reversed.

The physical significance of the foregoing adjustment procedure is that the upper stages of the Nolen network are adjusted to maximize the interference and minimize the desired signal observed at the output of each stage. This process is the same as maximizing the desired signal and minimizing the interference that proceeds downward to the lower stages. When the $n$th stage is reached where the useful output is desired, then the desired signal should be maximized and the interference minimized so the adjustment criterion is reversed.

### 8.2.2 *Recognition of the Desired Signal*

In using the performance criterion it is absolutely essential to be able to distinguish between the desired signal and interference. If the direction of arrival of the desired signal is not known *a priori* (as would usually be the case), it is necessary to be able to recognize some distinguishing characteristics of the signals themselves. Since communication systems designed to operate in a jamming environment commonly employ some form of spread spectrum modulation, it is reasonable to base the signal recognition scheme on the characteristics associated with spread spectrum signals.

With spread spectrum modulation, some form of pseudorandom coding is applied at the transmitter so as to spread the transmitted signal spectrum. When the receiver applies appropriate decoding to the spread spectrum signal, the original (unspread) desired signal is recovered with a narrow spectrum whereas any interference still emerges from the decoder having a wide spectrum. The demodulation process for spread spectrum signals thereby provides a basis for at least partially separating the desired signal from surrounding interference by the use of narrowband filters.

In determining whether the parameter adjustments in each stage of the Nolen network are succeeding in the diagonalization of the matrix $\mathbf{Z}$, it is useful to define a generalized correlation product [4]:

$$A^* \otimes B = E\{A_s^* B_s\} - \Gamma E\{A_I^* B_I\} \tag{8.43}$$

where $A$ is the complex envelope of one waveform having desired signal and interference components $A_s$ and $A_I$. Likewise $B$ represents the complex envelope of a second waveform having desired signal and interference components $B_s$ and $B_I$. If we assume that the desired signal is modulated with pseudorandom phase reversals and that a synchronized key generator is available for demodulation at the receiver, Figure 8.5 shows the block diagram of a generalized correlator that forms an estimate of $A^* \otimes B$. After passing the received signal through a synchronized demodulator to obtain the original unspread signal, narrow passband filters extract the desired signal while band reject filters extract the interference. The narrow passband outputs are applied to one correlator (the "signal correlator") that then forms estimates of $E\{A_s^* B_s\}$. The band reject outputs likewise are applied to a second correlator (the "interference correlator") that then forms estimates of $E\{A_I^* B_I\}$. A weighted combination of the outputs (signal component weighted by $K_s$ and interference component weighted by $K_I$) then forms an estimate of the complex quantity $A^* \otimes B$. The ratio of $K_s$ to $K_I$ determines the effective value of $\Gamma$.

If the correlation product operator $\otimes$ is taken to include operations on vector quantities, then we may write

$$\mathbf{M} = \mathbf{x}^* \otimes \mathbf{x}^T \tag{8.44}$$

and

$$\mathbf{Z} = \mathbf{h}^* \otimes \mathbf{h}^T \tag{8.45}$$

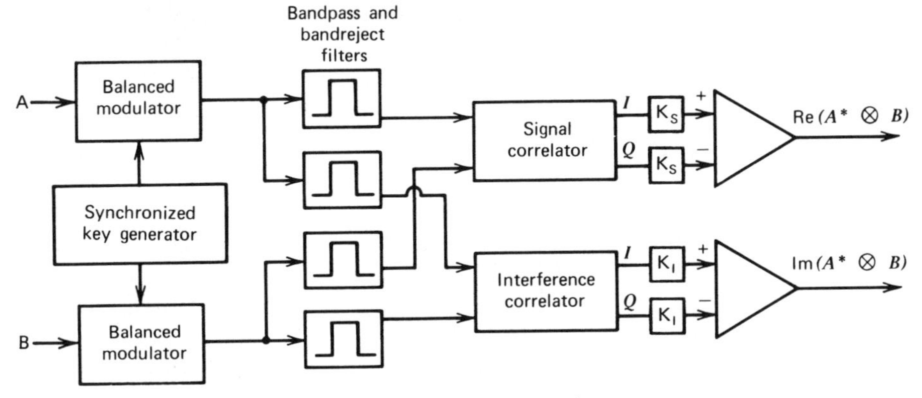


**Figure 8.5** Generalized correlator.

The generalized correlator network of Figure 8.5 thereby provides a basis for estimating the elements of the matrix **Z** and determining whether the diagonalization of this matrix is complete.

### 8.2.3 *Piecemeal Adjustment of the Nolen Network Stages*

The piecemeal adjustment procedure for Nolen network parameters described by (8.25) and (8.26) can no longer employ the correlation products of $\overline{y_k^l v_k^{l*}}$, $\overline{(y_k^l)^2}$, and $\overline{(v_k^l)^2}$ if the modified performance measure $\Omega_l$ is to be maximized (or minimized) for the $l$th stage, since it is now the generalized correlation product that is related to the elements of the matrix **Z** [4]. Consequently, (8.25) and (8.26) should now be modified to yield

$$\phi_{lk} = \arg\left\{ y_k^l \otimes v_k^{l*} \right\} \tag{8.46}$$

and

$$\psi_{lk} = \frac{1}{2}\tan^{-1}\left\{ \frac{2|y_k^l \otimes v_k^{l*}|}{|y_k^l \otimes y_k^{l*}| - |v_k^l \otimes v_k^{l*}|} \right\} \tag{8.47}$$

To minimize $\Omega_l$, the value of the arctangent function lying between $-\pi$ and zero should be taken, while to maximize $\Omega_l$ the value lying between zero and $\pi$ radians should be used. The complete piecemeal adjustment of a full Nolen cascade network for an $N$-element array using (8.46) and (8.47) requires $N(N-1)/2$ iterations, and this fact makes the practical use of a Nolen eigenvector component cascade processor less attractive when compared with the Gram-Schmidt cascade preprocessor that is described in the next section.

## 8.3 GRAM-SCHMIDT ORTHOGONALIZATION PREPROCESSOR

It is shown in Section 8.1 that use of a unitary transformation to obtain an orthogonal signal set enables a gradient-based algorithm to achieve accelerated convergence by circumventing the problem presented by eigenvalue spread. An orthogonal signal set can also be obtained by a transformation based on the Gram-Schmidt orthogonalization procedure [10]. While a normalized Gram-Schmidt preprocessor may be followed by a Howells-Applebaum adaptive processor to realize accelerated convergence as illustrated in Figure 8.6, it is also possible to simply utilize the preprocessor alone in a CSLC configuration to achieve interference cancellation. It will be seen that the Gram-Schmidt orthogonalization cascade preprocessor is very easily implemented, has excellent transient response characteristics, and therefore presents an attractive alternative to the eigenvector component preprocessor of Section 8.1.

To understand the coordinate transformation based on the Gram-Schmidt orthogonalization procedure, consider the five-element array of Figure 8.7 in which a transformation (indicated by a square) is introduced at each node in the cascade network to achieve independence (orthogonality) between the transformed output signal and a reference input signal [5].

Each transformation in the network has two input signals $v_k^k = y_k$ (which is the reference signal for level $k$) and $v_n^k$ where $n \geqslant k+1$. Every transformation achieves independence between $v_k^k$ and $v_n^k$. At the last level the output signals $y_1, y_2, \ldots, y_k$ are then mutually independent so that $E\{y_m y_n^*\} = 0$ for all $m \neq n$.

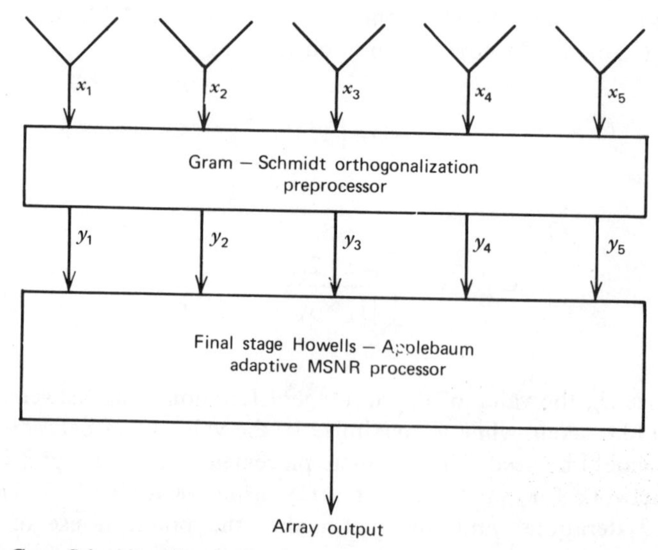


**Figure 8.6** Gram-Schmidt orthogonalization network with Howells-Applebaum adaptive processor for accelerated convergence.

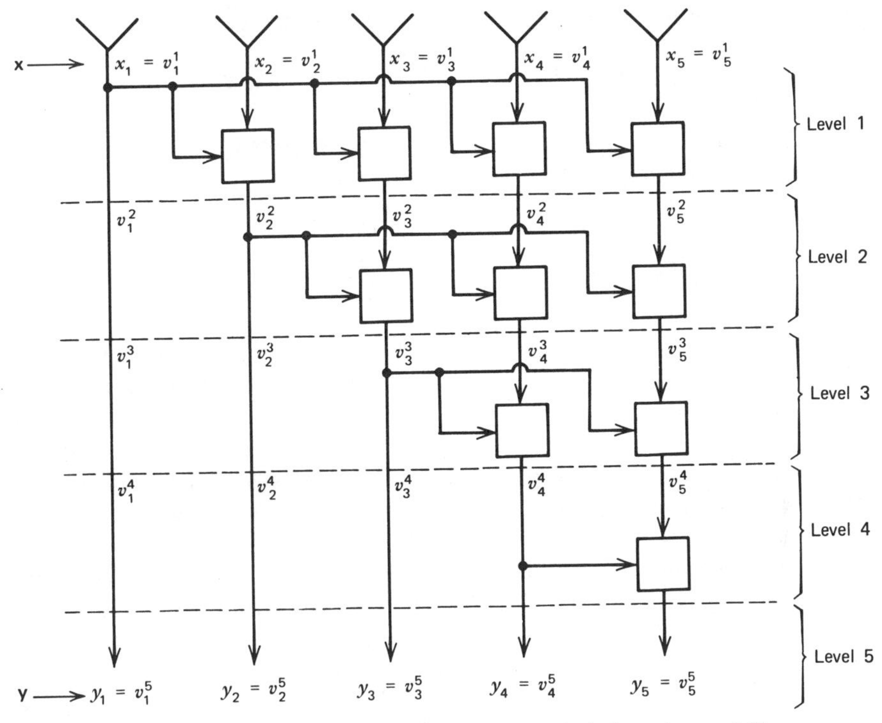


**Figure 8.7** Gram-Schmidt transformation to obtain independent variables.

Having achieved mutual independence among the output signals, a single Howells-Applebaum loop can then be applied to each signal where the loop gain is high for small signals and low for large signals, thereby equalizing the time response for the various signals in the interference environment. The first level of the network provides the necessary transformation of $v_n^1$, $n \geq 2$, to achieve independence from $v_1^1 = y_1$ for the signal set $v_n^2$, $n \geq 2$. The signal set $v_n^2$, $n \geq 3$ is then transformed in the second level of the network to achieve independence from the signal $v_2^2 = y_2$. This process continues until a complete transformation **G** is obtained, where

$$\mathbf{y} = \mathbf{G}\mathbf{x} \tag{8.48}$$

To obtain the Gram-Schmidt transformation **G** it is only necessary to ensure that the appropriate transformations take place at each node indicated in Figure 8.7. Recalling from Chapter 5 that a single Howells-Applebaum adaptive loop will yield an output signal that in the steady state is uncorrelated with a selected reference signal, we see that each transformation in Figure 8.7 may be realized with a single Howells-Applebaum adaptive control loop as shown in Figure 8.8. The transformation occurring at each node in

Figure 8.7 (using the weight indexing of Figure 8.8) can be expressed as

$$v_n^{k+1} = v_n^k - u_{k(n-1)} v_k^k, \qquad k+1 \leqslant n \leqslant N \tag{8.49}$$

where $N$ = number of elements in the array. In the steady state, the adaptive weights have values given by

$$u_{k(n-1)} = \frac{\overline{\left(v_k^{k*} v_n^{k+1}\right)}}{\overline{\left(v_k^{k*} v_k^k\right)}} \tag{8.50}$$

where the overbars denote expected values. The transformation represented by (8.48) and (8.49) is a close analog of the familiar Gram-Schmidt orthogonalization equations. For an $N$-element array, $N(N-1)/2$ adaptive weights are required to accomplish the Gram-Schmidt orthogonalization. Since $N$ adaptive weights are required for an adaptive Howells-Applebaum maximum SNR processor, a total of $N(N+1)/2$ adaptive weights are required for the configuration of Figure 8.6.

The transformation of the input signal vector $\mathbf{x}$ into a set of independent output signals $\mathbf{y}$ is not unique. Unlike the eigenvector transformation, which yields output signals in a normal coordinate system in which the maximum power signal is found to be the first output component, the Gram-Schmidt transformation can adopt any component of $\mathbf{x}$ as the first output component and any of the remaining components of $\mathbf{x}$ as the signal component $v_n^k$ to be transformed by way of (8.49).

The fact that the transformation network of Figure 8.7 yields a set of uncorrelated output signals suggests that this network may be capable of functioning in the manner of a CSLC system whose output signal (in the steady state) is uncorrelated with each of the auxiliary channel input signals. It will be recalled from the discussion of the SNR performance measure in Chapter 3 that an $N-1$ element CSLC is equivalent to an $N$-element adaptive array with a generalized signal vector given by

$$\mathbf{t} = \begin{bmatrix} 1 \\ 0 \\ \vdots \\ 0 \end{bmatrix} \tag{8.51}$$

It is shown in what follows that by selecting $x_5$ of Figure 8.7 as the main beam channel signal $b$ (so that $\mathbf{t}^T = [0, 0, \ldots, 0, 1]$) and $z = y_5$ as the output signal, then the cascade preprocessor yields an output that converges to $z = b - \mathbf{w}^T\mathbf{x}$. Here $\mathbf{x}$ is the auxiliary channel signal vector and $\mathbf{w}$ is the column vector of auxiliary channel weights for the equivalent CSLC system of Figure 8.9 that minimizes the output noise power. The CSLC system of Figure 8.9 and the cascade preprocessor of Figure 8.7 are then equivalent in the sense

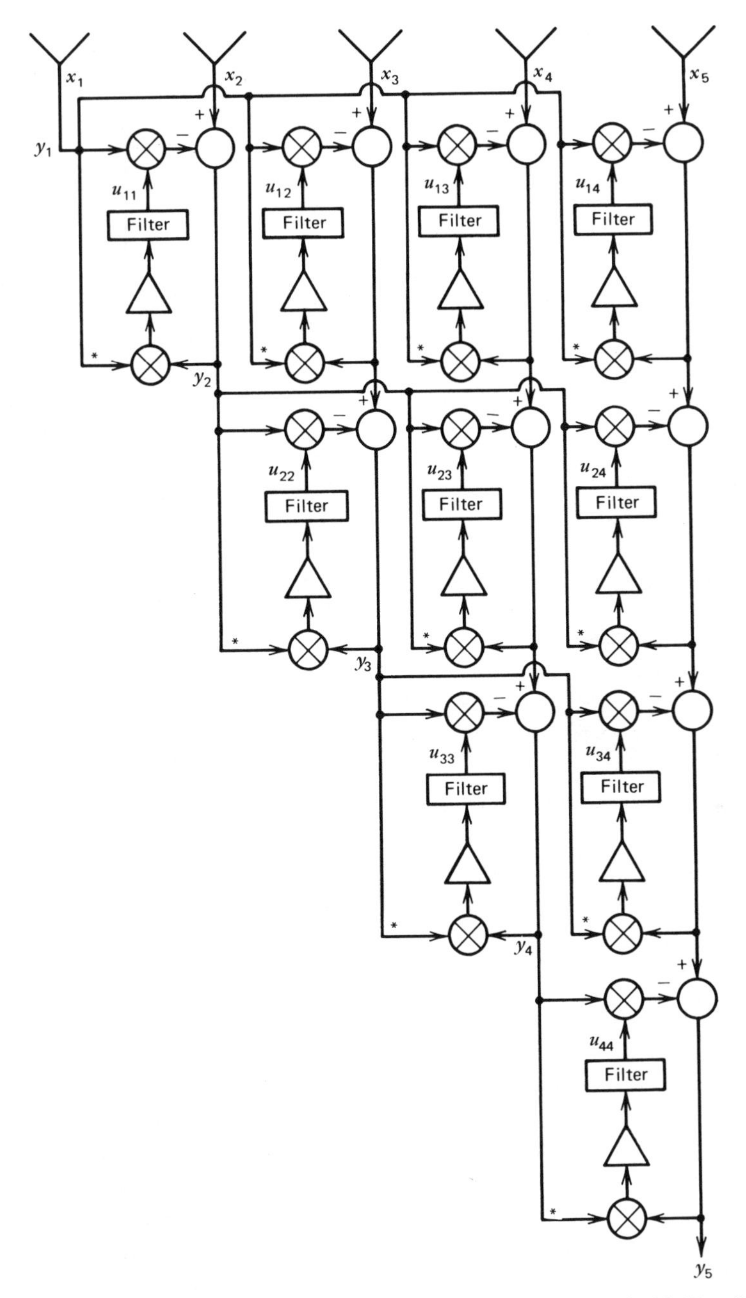


**Figure 8.8** Gram-Schmidt orthogonalization for five-element array realized with Howells-Applebaum adaptive loops.

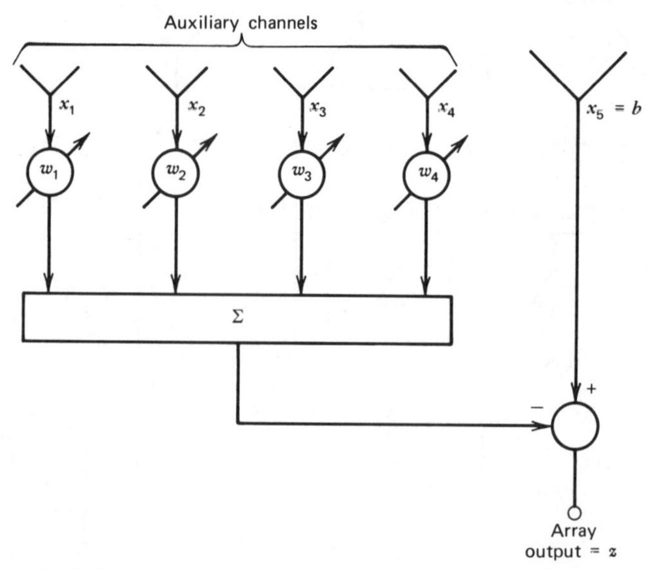


**Figure 8.9** Equivalent CSLC system for the cascade preprocessor system of Figure 8.7.

that the cascade network output is the same as the CSLC system output once the optimum solutions are reached. It will be recalled from Chapter 3 that the optimum weight vector for the CSLC system is given by

$$\mathbf{w}_{\text{opt}} = \mathbf{R}_{xx}^{-1}\overline{(\mathbf{x}^* b)} \tag{8.52}$$

If the main beam channel signal is replaced by a locally generated pilot signal $p(t)$, then

$$\mathbf{w}_{\text{opt}} = \mathbf{R}_{xx}^{-1}\overline{(\mathbf{x}^* p)} \tag{8.53}$$

and the array output is given by

### *8.3.1 Convergence of the Gram-Schmidt Cascade Preprocessor*

Let us first demonstrate that the cascade network of Figure 8.7 converges to the solution provided by a conventional CSLC system. Complete steady-state equivalence is shown for the three-element CSLC system of Figure 8.10 (in which two Howells-Applebaum SLC control loops yield weights $w_1$ and $w_2$) and the three-element cascade system of Figure 8.11 (in which three Howells-Applebaum control loops yield weights $u_{11}$, $u_{12}$, and $u_{22}$) [5]. The analysis for this case may then be easily extended to any arbitrary number of elements.

For the conventional three-element CSLC system of Figure 8.10, the steady-state solution for the optimum weight vector (whose components are

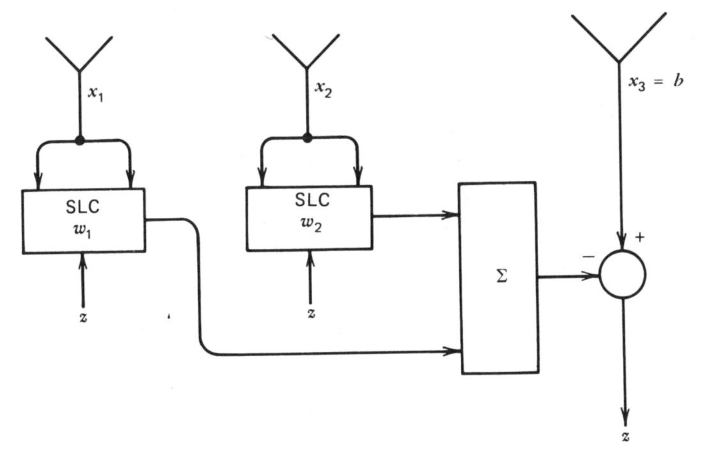


**Figure 8.10** Conventional three-element CSLC system having two Howells-Applebaum SLC control loops.

$w_1$ and $w_2$) is given by (8.52) where $\mathbf{R}_{xx} = \mathbf{R}_{nn}$ in the absence of a desired signal. From (8.52) it follows that the optimum steady-state weights of Figure 8.10 are given by

$$\begin{bmatrix} w_1 \\ w_2 \end{bmatrix} = \begin{bmatrix} \overline{x_1^* x_1} & \overline{x_1^* x_2} \\ \overline{x_2^* x_1} & \overline{x_2^* x_2} \end{bmatrix}^{-1} \begin{bmatrix} \overline{x_1^* b} \\ \overline{x_2^* b} \end{bmatrix}$$

$$= \frac{\begin{bmatrix} \overline{x_2^* x_2} & -\overline{x_1^* x_2} \\ -\overline{x_2^* x_1} & \overline{x_1^* x_1} \end{bmatrix} \begin{bmatrix} \overline{x_1^* b} \\ \overline{x_2^* b} \end{bmatrix}}{(\overline{x_1^* x_1})(\overline{x_2^* x_2}) - (\overline{x_1^* x_2})(\overline{x_2^* x_1})} \tag{8.55}$$

so that

$$w_1 = \frac{(\overline{x_2^* x_2})(\overline{x_1^* b}) - (\overline{x_1^* x_2})(\overline{x_2^* b})}{(\overline{x_1^* x_1})(\overline{x_2^* x_2}) - (\overline{x_1^* x_2})(\overline{x_2^* x_1})} \tag{8.56}$$

and

$$w_2 = \frac{(\overline{x_1^* x_1})(\overline{x_2^* b}) - (\overline{x_2^* x_1})(\overline{x_1^* b})}{(\overline{x_1^* x_1})(\overline{x_2^* x_2}) - (\overline{x_1^* x_2})(\overline{x_2^* x_1})} \tag{8.57}$$

For the cascade system of Figure 8.11, it follows from (8.49) and (8.50) that

$$y_2 = x_2 - u_{11} x_1 \tag{8.58}$$

$$v_3^2 = b - u_{12} x_1 \tag{8.59}$$

and

$$z = v_3^2 - u_{22} y_2 \tag{8.60}$$

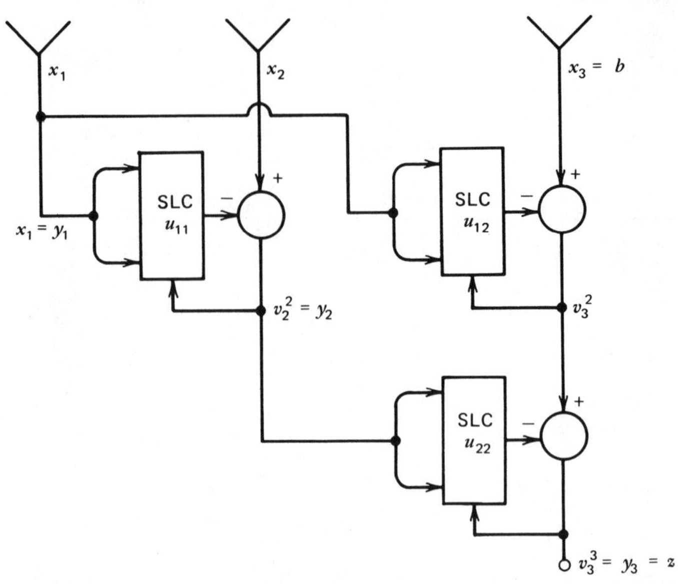


**Figure 8.11** Cascade Gram-Schmidt three-element CSLC system having three Howells-Applebaum SLC control loops.

The weight elements have steady-state values given by

$$u_{11} = \frac{\left(\overline{x_1^* x_2}\right)}{\left(\overline{x_1^* x_1}\right)} \tag{8.61}$$

$$u_{12} = \frac{\left(\overline{x_1^* b}\right)}{\left(\overline{x_1^* x_1}\right)} \tag{8.62}$$

and

$$u_{22} = \frac{\left(\overline{y_2^* v_3^2}\right)}{\left(\overline{y_2^* y_2}\right)} \tag{8.63}$$

From (8.58) to (8.60) it follows that

$$z = b - \left[ (u_{12} - u_{11}u_{22})x_1 + u_{22}x_2 \right] \tag{8.64}$$

Comparing (8.64) with Figure 8.10, we see that $w_2$ and $w_1$ fill the roles taken by $u_{22}$ and $u_{12} - u_{11}u_{22}$, respectively, in the cascade system.

If we substitute (8.58) and (8.59) into (8.63) it then follows that

$$u_{22} = \frac{\left(\overline{x_1^* x_1}\right)\left(\overline{x_2^* b}\right) - \left(\overline{x_2^* x_1}\right)\left(\overline{x_1^* b}\right)}{\left(\overline{x_1^* x_1}\right)\left(\overline{x_2^* x_2}\right) - \left(\overline{x_1^* x_2}\right)\left(\overline{x_2^* x_1}\right)} \tag{8.65}$$

which is identical with the solution (8.57) for $w_2$. Likewise if we substitute (8.61), (8.62), and (8.65) into $u_{12}-u_{11}u_{22}$ it follows that

$$u_{12}-u_{11}u_{22}=\frac{(\overline{x_1^* b})(\overline{x_2^* x_2})-(\overline{x_1^* x_2})(\overline{x_2^* b})}{(\overline{x_1^* x_1})(\overline{x_2^* x_2})-(\overline{x_1^* x_2})(\overline{x_2^* x_1})} \tag{8.66}$$

which is identical with the solution (8.56) for $w_1$. Therefore the steady-state solutions reached by the two CSLC of Figures 8.10 and 8.11 are identical with weight element equivalences given by

$$\left.\begin{aligned} w_2&=u_{22}\\ w_1&=u_{12}-u_{11}u_{22}\end{aligned}\right\} \tag{8.67}$$

Additional weight element equivalences may easily be obtained for an arbitrary number of sensor elements by considering the transformation (8.48). It can easily be shown by induction that the transformation **G** is a lower triangular matrix whose elements $g_{ij}$ are given by the following iterative relationships:

$$g_{ij}=0 \qquad \text{if } i<j \tag{8.68}$$

$$g_{ij}=1 \qquad \text{if } i=j \tag{8.69}$$

and

$$g_{ij}=-\sum_{k=j}^{i-1} u_{jk}^* \cdot g_{i(k+1)} \qquad \text{for } i>j \tag{8.70}$$

The $j$th row of the transformation matrix **G** then contains the desired weight element equivalences for a $j$-element CSLC system where the auxiliary channel weights are given by

$$\left.\begin{aligned} w_{j-1}&=-g_{j(j-1)}\\ w_{j-2}&=-g_{j(j-2)}\\ &\vdots\\ w_1&=-g_{j1}\end{aligned}\right\} \tag{8.71}$$

Having found the steady-state weight element equivalence relationships, it is now appropriate to consider the transient response of the networks of Figures 8.10 and 8.11.

The transient response of the two CSLC systems under consideration can be investigated by examining the system response to discrete signal samples of the main beam and the auxiliary channel outputs. The following analysis assumes that the adaptive weights reach their steady-state expected values on each iteration and therefore ignores errors that would be present due to loop noise. Let $x_{kn}$ denote the $n$th signal sample for the $k$th element channel. The main beam channel samples are then denoted by $b_n$. For a system having $N$ auxiliary channels and one main beam channel, after $N$ independent samples have been collected, a set of auxiliary channel weights can be computed using

$$\sum_{k=1}^{N} w_k x_{kn} = b_n \qquad \text{for } n = 1, 2, \ldots, N \tag{8.72}$$

The foregoing system of equations yields a unique solution since the matrix defined by the signal samples $x_{kn}$ is nonsingular provided that either thermal receiver noise or $N$ directional interference sources are present. Equation (8.72) is closely related to (8.52), since multiplying both sides of (8.72) by $x_{mn}^*$ and summing over the index $n$ yields

$$\sum_{n=1}^{N} \sum_{k=1}^{N} w_k x_{mn}^* x_{kn} = \sum_{n=1}^{N} x_{mn}^* b_n \tag{8.73}$$

which can be written in matrix form as

$$\hat{\mathbf{R}}_{xx}\mathbf{w} = \widehat{\overline{(\mathbf{x}^*\mathbf{b})}} \tag{8.74}$$

Comparing (8.74) with (8.52) reveals that these two equations yield similar solutions for the weight vector $\mathbf{w}$ in a stationary signal environment.

The iterative weight correction procedure for a single Howells-Applebaum SLC loop can be modeled as shown in Figure 8.12. On receipt of the $i$th signal sample, the resulting change in the weight $u_{k(n-1)}$ can be computed by applying (8.50) and assuming that the signal samples are approximately equal to their expected values to yield

$$\Delta u_{k(n-1)}(i) = \frac{v_k^{k*}\left[v_n^k(i) - u_{k(n-1)}(i-1)\cdot v_k^k(i)\right]}{\left(v_k^{k*} v_k^k\right)} \tag{8.75}$$

The element weight value is then updated in accordance with

$$u_{k(n-1)}(i) = u_{k(n-1)}(i-1) + \Delta u_{k(n-1)}(i) \tag{8.76}$$

With all weights in the cascade Gram-Schmidt network initially set to zero, it follows that on receipt of the first signal sample $v_2^2 = x_{21}$, and $v_3^2 = v_3^3 = b_1$ so the change in weight settings after the first iteration results in

$$u_{11}(1) = \frac{x_{11}^* v_2^1}{|x_{11}|^2} = \frac{x_{11}^* x_{21}}{|x_{11}|^2} \tag{8.77}$$

$$u_{12}(1) = \frac{x_{11}^* v_3^1}{|x_{11}|^2} = \frac{x_{11}^* b_1}{|x_{11}|^2} \tag{8.78}$$

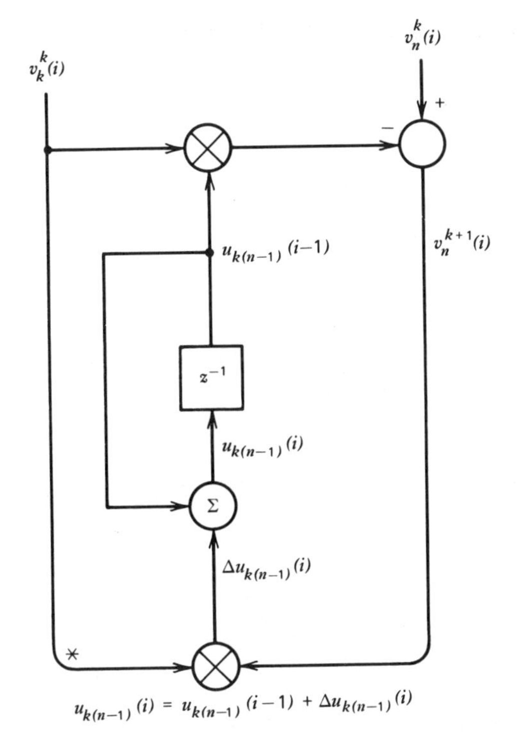


**Figure 8.12** Iterative weight correction model for Howells-Applebaum SLC loops.

and

$$u_{22}(1) = \frac{v_2^{2*} v_3^2}{|v_2^2|^2} = \frac{x_{21}^* b_1}{|x_{21}|^2} \tag{8.79}$$

On receipt of the second signal sample $v_2^2 = x_{22} - u_{11}(1)x_{12}$, $v_3^2 = b_2 - u_{12}(1)x_{12}$, and $v_3^3 = v_3^2 - u_{22}(1)v_2^2$ so that

$$\Delta u_{11}(2) = \frac{x_{12}^*[x_{22} - u_{11}(1)x_{12}]}{|x_{12}|^2} \tag{8.80}$$

$$\Delta u_{12}(2) = \frac{x_{12}^*[b_2 - u_{12}(1)x_{12}]}{|x_{12}|^2} \tag{8.81}$$

and

$$\Delta u_{22}(2) = \frac{(x_{22} - u_{11}(1)x_{12})^*[b_2 - u_{12}(1)x_{12} - u_{22}(1)(x_{22} - u_{11}(1)x_{12})]}{|x_{22} - u_{11}(1)x_{12}|^2} \tag{8.82}$$

Substitute (8.77) to (8.79) into (8.80) to (8.82); it then follows that

$$u_{11}(2)=u_{11}(1)+\Delta u_{11}(2)=\frac{x_{12}^{*}x_{22}}{|x_{12}|^2} \tag{8.83}$$

$$u_{12}(2)=u_{12}(1)+\Delta u_{12}(2)=\frac{x_{12}^{*}b_2}{|x_{12}|^2} \tag{8.84}$$

and

$$u_{22}(2)=u_{22}(1)+\Delta u_{22}(2)=\frac{x_{11}b_2-x_{12}b_1}{x_{11}x_{22}-x_{12}x_{21}} \tag{8.85}$$

By use of the weight equivalence relationships of (8.67) it follows that $w_2(2)=u_{22}(2)$ and

$$w_1(2)=u_{12}(2)-u_{11}(2)u_{22}(2)=\frac{x_{22}b_1-x_{21}b_2}{x_{11}x_{22}-x_{12}x_{21}} \tag{8.86}$$

Note, however, that $w_1(2)$ and $w_2(2)$ of (8.85) and (8.86) are exactly the weights that satisfy (8.72) for the case where $N=2$. Therefore for a two-auxiliary channel CSLC, the cascade Gram-Schmidt network converges to a set of near-optimum weights in only two iterations. The foregoing analysis can also be carried out for the case of an $N$-auxiliary channel CSLC and leads to the conclusion that $N$ iterations are required to converge to a set of near-optimum weights in the noise free case. When noise is present, then averaging of the input signals is required to mitigate the effects of noise induced errors.

Note that the use of (8.75) to update the weight settings results in values for $u_{11}$ and $u_{12}$ (the weights in the first level) that depend only on the current signal sample. The update setting for $u_{22}$, however, depends on the last two signal samples. Therefore an $N$-level cascade Gram-Schmidt network is always updating the weight settings based on the $N$ preceding signal samples.

The conventional two-weight element CSLC does not converge to a near-optimum weight solution with only two signal samples. To see this result note that for the configuration of Figure 8.10 after receipt of the first signal sample that

$$w_1(1)=\frac{x_{11}^{*}b_1}{|x_{11}|^2} \tag{8.87}$$

and

$$w_2(1) = \frac{x_{21}^* b_1}{|x_{21}|^2} \tag{8.88}$$

Following the second signal sample,

$$w_1(2) = w_1(1) + \frac{x_{12}^*(b_2 - w_1(1)x_{12} - w_2(1)x_{22})}{|x_{12}|^2} \tag{8.89}$$

and

$$w_2(2) = w_2(1) + \frac{x_{22}^*(b_2 - w_1(1)x_{12} - w_2(1)x_{22})}{|x_{12}|^2} \tag{8.90}$$

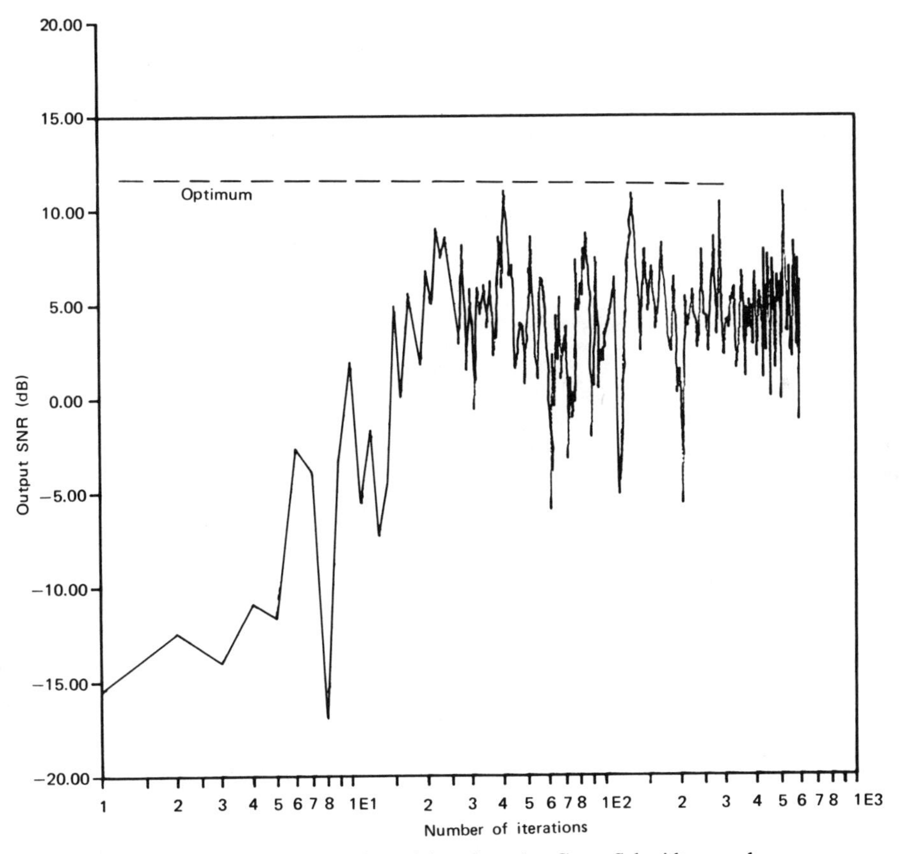


**Figure 8.13** Output SNR versus number of iterations for Gram-Schmidt cascade preprocessor with eigenvalue spread = 16,700. Algorithm parameters are $K=3$, $\alpha=\alpha_L=0.5$, and $\mathbf{w}(0)=\mathbf{0}$ with $s/n=10$ for which $\text{SNR}_{\text{opt}}=15$ (11.76 dB).

The weights given by (8.89) and (8.90) are not the same as the weights given by (8.85) and (8.86) and consequently do not satisfy (8.72). It will be seen from the simulation results that the configuration of Figure 8.10 yields a transient response that is considerably slower (even with a higher level of loop noise) than the transient response of a Gram-Schmidt cascade preprocessor.

## 8.4 SIMULATION RESULTS

To obtain an indication of the convergence rate that can be obtained through the use of a Gram-Schmidt orthogonalization preprocessor, a four-element network configuration like that of Figure 8.6 was simulated with a digital

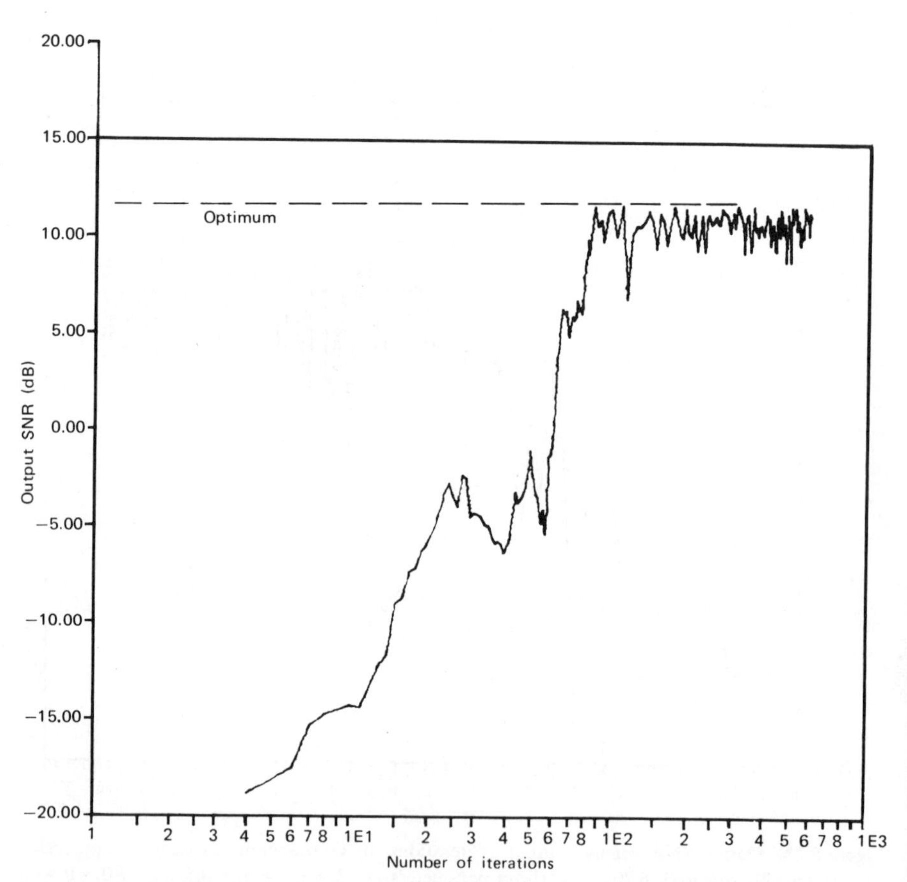


**Figure 8.14** Output SNR versus number of iterations for Gram-Schmidt cascade preprocessor with eigenvalue spread = 16,700. Algorithm parameters are $K=3$, $\alpha=\alpha_L=0.1$, and $\mathbf{w}(0)=\mathbf{0}$ with $s/n=10$ for which $\text{SNR}_{\text{opt}}=15$.

LMS adaptive loop used in place of each Howells-Applebaum loop in the final maximum SNR stage. The four loop gains of the LMS adaptive loops in the final maximum SNR stage of the selected configuration were set inversely proportional to an estimate of the input signal power to each loop $\hat{p}_i$ obtained by averaging the squared loop input signal over $K$ samples so that $\Delta_{s_i} = \alpha_L / \hat{p}_i$. Likewise, the loop gain of the digital version of each Howells-Applebaum loop in the cascade preprocessor was selected as $G_i = \alpha / \hat{p}_i$. Selecting the adaptive loop gains in this manner equalizes the transient response time to the signals $y_1 \ldots y_4$. Furthermore, the correlations (products) that are performed in the analog processor of Figure 8.8 are represented in the digital version as products of individual sample values averaged over $K$ samples. The averaging interval also permits signal power estimates to be obtained by averaging the instantaneous signal powers appearing at the loop inputs of the

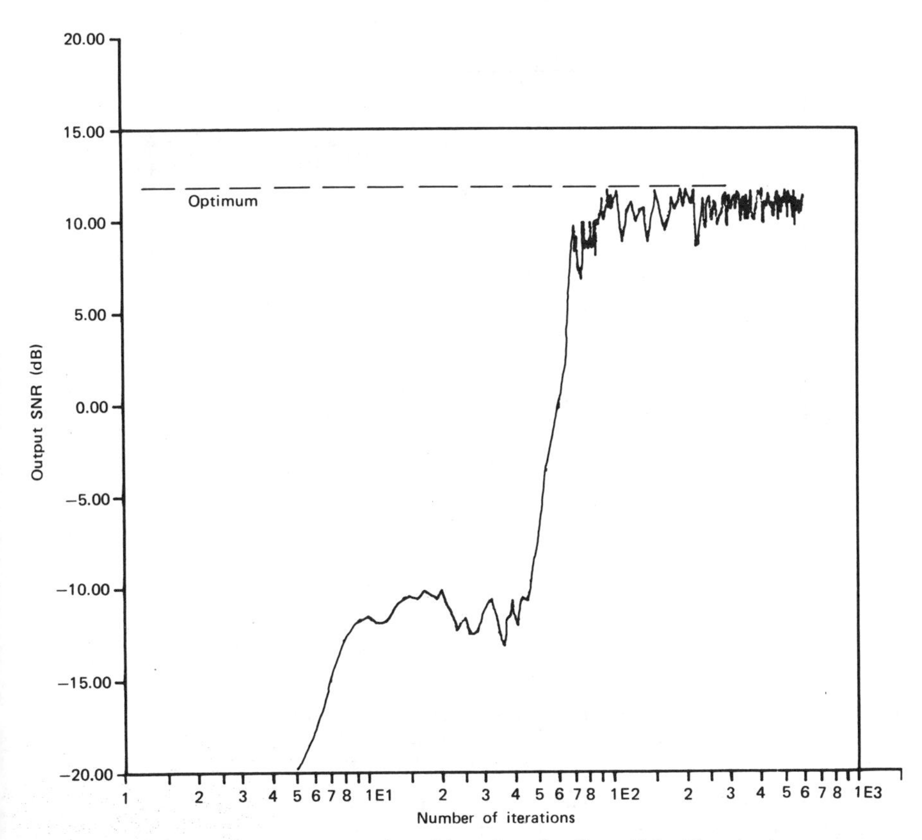


**Figure 8.15** Output SNR versus number of iterations for Gram-Schmidt cascade preprocessor with eigenvalue spread = 16,700. Algorithm parameters are $K = 9$, $\alpha = 0.3$, $\alpha_L = 0.25\alpha$, and $\mathbf{w}(0) = \mathbf{0}$ with $s/n = 10$ for which $\text{SNR}_{\text{opt}} = 15$.

final processing stage. The array configuration and signal environment correspond to that given in Figure 7.4, and the desired signal was again selected to be a biphase modulated signal having a 0° or 180° reference phase equally likely at each sample.

The algorithm performance in each case is given in terms of the resulting output SNR versus the number of iterations. The signal environment conditions are specified to represent two values of eigenvalue spread: $\lambda_{max}/\lambda_{min} = 16{,}700$ and 2440. The convergence results given in Figures 8.13 to 8.17 are for jammer-to-thermal noise ratios $J_1/n = 4000$, $J_2/n = 40$, and $J_3/n = 400$ for which the corresponding eigenvalues of the noise covariance matrix are $\lambda_1 = 1.67 \times 10^4$, $\lambda_2 = 10^3$, $\lambda_3 = 29$, and $\lambda_4 = 1.0$. The noise environment condi-

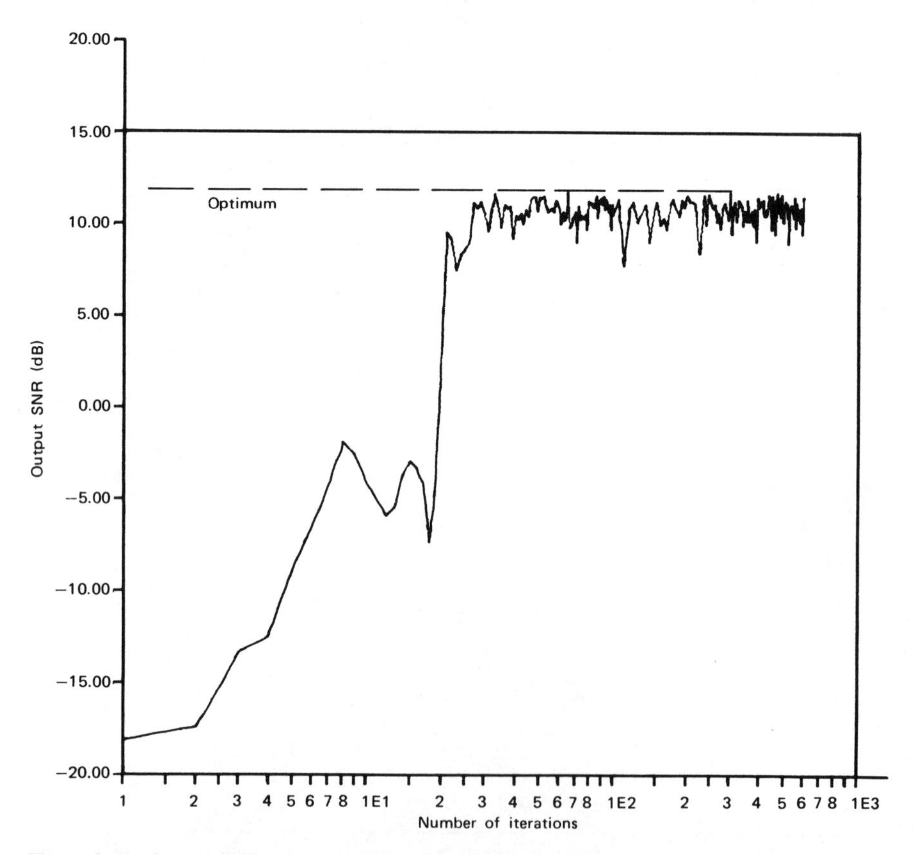


**Figure 8.16** Output SNR versus number of iterations for Gram-Schmidt cascade preprocessor with eigenvalue spread = 16,700. Algorithm parameters are $K = 9$, $\alpha = \alpha_L = 0.3$, and $\mathbf{w}(0) = \mathbf{0}$ with $s/n = 10$ for which $\text{SNR}_{opt} = 15$.

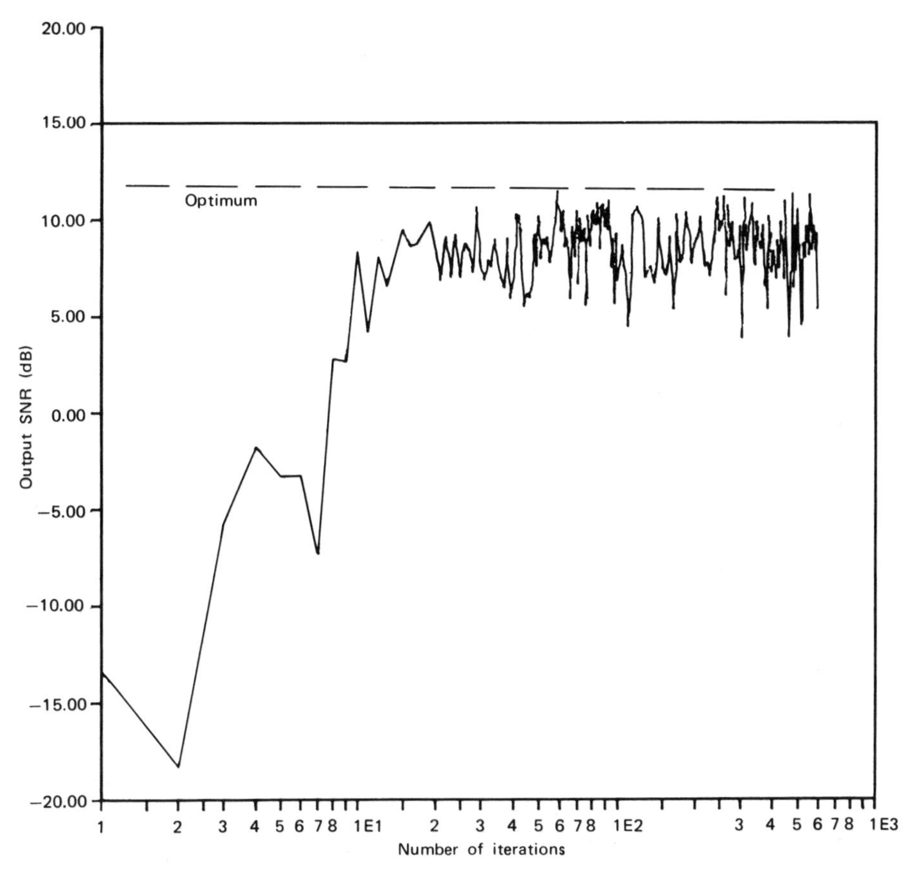


**Figure 8.17** Output SNR versus number of iterations for Gram-Schmidt cascade preprocessor with eigenvalue spread = 16,700. Algorithm parameters are $K=9$, $\alpha=\alpha_L=0.65$, and $\mathbf{w}(0)=\mathbf{0}$ with $s/n=10$ for which $\text{SNR}_{\text{opt}}=15$.

tions for Figures 8.18 to 8.20 are likewise given by the jammer-to-thermal noise ratios $J_1/n=500$, $J_2/n=40$, and $J_3/n=200$ for which the corresponding eigenvalues of the noise covariance matrix are $\lambda_1=2.44\times10^3$, $\lambda_2=4.94\times10^2$, $\lambda_3=25.62$, and $\lambda_4=1.0$. All other conditions are specified by the number of measurement samples averaged per iteration $K$, the Howells-Applebaum loop gain $\alpha$, the LMS loop gain $\alpha_L$, the desired signal-to-thermal noise ratio $s/n$, and the initial weight vector selection $\mathbf{w}(0)$. For $s/n=10$, the optimum output SNR is $\text{SNR}_{\text{opt}}=15$ (11.76 dB), while for $s/n=0.1$ the optimum output SNR is $\text{SNR}_{\text{opt}}=0.15$ ($-8.2$ dB).

On examining Figures 8.13 to 8.20, a number of observations can be made as follows:

1. A severe spread in covariance matrix eigenvalues affects the speed of the algorithm response only slightly (compare Figure 8.14 with Figure 8.20).
2. Comparable steady-state output variations (reflecting the same level of loop noise) are obtained for $K=3$, $\alpha=0.3/p$ and $K=9$, $\alpha=0.1/p$, thereby indicating that the product $K\cdot\alpha$ determines the loop noise level (compare Figure 8.14 with Figure 8.15).
3. Only a slight increase in the loop gain $\alpha$ is required before the weights become excessively noisy (compare Figure 8.16 with Figure 8.13 and Figure 8.17).
4. The appropriate value to which the product $K\cdot\alpha$ should be set for an acceptable level of loop noise depends on the value of $\text{SNR}_{\text{opt}}$ (com-

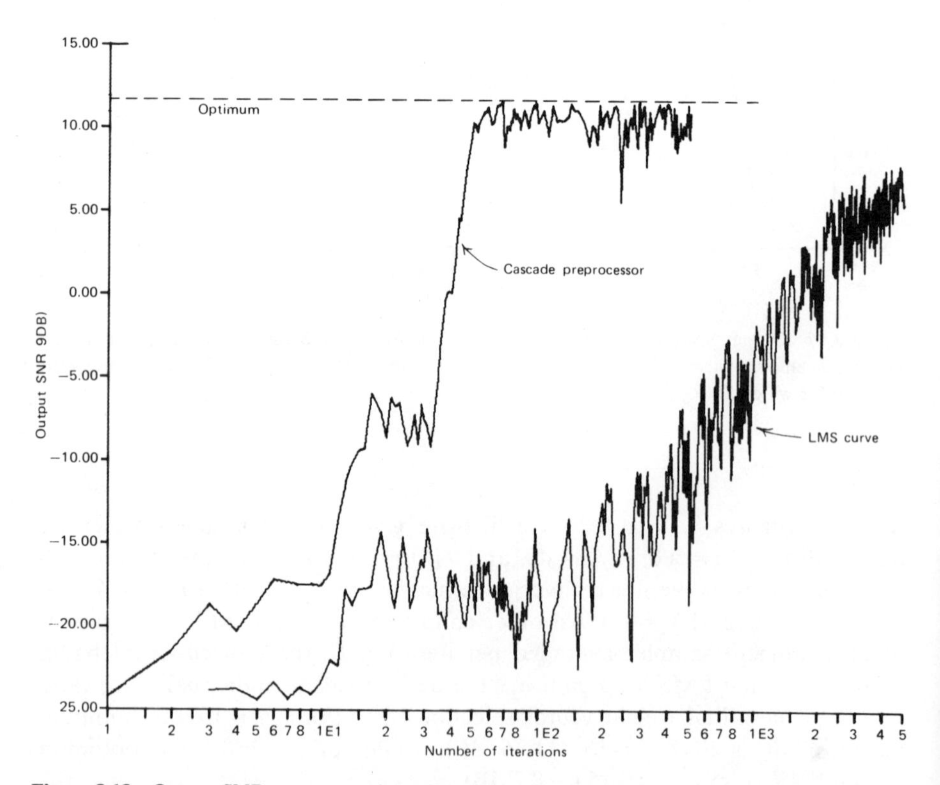


**Figure 8.18** Output SNR versus number of iterations for Gram-Schmidt cascade preprocessor and LMS processor with eigenvalue spread = 2440. Algorithm parameters are $K=3$, $\alpha=\alpha_L=0.1$, and $\mathbf{w}^T(0)=[1,0,0,0]$ with $s/n=10$ for which $\text{SNR}_{\text{opt}}=15$.

pare Figures 8.19 and 8.20). Smaller values of $SNR_{opt}$ require smaller values of the product $K \cdot \alpha$ to maintain the same loop noise level. It may also be seen that the adaptation time required with these parameter values for the Gram-Schmidt orthogonalization preprocessor is greater than that required for either a recursive or DMI algorithm. The question of relative adaptation times for the various algorithms is pursued further in Chapter 10.

5. The degree of transient response improvement that can be obtained with a cascade preprocessor is shown in Figure 8.18 where the Gram-Schmidt cascade preprocessor response is compared with the LMS processor response. The LMS curve in this figure was obtained simply by removing the cascade preprocessing stage in Figure 8.6 thereby leaving the final maximum SNR stage consisting of four LMS adaptive loops.

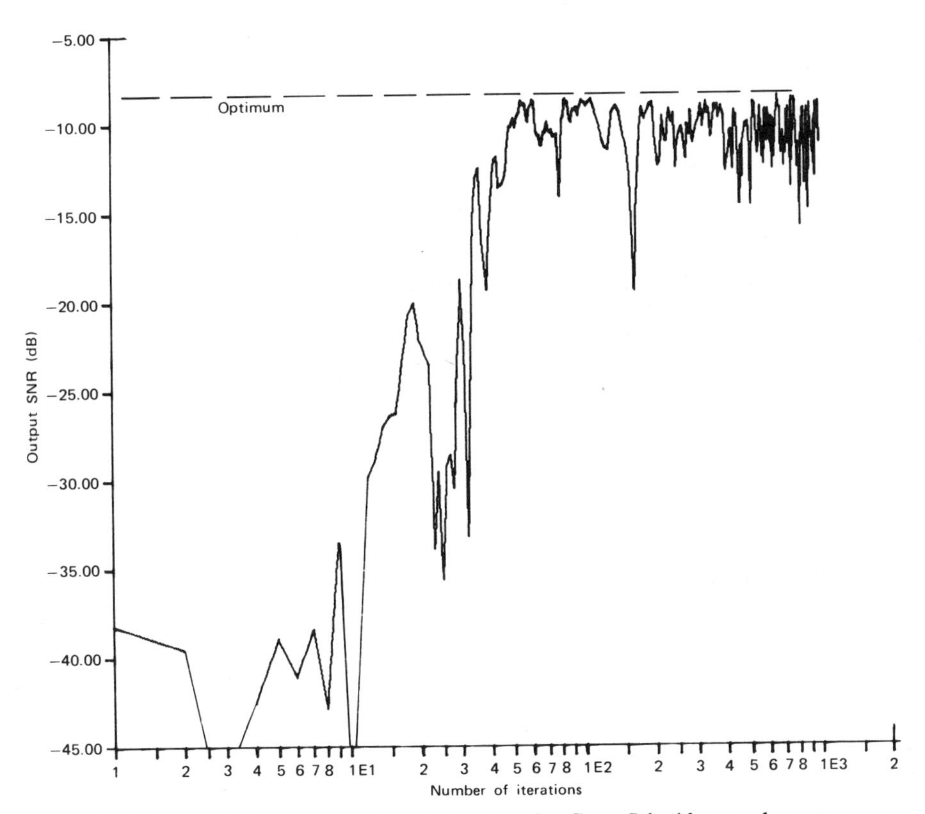


**Figure 8.19** Output SNR versus number of iterations for Gram-Schmidt cascade preprocessor with eigenvalue spread = 2440. Algorithm parameters are $K=3$, $\alpha=\alpha_L=0.1$, and $\mathbf{w}(0)=\mathbf{0}$ with $s/n=0.1$ for which $SNR_{opt}=0.15$ ($-8.2$ dB).

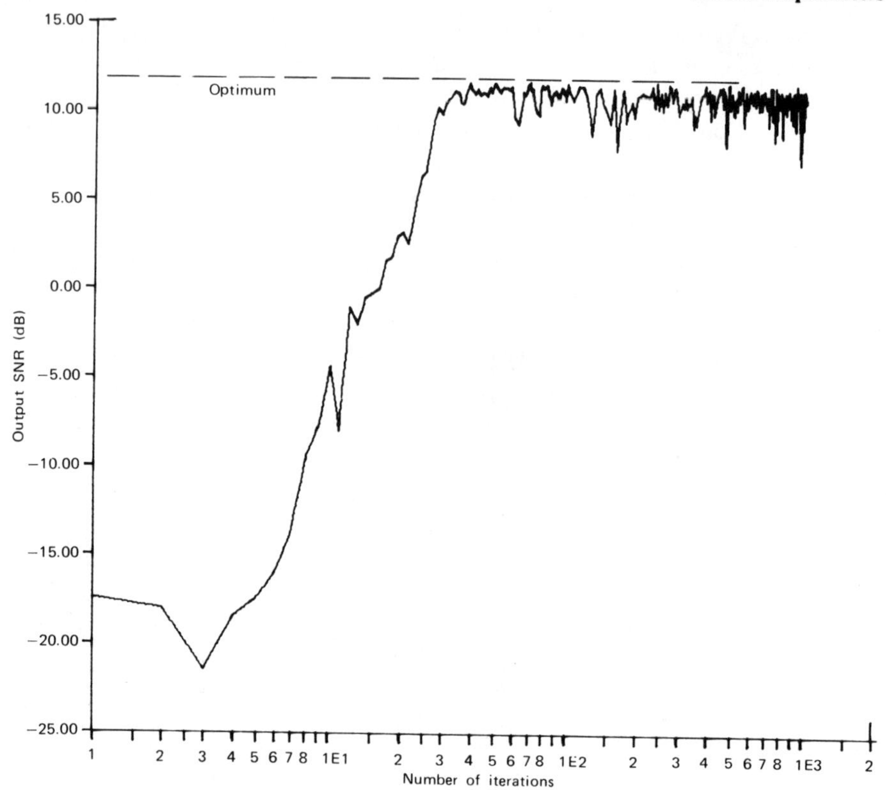


**Figure 8.20** Output SNR versus number of iterations for Gram-Schmidt cascade preprocessor with eigenvalue spread = 2440. Algorithm parameters are $K=3$, $\alpha=\alpha_L=0.1$, and $\mathbf{w}(0)=\mathbf{0}$ with $s/n=10$ for which $\text{SNR}_{\text{opt}}=15$.

The greater the eigenvalue spread of the $\mathbf{R}_{xx}$ matrix, then the greater is the degree of improvement that can be realized with a cascade preprocessor compared to the LMS processor.

## 8.5 SUMMARY AND CONCLUSIONS

The sensitivity of the convergence rate of the LMS and Howells-Applebaum adaptive processors to eigenvalue spread can be greatly reduced through the use of a cascade preprocessor that resolves the input signal vector into orthogonal components. A Nolen cascade network resolves the input signal vector into orthogonal eigenvector beam components but requires an excessive number of iterations to adjust the network parameters before rapid convergence can be achieved. The Gram-Schmidt cascade preprocessor resolves the input signal vector into orthogonal components (although these components are not eigenvector beam components), and succeeds in greatly reducing the sensitivity of the convergence rate to eigenvalue spread while requiring a relatively small number of iterations to adjust the preprocessor parameters to the desired steady-state values. The Gram-Schmidt preprocessor therefore is a highly attractive candidate for achieving fast algorithm convergence rates while retaining the implementation simplicity associated with the LMS and Howells-Applebaum adaptive processors.

## PROBLEMS

**1** Consider the cascade control loop configuration of Figure 8.21.

(a) Show that the weight equivalence between the configuration of Figure 8.21 and the standard CSLC configuration are given by

$$w_1 = u_1 + u_3 \qquad w_2 = -u_2 u_3$$

(b) Show that the steady-state weight values of Figure 8.21 are given by

$$u_1 = \frac{\overline{x_1^* b}}{\overline{x_1^* x_1}} \qquad u_2 = \frac{\overline{x_2^* x_1}}{\overline{x_2^* x_2}}$$

$$u_3 = \frac{(\overline{x_1^* x_2})}{(\overline{x_1^* x_1})}\left[\frac{(\overline{x_1^* x_1})(\overline{x_2^* b}) - (\overline{x_2^* x_1})(\overline{x_1^* b})}{(\overline{x_2^* x_2})(\overline{x_1^* x_1}) - (\overline{x_1^* x_2})(\overline{x_2^* x_1})}\right]$$

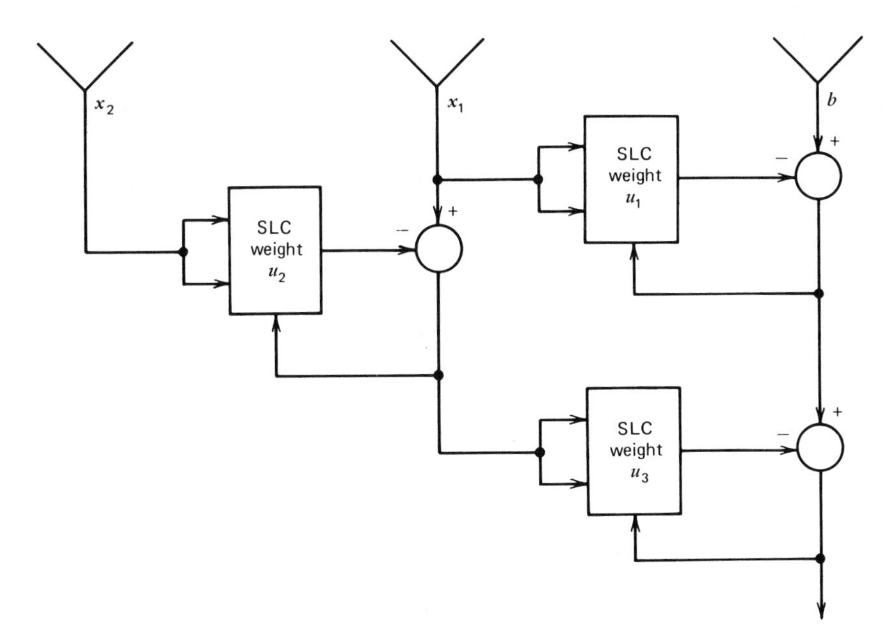


**Figure 8.21** Cascade arrangement of three Howells-Applebaum SLC control loops yielding convergence to incorrect solution.

Note that these steady-state weights do not correspond to the correct solution for the standard CSLC configuration.

*Note*: Problems 2 through 12 concern derivations that may be found in reference [7].

**2** Consider the three-element array cascade configuration of Figure 8.11.

(a) Show that the steady-state weight values for this configuration are given by (8.61), (8.62), and (8.65).

(b) Using (8.64) for the array output, show that

$$z = x_3 - u_{22}x_2 - \frac{(\overline{x_1^* x_3})(\overline{x_2^* x_2}) - (\overline{x_1^* x_2})(\overline{x_2^* x_3})}{(\overline{x_1^* x_1})(\overline{x_2^* x_2}) - (\overline{x_2^* x_1})(\overline{x_1^* x_2})} x_1$$

(c) The average noise output power is defined by $N_0 = \overline{|z|^2} = \overline{zz^*}$. For notational simplicity let $u_1 = u_{11}$, $u_2 = u_{12}$, $u_3 = u_{22}$ and show that

$$\begin{aligned} N_0 = {} & (\overline{x_3^* x_3}) + |\bar{u}_3|^2(\overline{x_2^* x_2}) + |\bar{u}_2 - \bar{u}_1\bar{u}_3|^2(\overline{x_1^* x_1}) \\ & - \bar{u}_3(\overline{x_3^* x_2}) - \overline{u_3^*}(\overline{x_2^* x_3}) \\ & - (\bar{u}_2 - \bar{u}_1\bar{u}_3)(\overline{x_3^* x_1}) - (\overline{u_2^*} - \overline{u_1^* u_3^*})(\overline{x_1^* x_3}) \\ & + \bar{u}_3(\overline{u_2^*} - \overline{u_1^* u_3^*})(\overline{x_1^* x_2}) + \overline{u_3^*}(\bar{u}_2 - \bar{u}_1\bar{u}_3)(\overline{x_2^* x_1}) \end{aligned}$$

**3** Using the weight notation of Problem 2(c) for the configuration of Figure 8.11, let

$$u_n = \bar{u}_n + \delta_n$$

where $\delta_n$ represents the fluctuation component of the weight. The total output noise power is then given by

$$\begin{aligned} N_{0_{\text{TOT}}} &= N_0 + N_u \\ &= \overline{|x_3 - (\bar{u}_3 + \delta_3)x_2 - [(\bar{u}_2 + \delta_2) - (\bar{u}_1 + \delta_1)(\bar{u}_2 + \delta_3)]\, x_1|^2} \end{aligned}$$

where $N_u$ represents the excess weight noise due to the fluctuation components. Note that $\bar{\delta}_n = 0$ and neglect third- and fourth-order terms in $\delta_n$ to show that

$$N_u = \overline{|\delta_3|^2}[(\overline{x_2^* x_2}) - (\overline{x_2^* x_1})\bar{u}_1] + (\overline{x_1^* x_1})\overline{|\delta_2 - \bar{u}_3\delta_1|^2}$$

**4** The weight $u_1 = u_{11}$ of Figure 8.11 satisfies the differential equation

$$\tau_1\left(\frac{\dot{u}_1}{G_1}\right) + \left(\frac{u_1}{G_1}\right) = x_1^*(x_2 - u_1 x_1)$$

where $G_1$ denotes amplifier gain and $\tau_1$ denotes time constant of the integrating filter in the Howells-Applebaum control loop for $u_1$. The foregoing equation can be rewritten as

$$\frac{1}{\alpha_1}\dot{u}_1+\left(x_1^* x_1+\frac{1}{G_1}\right)u_1=x_1^* x_2 \qquad \text{where } \alpha_1=\frac{G_1}{\tau_1}$$

so that

$$\frac{1}{\alpha_1}\dot{\bar{u}}_1+\left(\overline{x_1^* x_1}+\frac{1}{G_1}\right)\bar{u}_1=\overline{x_1^* x_2}$$

Subtracting the mean value differential equation from the instantaneous value differential equation and recalling that $u_1=\bar{u}_1+\delta_1$ then yields

$$\frac{1}{\alpha_1}\dot{\delta}_1+\left(\frac{1}{G_1}+\overline{x_1^* x_1}\right)\delta_1+\left(x_1^* x_1-\overline{x_1^* x_1}\right)u_1=\left(x_1^* x_2-\overline{x_1^* x_2}\right)$$

provided that the second-order term $\delta_1(x_1^* x_1-\overline{x_1^* x_1})$ is ignored. In the steady state where

$$\bar{u}_1=\frac{\overline{x_1^* x_2}}{\overline{x_1^* x_1}}$$

show that

$$\frac{1}{\alpha_1}\dot{\delta}_1+\left(\overline{x_1^* x_1}+\frac{1}{G_1}\right)\delta_1=\left(x_1^* x_2-\frac{(\overline{x_1^* x_2})}{(\overline{x_1^* x_1})}x_1^* x_1\right)=\frac{1}{\alpha_1}f_1(t)$$

where $f_1(t)$ may be regarded as a random variable.

**5** The solution of the differential equation obtained in Problem 4 is given by

$$\delta_1(t)=\int_0^t f_1(\tau)\exp\left[-\alpha_1\left(\overline{x_1^* x_1}+\frac{1}{G_1}\right)(t-\tau)\right]d\tau$$

so that

$$\dot{\delta}_1(t)=f_1(t)-\alpha_1\left(\overline{x_1^* x_1}+\frac{1}{G_1}\right)\delta_1(t)$$

Substitute $\dot{\delta}_1(t)$ into the differential equation for $\dot{\delta}_1(t)$ obtained in Problem 4 and show that this reduces the resulting expression to an identity thereby proving that $\delta_1(t)$ above is indeed a solution to the original differential equation.

**6** The second moment of the fluctuation $\delta_1$ is given by $\overline{\delta_1^* \delta_1}$.

(a) Use the expression for $\delta_1(t)$ developed in Problem 5 to show that

$$\overline{\delta_1^* \delta_1} = \exp\left[-2\alpha_1\left(\overline{x_1^* x_1} + \frac{1}{G_1}\right)\right] \int_0^t \int_0^t \overline{f_1^*(\tau) f_1(u)}$$
$$\cdot \exp\left[\alpha_1\left(\overline{x_1^* x_1} + \frac{1}{G_1}\right)(\tau + u)\right] d\tau\, du$$

(b) Assume that the random variable

$$f_1(t) = \alpha_1\left[x_1^* x_2 - \frac{(\overline{x_1^* x_2})}{(\overline{x_1^* x_1})} x_1^* x_1\right]$$

has a correlation interval denoted by $\varepsilon$, such that values of $f_1(t)$ separated by more than $\varepsilon$ are independent. Let

$$\int_0^t \overline{f_1^*(\tau) f_1(u)}\, du = \overline{f_1^* f_1}\varepsilon \qquad \text{for } t > \tau + \varepsilon.$$

Show that $\overline{\delta_1^* \delta_1}$ then reduces to

$$\overline{\delta_1^* \delta_1} = \frac{\varepsilon \overline{f_1^* f_1}}{2\alpha_1[(\overline{x_1^* x_1}) + 1/G_1]} \qquad \text{for } t > \tau + \varepsilon$$

**7** Show that the second moment of $f_1$ defined in Problem 6 (b) is given by

$$\overline{f_1^* f_1} = \alpha_1^2 \overline{\left[x_1^* x_2 - \frac{(\overline{x_1^* x_2})}{(\overline{x_1^* x_1})} x_1^* x_1\right]}$$
$$\times \overline{\left[x_1 x_2^* - \frac{(\overline{x_2^* x_1})}{(\overline{x_1^* x_1})} x_1^* x_1\right]}$$
$$= \alpha_1^2[(\overline{x_1^* x_1})(\overline{x_1^* x_2}) - (\overline{x_1^* x_2})(\overline{x_2^* x_1})]$$

**8** Substitute $\overline{f_1^* f_1}$ obtained in Problem 7 into $\overline{\delta_1^* \delta_1}$ obtained in Problem 6 (b) to show that

$$\overline{\delta_1^* \delta_1} = \frac{\varepsilon \alpha_1[(\overline{x_1^* x_1})(\overline{x_2^* x_2}) - (\overline{x_2^* x_1})(\overline{x_1^* x_2})]}{2[(\overline{x_1^* x_1}) + 1/G_1]}$$

Note that to obtain $\overline{\delta_2^* \delta_2}$, merely replace $x_2$ with $x_3$, $\alpha_1$ by $\alpha_2$, and $G_1$ by $G_2$ in the above expression.

**9** From Figure 8.11 it may be seen that the inputs to $u_3$ ($u_{22}$ of the figure) are $y_1$ and $y_2$. Consequently replace $x_1$ by $y_2$, $x_2$ by $y_1$, and $G_1$ by $G_3$ in the expression obtained in Problem 8. This replacement requires the second moments $\overline{y_1^* y_1}$ and $\overline{y_2^* y_2}$ and $\overline{y_2^* y_1}$ to be computed. Show that

$$\overline{y_1^* y_1} = \overline{|x_3 - \bar{u}_3 x_1|^2} = (\overline{x_3^* x_3}) - \frac{(\overline{x_1^* x_3})(\overline{x_3^* x_1})}{(\overline{x_1^* x_1})}$$

and

$$y_2^* y_2 = \overline{|x_2 - \bar{u}_1 x_1|^2} = \overline{\left| x_2 - \frac{(\overline{x_1^* x_2})}{(\overline{x_1^* x_1})} x_1 \right|^2}$$

$$= (\overline{x_2^* x_2}) - \frac{(\overline{x_1^* x_2})(\overline{x_2^* x_1})}{(\overline{x_1^* x_1})}$$

Furthermore

$$\overline{y_2^* y_1} = \overline{\left[ x_2 - \frac{(\overline{x_1^* x_2})}{(\overline{x_1^* x_1})} x_1 \right]^* \left[ x_3 - \frac{(\overline{x_1^* x_3})}{(\overline{x_1^* x_1})} x_1 \right]}$$

$$= (\overline{x_2^* x_3}) - \frac{(\overline{x_2^* x_1})(\overline{x_1^* x_3})}{(\overline{x_1^* x_1})}$$

**10** The computation of $N_u$ found in problem 3 requires the second moment $\overline{\delta_2^* \delta_1}$. Corresponding to the expression for $\delta_1(t)$ in Problem 5 we may write

$$\delta_2(t) = \int_0^t f_2(\tau) \exp - \alpha_2 \left( \overline{x_1^* x_1} + \frac{1}{G_2} \right)(t - \tau)\, d\tau$$

where

$$f_2(t) = \alpha_2 \left[ x_1^* x_3 - \frac{(\overline{x_1^* x_3})}{(\overline{x_1^* x_1})} x_1^* x_1 \right]$$

With the above results show that

$$\overline{\delta_2^* \delta_1} = \int_0^t \int_0^t \overline{f_2^*(\tau) f_1(u)}$$

$$\cdot \exp\left[ -\alpha_2 \left( \overline{x_1^* x_1} + \frac{1}{G_2} \right)(t - \tau) - \alpha_1 \left( \overline{x_1^* x_1} + \frac{1}{G_1} \right)(t - u)\, d\tau du \right.$$

Now assuming that

$$\int_0^t \overline{f_2^*(\tau) f_1(u)}\, du = \overline{f_2^* f_1}$$

show that

$$\overline{\delta_2^* \delta_1} = \frac{\varepsilon\, \overline{f_2^* f_1}}{\left[\alpha_2(\overline{x_1^* x_1}) + 1/G_2\right] + \alpha_1\left[(\overline{x_1^* x_1}) + 1/G_1\right]}$$

**11** Evaluate $\overline{f_2^* f_1}$ by retaining only terms of the form $\overline{x_m^* x_n}$ in the expansion:

$$\overline{f_2^* f_1} = \alpha_2 \alpha_1 \left[ x_1 x_3^* - \frac{(\overline{x_3^* x_1})}{(\overline{x_1^* x_1})} x_1^* x_1 \right] \cdot \left[ x_1^* x_2 - \frac{(\overline{x_1^* x_2})}{(\overline{x_1^* x_1})} x_1^* x_1 \right]$$

show that the resulting expression for $\overline{f_2^* f_1}$ is given by

$$\overline{f_2^* f_1} = \alpha_2 \alpha_1 \left[ (\overline{x_1^* x_1})(\overline{x_3^* x_2}) - (\overline{x_3^* x_1})(\overline{x_1^* x_2}) \right]$$

so that

$$\overline{\delta_2^* \delta_1} = \frac{\varepsilon \alpha_1 \alpha_2 \left[ (\overline{x_1^* x_1})(\overline{x_3^* x_2}) - (\overline{x_3^* x_1})(\overline{x_1^* x_2}) \right]}{\alpha_2\left[(\overline{x_1^* x_1}) + 1/G_2\right] + \alpha_1\left[(\overline{x_1^* x_1}) + 1/G_1\right]}$$

**12** Substitute the results obtained in Problem 11 into the expression for $N_u$, assuming that $\alpha_1 = \alpha_2$ and neglecting all $1/G$ terms, show that

$$\begin{aligned} N_u = \{ & \left[ (\overline{x_1^* x_1})(\overline{x_2^* x_2}) - (\overline{x_1^* x_2})(\overline{x_2^* x_1}) \right] \\ & \cdot \left[ (\overline{x_1^* x_1})(\overline{x_3^* x_3}) - (\overline{x_3^* x_1})(\overline{x_1^* x_2}) \right] \\ & - |(\overline{x_1^* x_1})(\overline{x_2^* x_3}) - (\overline{x_2^* x_1})(\overline{x_1^* x_2})|^2 \} \\ & \cdot \left[ \frac{\alpha_3 \varepsilon}{2(\overline{x_1^* x_1})^2} + \frac{\alpha_1 \varepsilon}{2(\overline{x_1^* x_1})(\overline{x_2^* x_2}) - (\overline{x_2^* x_1})(\overline{x_1^* x_2})} \right] \end{aligned}$$

Note that the portion of $N_u$ proportional to $\alpha_3$ is the component due to fluctuations in $u_3$. The remaining part of $N_u$ that is proportional to $\alpha_1$ is the contribution due to fluctuations in both $u_1$ and $u_2$. If $\tau_1$, $\tau_2$, and $\tau_3$ are adjusted so that all three circuits have the same effective time constant,

$$\alpha_1 = \alpha_2 = \alpha_3 \frac{(\overline{x_1^* x_1})^2}{(\overline{x_1^* x_1})(\overline{x_2^* x_2}) - (\overline{x_2^* x_1})(\overline{x_1^* x_2})}$$

then the above expression for $N_u$ shows that the control loop noise contribution due to fluctuations in $u_3$ is just equal to the sum of the contributions from fluctuations in $u_1$ and $u_2$.

## REFERENCES

[1] L. Hasdorff, *Gradient Optimization and Nonlinear Control*, Wiley, New York, 1976, Ch. 3.

[2] W. D. White, "Accelerated Convergence Techniques," Proceedings of the 1974 Adaptive Antenna Systems Workshop, March 11–13, Vol. 1, Naval Research Laboratory, Washington, DC, pp. 171–215.

[3] W. D. White, "Cascade Preprocessors for Adaptive Antennas," *IEEE Trans. Antennas Propag.*, Vol. AP-24, No. 5, September, 1976, pp. 670–684.

[4] W. D. White, "Adaptive Cascade Networks for Deep Nulling," *IEEE Trans. Antennas Propag.*, Vol. AP-26, No. 3, May 1978, pp. 396–402.

[5] L. E. Brennan, J. D. Mallett, I. S. Reed, "Convergence Rate in Adaptive Arrays," Technology Service Corporation Report No. TSC-PD-A177-2, July 15, 1977.

[6] R. C. Davis, "Convergence Rate in Adaptive Arrays," Technology Service Corporation Report No. TSC-PD-A177-3, October 18, 1977.

[7] L. E. Brennan and I. S. Reed, "Convergence Rate in Adaptive Arrays," Technology Service Corporation Report No. TSC-PD-177-4, January 13, 1978.

[8] W. F. Gabriel, "Adaptive Array Constraint Optimization," Program and Digest 1972 G-AP International Symposium, December 11–14, 1972; pp. 4–7.

[9] J. C. Nolen, "Synthesis of Multiple Beam Networks for Arbitrary Illuminations," Bendix Corporation, Radio Division, Baltimore, MD, April 21, 1960.

[10] P. R. Halmos, *Finite Dimensional Vector Spaces*, Princeton University Press, Princeton, NJ, 1948, p. 98.

# Chapter 9. Random Search Algorithms

Gradient-based algorithms are particularly suitable for performance measures that are either quadratic or at least unimodal. For some classes of problems [1]–[3] the mathematical relation of the variable parameters to the performance measure is either unknown or is too complex to be useful. In yet other problems, constraints may be placed on the variable parameters of the adaptive controller with the result that the performance surface may no longer be unimodal. When the performance surface of interest is multimodal and contains saddlepoints, then any gradient-based algorithm must be used with caution. One way of approaching such problems is to employ a search algorithm. Algorithms belonging to this class have global search capabilities for both unimodal and multimodal performance surfaces and work for any computable performance measure [4]–[11]. Until recently search algorithms were used only reluctantly, principally because little effort was made to use past information in guiding the search thereby resulting in slow convergence. Furthermore, search algorithms are not very powerful in unimodal applications. They do, however, have the advantage of being simple to implement in logical form, require little computation, are insensitive to discontinuities and exhibit a high degree of efficiency where little is known about the performance surface. The major types of search algorithms are systematic and random searches.

Systematic searches employ an exhaustive survey of the parameter space within the bounds of interest and therefore are easily capable of finding the global extremum of a multimodal performance measure. As a practical matter, however, this type of search can be very time consuming and incurs a high search loss since most of the search period is spent in regions of the parameter space having poor performance.

Random searches may be classified as either guided or unguided, depending on whether information is retained whenever the outcome of a trial step is learned. Furthermore, both the guided and unguided varieties of random search can be given accelerated convergence by increasing the adopted step size in a successful search direction. Three representative examples of random search algorithms that have been used for adaptive array applications are considered in this chapter: linear random search (LRS), accelerated random search (ARS), and guided accelerated random search (GARS).

## 9.1 LINEAR RANDOM SEARCH (LRS)

A random search algorithm having a single mode of operation that attempts to "learn" when a trial change in the adaptive weight vector is rejected is the "linear random search" (LRS) algorithm discussed by Widrow and McCool [12]. With this algorithm a random change $\Delta\mathbf{w}_k$ is tentatively added to the weight vector at the beginning of each interation, and the corresponding change in the performance measured is observed. A permanent weight vector change is then made that is proportional to the product of the change in the estimated performance measure and the tentative weight vector change, that is,

$$\mathbf{w}_{k+1}=\mathbf{w}_k+\mu_s\{\hat{\mathfrak{P}}[\mathbf{w}_k]-\hat{\mathfrak{P}}[\mathbf{w}_k+\Delta\mathbf{w}_k]\}\Delta\mathbf{w}_k \tag{9.1}$$

where $\mathfrak{P}[\cdot]$ denotes the selected array performance measure and $\mu_s$ is a step size constant. The random vector $\Delta\mathbf{w}_k$ has components generated from a normal probability density function having zero mean and variance $\sigma^2$. The constants $\mu_s$ and $\sigma^2$ must be selected so that stability and rate of adaptation of the algorithm are both satisfactory. The LRS algorithm is referred to as "linear" because the weight change is directly proportional to the change in the performance measure.

To analyze the LRS algorithm behavior, it is useful to introduce some definitions. The true change in the performance measure resulting from adding $\Delta\mathbf{w}_k$ to $\mathbf{w}_k$ is

$$(\Delta\mathfrak{P})_k \triangleq \mathfrak{P}[\mathbf{w}_k+\Delta\mathbf{w}_k]-\mathfrak{P}[\mathbf{w}_k] \tag{9.2}$$

When the performance measure value is estimated, then the corresponding estimated change in the performance measure is given by

$$(\Delta\hat{\mathfrak{P}})_k \triangleq \hat{\mathfrak{P}}[\mathbf{w}_k+\Delta\mathbf{w}_k]-\hat{\mathfrak{P}}[\mathbf{w}_k] \tag{9.3}$$

The error in the estimated change of the performance measure is then

$$\gamma_k \triangleq (\Delta\mathfrak{P})_k-(\Delta\hat{\mathfrak{P}})_k \tag{9.4}$$

and the variance in this error is given by

$$\begin{aligned}\operatorname{var}[\gamma_k]&=\operatorname{var}[(\Delta\hat{\mathfrak{P}})_k]\\&=\operatorname{var}\{\hat{\mathfrak{P}}[\mathbf{w}_k+\Delta\mathbf{w}_k]\}+\operatorname{var}\{\hat{\mathfrak{P}}[\mathbf{w}_k]\}\end{aligned} \tag{9.5}$$

To determine the variance of the estimate $\hat{\mathfrak{P}}[\mathbf{w}_k]$, it is necessary to consider a specific performance measure and the estimate of that measure to be employed.

To proceed with an analysis of the LRS algorithm, select the MSE as the performance measure of interest so that $\mathfrak{P}[\mathbf{w}]=\xi[\mathbf{w}]$ and let the estimate of the MSE be obtained by averaging $K$ independent samples as given by (4.98) of Section 4.3.2. With this choice of performance measure and its corresponding estimate, it follows that

$$\text{var}[\gamma_k]=\frac{2}{K}\{\xi^2[\mathbf{w}_k+\Delta\mathbf{w}_k]+\xi^2[\mathbf{w}_k]\} \tag{9.6}$$

where $K$= the number of independent samples on which the estimate $\hat{\xi}[\mathbf{w}]$ is based. In the steady state when the weight adjustment process is operating near the minimum point of the performance surface, then (9.6) is very nearly

$$\text{var}[\gamma_k]\approx\frac{4}{K}\xi_{\min}^2 \tag{9.7}$$

The tentative random changes in the weight vector produced by the LRS algorithm result in MSE perturbations. $K$ data samples are used at each iteration to obtain $\hat{\xi}[\mathbf{w}_k+\Delta\mathbf{w}_k]$. One way of decreasing the required convergence time for this algorithm is to employ parallel processors (with one processor using $\mathbf{w}_k$ and the other processor using $\mathbf{w}_k+\Delta\mathbf{w}_k$) so the samples required to obtain $\hat{\xi}[\mathbf{w}_k]$ and $\hat{\xi}[\mathbf{w}_k+\Delta\mathbf{w}_k]$ may be collected simultaneously. The value of the adaptive weight vector at the beginning of the next iteration is selected after the two $\hat{\xi}$ estimates are obtained. For any given iteration the average excess MSE resulting from the perturbation in the weight vector is given by

$$E\left\{\xi[\mathbf{w}_k]-\frac{\xi[\mathbf{w}_k]+\xi[\mathbf{w}_k+\Delta\mathbf{w}_k]}{2}\right\}=\frac{1}{2}E\{\xi[\mathbf{w}_k]-\xi[\mathbf{w}_k+\Delta\mathbf{w}_k]\} \tag{9.8}$$

The random weight perturbation vector $\Delta\mathbf{w}_k$ has zero mean and is uncorrelated with $\mathbf{w}_k$. Let $\Delta\mathbf{w}'_k$ denote the random weight perturbation vector in normal coordinates (in which the covariance matrix is diagonal so $\mathbf{R}'_{xx}=\mathbf{\Lambda}$), then $\text{cov}[\Delta\mathbf{w}_k]=\text{cov}[\Delta\mathbf{w}'_k]=\sigma^2\mathbf{I}$. Consequently, the average excess MSE can also be expressed as

$$\tfrac{1}{2}E\{\Delta\mathbf{w}_k^T\mathbf{R}_{xx}\Delta\mathbf{w}_k\}=\tfrac{1}{2}E\{\Delta\mathbf{w}'^T_k\mathbf{\Lambda}\Delta\mathbf{w}'_k\}=\tfrac{1}{2}\sigma^2\,\text{tr}(\mathbf{R}_{xx}) \tag{9.9}$$

Define the perturbation $P$ as the ratio of the average excess MSE (resulting from the random perturbations in the weight vector) to the minimum MSE, then

$$P=\frac{\sigma^2\,\text{tr}(\mathbf{R}_{xx})}{2\xi_{\min}} \tag{9.10}$$

### 9.1.1 LRS Algorithm Stability

The weight adjustment equation (9.1) can be rewritten in terms of the definitions given by (9.2), (9.3), and (9.4) for the MSE performance measure as

$$\mathbf{w}_{k+1} = \mathbf{w}_k + \mu_s\{-(\Delta\xi)_k + \gamma_k\}\Delta\mathbf{w}_k. \tag{9.11}$$

Recalling that $\mathbf{v}_k \triangleq \mathbf{w}_k - \mathbf{w}_{\text{opt}}$, then we can rewrite (9.11) as

$$\mathbf{v}_{k+1} = \mathbf{v}_k + \mu_s\{-(\Delta\xi)_k + \gamma_k\}\Delta\mathbf{w}_k \tag{9.12}$$

Specify that $\sigma^2$ be small so that $\Delta\mathbf{w}_k$ is always small, then

$$(\Delta\xi)_k = \Delta\mathbf{w}_k^T \nabla_k \tag{9.13}$$

where $\nabla_k$ is the gradient of the performance surface evaluated at $\mathbf{w}_k$. Since $\nabla_k = 2\mathbf{R}_{xx}\mathbf{v}_k$, (9.13) can be written as

$$(\Delta\xi)_k = 2\Delta\mathbf{w}_k^T \mathbf{R}_{xx}\mathbf{v}_k \tag{9.14}$$

Consequently, (9.12) can be rewritten as

$$\begin{aligned}\mathbf{v}_{k+1} &= \mathbf{v}_k + \mu_s\Delta\mathbf{w}_k\left[-2\Delta\mathbf{w}_k^T\mathbf{R}_{xx}\mathbf{v}_k + \gamma_k\right]\\ &= (\mathbf{I} - 2\mu_s\Delta\mathbf{w}_k\Delta\mathbf{w}_k^T\mathbf{R}_{xx})\mathbf{v}_k + \mu_s\gamma_k\Delta\mathbf{w}_k\end{aligned} \tag{9.15}$$

The alternate (but equivalent) form of (9.1) represented by (9.15) is more useful for analysis even though the algorithm is implemented in the form suggested by (9.1). Equation (9.15) emphasizes that the adaptive weight vector can be regarded as the solution of a first-order linear vector difference equation having a randomly varying coefficient $\mathbf{I} - 2\mu_s\Delta\mathbf{w}_k\Delta\mathbf{w}_k^T\mathbf{R}_{xx}$ and a random driving function $\mu_s\gamma_k\Delta\mathbf{w}_k$.

It is more convenient to deal with the foregoing linear vector difference equation in normal coordinates by premultiplying both sides of (9.15) by the transformation matrix $\mathbf{Q}$ of Section 4.1.3 to obtain

$$\mathbf{v}'_{k+1} = (\mathbf{I} - 2\mu_s\Delta\mathbf{w}'_k\Delta\mathbf{w}'^T_k\Lambda)\mathbf{v}'_k + \mu_s\gamma_k\Delta\mathbf{w}'_k \tag{9.16}$$

Although (9.16) is somewhat simpler than (9.15), the matrix coefficient of $\mathbf{v}'_k$ still contains cross-coupling and randomness, thereby rendering (9.16) a difficult equation to solve. Stability conditions for the LRS algorithm will therefore be obtained without an explicit solution to (9.16) by considering the behavior of the adaptive weight vector mean.

Taking the expected value of both sides of (9.16) and recognizing that $\Delta \mathbf{w}'_k$ is a random vector that is uncorrelated with $\gamma_k$ and $\mathbf{v}'_k$, we find that

$$\begin{aligned} E\{\mathbf{v}'_{k+1}\} &= E\{(\mathbf{I}-2\mu_s \Delta \mathbf{w}'_k \Delta \mathbf{w}'^T_k \boldsymbol{\Lambda})\mathbf{v}'_k\} + \mu_s E\{\gamma_k \Delta \mathbf{w}'_k\} \\ &= (\mathbf{I}-2\mu_s E\{\Delta \mathbf{w}'_k \Delta \mathbf{w}'^T_k\}\boldsymbol{\Lambda})E\{\mathbf{v}'_k\}+0 \\ &= (\mathbf{I}-2\mu_s \sigma^2 \boldsymbol{\Lambda})E\{\mathbf{v}'_k\} \end{aligned} \tag{9.17}$$

The solution to (9.17) is given by [12]

$$E\{\mathbf{v}'_k\} = (\mathbf{I}-2\mu_s\sigma^2\boldsymbol{\Lambda})^k \mathbf{v}'_0 \tag{9.18}$$

For the initial conditions $\mathbf{v}'_0$, (9.18) gives the expected value of the weight vector's transient response. If (9.18) is stable, then the mean of $\mathbf{v}'_k$ must converge. The stability condition for (9.18) is

$$\frac{1}{\lambda_{\max}} > \mu_s \sigma^2 > 0 \tag{9.19}$$

If we choose $\mu_s \sigma^2$ to satisfy (9.19), it then follows that

$$\lim_{k\to\infty} E\{\mathbf{v}'_k\} = 0 \tag{9.20}$$

Since the foregoing transient behavior is analogous to that of the method of steepest descent discussed in Section 4.1.2, it can be argued by analogy that the time constant of the $p$th mode of the expected value of the weight vector is given by

$$\tau_p = \frac{1}{2\mu_s \sigma^2 \lambda_p} \tag{9.21}$$

Furthermore the time constant of the $p$th mode of the MSE learning curve is one-half the above value so that

$$\tau_{p_{\text{mse}}} = \frac{1}{4\mu_s \sigma^2 \lambda_p} \tag{9.22}$$

Satisfying the stability condition (9.19) only implies that the mean of the adaptive weight vector will converge according to (9.20); variations in the weight vector about the mean value may be quite severe, however. It is therefore of interest to obtain an indication of the severity of variations in the weight vector when using the LRS algorithm by deriving an expression for the covariance of the weight vector. In obtaining such an expression, it will simply be assumed that the weight vector covariance is bounded and that the weight vector behaves as a stationary stochastic process after initial transients have died out. Assuming a bounded steady-state covariance matrix exists, we may calculate an expression for such a covariance by multiplying both sides of (9.16) by their respective transposes to obtain

$$\begin{aligned}\mathbf{v}'_{k+1}\mathbf{v}'^{T}_{k+1}=&(\mathbf{I}-2\mu_s\,\Delta\mathbf{w}'_k\Delta\mathbf{w}'^{T}_k\boldsymbol{\Lambda})\mathbf{v}'_k\mathbf{v}'^{T}_k(\mathbf{I}-2\mu_s\boldsymbol{\Lambda}\,\Delta\mathbf{w}'_k\,\Delta\mathbf{w}'^{T}_k)\\&+\mu_s^2\gamma_k^2\,\Delta\mathbf{w}'_k\,\Delta\mathbf{w}'^{T}_k\\&+(I-2\mu_s\,\Delta\mathbf{w}'_k\,\Delta\mathbf{w}'^{T}_k\boldsymbol{\Lambda})\mathbf{v}'_k\mu_s\gamma_k\,\Delta\mathbf{w}'^{T}_k\\&+\mu_s\gamma_k\,\Delta\mathbf{w}'_k\,\mathbf{v}'^{T}_k(\mathbf{I}-2\mu_s\boldsymbol{\Lambda}\,\Delta\mathbf{w}'_k\,\Delta\mathbf{w}'^{T}_k)\end{aligned}\tag{9.23}$$

Now we take expected values of both sides of (9.23) recalling that $\gamma_k$ and $\Delta\mathbf{w}'_k$ are zero mean uncorrelated stationary processes so that

$$\begin{aligned}E\{\mathbf{v}'_{k+1}\mathbf{v}'^{T}_{k+1}\}=&E\{(\mathbf{I}-2\mu_s\,\Delta\mathbf{w}'_k\,\Delta\mathbf{w}'^{T}_k\boldsymbol{\Lambda})\mathbf{v}'_k\mathbf{v}'^{T}_k\\&\cdot(\mathbf{I}-2\mu_s\boldsymbol{\Lambda}\,\Delta\mathbf{w}'_k\,\Delta\mathbf{w}'^{T}_k)\}\\&+\mu_s^2E\{\gamma_k^2\}E\{\Delta\mathbf{w}'_k\,\Delta\mathbf{w}'^{T}_k\}+0\end{aligned}\tag{9.24}$$

Since $\operatorname{var}[\gamma_k]\cong(4/K)\xi^2_{\min}$ and $\operatorname{cov}[\Delta\mathbf{w}'_k]=\sigma^2\mathbf{I}$, it follows that (9.24) can be expressed as

$$\begin{aligned}E\{\mathbf{v}'_{k+1}\mathbf{v}'^{T}_{k+1}\}=&E\{(\mathbf{I}-2\mu_s\,\Delta\mathbf{w}'_k\,\Delta\mathbf{w}'^{T}_k\boldsymbol{\Lambda})\\&\cdot\mathbf{v}'_k\mathbf{v}'^{T}_k(\mathbf{I}-2\mu_s\boldsymbol{\Lambda}\,\Delta\mathbf{w}'_k\,\Delta\mathbf{w}'^{T}_k)\}+\mu_s^2\frac{4}{K}\xi^2_{\min}\sigma^2\mathbf{I}\end{aligned}\tag{9.25}$$

In the steady state $\mathbf{v}'_k$ is also a zero mean stationary random process that is uncorrelated with $\Delta\mathbf{w}'_k$ so (9.25) can be written as

$$\begin{aligned}E\{\mathbf{v}'_{k+1}\mathbf{v}'^{T}_{k+1}\}=&E\{(\mathbf{I}-2\mu_s\,\Delta\mathbf{w}'_k\,\Delta\mathbf{w}'^{T}_k\boldsymbol{\Lambda})\\&\cdot E[\mathbf{v}'_k\mathbf{v}'^{T}_k](\mathbf{I}-2\mu_s\boldsymbol{\Lambda}\,\Delta\mathbf{w}'_k\,\Delta\mathbf{w}'^{T}_k)\}\\&+\mu_s^2\frac{4}{K}\xi^2_{\min}\sigma^2\mathbf{I}\end{aligned}\tag{9.26}$$

Consequently the steady state covariance of the adaptive weight vector is

$$\begin{aligned}\operatorname{cov}[\mathbf{v}'_k]=&E\{(\mathbf{I}-2\mu_s\,\Delta\mathbf{w}'_k\,\Delta\mathbf{w}'^{T}_k\boldsymbol{\Lambda})\\&\cdot\operatorname{cov}[\mathbf{v}'_k](\mathbf{I}-2\mu_s\boldsymbol{\Lambda}\,\Delta\mathbf{w}'_k\,\Delta\mathbf{w}'^{T}_k)\}+\mu_s^2\frac{4}{N}\xi^2_{\min}\sigma^2\mathbf{I}\\=&\operatorname{cov}[\mathbf{v}'_k]-2\mu_sE\{\Delta\mathbf{w}'_k\,\Delta\mathbf{w}'^{T}_k\}\boldsymbol{\Lambda}\operatorname{cov}[\mathbf{v}'_k]\\&-2\mu_s\operatorname{cov}[\mathbf{v}'_k]\boldsymbol{\Lambda}E\{\Delta\mathbf{w}'_k\,\Delta\mathbf{w}'^{T}_k\}\\&+4\mu_s^2E\{\Delta\mathbf{w}'_k\,\Delta\mathbf{w}'^{T}_k\boldsymbol{\Lambda}\operatorname{cov}[\mathbf{v}'_k]\boldsymbol{\Lambda}\,\Delta\mathbf{w}'_k\,\Delta\mathbf{w}'^{T}_k\}+\mu_s^2\frac{4}{N}\xi^2_{\min}\sigma^2\mathbf{I}\\=&\operatorname{cov}[\mathbf{v}'_k]-2\mu_s\sigma^2\boldsymbol{\Lambda}\operatorname{cov}[\mathbf{v}'_k]-2\mu_s\sigma^2\operatorname{cov}[\mathbf{v}'_k]\boldsymbol{\Lambda}\\&+4\mu_s^2E\{\Delta\mathbf{w}'_k\,\Delta\mathbf{w}'^{T}_k\boldsymbol{\Lambda}\operatorname{cov}[\mathbf{v}'_k]\boldsymbol{\Lambda}\,\Delta\mathbf{w}'_k\,\Delta\mathbf{w}'^{T}_k\}+\mu_s^2\frac{4}{K}\xi^2_{\min}\sigma^2\mathbf{I}\end{aligned}\tag{9.27}$$

Equation (9.27) is not easily solved to yield the covariance of $\mathbf{v}'_k$ because the matrices appearing in the equation cannot be factored. It can be plausibly argued (although not proven) that the steady state covariance matrix of $\mathbf{v}'_k$ should be diagonal. The results that have been obtained with such a simplifying assumption do, however, indicate that there is some merit in the plausibility argument.

The random driving function appearing in (9.16) consists of components that are uncorrelated with each other and uncorrelated over time. Furthermore the random coefficient $\mathbf{I}-2\mu_s \Delta\mathbf{w}'_k \Delta\mathbf{w}'^T_k \mathbf{\Lambda}$ is diagonal on the average (though generally not for every value of $k$), and uncorrelated both with $\mathbf{v}'_k$ and with itself over time. Consequently it is plausible that the covariance of $\mathbf{v}'_k$ is a diagonal matrix.

Assume that $\text{cov}[\mathbf{v}'_k]$ is in fact diagonal, then (9.27) can immediately be rewritten by merely rearranging terms as

$$4\mu_s\sigma^2\mathbf{\Lambda}\,\text{cov}[\mathbf{v}'_k]-4\mu_s^2 E\{\Delta\mathbf{w}'_k\Delta\mathbf{w}'^T_k\mathbf{\Lambda}\,\text{cov}[\mathbf{v}'_k]\mathbf{\Lambda}\Delta\mathbf{w}'_k\Delta\mathbf{w}'^T_k\}=\mu_s^2\frac{4}{K}\xi_{\min}^2\sigma^2\mathbf{I} \tag{9.28}$$

The case of greatest interest is that for which adaptation is slow, in which case

$$\mu_s\sigma^2\mathbf{\Lambda}\ll\mathbf{I} \tag{9.29}$$

Furthermore it may be noted that

$$\mu_s^2 E\{\Delta\mathbf{w}'_k\Delta\mathbf{w}'^T_k\mathbf{\Lambda}\,\text{cov}[\mathbf{v}'_k]\mathbf{\Lambda}\Delta\mathbf{w}'_k\Delta\mathbf{w}'^T_k\}\cong(\mu_s\sigma^2\mathbf{\Lambda})^2\,\text{cov}[\mathbf{v}'_k] \tag{9.30}$$

and from (9.29) it follows that

$$(\mu_s\sigma^2\mathbf{\Lambda})^2\,\text{cov}[\mathbf{v}'_k]\ll\mu_s\sigma^2\mathbf{\Lambda}\,\text{cov}[\mathbf{v}'_k] \tag{9.31}$$

With the result of (9.31) it follows that the term $-4\mu_s^2E\{\ \}$ appearing in (9.28) can be neglected. Consequently (9.28) can be rewritten as

$$\text{cov}[\mathbf{v}'_k]=\frac{\mu_s}{K}\xi_{\min}^2\mathbf{\Lambda}^{-1} \tag{9.32}$$

The steady state covariance matrix of $\mathbf{v}'_k$ given by (9.32) is based on a plausible assumption, but experience indicates that the predicted misadjustment obtained using this quantity generally yields accurate results [12].

The misadjustment experienced using the LRS algorithm can be obtained by considering the average excess MSE due to noise in the weight vector which is given by

$$E\{\mathbf{v}'^T_k\mathbf{\Lambda}\mathbf{v}'_k\}=\sum_{p=1}^{N}\lambda_p E\{(v'_{p_k})^2\} \tag{9.33}$$

where $N$ is the number of eigenvalues of $\mathbf{\Lambda}$. If we use (9.32), it follows that for the LRS algorithm,

$$E\left[\mathbf{v}_k'^T \mathbf{\Lambda} \mathbf{v}_k'\right] = \sum_{p=1}^{N} \lambda_p \left( \frac{\mu_s}{K} \xi_{\min}^2 \frac{1}{\lambda_p} \right) = \frac{N\mu_s}{K} \xi_{\min}^2 \tag{9.34}$$

Since the misadjustment $M$ is defined to be the average excess MSE divided by the minimum MSE,

$$M \triangleq \frac{E\left\{\mathbf{v}_k'^T \mathbf{\Lambda} \mathbf{v}_k'\right\}}{\xi_{\min}} \tag{9.35}$$

It follows that for the LRS algorithm,

$$M = \frac{N\mu_s}{K} \xi_{\min} \tag{9.36}$$

The result given by (9.36) can be usefully expressed in terms of the perturbation of the LRS process as

$$M = \frac{N\mu_s \sigma^2 \operatorname{tr}(\mathbf{R}_{xx})}{2KP} = \frac{N^2 \mu_s \sigma^2 \lambda_{\text{av}}}{2KP} \tag{9.37}$$

Now recall that the time constant of the $p$th mode of the learning curve for the LRS algorithm (in terms of the number of iterations required) is given by (9.21). Since one iteration of the weight vector requires two estimates of $\hat{\xi}$ to be obtained, $2K$ samples of data are used per iteration, and the learning curve time constant expressed in terms of the number of data samples is

$$T_{p_{\text{mse}}} \triangleq 2K\tau_{p_{\text{mse}}} = \frac{K}{2\mu_s \sigma^2 \lambda_p} \tag{9.38}$$

From (9.38) it follows immediately that

$$\lambda_p = \frac{K}{2\mu_s \sigma^2} \left( \frac{1}{T_{p_{\text{mse}}}} \right) \tag{9.39}$$

and

$$\lambda_{\text{av}} = \frac{K}{2\mu_s \sigma^2} \left( \frac{1}{T_{p_{\text{mse}}}} \right)_{\text{av}} \tag{9.40}$$

Substituting (9.40) into (9.37) then yields

$$M = \frac{N^2}{4P} \left( \frac{1}{T_{p_{\text{mse}}}} \right)_{\text{av}} \tag{9.41}$$

Since the total misadjustment consists of a stochastic component $M$ and a deterministic component $P$ from (9.41) we may write

$$M_{\text{tot}} = \frac{N^2}{4P}\left(\frac{1}{T_{p_{\text{mse}}}}\right)_{\text{av}} + P \tag{9.42}$$

If the deterministic component of the total misadjustment is optimally chosen, then both $M$ and $P$ are equal and $P$ is one-half the total misadjustment so that

$$(M_{\text{tot}})_{\min} = \frac{N^2}{2P_{\text{opt}}}\left(\frac{1}{T_{p_{\text{mse}}}}\right)_{\text{av}} = N\left[\left(\frac{1}{T_{p_{\text{mse}}}}\right)_{\text{av}}\right]^{1/2} \tag{9.43}$$

It is informative to compare this result with the corresponding result (4.83) for the LMS algorithm.

## 9.2 ACCELERATED RANDOM SEARCH (ARS)

Suppose a performance measure $\mathfrak{P}(\mathbf{w})$ has been adopted and it is desired to minimize this performance measure over some range of values for the complex weight vector $\mathbf{w}$. Let the adaptive processor change the complex weight vector elements using a simplified version of accelerated random search reported by Baird and Rassweiler [13] as follows:

$$\mathbf{w}(k+1) = \mathbf{w}(k) + \mu_s(k)[\Delta\mathbf{w}(k)] \tag{9.44}$$

where $\mu_s(k)$ is the step size initially set at $\mu_s(0) = \mu_0$ and $\Delta\mathbf{w}(k)$ is a random vector whose components are given by

$$\Delta w_i(k) = \cos\theta_i + j\sin\theta_i, \qquad i = 1, 2, \ldots, m \tag{9.45}$$

where $\theta_i$ is a uniformly distributed random angle on the interval $\{0, 2\pi\}$ so that $|\Delta w_i(k)| = 1$ and $\Delta\mathbf{w}(k)$ controls the direction of the weight vector change while $\mu_s$ controls the magnitude.

Initially the weight vector is set to $\mathbf{w}(0)$, and the corresponding performance measure $\mathfrak{P}[\mathbf{w}(0)]$, or an estimate thereof $\hat{\mathfrak{P}}[\mathbf{w}(0)]$, is evaluated. The weight vector is then changed in accordance with (9.44) using $\mu_s(0) = \mu_0$. The performance index $\mathfrak{P}[\mathbf{w}(1)]$ is evaluated and compared with $\mathfrak{P}[\mathbf{w}(0)]$. If this comparison indicates improved performance, then the weight direction change vector $\Delta\mathbf{w}$ is retained, and the step size $\mu_s$ is doubled (resulting in "accelerated" convergence). If, however, the resulting performance is not improved, then the previous value of $\mathbf{w}$ is retained as the starting point, a new value of $\Delta\mathbf{w}$ is selected, and $\mu_s$ is reset to $\mu_0$. As a consequence of always returning to the previous value of $\mathbf{w}$ as the starting point for a new weight perturbation in the event the performance measure is not improved, the ARS

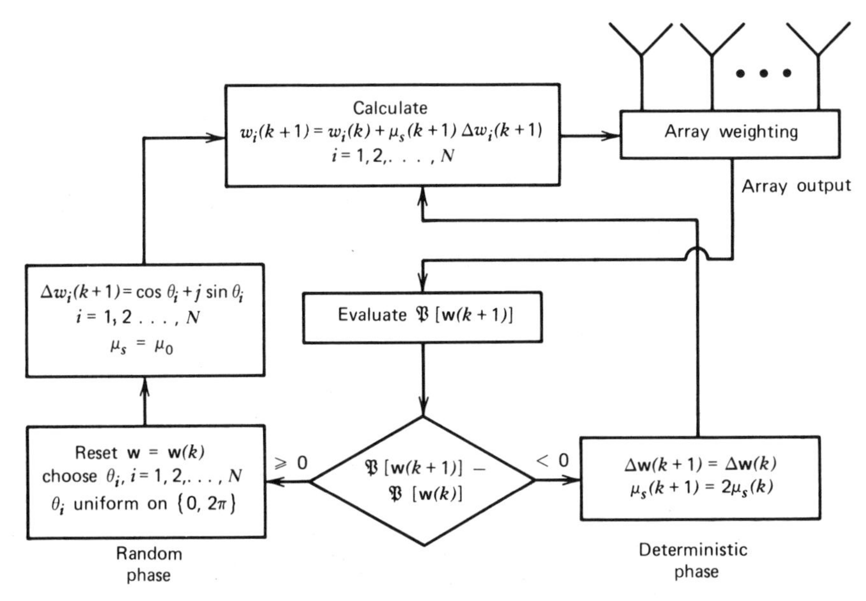


**Figure 9.1** Block diagram for ARS algorithm.

approach is inherently stable, and stability considerations do not play a role in step size selection. A block diagram of this simplified version of accelerated random search is given in Figure 9.1.

Consider a single component of the complex weight vector for which $v_i = w_i - w_{\text{opt}_i}$. If $v_i(k)$ lies within $\mu_0/2$ of $w_{\text{opt}}$, then any further perturbation in that component of the weight vector of step size $\mu_0$ will result in $v_i(k+1) \geqslant v_i(k)$ as shown in Figure 9.2. Consequently if all components of the weight vector lie within or on the best performance surface contour contained within the circle of radius $\mu_0/2$ about $w_{\text{opt}}$, then no further improvement in the performance measure can possibly occur using step size $\mu_0$. The condition where all weight vector components lie within this best performance surface contour therefore represents a lower limit on the possible improvement that can be achieved with the ARS procedure, and this is the ultimate condition to which the weight vector is driven in the steady state.

To simplify the development, assume it is equally likely that the weight vector component $w_i(k)$ will lie anywhere within or on the circle for which $v_i = \mu_0/2$, it then follows that the steady-state expected value of $v_i$ is $v_{ss}(k) = 0$ and the average excess MSE in this steady-state condition is (adopting $\mathfrak{P} = \xi$ and noting that $E\{|v_i|^2\} = \mu_0^2/8$)

$$E\{\mathbf{v}'^T(k)\boldsymbol{\Lambda}\mathbf{v}'(k)\} = \frac{\mu_0^2}{8}\xi_{\min}\,\text{tr}(\mathbf{R}_{xx}) \tag{9.46}$$

The average misadjustment for this steady-state condition is therefore

$$M_{\text{av}} = \frac{\mu_0^2}{8}\text{tr}(\mathbf{R}_{xx}) \tag{9.47}$$

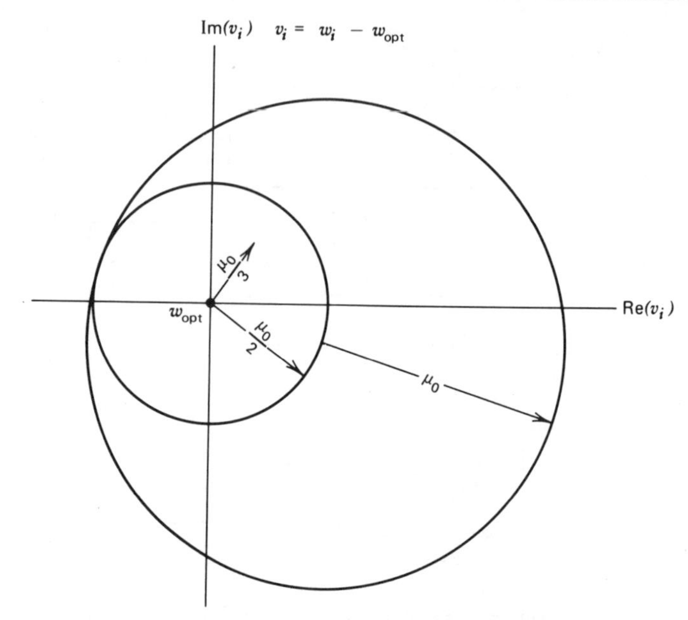


**Figure 9.2** Complex perturbation of step size $\mu_0$ for a single weight vector component.

On each successive iteration the weight vector components are perturbed by $\mu_0$ from their steady-state values. Assume that the perturbation is taken from $v_{ss} = \rho$ as shown in Figure 9.3, then $E\{v_p\} = \mu_0, E\{|v_p|^2\} = 9\mu_0^2/8$, and the average total misadjustment for the random search perturbation is therefore

$$M_{\text{tot}} = \frac{9\mu_0^2}{8}\text{tr}(\mathbf{R}_{xx}) \tag{9.48}$$

From the foregoing discussion, it follows that in the steady state as long as a correct decision is made concerning $\xi[\mathbf{w}(k+1)] - \xi[\mathbf{w}(k)]$ the average total misadjustment is given by (9.48).

In practice the ARS algorithm examines the statistic $\hat{\mathfrak{P}}[\mathbf{w}(k+1)] - \hat{\mathfrak{P}}[\mathbf{w}(k)]$ instead of $\mathfrak{P}[\mathbf{w}(k+1)] - \mathfrak{P}[\mathbf{w}(k)]$, and the measured statistic contains noise that may yield a misleading indication of the performance measure difference. It is therefore essential to guarantee that the performance measure difference due to the weight vector perturbation is significantly larger than the standard deviation of the error in the estimated change in the performance measure: this may be done by selecting $\Delta\mathfrak{P} = \mathfrak{P}[\mathbf{w}(k+1)] - \mathfrak{P}[\mathbf{w}(k)] > \sigma_\gamma$ where $\sigma_\gamma^2$ is given by (9.7). Having selected $K$ and $\mu_s$ so that $\Delta\mathfrak{P} > \sigma_\gamma$, we find that the average steady-state misadjustment is approximated by (9.48). Furthermore, it is desirable that the performance measure difference due to the weight vector perturbation be less than $E\{\mathfrak{P}[\mathbf{w}(k)]\}$. Therefore for $\mathfrak{P}$ selected

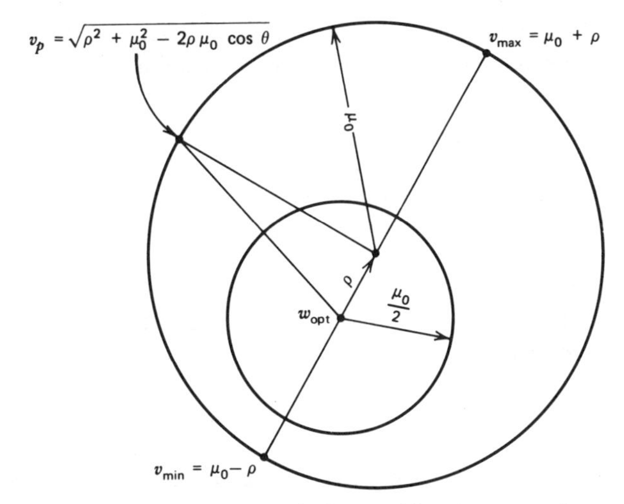


**Figure 9.3** Weight vector component perturbation resulting from step size $\mu_0$ starting from $v_{ss} = \mu_0/3$.

to be $\xi$, in the steady state the constants $K$ and $\mu_s$ should be selected to satisfy

$$\xi_{min} > \tfrac{9}{8}\mu_0^2 \operatorname{tr}(\mathbf{R}_{xx}) > \frac{2\xi_{min}}{\sqrt{K}} \tag{9.49}$$

Even with $K$ and $\mu_s$ selected to satisfy $\Delta\mathfrak{P} > \sigma_\gamma$, it is still possible for noise present in the measurements of system performance to produce deceptively good results for any one experiment. Such spurious results will on the average be corrected on successive trials.

## 9.3 GUIDED ACCELERATED RANDOM SEARCH (GARS)

The GARS introduced by Barron [14]–[16] consists of two phases: a random phase and a deterministic phase. Control of the parameter space search is passed back and forth between these two phases as the search finds parameter space regions having better or worse performance. In the initial random phase (which is an information gathering phase), the parameter values in the adaptive weight vector are perturbed randomly according to a multivariate probability density function *pdf*. Once a direction is found in which performance improvement occurs, the deterministic phase is entered, and the information acquired from the initial random phase is exploited as larger step sizes are taken in the direction of improved performance. Whenever an accelerated step in the deterministic phase produces an unsuccessful outcome, the random phase is reentered, but now the *pdf* governing the random search assigns smaller excursions to the parameter values than occurred in the initial random phase.

A block diagram representation of a simplified version of GARS is given in Figure 9.4. Starting with an initial weight vector $\mathbf{w}_0$ a corresponding performance measure $\mathfrak{P}[\mathbf{w}_0]$, is evaluated. At all times the minimum value of the performance measure attained is stored and denoted by $\mathfrak{P}^*$ so initially $\mathfrak{P}^* = \mathfrak{P}[\mathbf{w}_0]$. The GARS algorithm begins in its random phase by generating a random weight vector perturbation $\Delta\mathbf{w}$ each of whose elements are drawn from a normal probability density function having zero mean and a variance $\sigma^2$. The variance is selected according to

$$\sigma^2 = K_1 + K_2\mathfrak{P}^* \tag{9.50}$$

where $K_1$ and $K_2$ are design constants for the GARS algorithm selected so the

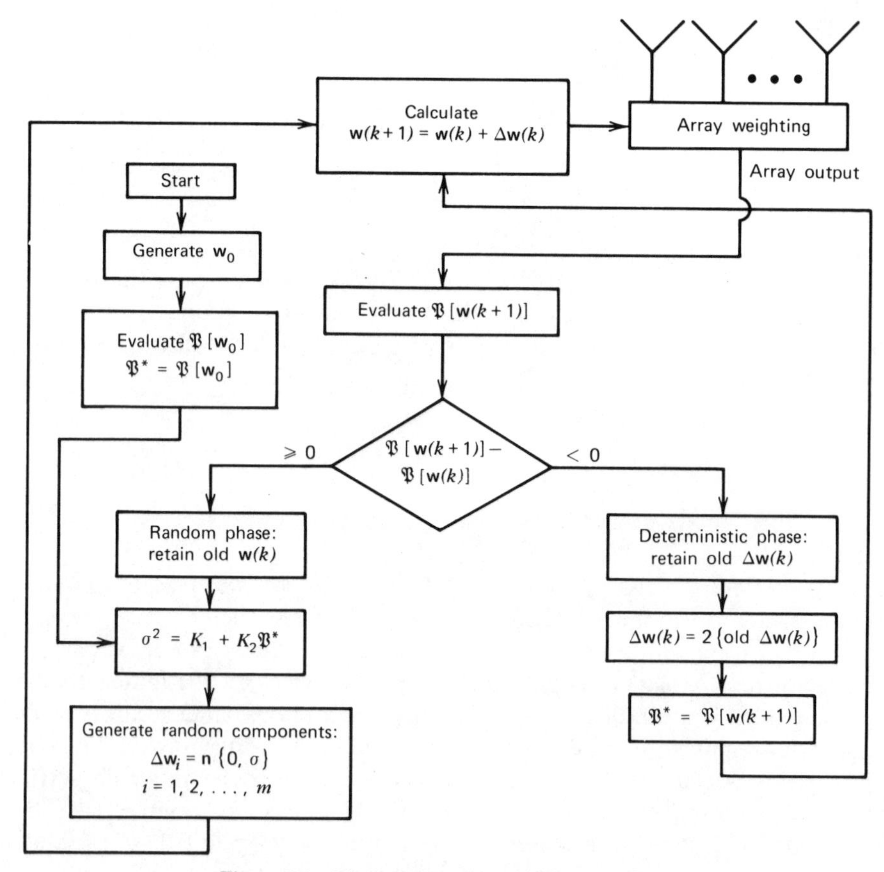


**Figure 9.4** Block diagram for GARS algorithm.

resulting algorithm step size is small enough when the optimum performance is realized, yet big enough to gain useful performance surface information when the trial weight vector is far from optimum.

The next trial adaptive weight vector is then computed using

$$\mathbf{w}(k+1)=\mathbf{w}(k)+\Delta\mathbf{w}(k) \tag{9.51}$$

and the corresponding performance is evaluated $\mathfrak{P}[\mathbf{w}(k+1)]$. If no improvement in the performance measure is realized, the algorithm remains in the random phase for the next trial weight vector, returning to the previous value of $\mathbf{w}$ as the starting point for the next weight perturbation. Once a direction in which to move for improved performance is determined, the deterministic phase of the algorithm is entered, and convergence is accelerated by continuing to travel in the direction of improved performance with twice the previous step size. The weight vector step $\Delta\mathbf{w}$ is continually doubled as long as performance measure improvements are realized. Once the performance measure begins to degrade, the search is returned to the random phase where the adaptive weight vector perturbations $\Delta\mathbf{w}$ will now be considerably smaller than before due to the smaller value of $\sigma$ used in generating new search directions.

From the foregoing description of the simplified version of GARS, it is seen that the principal difference between GARS and ARS lies in how the random phase of the search is conducted. Not only is the search direction random (as it was before), but the step size is also random and governed by the parameter $\sigma$ whose assigned value depends on the minimum value that the selected performance measure has attained. Consequently the search step size tends to be reduced as performance measure improvements are realized. The observations made in the previous section for the ARS algorithm with minimum step size $\mu_0$ now apply in a statistical sense to the GARS algorithm. When $\mathfrak{P}$ is selected to be $\xi$ the condition expressed by (9.49) should also be satisfied where now the expected change in performance measure due to weight perturbations is given by

$$E\{\Delta\xi\}=\sigma^2\,\mathrm{tr}(\mathbf{R}_{xx}) \tag{9.52}$$

Noise in the measurements of the performance measure can produce deceptively good results for any one experiment, thereby creating the risk of spurious measurements locking the search to a false solution. To avoid the risk of locking onto a false solution, it is only necessary to provide for the random phase of the algorithm to periodically reexamine the performance measure achieved at the supposed best-to-date setting of the weight vector. Furthermore, to provide the capability to handle nonstationary operating conditions, periodically require the algorithm to use large step sizes and conduct the exploration perturbations uniformly throughout the parameter space.

The general GARS algorithm incorporates features that will not be simulated here. The most important such feature is the provision for a long-term memory that is obtained by employing a nonuniform multivariate probability distribution function [17] to generate search directions and thereby guide the search to increase the probability that future trials will yield better performance scores than past trials. This multivariate *pdf* is shaped according to the results of a series of initial trials conducted during the opening stage of the search when no preference is given to any search direction. During the middle stage of the search the multivariate *pdf* formed during the opening stage guides the search by generating new search directions. In the final search stage the dimensionality of the parameter space search can be reduced by converting from a simultaneous search involving all the parameters to a nearly sequential search involving only a small fraction of the parameters at any step. This selected fraction of the parameters to search is chosen randomly for each new iteration.

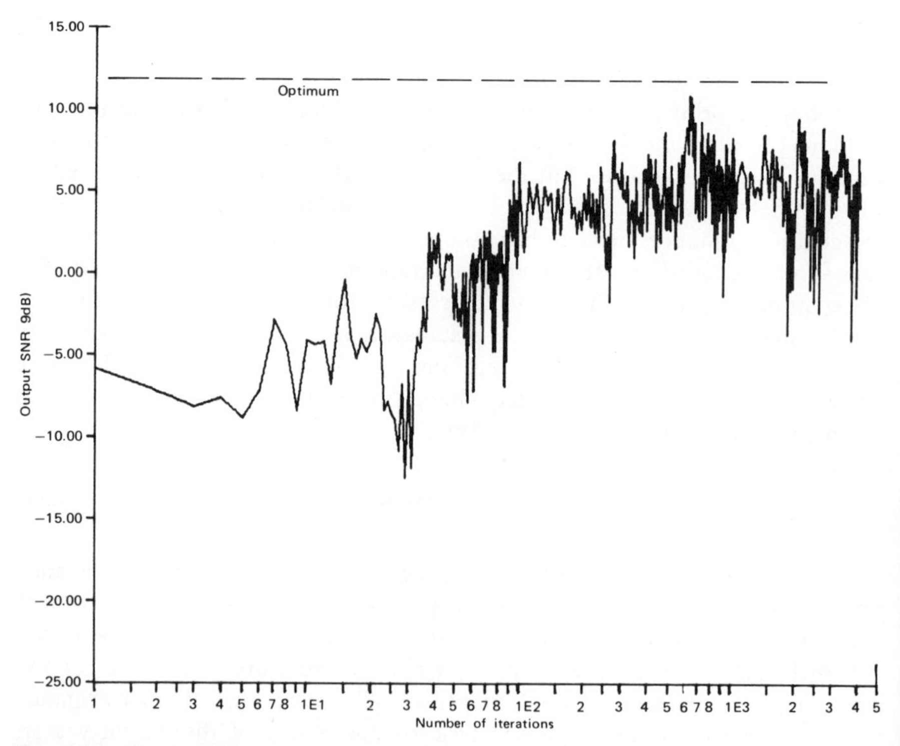


**Figure 9.5** Output SNR versus number of iterations for the LRS algorithm with eigenvalue spread = 153.1.

## 9.4 SIMULATION RESULTS

A four-element array was simulated having the array configuration and signal environment represented in Figure 7.4. The random search algorithms have convergence speeds that are much slower than that for the LMS algorithm when the MSE performance measure is used, and the selected three-jammer scenario was simulated under two different jamming conditions: one condition having a moderate eigenvalue spread of $\lambda_{max}/\lambda_{min}=153.1$ and another condition with a more severe eigenvalue spread of $\lambda_{max}/\lambda_{min}=2440$. When we select jammer-to-thermal noise ratios of $J_1/n=25$, $J_2/n=4$, $J_3/n=20$ and a signal-to-thermal noise ratio of $s/n=10$, the corresponding eigenvalues are $\lambda_1=153.1$, $\lambda_2=42.6$, $\lambda_3=3.34$, and $\lambda_4=1$ for which $\text{SNR}_{opt}=15.9$ (12 dB). Figures 9.5–9.8 give typical convergence results for the case where $\lambda_{max}/\lambda_{min}=153.1$. Likewise, when we select jammer-to-thermal noise ratios of $J_1/n=500$, $J_2/n=40, J_3/n=200$ and a signal-to-thermal noise ratio of $s/n=10$, the

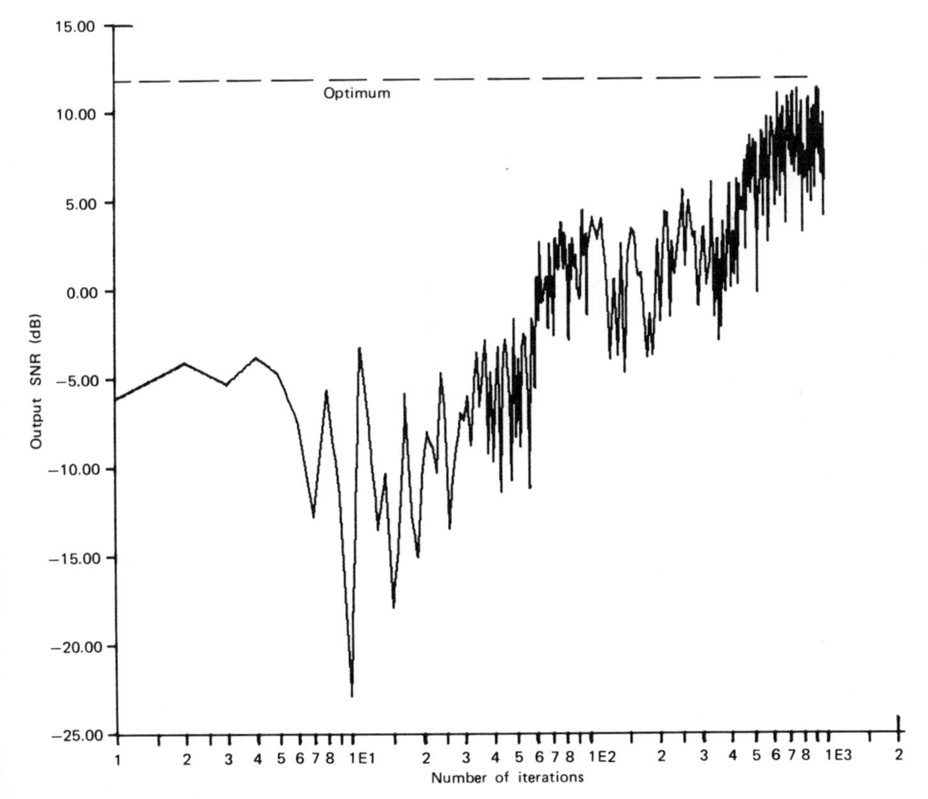


**Figure 9.6** Output SNR versus number of iterations for the ARS algorithm with eigenvalue spread = 153.1.

corresponding eigenvalues are then $\lambda_1 = 2440$, $\lambda_2 = 494$, $\lambda_3 = 25.6$, and $\lambda_4 = 1$ for which $SNR_{opt} = 15.08$ (11.8 dB). Figures 9.9–9.12 then give typical convergence results for the case where $\lambda_{max}/\lambda_{min} = 2440$.

In simulating the ARS algorithm, it was found that the weight adjustment scheme depicted in Figure 9.1 should be modified in order to obtain satisfactory results. The difficulty with the scheme of Figure 9.1 lies in the fact that the farther away from the solution $\mathbf{w}_{opt}$ that $\mathbf{w}(k+1)$ lies, the greater is the variance in the estimate $\hat{\xi}[\mathbf{w}(k+1)]$. Consequently if the step size $\mu_0$ is selected to obtain an acceptable steady-state error in the neighborhood of $\mathbf{w}_{opt}$, it may well be that the changes in $\xi[\mathbf{w}(k+1)]$ occurring as a consequence of the perturbation $\Delta\mathbf{w}(k+1)$ are overwhelmed by the random fluctuations experienced in $\hat{\xi}[\mathbf{w}(k+1)]$ when $\mathbf{w}(k+1)$ is far removed from $\mathbf{w}_{opt}$. When this situation occurs, the adjustment algorithm will merely yield a succession of weights that slowly meander aimlessly with step size $\mu_0$.

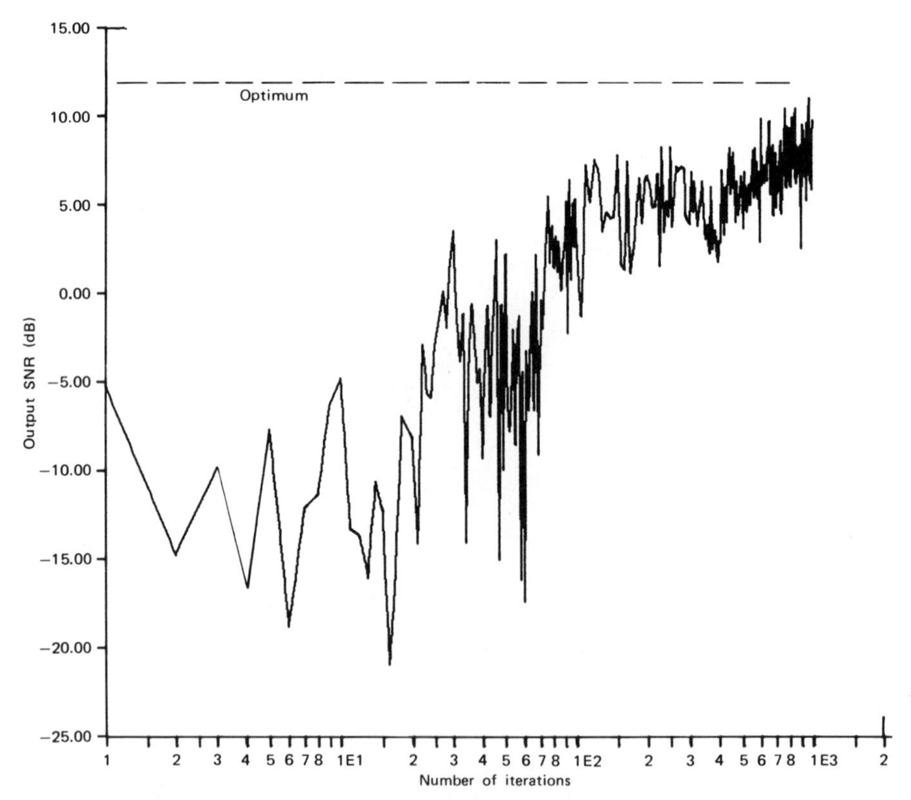


**Figure 9.7** Output SNR versus number of iterations for the GARS algorithm with eigenvalue spread = 153.1.

What is required to correct the foregoing situation is to recognize that the step size $\mu_0$ should be selected to reflect the changes in the variance of $\hat{\xi}[\mathbf{w}(k+1)]$ that occur when $\mathbf{w}(k+1)$ is well removed from $\mathbf{w}_{\text{opt}}$. This correction can be accomplished by incorporating a step size $\mu_s = \sqrt{K_1 + K_2\mathfrak{P}^*}$ into the ARS algorithm in accordance with the philosophy expressed by the GARS algorithm in Figure 9.4. Of course it would be preferable to use $\mu_s = \sqrt{K_1 + K_2(\mathfrak{P}^* - \mathfrak{P}_{\min})}$, but in general $\mathfrak{P}_{\min}$ is unknown.

The LRS, ARS, and GARS algorithms were all simulated using $K=90$ to obtain the estimate of MSE, which was the performance measure used in all cases. To satisfy the condition imposed by (9.49), the GARS algorithm was simulated using

$$\sigma^2 \operatorname{tr}(\mathbf{R}_{xx}) = \mathfrak{k}_1 + \mathfrak{k}_2\xi^* \tag{9.53}$$

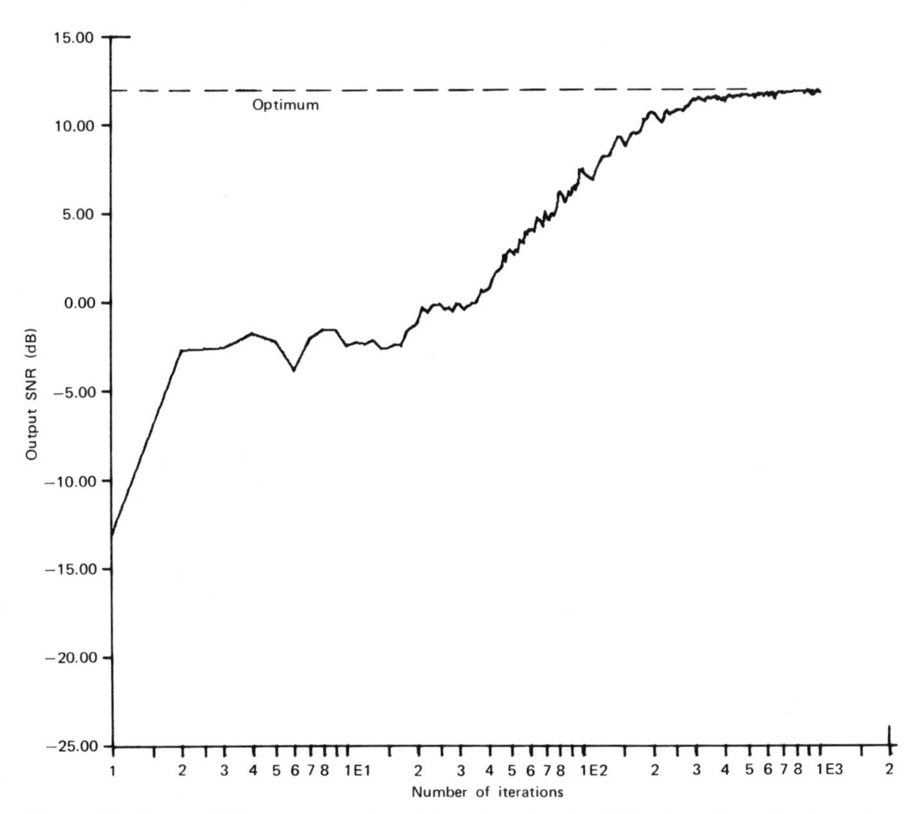


**Figure 9.8** Output SNR versus number of iterations for the LMS algorithm with eigenvalue spread = 153.1.

where $\mathfrak{k}_1 = \frac{1}{160}$ and $\mathfrak{k}_2 = 0.1$. Likewise the ARS algorithm was simulated using

$$\mu_s^2 \operatorname{tr}(\mathbf{R}_{xx}) = \mathfrak{k}_1 + \mathfrak{k}_2 \xi^* \tag{9.54}$$

with $\mathfrak{k}_1$ and $\mathfrak{k}_2$ assigned the same values as for the GARS algorithm. The LRS algorithm was simulated using the constants $\mu_s = 1.6$ and $\sigma^2 \operatorname{tr}(\mathbf{R}_{xx}) = 0.05$, thereby yielding a greater misadjustment error than either the ARS or GARS algorithms. The LMS algorithm was also simulated for purposes of comparison with step size corresponding to $\mu_s \operatorname{tr}(\mathbf{R}_{xx}) = 0.1$ and using an estimated gradient derived from the average value of three samples of $e(k)\mathbf{x}(k)$ so $K=3$ instead of the more common $K=1$. In all cases the initial weight vector was taken to be $\mathbf{w}^T(0) = [0.1, 0, 0, 0]$.

The results of Figures 9.5–9.7 show that both the ARS and GARS algorithms are within 3 dB of the optimum SNR after about 800 iterations, while the LRS algorithm has not succeeded in reaching this point after 4000 iterations, even though the misadjustment is more severe than for the ARS and GARS algorithms. This result indicates that the misadjustment versus

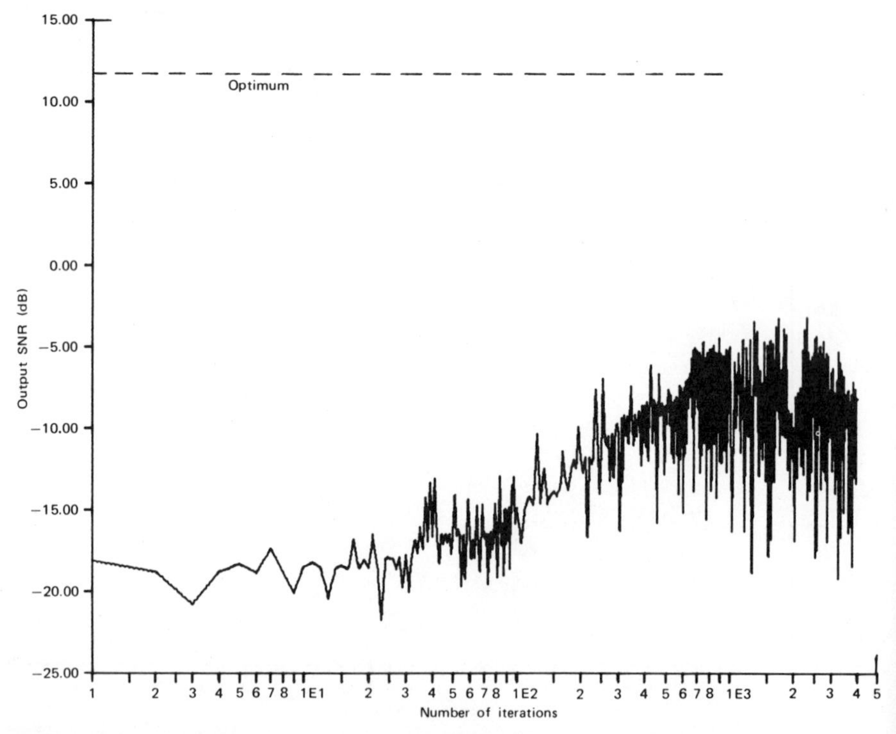


**Figure 9.9** Output SNR versus number of iterations for the LRS algorithm with eigenvalue spread = 2440.

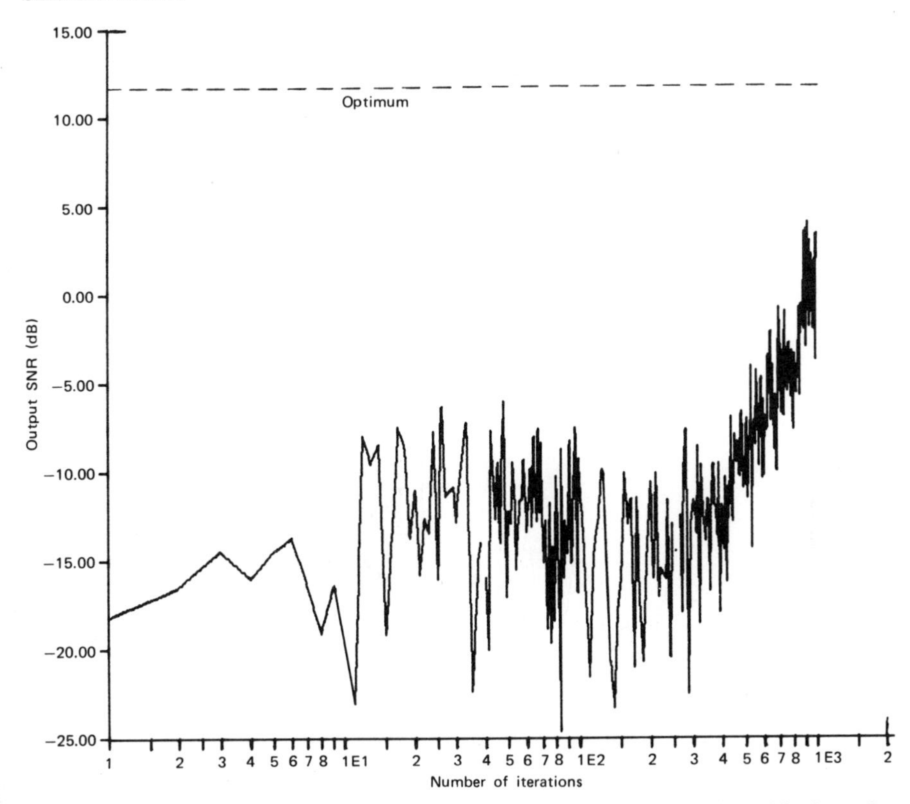


**Figure 9.10** Output SNR versus number of iterations for the ARS algorithm with eigenvalue spread = 2440.

speed of adaptation trade-off is more favorable for the ARS and GARS algorithms than it is for the LRS algorithm. The LMS algorithm by contrast is within 3 dB of the optimum output SNR after only 150 iterations with only a small degree of misadjustment. The extreme disparity in speed of convergence between the LMS algorithm and the three random search algorithms is actually more pronounced than the comparison of number of iterations indicates because each iteration in the random search algorithms represents 90 samples, whereas each iteration in the LMS algorithm represents only three samples. Consequently the time scale on Figures 9.5–9.7 is 30 times greater than the corresponding time scale on Figure 9.8.

The results given in Figures 9.8–9.12 for the case where eigenvalue spread = 2440 confirm the previous results obtained with the less severe eigenvalue spread of only 153.1. These results also show that both the random search algorithms and the LMS algorithm are sensitive to eigenvalue spread in the $\mathbf{R}_{xx}$ matrix.

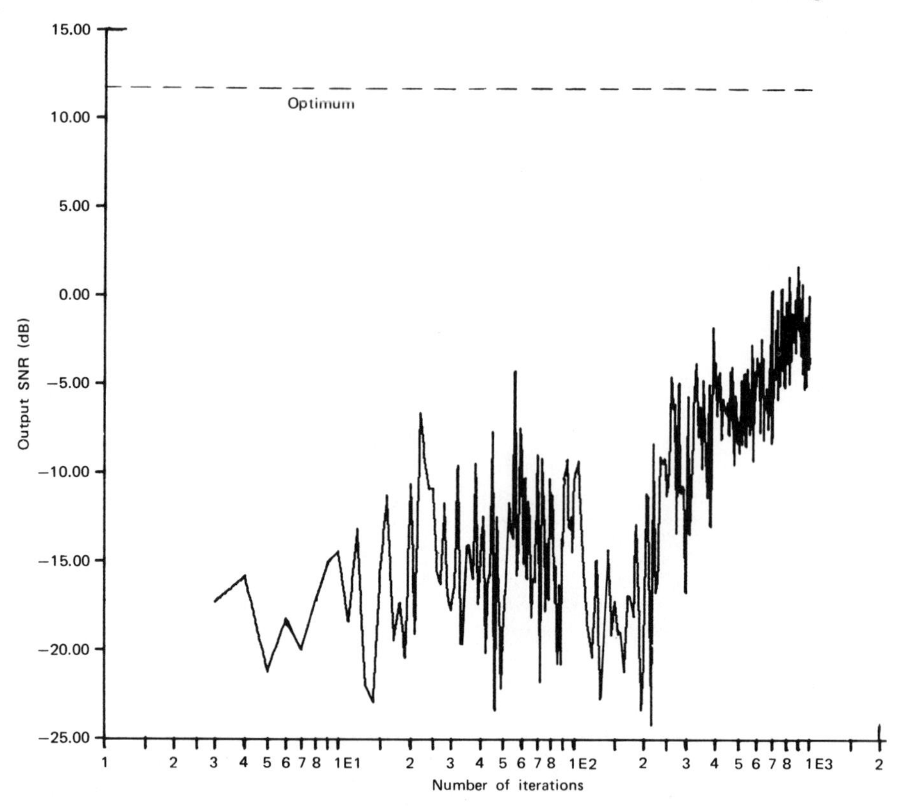


**Figure 9.11** Output SNR versus number of iterations for the GARS algorithm with eigenvalue spread = 2440.

## 9.5 SUMMARY AND CONCLUSIONS

Random search algorithms are useful in searching irregular and multimodal performance surfaces and only require a direct evaluation of the selected performance measure in order to implement. The weight adjustment computation for the LRS, ARS, and GARS algorithms is extremely simple, requiring only modest computational power. More elaborate and complicated random search algorithms have been applied to adaptive control and pattern recognition systems [12], but the introduction of more sophisticated measures into random searches removes their simplicity (which is a primary virtue for adaptive array applications).

The price to be paid for simple computation and implementation requirements is longer required convergence time to reach the optimal weight vector solution, although both the ARS and GARS algorithms have more favorable misadjustment versus speed of convergence trade-offs than the LRS algo-

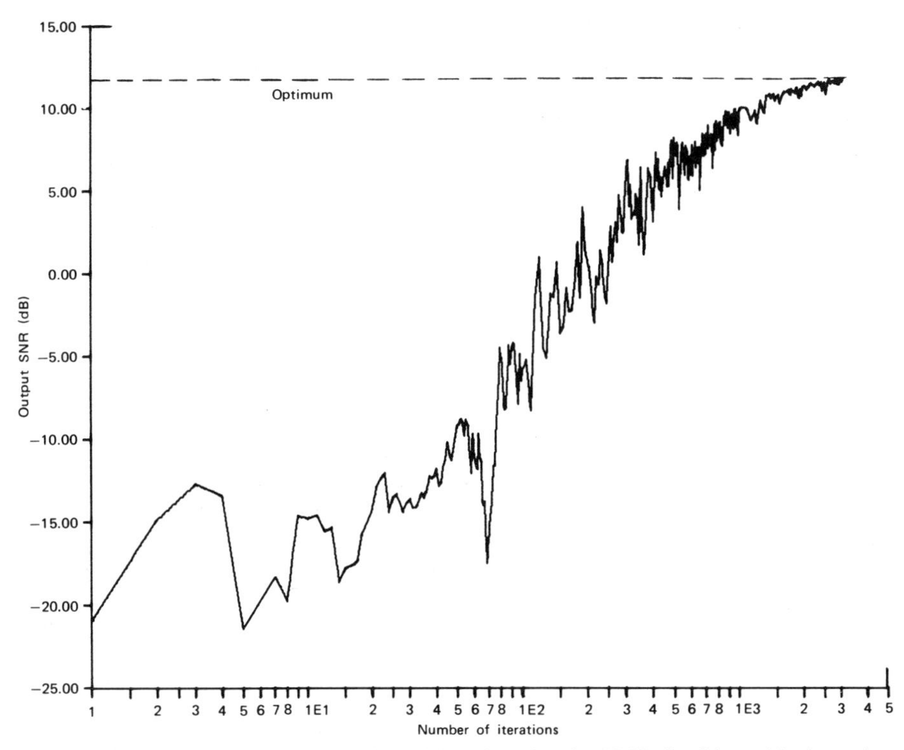


**Figure 9.12** Output SNR versus number of iterations for the LMS algorithm with eigenvalue spread = 2440.

rithm. Nevertheless, the convergence speed that can be realized with the LMS algorithm on unimodal performance surfaces is orders of magnitude faster than that which can be realized by a random search algorithm. Furthermore, the three random search algorithms considered here all exhibited the same degree of convergence speed sensitivity to eigenvalue spread as the LMS algorithm. Finally, all random searches exhibit relatively slow convergence in high dimensional spaces, a characteristic that reflects the fact that as the number of possible directions to be searched increases, then the convergence time also increases. This performance characteristic leads to the suggestion that reducing the dimensionality of the parameter space search over the final stages of convergence can be helpful. Note that the brief discussion given here is not a complete survey of search algorithms that employ only measurements of the output power. In particular, consideration should be given to the "directed search" techniques like that proposed by Hooke and Jeeves [20] which may well exhibit superior convergence properties under certain conditions.

## PROBLEMS

**1** [12] ***Misadjustment vs. Speed of Adaptation Trade-off for the LRS Algorithm.*** Assuming all eigenvalues are equal so that $(T_{p_{mse}})_{av} = T_{mse}$, plot $T_{mse}$ versus $N$ for the LRS algorithm assuming $(M_{tot})_{min} = 10\%$ in (9.43) and compare this result with the corresponding plots obtained for the LMS and DSD algorithms in Problem 1 of Chapter 4.

**2** [18] ***Search Loss for a Simple Random Search with Reversing Step.*** Consider the simple random search algorithm described by

$$\mathbf{x}(i+1) = \mathbf{x}(i) + \Delta\mathbf{x}(i+1)$$

where $\Delta\mathbf{x}$ is a random displacement vector satisfying $|\Delta\mathbf{x}|^2 = 1$ and

$$\Delta\mathbf{x}(i+1) = \begin{cases} \xi \text{ if performance improvement is observed} \\ -\Delta\mathbf{x}(i) \text{ if no performance improvement is observed} \end{cases}$$

where $\xi$ is the random direction of the displacement vector.

(a) Consider the parameter space of Figure 9.13 in which a performance improvement is realized for any angle in the range $-\pi/2 < \phi < \pi/2$ and no performance improvement is realized for any angle in the range $\pi/2 \leqslant \phi \leqslant 3\pi/2$. The mean displacement in the direction of a successful random step is given by

$$U(n) = \int_0^{\pi/2} \cos\phi p(\phi)\, d\phi$$

where $n$ is the number of degrees of freedom and $p(\phi)$ is the probability density function of the angle $\phi$ for a uniform distribution of directions of the random step in the $n$-dimensional space. Show that

$$p(\phi) = \frac{\sin^{n-2}\phi}{2\int_0^{\pi/2} \sin^{n-2}\phi\, d\phi} = \frac{\Gamma(n-1)}{2^{n-2}\left[\Gamma\left(\frac{n-1}{2}\right)\right]^2} \sin^{n-2}\phi$$

where $\Gamma(\cdot)$ is the Gamma function.

*Hint.* Note that the area of a ring-shaped zone on the surface of an $n$-dimensional sphere corresponding to the angle $d\phi$ is $A_{n-2} \times \sin^{n-2}\phi\, d\phi$. Consequently the area of the surface of the hypersphere included in the hypercone with vertical angle $2\phi$ is given by

$$S(\phi) = A_{n-2}\int_0^{\phi} \sin^{n-2}\phi\, d\phi$$

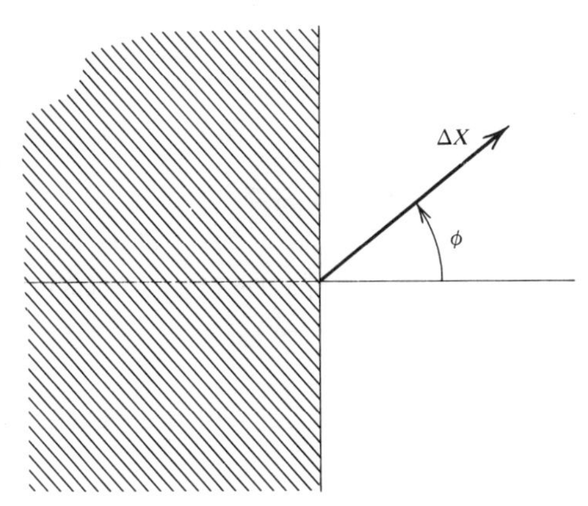


**Figure 9.13** Parameter space section showing displacement vector ΔX and direction $\phi$.

The probability that a random vector lies in this cone for a uniform probability choice of the random direction is equal to the ratio of the "areas" $S(\phi)$ and $S(\pi)$. The desired probability density is then the derivative of this ratio with respect to the angle $\phi$.

(b) Show that $U(n)$ defined in part (a) is given by

$$U(n)=\frac{\int_0^{\pi/2}\cos\phi\sin^{n-2}\phi\,d\phi}{\int_0^{\pi/2}\sin^{n-2}\phi\,d\phi}=\frac{\Gamma(n-1)}{2^{n-3}(n-1)\left[\Gamma\left(\frac{n-1}{2}\right)\right]^2}$$

Since the probability of successful and unsuccessful steps is the same for the parameter space of Figure 9.13, then on the average for one successful step there is one unsuccessful step and a corresponding reverse step—three steps in all. Therefore the mean displacement for one successful step is reduced by two-thirds, and is only $\frac{1}{3}$ $U(n)$.

Defining search loss to be the number of steps required by the search such that the vector sum of these steps has the same length as one operating step in the successful direction, it follows that the mean search loss for the above algorithm is just $3/U(n)$.

**3** [19] ***Relative Search Efficiency Using the Search Loss Function.*** Consider a fixed step size gradient search defined by

$$\mathbf{x}(i+1)=\mathbf{x}(i)-a(i)\mu_s\frac{\Delta(i)}{|\Delta(i)|}$$

where

$$\mu_s = \text{step size}$$

$$\Delta(i) = \left[ \frac{\partial F}{\partial x_1}, \ldots, \frac{\partial F}{\partial x_n} \right]_{\mathbf{x}(i)}$$

$$a(i) = \begin{cases} 1 & \text{if } F[\mathbf{x}(i+1)] < F[\mathbf{x}(i)] \\ 0 & \text{otherwise} \end{cases}$$

and where $F[\cdot]$ denotes a known performance measure. Likewise consider a fixed step size random search defined by

$$\mathbf{x}(i+1) = \mathbf{x}(i) - b(i)\Delta\mathbf{x}(i) + \Delta\mathbf{x}(i+1)$$

where $\Delta\mathbf{x}(i)$ is a random vector having a uniform distribution on a hypersphere of radius $\mu_s$ and centered at the origin, and

$$b(i) = \begin{cases} 1 & \text{if } F[\mathbf{x}(i)] \geqslant F[\mathbf{x}(i-1)] \\ 0 & \text{otherwise} \end{cases}$$

Define the search loss function to be the performance measure $F[x]$ divided by the ratio of performance improvement per performance evaluation, that is,

$$SL(\mathbf{x}) \triangleq \frac{F(\mathbf{x})}{\left( \dfrac{F(\mathbf{x}) - F(\mathbf{x} + \Delta\mathbf{x})}{N} \right)}$$

where $N$ = total number of performance measure evaluations required at each search point.

(a) For the quadratic performance measure $F(\mathbf{x}) = \rho^2$, show that for the gradient search algorithm $F(\mathbf{x} + \Delta\mathbf{x}) = \rho^2 - 2\rho\mu_s + \mu_s^2$.

(b) To empirically evaluate the gradient $\Delta$ at some point $\mathbf{x}$ requires the evaluation of $F[\cdot]$ at $\mathbf{x}$ and at points along each coordinate direction separated from $\mathbf{x}$ by some distance $\epsilon$. Therefore a total of $n+1$ function evaluations are required (where $n$ = number of degrees of freedom). Show for the gradient search defined above that

$$SL(\mathbf{x}) = \frac{\rho^2(n+1)}{(2\rho\mu_s - \mu_s^2)}$$

(c) For a given base point $\mathbf{x}(i)$, the successor trial state $\mathbf{x}(i+1)$ for the random search defines an angle $\phi$ with respect to a line connecting

$\mathbf{x}(i)$ with the extremum point of the performance measure for which the probability density function $p(\phi)$ was obtained in Problem 2. Show that the expected value of performance measure improvement using the random search defined above is

$$E\{-\Delta F\} = E\{\Delta\rho^2\} = \int_0^{\phi_0} \Delta\rho^2 p(\phi)\,d\phi$$

where

$$\phi_0 = \cos^{-1}\left\{\frac{\mu_s}{2\rho}\right\} \qquad \text{and} \qquad \Delta\rho^2 = 2\rho\mu_s\cos\phi - \mu_s^2$$

(d) Show for the random search algorithm above that

$$SL(\mathbf{x}) = \frac{2\rho^2 \int_0^{\pi/2} \sin^{n-2}\phi\,d\phi}{\int_0^{\phi_0} 2\mu_s\rho\cos\phi\sin^{n-2}\phi\,d\phi - \int_0^{\phi_0} \mu_s^2 \sin^{n-2}\phi\,d\phi}$$

The search loss function of parts (b) and (d) can be compared for specific values of $\rho$, $\mu_s$, and $n$ to determine whether the gradient search or the random search is more efficient.

**4** [19] ***Search Loss Function Improvement Using Step Reversal.*** The relative efficiency of the fixed step size random algorithm introduced in Problem 3 can be significantly improved merely by adding a "reversal" feature to the random search. The fixed step size random search algorithm with reversal is described by

$$\mathbf{x}(i+1) = \mathbf{x}(i) + c(i)\Delta\mathbf{x}(i+1) + 2[c(i)-1]\Delta\mathbf{x}(i)$$

where

$$c(i) = \begin{cases} 1 & \text{if } F[\mathbf{x}(i)] \leq F[\mathbf{x}(i-1)] \\ 0 & \text{otherwise} \end{cases}$$

The above modification of the random search in effect searches at $\mathbf{y} - \Delta\mathbf{y}$ if the initial search at $\mathbf{y} + \Delta\mathbf{y}$ failed to produce an improvement in the performance measure.

(a) Show that

$$E\{-\Delta F\} = E\{\Delta\rho^2\} = \frac{3\int_0^{\phi_0} \Delta\rho^2 \sin^{n-2}\phi\,d\phi}{2\int_0^{\pi} \sin^{n-2}\phi\,d\phi}$$

where

$$\phi_0 = \cos^{-1}\left\{\frac{\mu_s}{2}\right\} \qquad \text{and } \Delta\rho^2 = 2\rho\mu_s\cos\phi - \mu_s^2$$

(b) Show that

$$SL(\mathbf{x}) = \frac{4\mu_s^2 \int_0^{\pi/2} \sin^{n-2}\phi \, d\phi}{\int_0^{\phi_0} 6\mu_s \rho \cos\phi \sin^{n-2}\phi \, d\phi - \int_0^{\phi_0} 3\mu_s^2 \sin^{n-2}\phi \, d\phi}$$

This search loss can be compared with that obtained in Problem 3(d) for specified $\rho$, $\mu_s$, and $n$ to determine exactly the improvement that can be realized by addition of the reversal feature.

## REFERENCES


[1] M. K. Leavitt, "A Phase Adaptation Algorithm," *IEEE Trans. Antennas Propag.*, Vol. AP-24, No. 5, September 1976, pp. 754–756.

[2] P. A. Thompson, "Adaptation by Direct Phase-Shift Adjustment in Narrow-Band Adaptive Antenna Systems," *IEEE Trans. Antennas Propag.*, Vol. AP-24, No. 5, September 1976, pp. 756–760.

[3] G. J. McMurty, "Search Strategies in Optimization," Proceedings of the 1972 International Conference on Cybernetics and Society, October, Washington, DC, pp. 436–439.

[4] C. Karnopp, "Random Search Techniques for Optimization Problems," *Automatica*, Vol. 1, August 1963, pp. 111–121.

[5] G. J. McMurty and K. S. Fu, "A Variable Structure Automation Used as a Multimodal Searching Technique," *IEEE Trans. Autom. Control*, Vol. AC-11, July 1966, pp. 379–387.

[6] R. L. Barton, "Self-Organizing Control: The Elementary SOC—Part I," *Control Eng.*, February 1968.

[7] ———, "Self-Organizing Control: The General Purpose SOC—Part II," *Control Eng.*, March 1968.

[8] M. A. Schumer and K. Steiglitz, "Adaptive Step Size Random Search," *IEEE Trans. Autom. Control*, Vol. AC-13, June 1968, pp. 270–276.

[9] R. A. Jarvis, "Adaptive Global Search in a Time-Variant Environment Using a Probabilistic Automaton with Pattern Recognition Supervision," *IEEE Trans. Syst. Sci. Cybern.*, Vol. SSC-6, July 1970, pp. 209–217.

[10] A. N. Mucciardi, "Self-Organizing Probability State Variable Parameter Search Algorithms for the Systems that Must Avoid High-Penalty Operating Regions," *IEEE Trans. Syst., Man, Cybern.*, Vol. SMC-4, July 1974, pp. 350–362.

[11] R. A. Jarvis, "Adaptive Global Search by the Process of Competitive Evolution," *IEEE Trans. Syst., Man, Cybern.* Vol. SMC-5, May 1975, pp. 297–311.

[12] B. Widrow and J. M. McCool, "A Comparison of Adaptive Algorithms based on the Methods of Steepest Descent and Random Search," *IEEE Trans. Antennas Propag.*, Vol. AP-24, No. 5, September 1976, pp. 615–637.

[13] C. A. Baird and G. G. Rassweiler, "Search Algorithms for Sonobuoy Communication," Proceedings of the Adaptive Antenna Systems Workshop, March 11–13, 1974, NRL Report 7803, Vol. I September 27, 1974, pp. 285–303.

[14] R. L. Barron, "Inference of Vehicle and Atmosphere Parameters from Free-Flight Motions," *AIAA J. Spacecr. Rockets*, Vol. 6, No. 6, June 1969, pp. 641–648.

[15] ———, "Guided Accelerated Random Search as Applied to Adaptive Array AMTI Radar," Proceedings of the Adaptive Antenna Systems Workshop, March 11–13, 1974, Vol. I, NRL Report 7803, September 27, 1974, pp. 101–112.

[16] A. E. Zeger and L. R. Burgess, "Adaptive Array AMTI Radar," Proceedings of the Adaptive Antenna Systems Workshop, March 11–13, 1974, NRL Report 7803, Vol. I, September 27, 1974, pp. 81–100.

[17] A. N. Mucciardi, "A New Class of Search Algorithms for Adaptive Computation," Proc. 1973 IEEE Conference on Decision and Control, December, San Diego, paper No. WA5-3, pp. 94–100.

[18] L. A. Rastrigin, "The Convergence of the Random Search Method in the External Control of a Many-Parameter System," *Autom. Remote Control*, Vol. 24, No. 11, April 1964, pp. 1337–1342.

[19] J. P. Lawrence, III and F. P. Emad, "An Analytic Comparison of Random Searching and Gradient Searching for the Extremum of a Known Objective Function," *IEEE Trans. Autom. Control*, Vol. AC-18, No. 6, December 1973, pp. 669–671.

[20] R. Hooke and T. A. Jeeves, "Direct Search Solution of Numerical and Statistical Problems," J. of the Assoc. for Comput. Mach., Vol. 8, April 1961, pp. 212–229.

# Chapter 10. Adaptive Algorithm Performance Summary

Chapters 4 through 9 have considered the transient response characteristics and implementation considerations associated with different classes of adaptive algorithms that are widely used for adaptive array applications. Before considering some practical problems associated with adaptive array system design, it is appropriate to summarize the principal characteristics of each algorithm class.

In each chapter of Part Two an algorithm representing a distinct adaptation philosophy has been directly compared with the LMS algorithm to determine the convergence speed relative to LMS adaptation. It is now convenient to compare directly the transient responses of the various algorithms for a selected example to determine the relative convergence speeds. Since the misadjustment versus rate of adaptation trade-offs for the random search algorithms (LRS, ARS, and GARS) and for the DSD algorithm of Chapter 4 are unfavorable compared with the LMS algorithm, recourse to these methods would be taken only if the meager instrumentation required was regarded as a cardinal advantage or non-unimodal performance surfaces were of concern. Furthermore, the Howells-Applebaum maximum SNR algorithm has a misadjustment versus convergence speed trade-off that is nearly identical with the LMS algorithm. Attention for the direct comparison consequently is focused on the following adaptive algorithms:

1. Least mean square (LMS) error algorithm (Section 4.2 of Chapter 4).
2. Powell's accelerated gradient (PAG) algorithm (Section 4.4.1 of Chapter 4).
3. Direct matrix inversion (DMI) [Version (6.24) from Section 6.1.2 of Chapter 6].
4. Recursive (R) algorithm (Section 7.5 of Chapter 7).
5. Gram-Schmidt cascade preprocessor (GSCP) (Section 8.4 of Chapter 8).

Each algorithm is started with the same initial weight vector $\mathbf{w}^T(0) =$

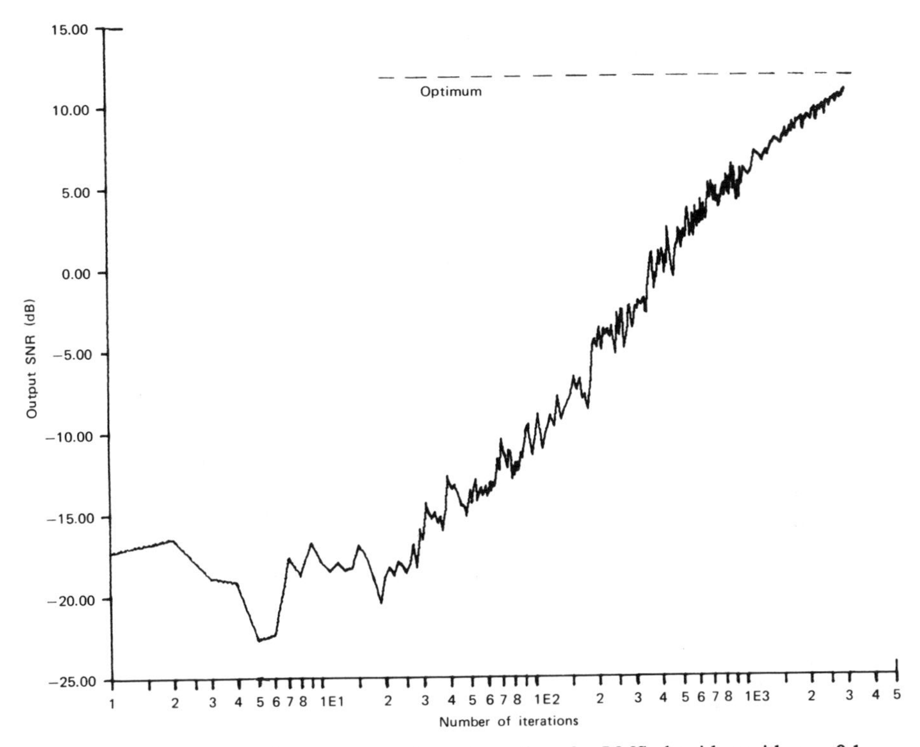


**Figure 10.1** Output SNR versus number of iterations for LMS algorithm with $\alpha_L = 0.1$.

[0.1, 0, 0, 0], for the array geometry and signal configuration of Figure 4.19 with signal conditions corresponding to an eigenvalue spread of 2440. The convergence results for each algorithm are plotted in Figures 10.1–10.5, where the corresponding algorithm parameters are also given. The results given in these figures represent the average of 10 separate adaptation trials obtained with each algorithm, thereby diminishing the effects of any highly unusual noise sequences that may occur.

The results of Figures 10.1–10.5 show the LMS algorithm requires about 1750 data samples to converge within 3 dB of the optimum output SNR, the PAG algorithm requires 110 iterations (990 data samples), the DMI and recursive algorithms both require 8 data samples, and the GSCP algorithm requires 40 iterations (120 data samples). Algorithm parameters were selected so the degree of steady-state misadjustment for the LMS, PAG, and GSCP algorithms is comparable. Misadjustment of the DMI and recursive algorithms decreases as the number of iterations increases and cannot be modified by altering algorithm parameters. These results indicate that the DMI and recursive algorithms offer by far the best misadjustment versus speed of

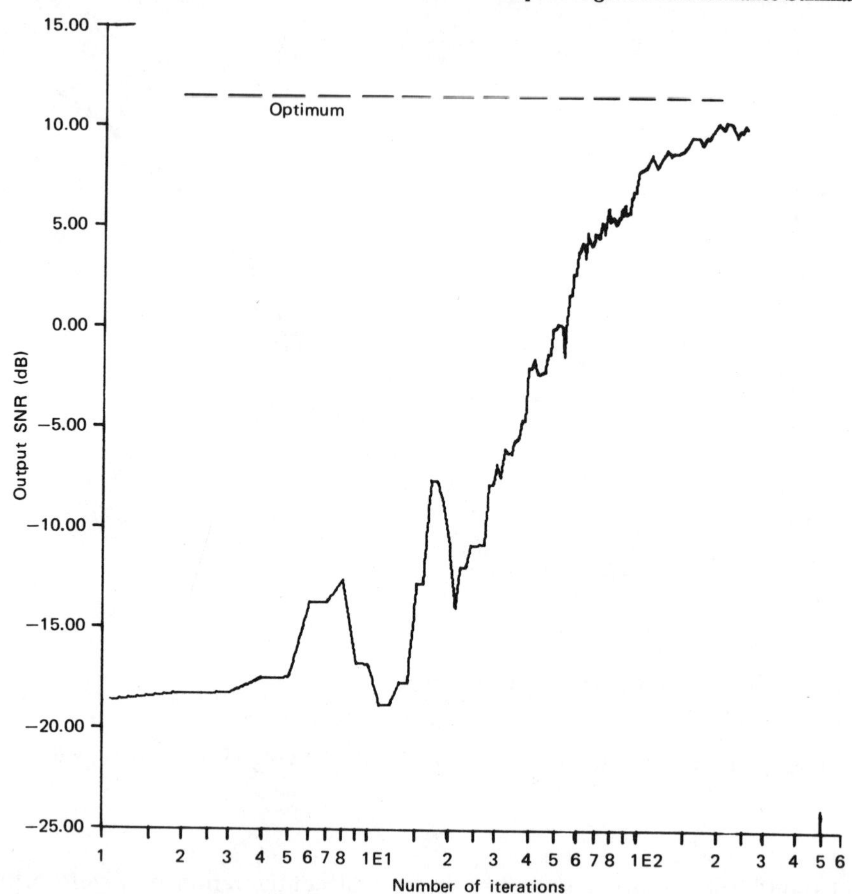


**Figure 10.2** Output SNR versus number of iterations for PAG algorithm with $K=9$ samples per iteration.

convergence trade-off, followed (in order) by the GSCP algorithm, the PAG algorithm, and the LMS algorithm for this moderate eigenvalue spread condition of $\lambda_{max}/\lambda_{min}=2440$.

It is now convenient to summarize the principal operational characteristics associated with the adaptive algorithms considered throughout Part Two. This summary is given in Table 10.1.

Those algorithms that achieve the simplest possible instrumentation by only requiring direct measurement of the selected performance measure pay a severe penalty in terms of the increased convergence time required to reach the steady-state solution for a given degree of misadjustment. Accepting the instrumentation necessary to incorporate one correlator for each controlled

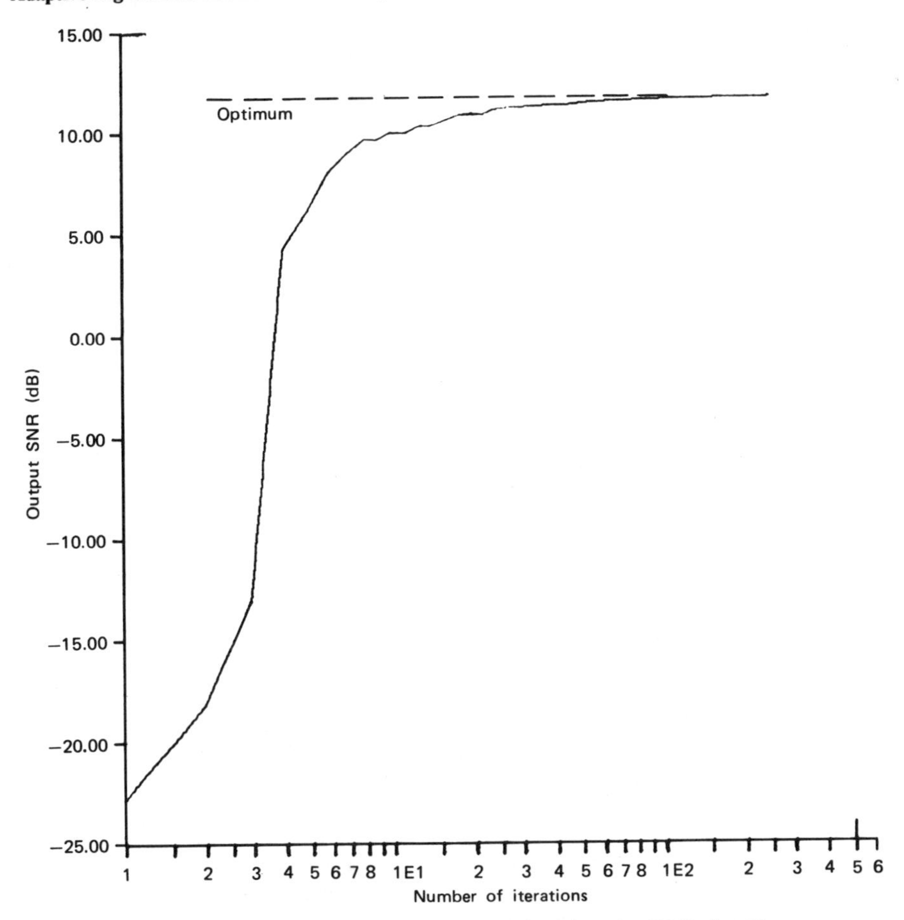


**Figure 10.3** Output SNR versus number of iterations for DMI algorithm.

array element enables the misadjustment versus convergence speed trade-off for the LMS and Howells-Applebaum interference suppression loops to be achieved. Further improvement in the misadjustment versus convergence speed trade-off can be obtained where sensitivity to eigenvalue spread is a concern by paying the price of additional instrumentation and/or additional computational power. As the shift to digital processing continues, algorithms requiring more sophisticated computation become not only practicable but also preferable in many cases. Not only can high performance be achieved that was impractical before, but also the low cost of the increased computational power may in some cases render a sophisticated algorithm more economical.

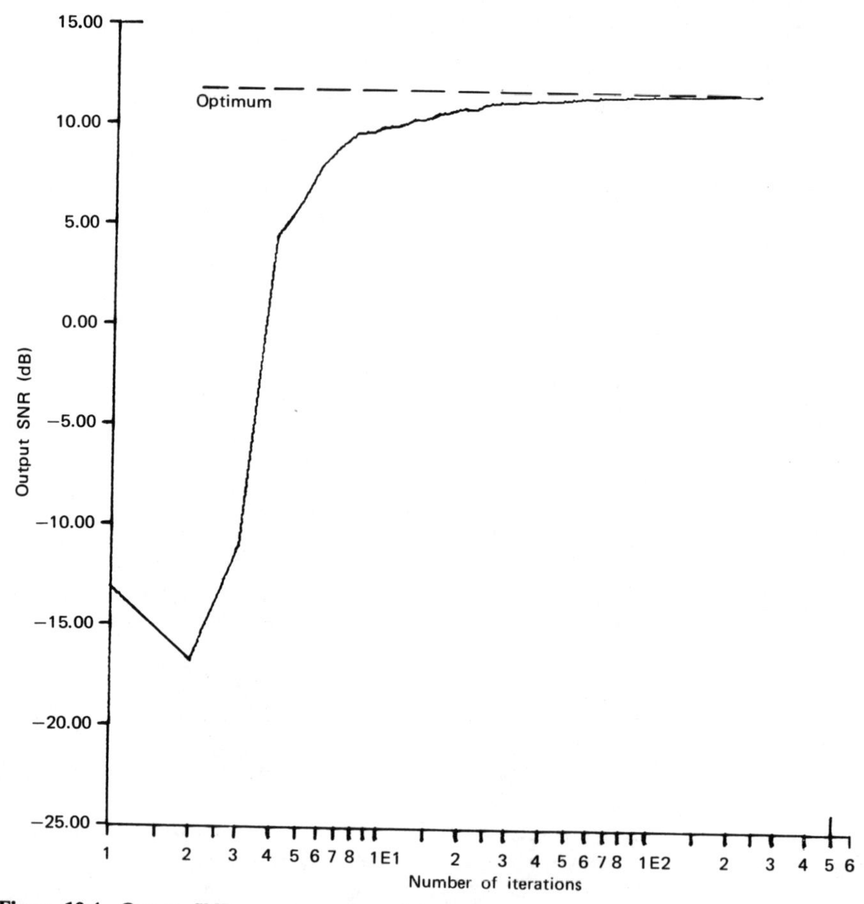


**Figure 10.4** Output SNR versus number of iterations for recursive algorithm with $\alpha = 1$ and $\mathbf{P}(0) = \mathbf{I}$.

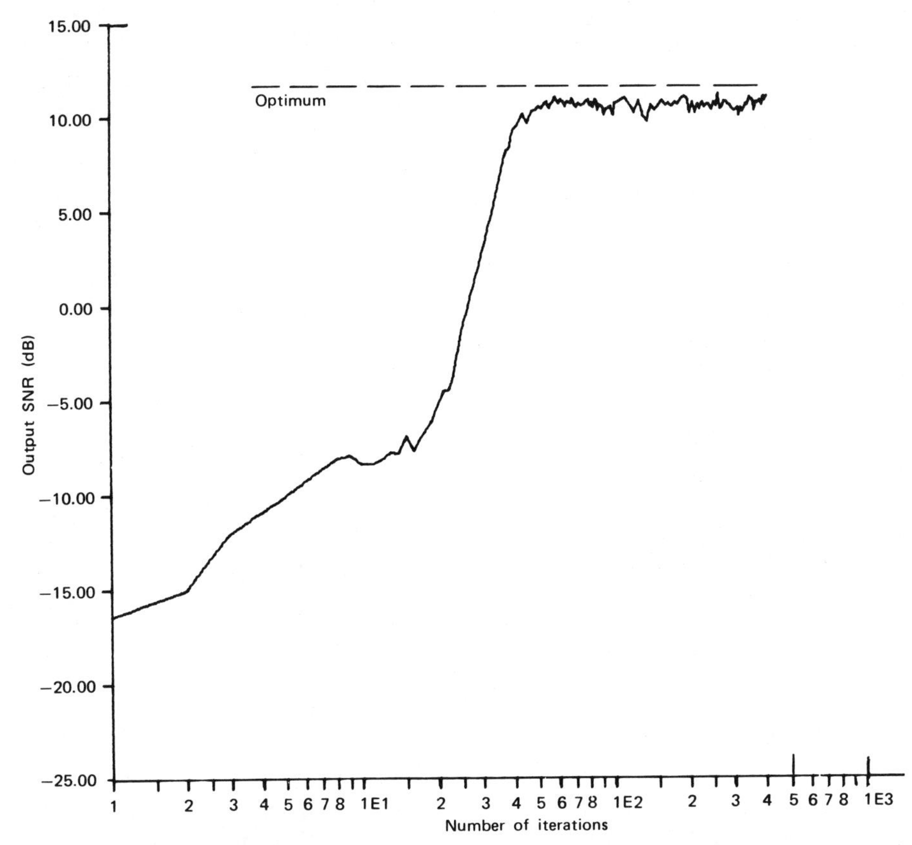


**Figure 10.5** Output SNR versus number of iterations for GSCP with $\alpha = \alpha_L = 0.1$ and $K = 3$ samples per iteration.

**Table 10.1 Operational Characteristics Summary of Selected Adaptive Algorithms**

| | LMS | MSNR | PAG | DSD | DMI | R | GSCP | RS |
|---|---|---|---|---|---|---|---|---|
| Algorithm philosophy | Steepest descent | Similar to LMS | Conjugate gradient descent | Perturbation technique | Direct estimate of covariance; open-loop. | Data weighting similar to Kalman filter; closed-loop. | Orthogonalize input signals | Trial and error |
| Transient response characteristic | Misadjustment vs. convergence speed trade-off is acceptable for numerous applications | Nearly the same as LMS. | More favorable misadjustment vs. convergence speed trade-off than LMS | Unfavorable misadjustment vs. convergence speed trade-off compared to LMS | Achieves the fastest convergence with most favorable misadjustment vs. convergence speed trade-off | Same as DMI | More favorable misadjustment vs. convergence speed trade-off than PAG. Convergence speed approaches DMI | Unfavorable misadjustment vs. convergence speed trade-off compared to LMS; accelerated steps improve speed |

**Table 10.1 cont'd**

| | | | | | | | | |
|---|---|---|---|---|---|---|---|---|
| Algorithm strengths | Easy to implement, requiring $N$ correlators and integrators. Tolerant of hardware errors. | Same as LMS. | Convergence speed less sensitive to eigenvalue spread than LMS. Fast convergence for small number of degrees-of-freedom. | Easy to implement. Only requires instrumentation to directly measure the performance index | Very fast convergence speed independent of eigenvalue spread. | Same as DMI; different data weighting schemes are easily incorporated | Convergence speed enjoys reduced sensitivity to eigenvalue spread compared to LMS. Tolerant of hardware errors. | Can be applied to any directly measurable performance index; easy to implement, with meager instrumentation and computation requirements |
| Algorithm weaknesses | Convergence speed sensitive to eigenvalue spread. | Same as LMS. | Relatively difficult to implement and requires parallel processors; convergence speed sensitive to number of degrees-of-freedom | Rate of convergence sensitive to eigenvalue spread, with speed comparable to that of RS with accelerated step | Requires $N(N+1)/2$ correlators to implement; matrix inversion requires adequate precision and $N^3/2+N^2$ complex multiplies | Requires $N(N+1)/2$ correlators and heavy computational load. | Requires a respectable amount of hardware—$N(N+1)/2$ adaptive loops to implement | Convergence speed sensitive to eigenvalue spread and the slowest of all algorithms considered |
| Chapter discussed in | 4 | 5 | 4 | 4 | 6 | 7 | 8 | 9 |

# Part III. Adaptive Array Compensation and Current Research Trends

# Chapter 11. Compensation of Adaptive Arrays

The presence of a single complex adaptive weight in each element channel of an adaptive array is sufficient for processing narrowband signals. To process broadband signals, however, requires that tapped-delay line (transversal filter) processing (or its frequency domain equivalent) be employed in each element channel because this permits frequency dependent amplitude and phase adjustments to be made. The analysis presented so far has assumed that each element channel consists of identical electronics so that except for the adaptive weight settings each channel is electrically "matched". As a practical matter, the electrical characteristics of each channel are slightly different and lead to "channel mismatching" in which significant differences in frequency response characteristics from channel to channel may severely degrade an array's performance unless some form of compensation is employed. This chapter addresses frequency dependent mismatch compensation by means of tapped-delay line processing, which is important if a practical broadband adaptive array design is to be realized.

Whether designing a tapped-delay line processor to accommodate broadband signals, to compensate for channel mismatch effects, or to compensate for the effects of multipath and finite array propagation delay, it is necessary to determine the number of taps that will be required to achieve a desired level of compensation. Since each additional tap (and associated weighting element) incorporated into the design increases the cost and complexity of the resulting adaptive array system, the question of how many taps are required for a specified set of conditions is an important practical design consideration.

## 11.1 BROADBAND SIGNAL PROCESSING CONSIDERATIONS

Since tapped-delay lines have frequency dependent transfer functions, their use has the potential for allowing much wider bandwidth performance in an adaptive array than would otherwise obtain. By examining adaptive array performance as a function of the number of taps, the tap spacing, and the total delay in each channel, one may determine the minimum number of taps required to obtain satisfactory performance for a given bandwidth.

Although the amplitude and phase adjustments required for broadband signals are frequency dependent, only certain kinds of frequency dependence will result in passing the desired signal through the array without distortion. The discussion of broadband signal processing considerations given here follows the treatment of this subject given by Rodgers and Compton [1]–[3]. The ideal (distortionless) channel transfer functions are derived, adaptive array performance using quadrature hybrid processing and two, three, and five-tap-delay line processing are considered, and results and conclusions for broadband signal processing are then discussed.

### *11.1.1 Distortionless Channel Transfer Functions*

Consider the two-element array depicted in Figure 11.1 in which the element channels are represented by the transfer functions $H_1(\omega)$ and $H_2(\omega)$, and a desired signal and an interference signal are incident on the array. Let the desired signal arrive from spatial angle $\theta_s$, measured relative to the array face normal. Furthermore assume that the array elements are omnidirectional and that the spacing between elements is one-half wavelength at the carrier frequency $\omega_0$ so that $d=\lambda_0/2=\pi\mathfrak{v}/\omega_0$ where $\mathfrak{v}$ is the wavefront propagation velocity.

From the point of view of the desired signal, the overall transfer function encountered in passing through the array of Figure 11.1 is

$$H_d(\omega)=H_1(\omega)+H_2(\omega)\exp\left(-j\frac{\omega d}{\mathfrak{v}}\sin\theta_s\right) \tag{11.1}$$

and the overall transfer function seen by the interference signal is

$$H_I(\omega)=H_1(\omega)+H_2(\omega)\exp\left(-j\frac{\omega d}{\mathfrak{v}}\sin\theta_i\right) \tag{11.2}$$

Now require that

$$H_d(\omega)=\exp(-j\omega T_1) \tag{11.3}$$

and

$$H_I(\omega)=0 \tag{11.4}$$

By choosing $H_d(\omega)$ according to (11.3), the desired signal is permitted to experience a time delay $T_1$ in passing through the array but otherwise remains undistorted. Choosing $H_I(\omega)=0$ results in complete suppression of the interference signal from the array output. To determine whether it is possible to select $H_1(\omega)$ and $H_2(\omega)$ to satisfy (11.3) and (11.4), solve (11.3) and (11.4) for $H_1(\omega)$ and $H_2(\omega)$. Setting $H_1(\omega)=|H_1(\omega)|\exp[j\alpha_1(\omega)]$ and $H_2(\omega)=$

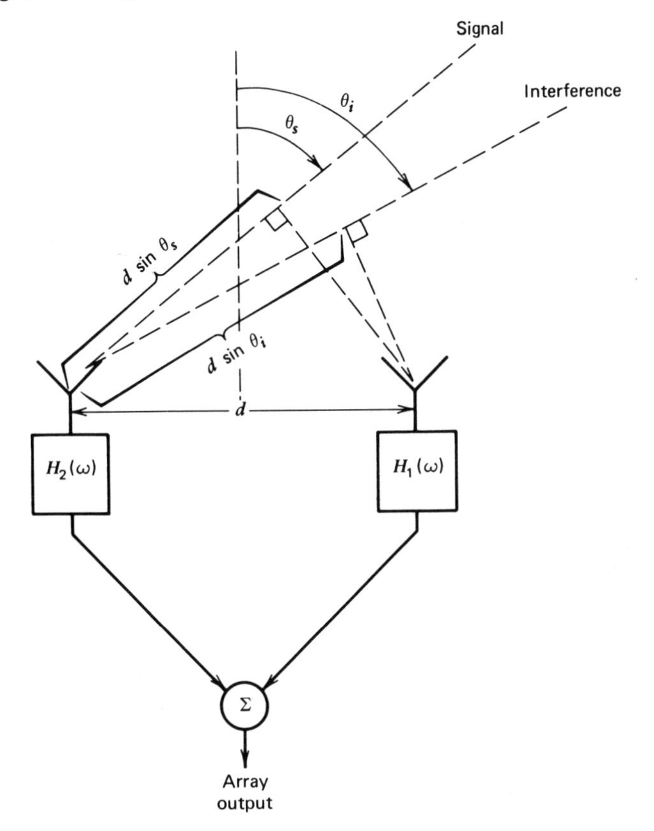


**Figure 11.1** Two-element array.

$|H_2(\omega)|\exp[j\alpha_2(\omega)]$ results in

$$|H_1(\omega)|\exp[j\alpha_1(\omega)] + |H_2(\omega)|\exp\left\{j\left[\alpha_2(\omega) - \frac{\pi\omega}{\omega_0}\sin\theta_s\right]\right\} = \exp(-j\omega T_1) \tag{11.5}$$

$$|H_1(\omega)|\exp[j\alpha_1(\omega)] + |H_2(\omega)|\exp\left\{j\left[\alpha_2(\omega) - \frac{\pi\omega}{\omega_0}\sin\theta_i\right]\right\} = 0 \tag{11.6}$$

From (11.5) and (11.6) it follows (as shown by the development outlined in the problems section) that the following conditions must obtain if (11.3) and (11.4) are to be satisfied:

$$H_1(\omega) = H_2(\omega) = \frac{1}{\sqrt{2\left(1 - \cos\left[\frac{\pi\omega}{\omega_0}(\sin\theta_i - \sin\theta_s)\right]\right)}} \tag{11.7}$$

$$\alpha_2(\omega) = \frac{\pi}{2}\left(\frac{\omega}{\omega_0}\right)[\sin\theta_s + \sin\theta_i] \mp n\frac{\pi}{2} - \omega T_1 \tag{11.8}$$

and

$$\alpha_1(\omega) = \frac{\pi}{2}\left(\frac{\omega}{\omega_0}\right)[\sin\theta_s - \sin\theta_i] \pm n\frac{\pi}{2} - \omega T_1 \tag{11.9}$$

where $n$ is any odd integer.

It may therefore be concluded that the amplitude of the ideal transfer functions must be equal and frequency dependent. Equations (11.8) and (11.9) furthermore show that the phase of each filter is a linear function of frequency with the slope dependent on the spatial arrival angles of the signals as well as on the time delay $T_1$ of the desired signal.

It is instructive to plot the amplitude function given by (11.7) for several different arrival angles. The results for the amplitude function are shown in Figure 11.2 for two choices of arrival angles ($\theta_s = 0°$ and $\theta_s = 80°$), where it is seen that the amplitude of the distortionless transfer function is nearly flat over a 40% bandwidth when the desired signal is at broadside ($\theta_s = 0°$) and the interference signal is 90° from broadside ($\theta_i = 90°$). Examination of (11.7) shows that whenever $(\sin\theta_i - \sin\theta_s)$ is in the neighborhood of $\pm 1$, then the resulting amplitude function will be nearly flat over the 40% bandwidth region. If, however, both the desired and interference signals are far from broadside (as when $\theta_d = 80°$ and $\theta_i = 90°$), then the amplitude function is no longer flat.

The degree of "flatness" of the distortionless filter amplitude function can be interpreted in terms of the signal geometry with respect to the array sensitivity pattern. In general, when the phases of $H_1(\omega)$ and $H_2(\omega)$ are

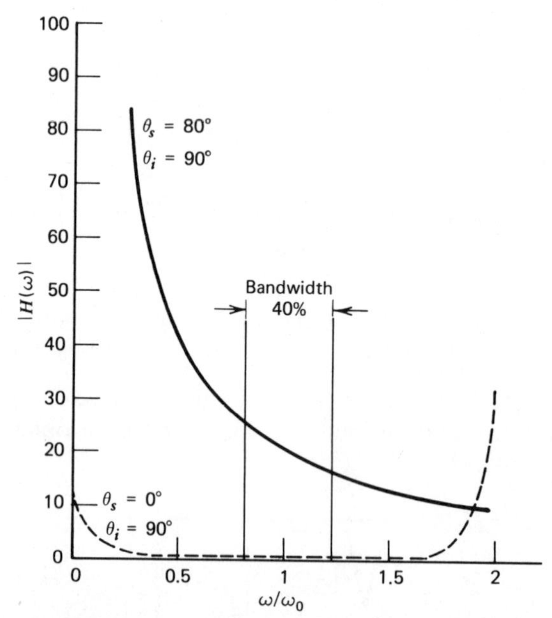


**Figure 11.2** Distortionless transfer function amplitude versus normalized frequency for $d = \lambda_0/2$. From Rodgers and Compton, Technical Report ESL 3832-3, 1975 [2].

adjusted to yield the maximum undistorted response to the desired signal, the corresponding array sensitivity pattern will have certain nulls. The distortionless filter amplitude function will then be the most flat when the interference signal falls into one of these pattern nulls.

Equation (11.7) furthermore shows that singularities occur in the distortionless channel transfer functions whenever $(\omega/\omega_0)\pi(\sin\theta_i - \sin\theta_s) = n2\pi$ where $n = 0, 1, 2, \ldots$. The case when $n = 0$ occurs when the desired and interference signals arrive from exactly the same direction, so it is hardly surprising that the array would experience difficulty trying to receive one signal while nulling the other in this case. The other cases when $n = 1, 2, \ldots,$ occur when the signals arrive from different directions, but the phase shifts between elements differ by a multiple of $2\pi$ radians at some frequency $\omega$ in the signal band.

The phase functions $\alpha_1(\omega)$ and $\alpha_2(\omega)$ of (11.8) and (11.9) are linear functions of frequency. When $T_1 = 0$, the phase slope of $H_1(\omega)$ is proportional to $\sin\theta_s - \sin\theta_i$, while that of $H_2(\omega)$ is proportional to $\sin\theta_i + \sin\theta_s$. Consequently when the desired signal is broadside, $\alpha_1(\omega) = -\alpha_2(\omega)$. Furthermore the phase difference between $\alpha_1(\omega)$ and $\alpha_2(\omega)$ is also a linear function of frequency, a result that would be expected since this allows the interelement phase shift (which is also a linear function of frequency) to be cancelled.

### *11.1.2 Quadrature Hybrid and Tapped-Delay Line Processing for an LMS Array*

Consider a two-element adaptive array using the LMS algorithm to provide the adaptive weight adjustment. If $\mathbf{w}$ is the column vector of array weights, $\mathbf{R}_{xx}$ is the correlation matrix of input signals to each adaptive weight, and $\mathbf{r}_{xd}$ is the cross-correlation vector between the received signal vector $\mathbf{x}(t)$ and the reference signal $d(t)$, then as shown in Chapter 3, the optimum array weight vector that minimizes $E\{\epsilon^2(t)\}$ (where $\epsilon(t) = d(t) -$ array output) is given by

$$\mathbf{w}_{\text{opt}} = \mathbf{R}_{xx}^{-1}\mathbf{r}_{xd} \tag{11.10}$$

If the signal appearing at the output of each sensor element consists of a desired signal, an interference signal, and a thermal noise component (where each component is statistically independent of the others and has zero mean), then the elements of $\mathbf{R}_{xx}$ can readily be evaluated in terms of these component signals.

Consider the tapped-delay line employing real (instead of complex) weights shown in Figure 11.3. Since each signal $x_i(t)$ is just a time-delayed version of $x_1(t)$, it follows that

$$\left.\begin{aligned} x_2(t) &= x_1(t-\Delta) \\ x_2(t) &= x_1(t-2\Delta) \\ &\vdots \\ x_L(t) &= x_1[t-(L-1)\Delta] \end{aligned}\right\} \tag{11.11}$$

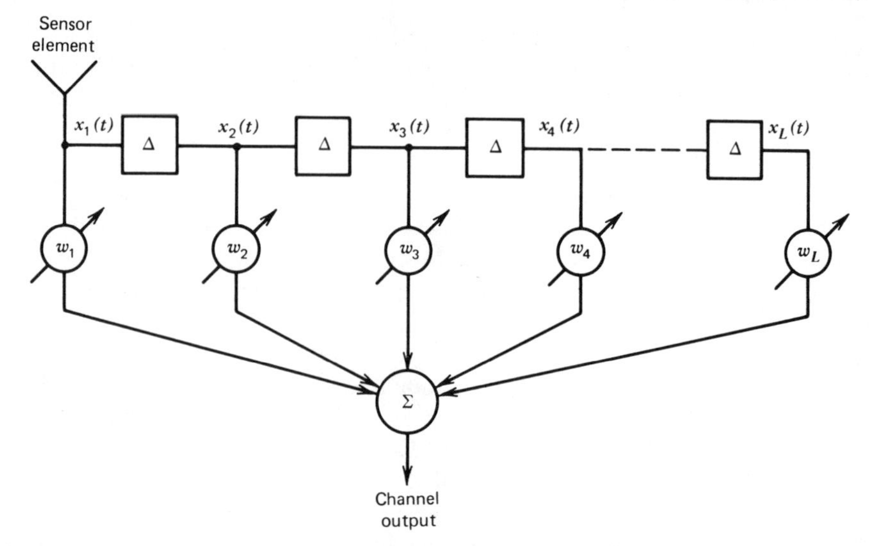


**Figure 11.3** Tapped-delay line processor for a single-element channel having real adaptive weights.

Now since the elements of $\mathbf{R}_{xx}$ are given by

$$r_{x_i x_j} \overset{\Delta}{=} E\{x_i(t)x_j(t)\} \tag{11.12}$$

it follows from (11.11) that

$$r_{x_i x_j} = r_{x_1 x_1}(\tau_{ij}) \tag{11.13}$$

where $r_{x_1 x_1}(\tau_{ij})$ is just the autocorrelation function of $x_1(t)$, and $\tau_{ij}$ is the time delay between $x_i(t)$ and $x_j(t)$. Furthermore $r_{x_i x_i}(\tau_{ij})$ is comprised of the sum of three autocorrelation functions—those of the desired signal, the interference, and the thermal noise so that:

$$r_{x_1 x_1}(\tau_{ij}) = r_{dd}(\tau_{ij}) + r_{II}(\tau_{ij}) + r_{nn}(\tau_{ij}) \tag{11.14}$$

For those elements of $\mathbf{R}_{xx}$ corresponding to $x_i(t)$ and $x_j(t)$ from different element channels, $r_{x_i x_j}$ only consists of the sum of the autocorrelation functions of the desired signal and the interference signal (with appropriate delays) but not the thermal noise since the element noise from channel to channel is uncorrelated. Consequently for signals in different element channels,

$$r_{x_i x_j}(\tau_{ij}) = r_{dd}(\tau_{d_{ij}}) + r_{II}(\tau_{I_{ij}}) \tag{11.15}$$

where $\tau_{d_{ij}}$ denotes the time delay between $x_i(t)$ and $x_j(t)$ for the desired signal, and $\tau_{I_{ij}}$ denotes the time delay between $x_i(t)$ and $x_j(t)$ for the interference signal (these two time delays will in general be different due to the different angles of arrival of the two signals). Only when $x_i(t)$ and $x_j(t)$ are from the same array element channel will $\tau_{d_{ij}}=\tau_{I_{ij}}$ (which may then be denoted by $\tau_{i_j}$).

Next consider the quadrature hybrid array processor depicted in Figure 11.4. Let $x_1(t)$ and $x_3(t)$ denote the in-phase signal components while $x_2(t)$ and $x_4(t)$ denote the quadrature-phase signal components of each of the elements output signals. Then the in-phase and quadrature components are related by

$$\left.\begin{aligned} x_2(t)&=\check{x}_1(t)\\ x_4(t)&=\check{x}_3(t)\end{aligned}\right\} \tag{11.16}$$

The symbol $\check{}$ denotes the Hilbert transform:

$$\check{x}(t) \triangleq \frac{1}{\pi}\int_{\infty-}^{\infty} \frac{x(\tau)}{t-\tau}\,d\tau \tag{11.17}$$

where the above integral is regarded as a Cauchy principal value integral. The various elements of the correlation matrix

$$r_{x_ix_j}=E\{x_i(t)x_j(t)\} \tag{11.18}$$

can then be found by making use of certain Hilbert transform relations as

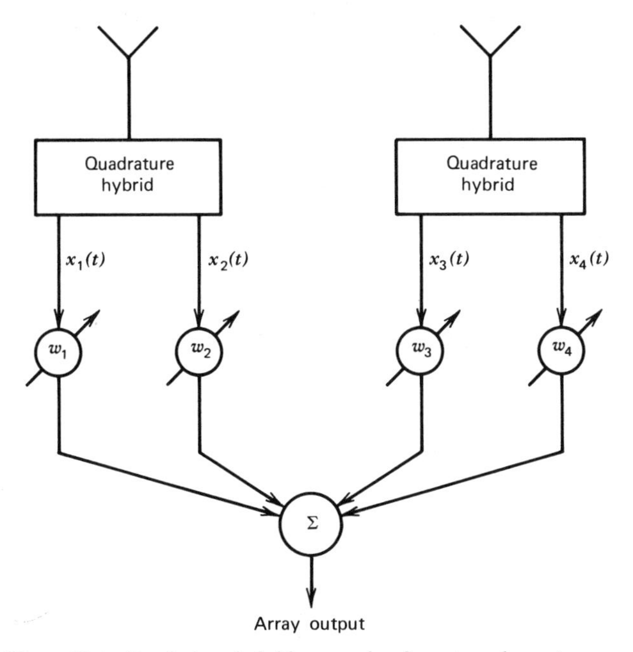


**Figure 11.4** Quadrature hybrid processing for a two-element array.

follows [4], [7]:

$$E\{\check{x}(t)\check{y}(s)\}=E\{x(t)y(s)\} \tag{11.19}$$

$$E\{\check{x}(t)y(s)\}=-E\{x(t)\check{y}(s)\} \tag{11.20}$$

so that

$$E\{\check{x}(t)x(t)\}=0 \tag{11.21}$$

and

$$E\{x(t)\check{y}(s)\}=\check{E}\{x(t)y(s)\} \tag{11.22}$$

where $\check{E}\{x(t)y(s)\}$ denotes the Hilbert transform of $r_{xy}(\tau)$ where $\tau=s-t$. With the above relations and from (11.16) it then follows that

$$r_{x_1x_1}=E\{x_1(t)x_1(t)\}=r_{x_1x_1}(0) \tag{11.23}$$

$$r_{x_1x_2}=E\{x_1(t)x_2(t)\}=E\{x_1(t)\check{x}_1(t)\}=0 \tag{11.24}$$

and

$$\begin{aligned} r_{x_2x_2}&=E\{x_2(t)x_2(t)\}=E\{\check{x}_1(t)\check{x}_1(x)\}\\ &=E\{x_1(t)x_1(t)\}=r_{x_1x_1}(0) \end{aligned} \tag{11.25}$$

where $r_{x_1x_1}(\tau)$ is the autocorrelation function of $x_1(t)$ given by (11.14).

When two different sensor element channels are involved [as with $x_1(t)$ and $x_3(t)$, for example], then

$$E\{x_1(t)x_3(t)\}=r_{dd}(\tau_{d_{13}})+r_{II}(\tau_{I_{13}}) \tag{11.26}$$

where $\tau_{d_{13}}$ and $\tau_{I_{13}}$ represent the spatial time delays between the sensor elements of Figure 11.4 for the desired and interference signals, respectively. Similarly,

$$\begin{aligned} E\{x_1(t)x_4(t)\}&=E\{x_1(t)\check{x}_3(t)\}=\check{E}\{x_1(t)x_3(t)\}\\ &=\check{r}_{dd}(\tau_{d_{13}})+\check{r}_{II}(\tau_{I_{13}}) \end{aligned} \tag{11.27}$$

$$\begin{aligned} E\{x_2(t)x_3(t)\}&=E\{\check{x}_1(t)x_3(t)\}=-E\{\check{x}_1(t)x_3(t)\}\\ &=-\check{E}\{x_1(t)x_3(t)\}=-\check{r}_{dd}(\tau_{d_{13}})-\check{r}_{II}(\tau_{I_{13}}) \end{aligned} \tag{11.28}$$

and

$$\begin{aligned} E\{x_2(t)x_4(t)\}&=E\{\check{x}_1(t)\check{x}_3(t)\}=x\{x_1(t)x_3(t)\}\\ &=r_{dd}(\tau_{d_{13}})+r_{II}(\tau_{I_{13}}) \end{aligned} \tag{11.29}$$

Now consider the cross-correlation vector $\mathbf{r}_{xd}$ defined by

$$\mathbf{r}_{xd} \triangleq E\begin{bmatrix} x_1(t)d(t) \\ x_2(t)d(t) \\ \vdots \\ x_{2N}(t)d(t) \end{bmatrix} \tag{11.30}$$

where $N$ is the number of sensor elements. Each element of $\mathbf{r}_{xd}$, denoted by $r_{x_i d}$, is just the cross-correlation between the reference signal $d(t)$ and signal $x_i(t)$. Since the reference signal is just a replica of the desired signal and is statistically independent of the interference and thermal noise signals, the elements of $\mathbf{r}_{xd}$ consist only of the autocorrelation function of the desired signal so that

$$r_{x_i d} = E\{x_i(t)d(t)\} = r_{dd}(\tau_{d_i}) \tag{11.31}$$

where $\tau_{d_i}$ represents the time delay between the reference signal and the desired signal component of $x_i(t)$. For an array with tapped-delay line processing, each element of $\mathbf{r}_{xd}$ is the autocorrelation function of the desired signal evaluated at a time delay value that reflects both the spatial delay between sensor elements and the delay-line delay to the tap of interest. For an array with quadrature hybrid processing, those elements of $\mathbf{r}_{xd}$ corresponding to an in-phase channel yield the autocorrelation function of the desired signal evaluated at the spatial delay appropriate for that element as follows:

$$r_{x_i d}(\text{in-phase channel}) = E\{x_i(t)d(t)\} = r_{dd}(\tau_{d_i}) \tag{11.32}$$

Those elements of $\mathbf{r}_{xd}$ corresponding to quadrature-phase channels can be evaluated using (11.21) and (11.22) as follows:

$$\begin{aligned} r_{x_{i+1} d}(\text{quadrature-phase channel}) &= E\{x_{i+1}(t)d(t)\} \\ &= E\{\check{x}_i(t)d(t)\} = -E\{x_i(t)\check{d}(t)\} \\ &= -\check{E}\{x_i(t)d(t)\} = -\check{r}_{x_i d}(\tau_{d_i}) \end{aligned} \tag{11.33}$$

Once $\mathbf{R}_{xx}$ and $\mathbf{r}_{xd}$ have been evaluated for a given signal environment, the optimal LMS weights can be computed from (11.10) and the steady-state response of the entire array can then be evaluated.

For the tapped-delay line of Figure 11.3 the element channel may be regarded as a processor having a channel transfer function given by

$$\begin{aligned} H_1(\omega) &= w_1 + w_2 e^{-j\omega\Delta} + w_3 e^{-j2\omega\Delta} \\ &\quad + \dots + w_L e^{-j(L-1)\omega\Delta} \end{aligned} \tag{11.34}$$

Likewise the quadrature hybrid processor of Figure 11.4 has a channel transfer function:

$$H_1(\omega) = w_1 - jw_2 \tag{11.35}$$

The overall array transfer function that results for the desired signal and the interference may now be evaluated by properly accounting for the effects of spatial delays between array elements. With a two-element array having channel transfer functions $H_1(\omega)$ and $H_2(\omega)$, the transfer function presented to the desired signal is

$$H_d(\omega) = H_1(\omega) + H_2(\omega)e^{-j\omega\tau_d} \tag{11.36}$$

while the transfer function for the interference is

$$H_I(\omega) = H_1(\omega) + H_2(\omega)e^{-j\omega\tau_I} \tag{11.37}$$

The spatial time delays associated with the desired and interference signals are represented by $\tau_d$ and $\tau_I$, respectively, between element 1 [with channel transfer function $H_1(\omega)$] and element 2 [with channel transfer function $H_2(\omega)$]. With two sensor elements spaced apart by a distance $d$ as in Figure 11.1, the two spatial time delays are given by

$$\tau_d = \frac{d}{\mathfrak{v}} \sin\theta_s \tag{11.38}$$

and

$$\tau_I = \frac{d}{\mathfrak{v}} \sin\theta_I \tag{11.39}$$

It is desired to compare tapped-delay line processing (employing real weights) with quadrature hybrid processing by evaluating the output signal-to-total noise ratio for a two-element array. The output signal-to-total noise ratio is defined as

$$\text{SNR} \triangleq \frac{P_d}{P_I + P_n} \tag{11.40}$$

where $P_d$, $P_I$, and $P_n$ represent the output desired signal power, interference signal power, and thermal noise power, respectively.

The array output power for each of the foregoing three signals may now be evaluated. Let $\phi_{dd}(\omega)$ and $\phi_{II}(\omega)$ represent the power spectral densities of the desired signal and the interference signal respectively, then the desired signal output power is given by

$$P_d = \int_{-\infty}^{\infty} \phi_{dd}(\omega)|H_d(\omega)|^2 d\omega \tag{11.41}$$

where $H_d(\omega)$ is the overall transfer function seen by the desired signal, and the interference signal output power is

$$P_I = \int_{-\infty}^{\infty} \phi_{II}(\omega)|H_I(\omega)|^2 d\omega \tag{11.42}$$

where $H_I(\omega)$ is the overall transfer function seen by the interference signal. The thermal noise present in each element output is statistically independent from one element to the next. Let $\phi_{nn}(\omega)$ denote the thermal noise power spectral density, then the noise power contributed to the array output by element 1 is

$$P_{n_1} = \int_{-\infty}^{\infty} \phi_{nn}(\omega)|H_1(\omega)|^2 d\omega \tag{11.43}$$

while that contributed by element 2 is

$$P_{n_2} = \int_{-\infty}^{\infty} \phi_{nn}(\omega)|H_2(\omega)|^2 d\omega \tag{11.44}$$

Consequently the total thermal noise output power from a two element array is

$$P_n = \int_{-\infty}^{\infty} \phi_{nn}(\omega)\left[|H_1(\omega)|^2 + |H_2(\omega)|^2\right] d\omega \tag{11.45}$$

The foregoing expressions may now be used in (11.40) to obtain the output signal-to-total noise ratio.

### *11.1.3 Performance Comparison of Four Array Processors*

The performance of four adaptive arrays—one with quadrature hybrid processing and three with tapped-delay line processing (using real weights)—can now be determined and compared. The comparisons are made for signal bandwidths of 4, 10, 20, and 40%. The objective of this comparison is to determine the relative desirability of tapped-delay line processing compared to quadrature hybrid processing when broadband signals are present in the signal environment. Real weights are employed in the tapped-delay line to preserve as much simplicity as possible in the hardware implementation, although this sacrifices the available degrees of freedom with a consequent degradation in tapped-delay line performance relative to combined amplitude and phase weighting. The results obtained will nevertheless serve as an indication of the relative effectiveness of tapped-delay line processing compared to quadrature hybrid processing for broadband signals.

The four array processors to be compared are shown in Figure 11.5 where each array has two sensor elements and the elements are spaced one half wavelength apart at the center frequency of the desired signal bandwidth. Figure 11.5*a* shows an array having quadrature hybrid processing while

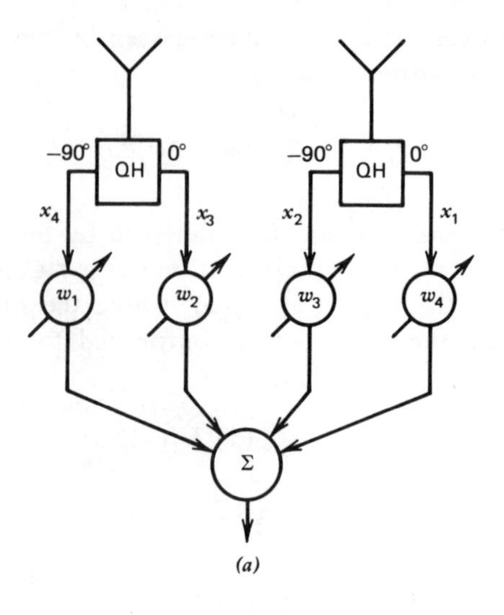


*(a)*

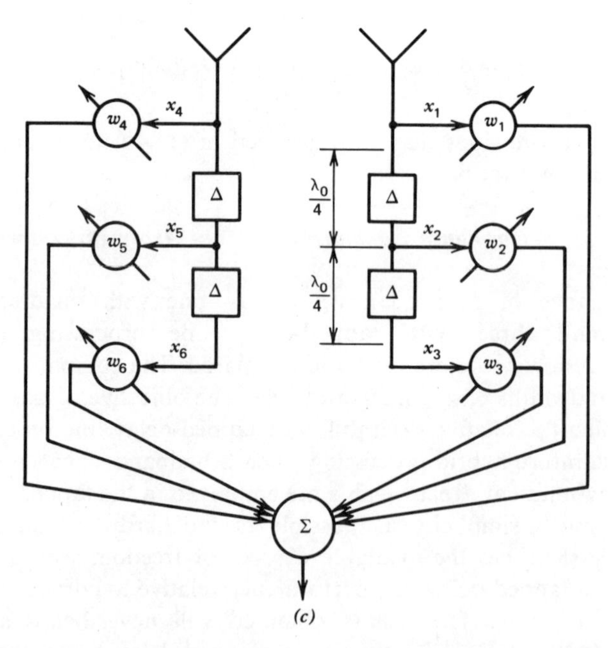


*(c)*

**Figure 11.5** Four adaptive array processors for broadband signal processing comparison. (*a*) Quadrature hybrid. (*b*) Two-tap delay line. (*c*) Three-tap delay line. (*d*) Five-tap delay line. From Rodgers and Compton, *IEEE Trans. Aerosp. Electron. Syst.*, January 1979 [3].

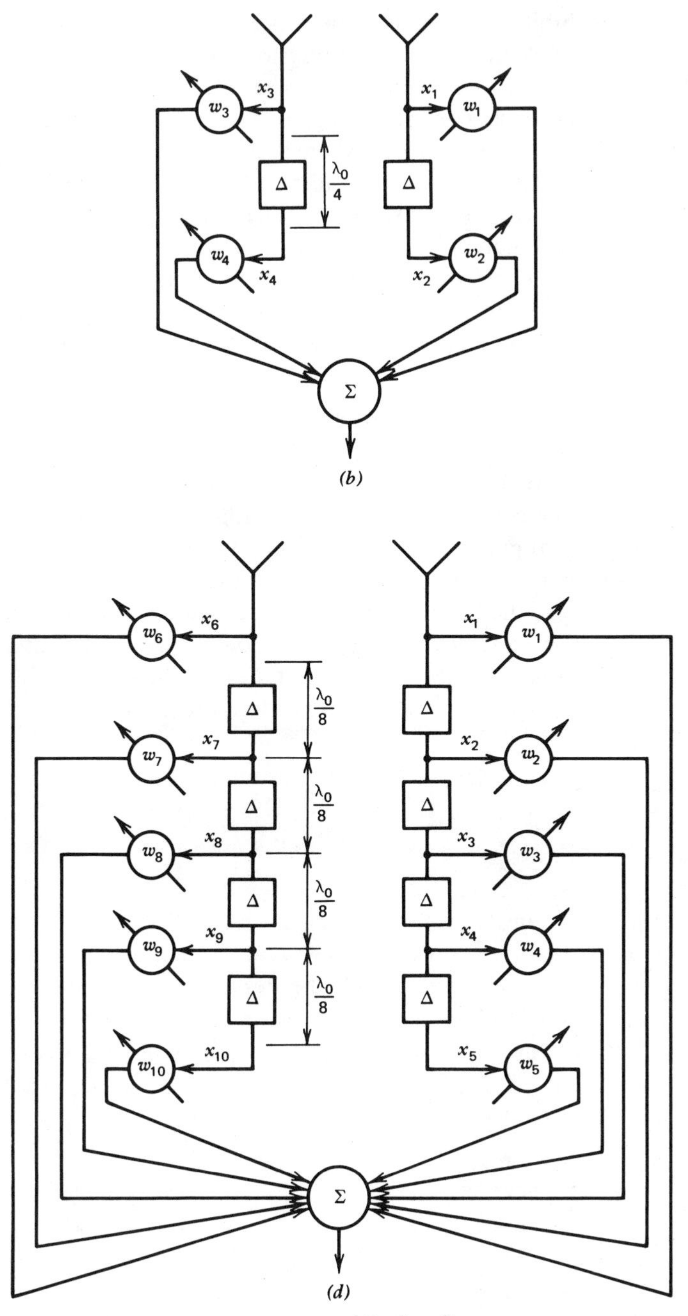


**Figure 11.5** (*Continued*).

Figure 11.5*b–d* exhibit tapped-delay line processing. The processor of Figure 11.5*b* has one delay element corresponding to one quarter wavelength at the center frequency and two associated taps. The processor of Figure 11.5*c* has two delay elements, each corresponding to one quarter wavelength at the center frequency, and three associated taps. The processor of Figure 11.5*d* has four delay elements, each corresponding to one-eighth wavelength at the center frequency, and five associated taps. Note that the total delay present in tapped-delay line of Figure 11.5*d* is the same as that of Figure 11.5*c*, so the processor in Figure 11.5*d* may be regarded as a more finely subdivided version of the processor in Figure 11.5*c*.

Assume that the desired signal is biphase modulated of the form

$$s_d(t) = A\cos[\omega_0 t + \phi(t) + \theta] \tag{11.46}$$

where $\phi(t)$ denotes a phase angle that is either zero or $\pi$ over each bit interval, and $\theta$ is an arbitrary constant phase angle (within the range $[0, 2\pi]$) for the duration of any signal pulse. The $n$th bit interval is defined over $T_0 + (n-1)T \leqslant t \leqslant T_0 + nT$ where $n$ is any integer, $T$ is the bit duration, and $T_0$ is a constant that determines where the bit transitions occur, as shown in Figure 11.6.

Assume that $\phi(t)$ is statistically independent over different bit intervals and is zero or $\pi$ with equal probability, and that $T_0$ is uniformly distributed over one bit interval, then $s_d(t)$ is a stationary random process with power spectral density given by

$$\phi_{dd}(\omega) = \frac{A^2 T}{2}\left[\frac{\sin(T/2)(\omega - \omega_0)}{(T/2)(\omega - \omega_0)}\right]^2 \tag{11.47}$$

This power spectral density is shown in Figure 11.7.

The reference signal is assumed to be equal to the desired signal component of $x_1(t)$, and is time aligned with the desired component of $x_2(t)$. The desired signal "bandwidth" will be taken to be the frequency range defined by the first nulls of the spectrum given by (11.47). With this definition, the

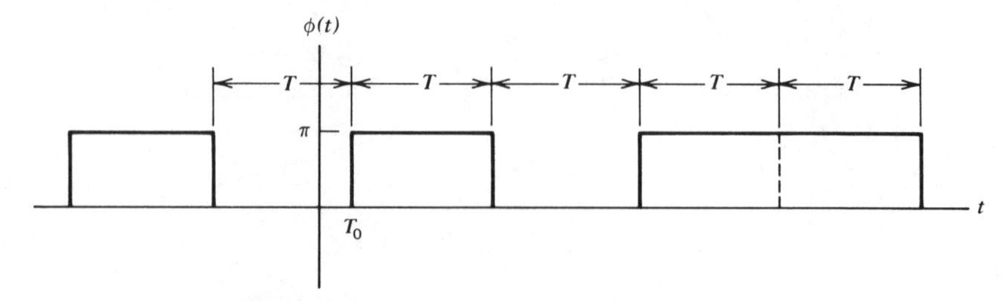


**Figure 11.6** Bit transitions for biphase modulated signal.

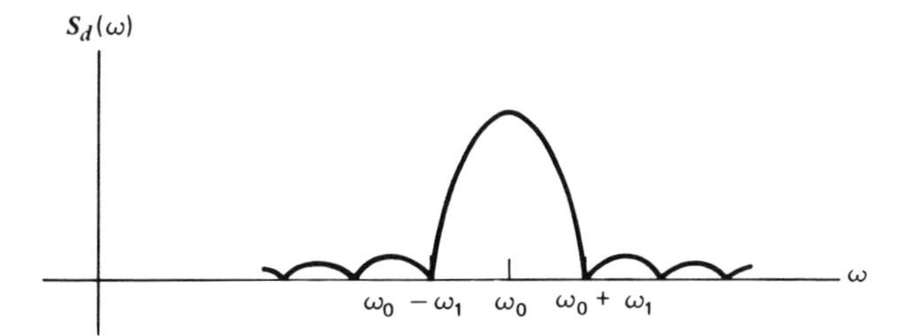


**Figure 11.7** Desired signal power spectral density.

fractional bandwidth then becomes

$$\text{desired signal bandwidth} = \frac{2\omega_1}{\omega_0} \tag{11.48}$$

where $\omega_1$ is the frequency separation between the center frequency $\omega_0$ and the first null

$$\omega_1 = \frac{2\pi}{T} \tag{11.49}$$

Assume the interference signal is a Gaussian random process with a flat, bandlimited power spectral density over the range $\omega_0 - \omega_1 < \omega < \omega_0 + \omega_1$, then the interference signal spectrum appears as shown in Figure 11.8. Finally, the thermal noise signals present at each element are assumed to be statistically independent between elements, having a flat, bandlimited, Gaussian spectral density over the range $\omega_0 - \omega_1 < \omega < \omega_0 + \omega_1$ (identical with the interference spectrum of Figure 11.8).

With the foregoing definitions of signal spectra, the integrals of (11.42) and (11.45) yielding interference and thermal noise power are taken only over the frequency range $\omega_0 - \omega_1 < \omega < \omega_0 + \omega_1$. The desired signal power also is considered only over the frequency range $\omega_0 - \omega_1 < \omega < \omega_0 + \omega_1$ to obtain a consistent

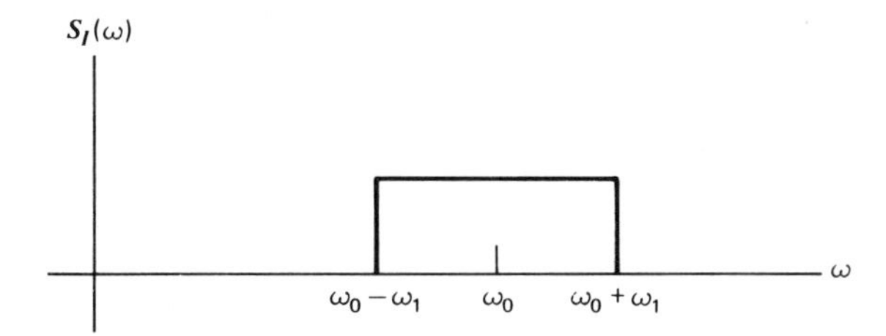


**Figure 11.8** Interference signal power spectral density.

definition of SNR. Consequently the integral of (11.41) is only carried out over $\omega_0-\omega_1<\omega<\omega_0+\omega_1$.

To compare the four adaptive array processors of Figure 11.5, the output SNR performance is evaluated for the above signal conditions. Assume the element thermal noise power $p_n$ is 10 dB below the element desired signal power $p_s$ so that $p_s/p_n=10$ dB. Furthermore suppose that the element interference signal power $p_i$ is 20 dB stronger than the element desired signal power so that $p_s/p_i=-20$ dB. Now assume that the desired signal is incident on the array from broadside. The output SNR given by (11.40) can be evaluated from (11.41), (11.42), and (11.43) by assuming the processor weights satisfy (11.10) for each of the four processor configurations. The resulting output signal-to-total noise ratio that results using each processor is plotted in Figures 11.9–11.12 as a function of the interference angle of arrival for 4, 10, 20, and 40% bandwidth signals, respectively.

It may be seen in all cases that, regardless of the signal bandwidth, when the interference approaches broadside (near the desired signal) the SNR degrades rapidly, and the performance of all four processors becomes identical. This SNR degradation is to be expected since when the interference approaches the desired signal, the desired signal falls into the null provided to cancel the interference; and the output SNR consequently falls. Furthermore,

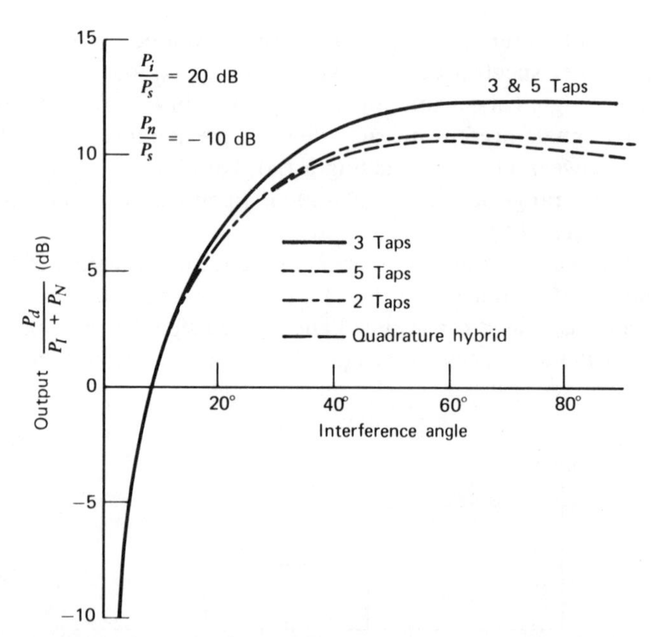


**Figure 11.9** Output signal to interference plus noise ratio interference angle for four adaptive processors with 4% bandwidth signal. From Rodgers and Compton, *IEEE Trans. Aerosp. Electron. Syst.*, January 1979 [3].

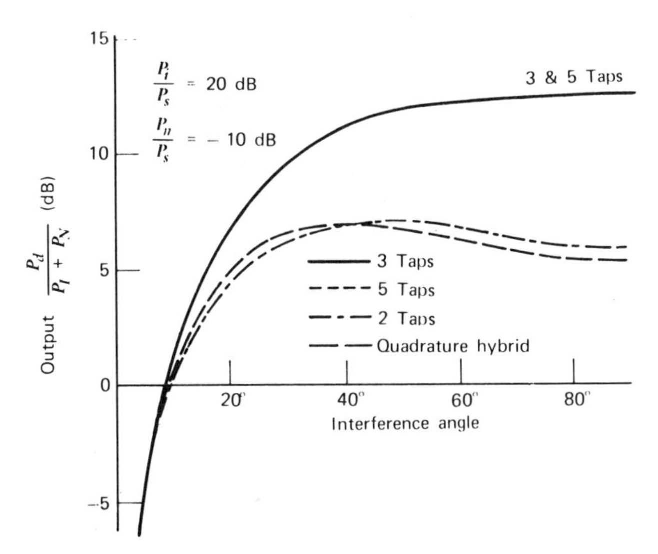


**Figure 11.10** Output signal to interference plus noise ratio versus interference angle for four adaptive processors with 10% bandwidth signal. From Rodgers and Compton, *IEEE Trans. Aerosp. Electron. Syst.*, January 1979 [3].

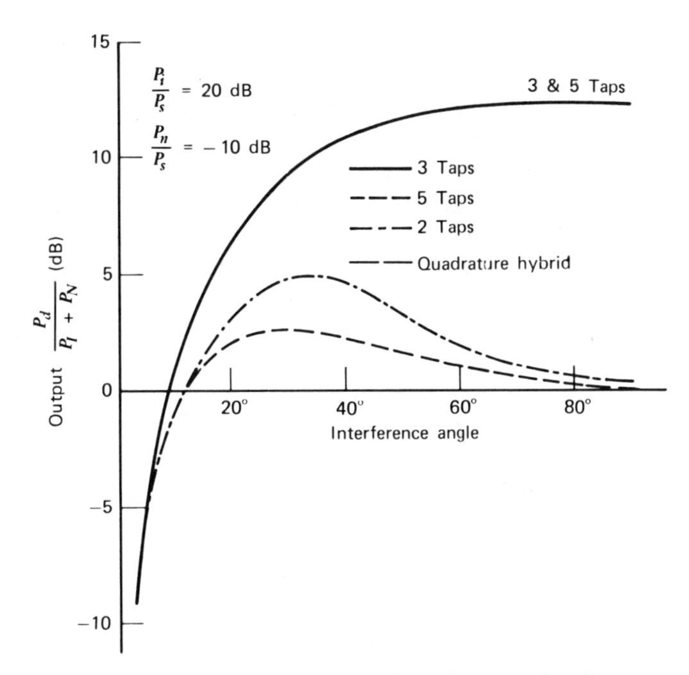


**Figure 11.11** Output signal to interference plus noise ratio versus interference angle for four adaptive processors with 20% bandwidth signal. From Rodgers and Compton, *IEEE Trans. Aerosp. Electron. Syst.*, January 1979 [3].

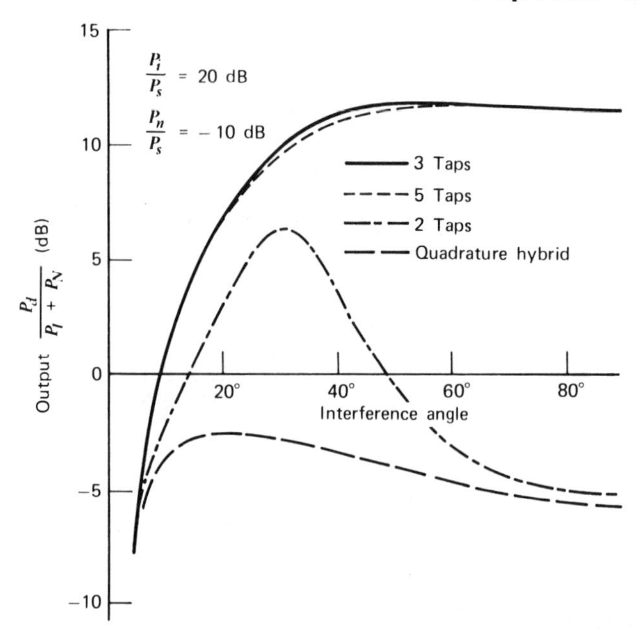


**Figure 11.12** Output signal to interference plus noise ratio versus interference angle for four adaptive processors with 40% bandwidth signal. From Rodgers and Compton, *IEEE Trans. Aerosp. Electron. Syst.*, January 1979 [3].

as the interference approaches broadside, the interelement phase shift for this signal approaches zero. Consequently the need to provide a frequency dependent phase shift behind each array element to deal with the interference signal is less, and the performance of all four processors becomes identical.

When the interference signal is widely separated from the desired signal, then the output SNR is different for the four processors being considered, and this difference becomes more pronounced as the bandwidth increases. For 20 and 40% bandwidth signals for example, neither the quadrature hybrid processor nor the two-tap delay line processor is capable of providing good performance as the interference signal approaches endfire. The performance of both the three- and five-tap delay line processors remains quite good in the endfire region, however. If 20% or more bandwidth signals must be accommodated, then tapped-delay line processing becomes a necessity. Figure 11.12 shows that there is no significant performance advantage provided by the five-tap processor compared to the three-tap processor, so a three-tap processor is adequate for up to 40% bandwidth signals in the case of a two-element array.

It may be noted that in Figures 11.11 and 11.12, the output SNR performance of the two-tap delay line processor peaks when the interference signal is located at 30° off broadside. This result occurs because when the interference is 30° off broadside, the interelement delay time is $\lambda/4$ (since the elements are spaced apart by $\lambda/2$). Consequently the single delay element value of $\lambda/4$ provides just the right amount of time delay to compensate exactly for the interelement time delay and to produce an improvement in the output SNR.

The three-tap and five-tap delay line processors both produce a maximum SNR of about 12.5 dB at wide interference angles of 70° or greater. For ideal channel processing, the interference signal is eliminated, the desired signal in each channel is added coherently to produce $P_d = 4p_s$, and the thermal noise is added noncoherently to yield $P_N = 2p_n$. Consequently the best possible theoretical output SNR for a two-element array with thermal noise 10 dB below the desired signal and no interference is 13 dB. Therefore the three-tap and five-tap delay line processors are successfully rejecting nearly all the interference signal power at wide off-boresight angles.

### *11.1.4 Processor Transfer Functions*

It is informative to examine the transfer function of the array as seen by the desired signal and the interference for each of the four processors considered

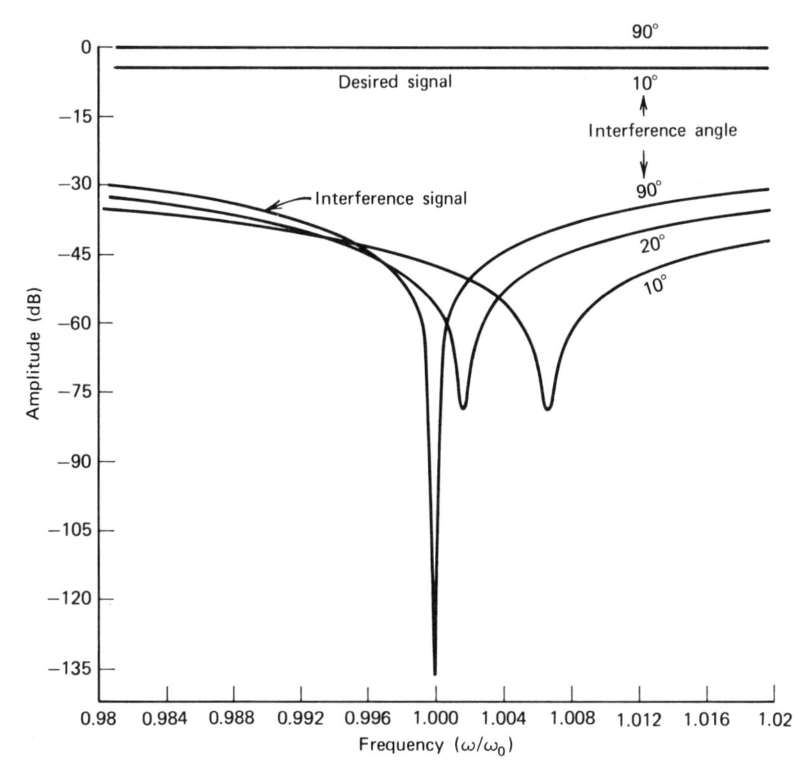


**Figure 11.13** Quadrature hybrid transfer functions at 4% bandwidth. From Rodgers and Compton, Technical Report ESL 3832-3, 1975 [2].

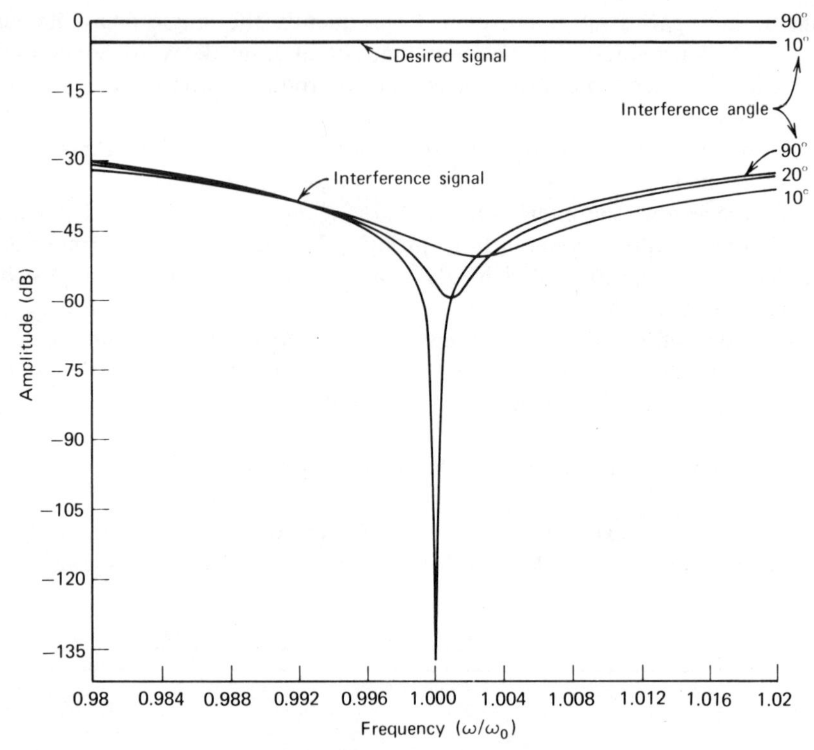


**Figure 11.14** Two-tap delay line transfer functions at 4% bandwidth. From Rodgers and Compton, Technical Report ESL 3832-3, 1975 [2].

in the preceding section. The array transfer function presented to the desired signal should ideally be constant across the desired signal bandwidth, thereby preventing desired signal distortion. For the interference signal it is only desired to maintain a suitably low array response over the interference bandwidth.

The transfer functions for the four processors and the two-element array can be evaluated using (11.34)–(11.39). Using the same conditions adopted in computing the SNR performance, Figures 11.13–11.16 show $|H_d(\omega)|$ and $|H_I(\omega)|$ for the four processors of Figure 11.5 with a 4% signal bandwidth and various interference signal angles. The results shown in these figures indicate that for all four processors and for all interference angles, the desired signal response is quite flat over the signal bandwidth. As the interference approaches the desired signal angle at broadside, however, the (constant) response level of the array to the desired signal drops as a consequence of the desired signal partially falling within the array pattern interference null.

The results in Figure 11.13 for quadrature hybrid processing show that the array response to the interference signal has a deep notch at the center frequency when the interference signal is well separated ($\theta_i > 20°$) from the

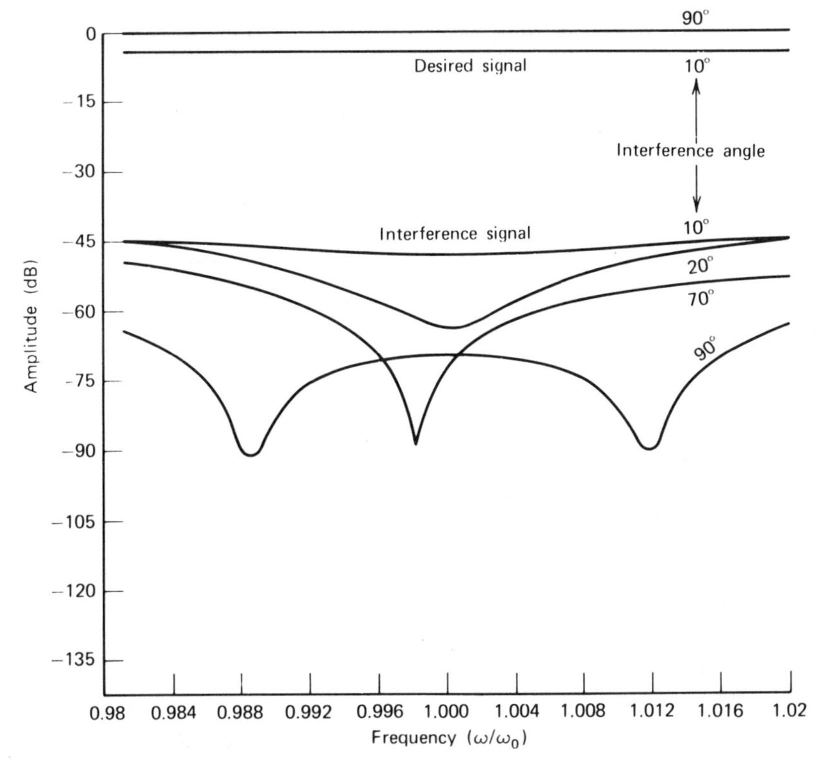


**Figure 11.15** Three-tap delay line transfer functions at 4% bandwidth. From Rodgers and Compton, Technical Report ESL 3832-3, 1975 [2].

desired signal. As the interference signal is permitted to approach the desired signal ($\theta_i < 20°$), the notch migrates away from the center frequency because the processor weights must compromise between rejection of the interference signal and enhancement of the desired signal when the two signals are close. Migration of the notch improves the desired signal response (since the desired signal power spectral density peaks at the center frequency) while affecting interference rejection only slightly (since the interference signal power spectral density is constant over the signal band).

The array response for the two-tap processor is shown in Figure 11.14. It is seen that the response to both the desired and interference signals is very similar to that obtained for quadrature hybrid processing. The most notable change is the slightly different shape of the transfer function notch presented to the interference signal by the two-tap delay line processor compared with the quadrature hybrid processor.

The three-tap processor array response is presented in Figure 11.15. The most significant difference is that the interference signal response is reduced considerably, with a minimum rejection of the interference signal of about 45 dB. When the interference signal is close to the desired signal, the array

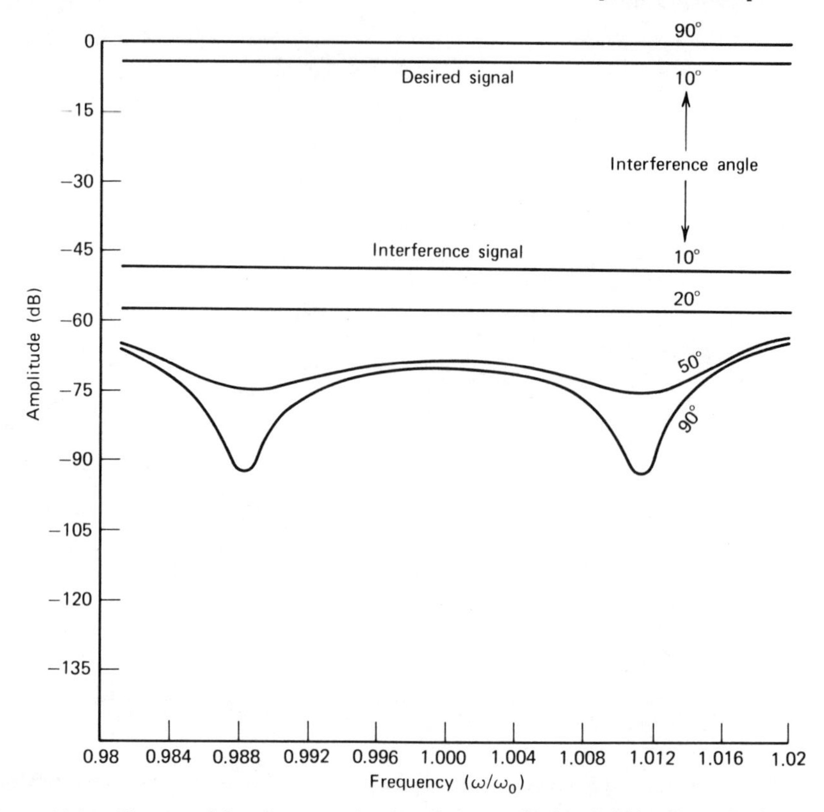


**Figure 11.16** Five-tap delay line transfer functions at 4% bandwidth. From Rodgers and Compton, Technical Report ESL 3832-3, 1975 [2].

response is characterized by a single mild dip. As the separation angle between the interference signal and the desired signal increases, the single dip becomes more pronounced and finally develops into a double dip at very wide angles. It is difficult to attribute much significance to the double-dip behavior since it occurs at such a low response level (of more than 75 dB attenuation).

The five-tap processor response of Figure 11.16 is very similar to the three-tap processor response. The most significant difference is that slightly more interference signal rejection is achieved.

As the signal bandwidth increases, the nature of the processor response curves remains the same as in Figures 11.13–11.16 with only two following significant developments occurring:

1. As the interference signal bandwidth increases, it becomes more difficult to reject the interference signal over the entire bandwidth, so the minimum rejection level increases.

2. The desired signal response decreases because the array feedback reduces all weights to compensate for the presence of a greater interference signal component at the array output, thereby resulting in greater desired signal attenuation.

The net result is that as the signal bandwidth increases, the output SNR performance degrades, a fact that is confirmed by the results of Figures 11.9–11.12.

## 11.2 MULTIPATH COMPENSATION

In many operating environments, multipath rays impinge on the array face slightly after the direct path signal wavefront arrives at the sensors. The presence of such multipath rays has the effect of distorting any interference signal that may appear in the various element channels, thereby severely limiting the interference cancellation that can be achieved by the adaptive array without compensation. A tapped-delay line processor combines delayed and weighted replicas of the input signal to form the filtered output signal and thereby has the potential for providing compensation for multipath effects since multipath rays also consist of delayed and weighted replicas of the direct path ray.

### *11.2.1 Two-Channel Interference Cancellation Model*

Consider an ideal two-element adaptive array system in which it is desired that one channel's (called the "auxiliary" channel) response be adjusted so that any jamming signal entering the other channel through the sidelobes (termed the "main" channel) is cancelled at the array output. A system designed to suppress sidelobe jamming in this manner is called a coherent sidelobe canceller (CSLC), and Figure 11.17 depicts a two-channel CSLC system in which the auxiliary channel employs tapped-delay line compensation involving $L$ weights and $L-1$ delay elements of value $\Delta$ seconds each. A delay element of value $D=(L-1)\Delta/2$ is included in the main channel so the center tap of the auxiliary channel corresponds to the output of the delay $D$ in the main channel thereby permitting compensation for both positive and negative values of the off-broadside angle $\theta$. This ideal two-element CSLC system model exhibits all the salient characteristics that a more complex system involving several auxiliary channels would also have, so the two-element system serves as a convenient model for performance evaluation of multipath cancellation [5].

The system performance is to be evaluated in terms of the ability of the CSLC to cancel an undesired interference signal through proper design of the tapped-delay line. In actual practice, the weight adjustment is achieved by means of an adaptive algorithm that modifies the weight settings. To

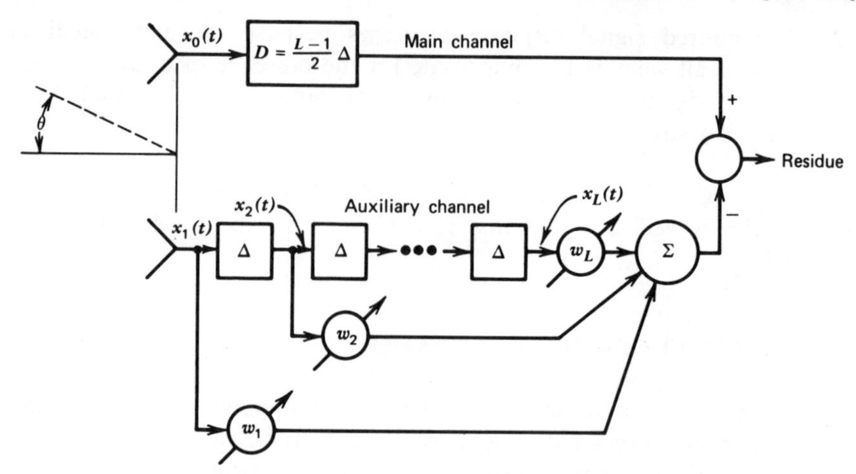


**Figure 11.17** Ideal two-element CSLC model with auxiliary channel compensation involving $L$ weights and $L-1$ delay elements.

eliminate the effect of algorithm selection from consideration, only the steady-state performance will be evaluated. Since the steady-state solution can be found analytically, it is necessary to determine only the resulting solution for the output residue power. This residue power is then a direct measure of the interference cancellation ability of the two-element CSLC model.

Let $x_0(t), x_1(t)$, and $e(t)$ represent the complex envelope signals of the main channel input signal, the auxiliary channel input signal, and the output residue signal, respectively. Define the complex signal vector

$$\mathbf{x}^T \triangleq \left[x_1(t), x_2(t), \ldots, x_L(t)\right] \tag{11.50}$$

where

$$x_2(t) \triangleq x_1(t-\Delta)$$

$$\vdots$$

$$x_L(t) \triangleq x_1\left[t-(L-1)\Delta\right]$$

Also define the complex weight vector

$$\mathbf{w}^T \triangleq \left[w_1, w_2, \ldots, w_L\right] \tag{11.51}$$

The output of the tapped-delay line may then be expressed as

$$\text{filter output} = \sum_{i=1}^{L} x_1\left[t-(i-1)\Delta\right] w_i^* = \mathbf{w}^\dagger \mathbf{x}(t) \tag{11.52}$$

The residue (complex envelope) signal is then given by

$$e(t)=x_0(t-D)+\mathbf{w}^\dagger\mathbf{x}(t) \tag{11.53}$$

It is desired to select the weight vector **w** to minimize the residue signal in a MSE sense. For stationary random processes, this is equivalent to minimizing the expression

$$R_{ee}(0)=E\{e(t)e^*(t)\} \tag{11.54}$$

From (11.53) and the fact that

$$E\{x_0(t-D)x_0^*(t-D)\}=r_{x_0x_0}(0) \tag{11.55}$$

$$E\{\mathbf{x}(t)x_0^*(t-D)\}=r_{xx_0}(-D) \tag{11.56}$$

and

$$E\{\mathbf{x}(t)\mathbf{x}^\dagger(t)\}=\mathbf{R}_{xx}(0) \tag{11.57}$$

it follows that

$$R_{ee}(0)=r_{x_0x_0}(0)-\mathbf{r}_{xx_0}^\dagger(-D)\mathbf{R}_{xx}^{-1}(0)\mathbf{r}_{xx_0}(-D) + \left[\mathbf{r}_{xx_0}^\dagger(-D)+\mathbf{w}^\dagger\mathbf{R}_{xx}(0)\right]\cdot\mathbf{R}_{xx}^{-1}(0)\left[\mathbf{r}_{xx_0}(-D)+\mathbf{R}_{xx}(0)\mathbf{w}\right] \tag{11.58}$$

It is desired to minimize (11.58) by appropriately selecting the complex weight vector **w**. Assume the matrix $\mathbf{R}_{xx}(0)$ is nonsingular: the value of **w** for which this minimum occurs is given by

$$\mathbf{w}_{\text{opt}}=\mathbf{R}_{xx}^{-1}(0)\mathbf{r}_{xx_0}(-D) \tag{11.59}$$

The corresponding minimum residue signal power then becomes

$$R_{ee}(0)_{\min}=r_{x_0x_0}(0)-\mathbf{r}_{xx_0}^\dagger(-D)\mathbf{R}_{xx}^{-1}(0)\mathbf{r}_{xx_0}(-D) \tag{11.60}$$

Interference cancellation performance of the CSLC model of Figure 11.17 can now be determined by evaluating (11.60) using selected signal environment assumptions.

### *11.2.2 Signal Environment Assumptions*

Let $s_1(t,\theta_1)$ represent the interference signal arriving from direction $\theta_1$, and let $s_m(t,\rho_m,D_m,\theta_{m+1})$ for $m=2,\ldots,M$ represent the multipath structure associated with the interference signal that is assumed to consist of a collection of $M-1$ correlated plane-wave signals of the same frequency arriving from different directions so that $\theta_{m+k} \neq \theta_1$ and $\theta_{m+k} \neq \theta_{m+l}$ for $k \neq l$. The multipath rays each have an associated reflection coefficient $\rho_m$ and a time delay with respect to the direct ray $D_m$. The structure of the covariance matrix for this multipath model can then be expressed as [6]

$$\mathbf{R}_{ss} = \mathbf{V}_s \mathbf{A} \mathbf{V}_s^{\dagger} \tag{11.61}$$

where $\mathbf{V}_s$ is the $N \times M$ signal matrix given by

$$\mathbf{V}_s = \left[ \mathbf{v}_{s_1} \,\vdots\, \mathbf{v}_{s_2} \,\vdots\, \cdots \,\vdots\, \mathbf{v}_{s_M} \right] \tag{11.62}$$

whose components are given by the $N \times 1$ vectors

$$\mathbf{v}_{s_m} = \sqrt{P_{s_m}} \begin{bmatrix} 1 \\ \exp[j2\pi(d/\lambda_0)\sin\theta_m] \\ \exp[j2\pi(d/\lambda_0)2\sin\theta_m] \\ \vdots \\ \exp[j2\pi(d/\lambda_0)(N-1)\sin\theta_m] \end{bmatrix} \tag{11.63}$$

where $P_{s_m} = \rho_m^2$ denotes the power associated with the signal $s_m$, and $\mathbf{A}$ is the multipath correlation matrix. When $\mathbf{A} = \mathbf{I}$, the various signal components are uncorrelated whereas for $\mathbf{A} = \mathbf{U}$ (the $M \times M$ matrix of unity elements) the various components are perfectly correlated. For purposes of numerical evaluation the correlation matrix model may be selected as [6]

$$\mathbf{A} = \begin{bmatrix} 1 & \alpha & \alpha^2 & \cdots & \alpha^{M-1} \\ \alpha & 1 & \alpha & \cdots & \alpha^{M-1} \\ \vdots & & & & \\ \alpha^{M-1} & \cdot & \cdot & \cdots & 1 \end{bmatrix} \qquad 0 \leqslant \alpha \leqslant 1 \tag{11.64}$$

Note that channel-to-channel variations in $\theta_m$, $D_m$, and $\rho_m$ cannot be accommodated by this simplified model. Consequently, a more general model must be developed to handle such variations which tend to occur where near-field scattering effects are significant.

The input signal covariance matrix may be written as

$$\mathbf{R}_{xx} = \mathbf{R}_{nn} + \mathbf{V}_s \mathbf{A} \mathbf{V}_s^{\dagger} \tag{11.65}$$

where $\mathbf{R}_{nn}$ denotes the noise covariance matrix.

For simplicity it is convenient to assume that only a single multipath ray is present. In this case let $s(t, \theta_1)$ denote the direct interference signal, and let $s_m(t, \rho_m, D_m, \theta_2)$ represent the multipath ray associated with the direct interference signal. The received signal at the main channel element is then given by

$$x_0(t) = s(t,\theta_1) + s_m(t,\rho_m,D_m,\theta_2) \tag{11.66}$$

Denote $s(t,\theta_1)$ by $s(t)$, then $s_m(t,\rho_m,D_m,\theta_2)$ can be written as $\rho_m s(t-D_m)\times \exp(-j\omega_0 D_m)$ so that

$$x_0(t) = s(t) + \rho_m s(t-D_m)\exp(-j\omega_0 D_m) \tag{11.67}$$

where $\omega_0$ is the center frequency of the interference signal. It then follows that

$$x_1(t) = s(t-\tau_{12})\exp(-j\omega_0\tau_{12}) + \rho_m s(t-D_m-\tau_{22})\exp[-j\omega_0(D_m+\tau_{22})] \tag{11.68}$$

where $\tau_{12}$ and $\tau_{22}$ represent the propagation delay between the main channel element and the auxiliary channel element for the wavefronts of $s(t,\theta_1)$ and $s_m(t,\rho_m,d_m,\theta_2)$, respectively.

Assume the signals $s(t,\theta_1)$ and $s_m(t,\rho_m,D_m,\theta_2)$ possess flat spectral density functions over the bandwidth $B$, as shown in Figure 11.18*a*, then the corresponding auto- and cross-correlation functions of $x_0(t)$ and $x_1(t)$ can be evaluated by recognizing that

$$\mathbf{R}_{xx}(\tau) = \mathfrak{F}^{-1}\{\mathbf{\Phi}_{xx}(\omega)\} \tag{11.69}$$

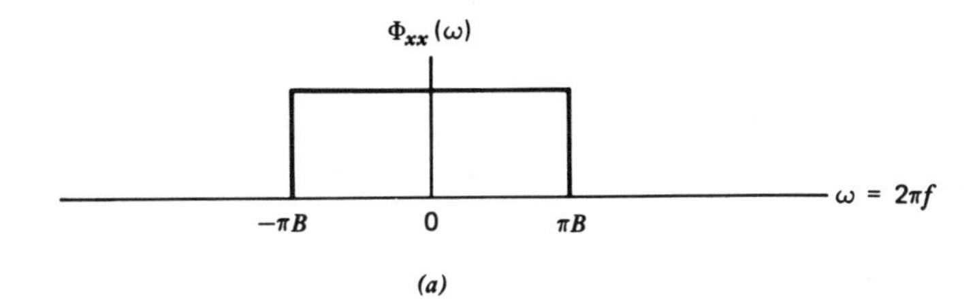


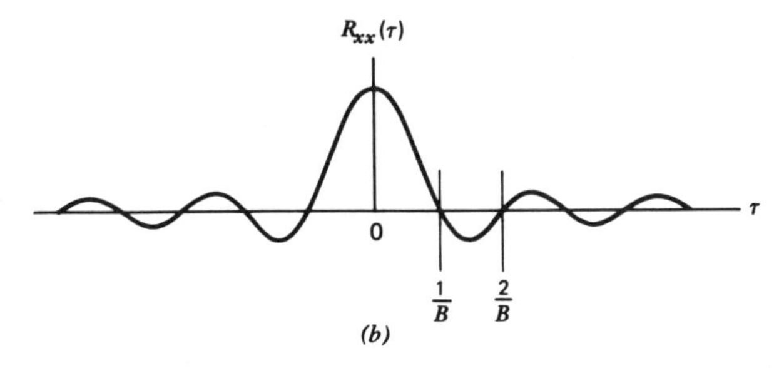


**Figure 11.18** Flat spectral density function and corresponding autocorrelation function for interference signal. (*a*) Spectral Density Function. (*b*) Autocorrelation Function.

where $\mathfrak{F}^{-1}\{\cdot\}$ denotes "inverse Fourier transform," and $\mathbf{\Phi}_{xx}(\omega)$ denotes the cross-spectral density matrix of $\mathbf{x}(t)$.

From (11.66), (11.68), and (11.69) it immediately follows that

$$r_{x_0x_0}(0)=1+|\rho_m|^2 + \frac{\sin \pi BD_m}{\pi BD_m}\left(\rho_m e^{-j\omega_0 D_m}+\rho_m^* e^{j\omega_0 D_m}\right) \tag{11.70}$$

Likewise defining $f[\psi,\text{sgn}1,\text{sgn}2] \triangleq \dfrac{\sin \pi B[\psi+\text{sgn}1\cdot(i-1)\Delta+\text{sgn}2\cdot D]}{\pi B[\psi+\text{sgn}1\cdot(i-1)\Delta+\text{sgn}2\cdot D]}$

and $g[\psi,\text{sgn}] \triangleq \dfrac{\sin \pi B[\psi+\text{sgn}\cdot(i-k)\Delta]}{\pi B[\psi+\text{sgn}\cdot(i-k)\Delta]}$, then

$$\begin{aligned} r_{x_ix_0}(-D)=&f[\tau_{12},+,-]\exp\{-j\omega_0[\tau_{12}+(i-1)\Delta]\}\\ &+f[D_m+\tau_{22},+,-]\rho_m\exp\{-j\omega_0[\tau_{22}+(i-1)\Delta+D_m]\}\\ &+f[D_m-\tau_{12},-,+]\rho_m^*\exp\{-j\omega_0[\tau_{12}+(i-1)\Delta-D_m]\}\\ &+f[\tau_{22},+,-]|\rho_m|^2\exp\{-j\omega_0[\tau_{22}+(i-1)\Delta]\} \end{aligned} \tag{11.71}$$

$$\begin{aligned} r_{x_ix_k}(0)=&g[0,+][1+|\rho_m|^2]\exp[-j\omega_0(i-k)\Delta]\\ &+g[\tau_{12}-\tau_{22}-D_m,-]\rho_m\exp\{j\omega_0[\tau_{12}-\tau_{22}-D_m-(i-k)\Delta]\}\\ &+g[\tau_{12}-\tau_{22}-D_m,+]\rho_m^*\exp\{-j\omega_0[\tau_{12}-\tau_{22}-D_m+(i-k)\Delta]\} \end{aligned} \tag{11.72}$$

The vector $\mathbf{r}_{xx_0}(-D)$ is then given by

$$\mathbf{r}_{xx_0}(-D)=\begin{bmatrix} r_{x_1x_0}(-D)\\ r_{x_2x_0}(-D)\\ \vdots\\ r_{x_Nx_0}(-D)\end{bmatrix} \tag{11.73}$$

and the matrix $\mathbf{R}_{xx}(0)$ is given by

$$\mathbf{R}_{xx}(0)=\begin{bmatrix} r_{x_1x_1}(0) & r_{x_1x_2}(0) & \cdots & r_{x_1x_N}(0)\\ \vdots & r_{x_2x_2}(0) & & \\ & & \ddots & \\ r_{x_1x_N}(0) & \cdots & & r_{x_Nx_N}(0)\end{bmatrix} \tag{11.74}$$

In order to evaluate (11.60) for the minimum possible value of output residue power (11.70), (11.71), and (11.72) show that it is necessary to specify the following parameters:

$N$ = number of taps in the transversal filter
$\rho_m$ = multipath reflection coefficient
$\omega_0$ = (radian) center frequency of interference signal
$D_m$ = multipath delay time with respect to direct ray
$\tau_{12}$ = propagation delay between the main antenna element and the auxiliary antenna element for the direct ray
$\tau_{22}$ = propagation delay between the main antenna element and the auxiliary antenna element for the multipath ray
$\Delta$ = transversal filter intertap delay
$B$ = interference signal bandwidth
$D$ = main channel receiver time delay

The quantities $\tau_{12}$ and $\tau_{22}$ are related to the CSLC array geometry by

$$\left.\begin{aligned} \tau_{12} &= \frac{d}{\mathfrak{v}} \sin\theta_1 \\ \tau_{22} &= \frac{d}{\mathfrak{v}} \sin\theta_2 \end{aligned}\right\} \tag{11.75}$$

where

$d$ = interelement array spacing
$\mathfrak{v}$ = wavefront propagation speed
$\theta_1$ = angle of incidence of direct ray
$\theta_2$ = angle of incidence of multipath ray

### *11.2.3 Example: Results for Compensation of Multipath Effects*

Consider an interference signal for which the direct ray angle-of-arrival is $\theta_1 = 30°$, the multipath ray angle-of-arrival is $\theta_2 = -30°$, and the interelement spacing is $d = 2.25\lambda_0$. In order to specify all the parameters defined in the previous section, it is necessary to characterize further the interference signal and multipath ray as follows:

$$\begin{aligned} &\text{center frequency} \quad f_0 = 237\text{ MHz} \\ &\text{signal bandwidth} \quad B = 3\text{ MHz} \\ &\text{multipath reflection coefficient} \quad \rho_m = 0.5 \end{aligned} \tag{11.76}$$

Referring to (11.68), (11.71), and (11.72), we see that the parameters $\omega_0$, $\tau_{12}$, $\tau_{22}$, $D_m$, and $\Delta$ enter the evaluation of the output residue power in the form of the products $\omega_0\tau_{12}$, $\omega_0\tau_{22}$, $\omega_0 D_m$, and $\omega_0\Delta$. These products represent the phase shift experienced at the center frequency $\omega_0$ as a consequence of the four corresponding time delays. Likewise it may also be seen that the parameters $B$, $D$, $D_m$, $\tau_{12}$, $\tau_{22}$, and $\Delta$ enter the evaluation of the output residue power in the form of the products $BD$, $BD_m$, $B\tau_{12}$, $B\tau_{22}$, and $B\Delta$; these time-bandwidth products may be regarded as phase shifts experienced by the highest frequency component of the complex envelope interference signal as a consequence of the five corresponding time delays. In the results to be obtained, both the intertap delay $\Delta$ and the multipath delay $D_m$, are important parameters that affect the CSLC system performance through their corresponding time-bandwidth products: consequently the results are given here with the time-bandwidth products taken as the fundamental quantity of interest.

Since for this example $\theta_1 = -\theta_2$, it follows that if the product $\omega_0\tau_{12}$ is specified as

$$\left.\begin{aligned} &\omega_0\tau_{12} = \frac{\pi}{4} \\ \text{then the product} \quad & \\ &\omega_0\tau_{22} = -\frac{\pi}{4} \end{aligned}\right\} \tag{11.77}$$

Furthermore let the products $\omega_0 D_m$ and $\omega_0\Delta$ be given by

$$\left.\begin{aligned} \omega_0 D_m &= 0 \pm 2k\pi, \quad && k \text{ any integer} \\ \omega_0\Delta &= 0 \pm 2l\pi, \quad && l \text{ any integer} \end{aligned}\right\} \tag{11.78}$$

For the element spacing $d = 2.25\lambda_0$ and $\theta_1 = 30°$, then specify

$$B\tau_{12} = -B\tau_{22} = \frac{1}{P}, \qquad P = 72 \tag{11.79}$$

Finally, specifying the multipath delay time to correspond to 46 meters yields

$$BD_m = 0.45 \tag{11.80}$$

Since

$$D=\frac{N-1}{2}\Delta \tag{11.81}$$

it only remains to specify the parameters $N$ and $B\Delta$ in order to evaluate the output residue power by way of (11.60).

To evaluate the output residue power by way of (11.60) resulting from the array geometry and multipath conditions specified by (11.76)–(11.81) requires that the cross-correlation vector $\mathbf{r}_{xx_0}(-D)$, the $N\times N$ autocorrelation matrix $\mathbf{R}_{xx}(0)$, and the autocorrelation function $r_{x_0x_0}(0)$ be evaluated by way of (11.70)–(11.72). A computer program to evaluate (11.60) for the multipath conditions specified was written in complex, double precision arithmetic.

The results of the computer evaluation of output residue power are summarized in Figure 11.19 where the resulting minimum possible value of cancelled power output in dB is plotted as a function of $B\Delta$ for various specified values of $N$. It will be noted in Figure 11.19 that for $N=1$, the cancellation performance is independent of $B\Delta$ since no intertap delays are present with only a single tap. As explained in Appendix B the transfer function of the tapped-delay line transversal filter has a periodic structure with (radian) frequency period $2\pi B_f$ which is centered at the frequency $f_0$. It should be noted that the transversal filter frequency bandwidth $B_f$ is not necessarily the same as the signal frequency bandwidth $B$. The transfer

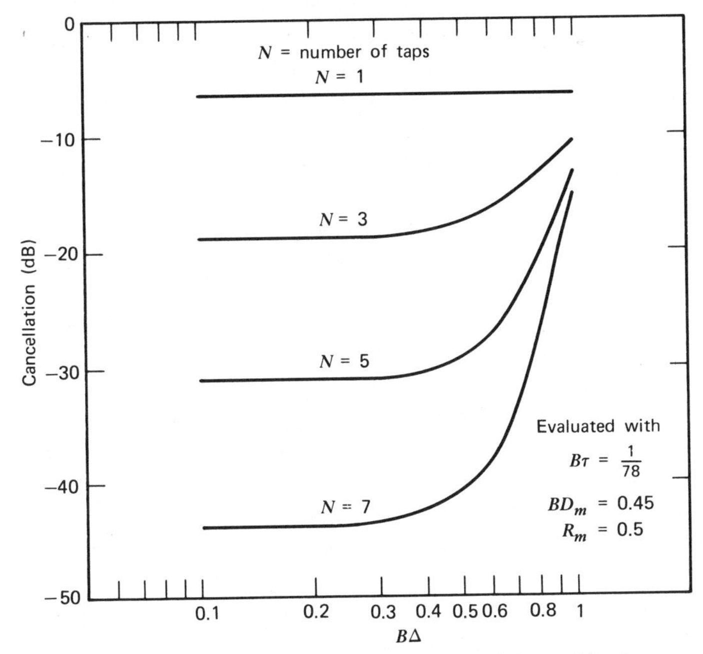


**Figure 11.19** Decibel cancellation versus $B\Delta$ for multipath.

function of a transversal filter within the primary frequency band ($|f-f_0|<B_f/2$) may be expressed as

$$F(f)=\sum_{k=1}^{N}\left[A_k e^{j\phi_k}\right]\exp\left[-j2\pi(k-1)\delta f\Delta\right] \tag{11.82}$$

where $A_k e^{j\phi_k}$ represents the $k$th complex weight, $\delta f=f-f_0, f_0=$center frequency; and the transversal filter frequency bandwidth is

$$B_f=\frac{1}{\Delta} \tag{11.83}$$

Since the transversal filter should be capable of adjusting the complex weights to achieve appropriate amplitude and phase values over the entire signal bandwidth $B$, it follows that $B_f$ should satisfy

$$B_f \geqslant B \tag{11.84}$$

Consequently, the maximum intertap delay spacing is given by

$$\Delta_{\max}=\frac{1}{B} \tag{11.85}$$

It follows that values of $B\Delta$ that are greater than *unity* should not be considered for practical compensation designs; however, values of $B_f>B$ (resulting in $0<B\Delta<1$) are *sometimes desirable.*

It is seen in Figure 11.19 that as $B\Delta$ decreases from 1, then for values of $N>1$ the cancellation performance rapidly improves (the minimum cancelled residue power decreases) until $B\Delta=BD_m$(0.45 for this example), after which very little significant improvement occurs. As $B\Delta$ is permitted to become very much smaller than $BD_m$ (approaching zero), the cancellation performance will once more degrade since the intertap delay is in effect being removed. The simulation could not compute this result since as $B\Delta$ approaches zero, the matrix $\mathbf{R}_{xx}(0)$ becomes singular and matrix inversion becomes impossible. Cancellation performance of $-30$ dB is virtually assured if the transversal filter has at least five taps and $\Delta$ is selected so that $\Delta=D_m$.

Suppose for example that the transversal filter is designed with $B\Delta=0.45$. Using the same set of selected constants as for the example above, we find it useful to consider what results would be obtained when the actual multipath delay is different from the anticipated value corresponding to $BD_m=0.45$. From the results already obtained in Figure 11.19, it may be anticipated that if $BD_m>B\Delta$, then the cancellation performance would degrade. If, however, $BD_m \ll B\Delta$, then the cancellation performance would improve since in the limit as $D_m \rightarrow 0$ the system performance with no multipath present would result.

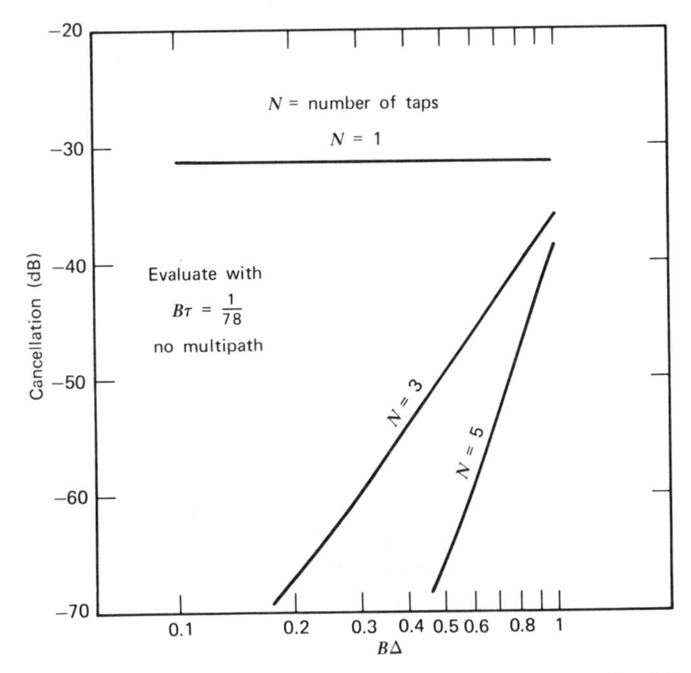


**Figure 11.20** Decibel cancellation versus $B\Delta$ for array propagation delay.

### *11.2.4 Results for Compensation of Array Propagation Delay*

In the absence of a multipath ray, the analysis presented in the preceding section includes all the features necessary to account for array propagation effects. When we set $\rho_m = 0$ and let $\tau_{12} = \tau$ represent the element-to-element array propagation delay, (11.70)–(11.72) permit (11.60) to be used to investigate the effects of array propagation delay on cancellation performance. On the basis of the behavior already found for multipath compensation, it would be reasonable to anticipate that with $B\Delta = B\tau$ then maximum cancellation performance would obtain, while if $B\Delta > B\tau$ then the cancellation performance would degrade. Figure 11.20 gives the resulting cancellation performance as a function of $B\Delta$ for fixed $B\tau$. The number of taps $N$ is an independent parameter, and all other system constants are the same as those in the example of Section 11.2.3. It is seen that the results confirm the anticipated performance noted above.

## 11.3 ANALYSIS OF INTERCHANNEL MISMATCH EFFECTS

Any adaptive array processor is susceptible to unavoidable frequency dependent variations in gain and phase between the various element channels. To compensate for such frequency dependent "channel mismatch" effects, it is possible to take advantage of the additional degrees of freedom provided by a tapped-delay line. Since a simple two-element CSLC system exhibits all the salient characteristics of channel mismatching present in more complex systems, the two-element model is again adopted as the example for performance evaluation of channel mismatch compensation.

Figure 11.21 is a simplified representation of a single auxiliary channel CSLC system in which the single complex weight is a function of frequency. The transfer function $T_0(\omega,\theta)$ reflects all amplitude and phase variations in the main beam sidelobes as a function of frequency as well as any tracking errors in amplitude and phase between the main and auxiliary channel electronics. Likewise the equivalent transfer function for the auxiliary channel (including any auxiliary antenna variations) is denoted by $T_1(\omega,\theta)$. The spectral power density of a wideband jammer is given by $\phi_{JJ}(\omega)$. The signal from the auxiliary channel is "multiplied" by the complex weight $w_1=\alpha e^{j\phi}$ and the "cancelled" output of residue power spectral density is represented by $\phi_{rr}(\omega,\theta)$.

The objective of the CSLC is to minimize the residue power, appropriately weighted, over the bandwidth. Since the integral of the power spectral density over the signal frequency spectrum yields the signal power, the requirement to minimize the residue power can be expressed as

$$\underset{w_1}{\text{Min}} \int_{-\infty}^{\infty} \phi_{rr}(\omega,\theta)\, d\omega \tag{11.86}$$

where

$$\phi_{rr}(\omega,\theta)=|T_0(\omega,\theta)-w_1 T_1(\omega,\theta)|^2\phi_{JJ}(\omega) \tag{11.87}$$

Now replace the complex weight $w_1$ in Figure 11.21 by a tapped-delay line having $2N+1$ adaptively controlled complex weights each separated by a

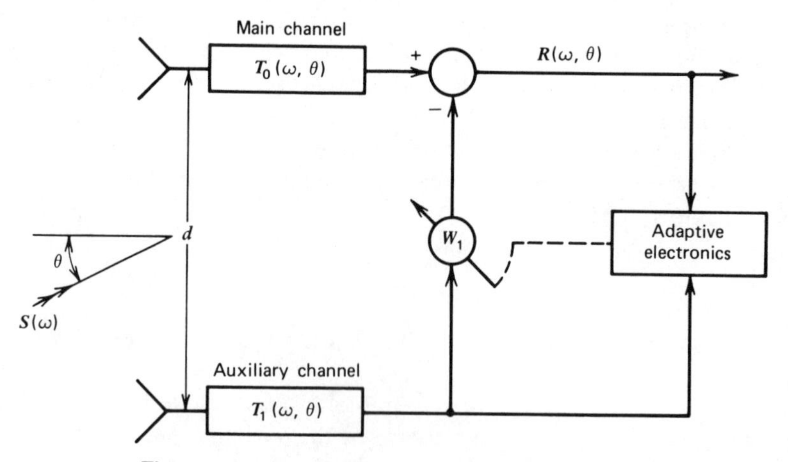


**Figure 11.21** Simplified model of single-channel CSLC.

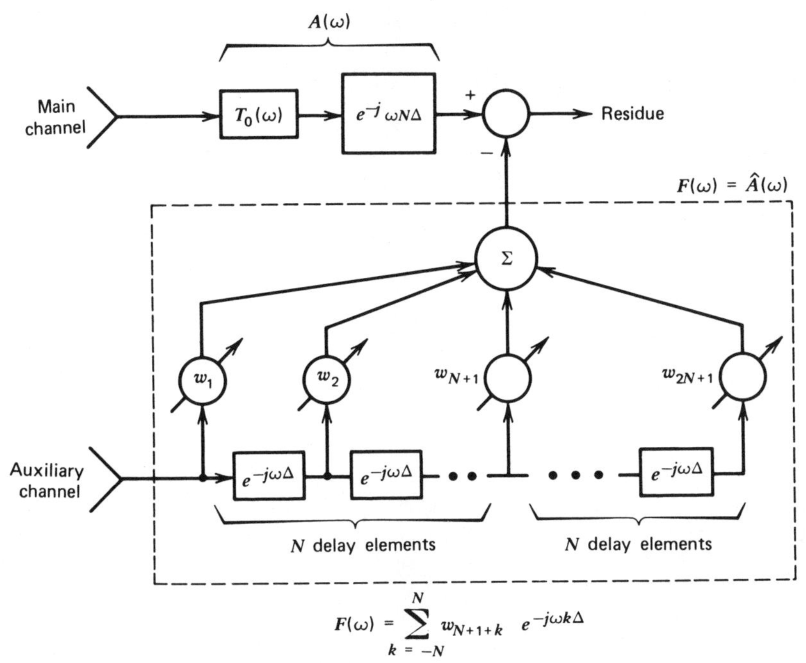


$$F(\omega) = \sum_{k=-N}^{N} w_{N+1+k}\, e^{-j\omega k\Delta}$$

**Figure 11.22** Single-channel CSLC having main channel distortion and tapped-delay line auxiliary channel compensation.

time delay $\Delta$ as in Figure 11.22. A delay element of value $N\Delta$ is included in the main channel (just as in the preceding section) so that compensation for both positive and negative angles of arrival can be provided. It is convenient to write the main and auxiliary channel transfer functions with respect to the output of the main channel so no delay terms occur in the resulting main channel transfer function denoted by $A(\omega)$. It is likewise convenient to assume for analysis purposes that all channel distortion is confined to the main channel and that $T_1(\omega,\theta)=1$. Denote the transversal filter transfer function by $F(\omega)$, it then follows that

$$F(\omega)=\sum_{k=-N}^{N} w_{N+1+k}e^{-j\omega k\Delta} \tag{11.88}$$

where the $w_{N+1+k}$'s are nonfrequency dependent complex weights.

It is desired to minimize the output residue power over the signal bandwidth by appropriately selecting the weight vector $\mathbf{w}$. Assuming the jammer power spectral density is a constant over the frequency region of interest, then minimizing the output residue power is equivalent to selecting the $F(\omega)$ that provides the "best" estimate (denoted by $\hat{A}(\omega)$) of the main channel transfer function over that frequency range. If the estimate $\hat{A}(\omega)$ is to be optimal in the MSE sense, then the error in this estimate $e(\omega)=A(\omega)-F(\omega)$

must be orthogonal to $\hat{A}(\omega)=F(\omega)$, that is,

$$E\{[A(\omega)-F(\omega)]F^*(\omega)\}=0 \tag{11.89}$$

where the expectation $E\{\cdot\}$ is taken over frequency and is therefore equivalent to

$$E\{\cdot\}=\frac{1}{2\pi B}\int_{-\pi B}^{\pi B}\{\ \}\,d\omega \tag{11.90}$$

where all frequency dependent elements in the integrand of (11.90) have been reduced to baseband. Letting $A(\omega)=A_0(\omega)e^{-j\phi_0(\omega)}$, substituting (11.88) into (11.89), and requiring the error to be orthogonal to all tap outputs to obtain the minimum MSE estimate $\hat{A}(\omega)$ then yields the condition

$$E\{[A_0(\omega)\exp[-j\phi_0(\omega)]-F(\omega)]\exp(j\omega k\Delta)\}=0 \qquad \text{for } k=-N,\ldots,0,\ldots,N \tag{11.91}$$

Equation (11.91) can be rewritten as

$$E\{A_0(\omega)\exp[j(\omega k\Delta-\phi_0(\omega))]\}-E\left\{\left[\sum_{l=-N}^{N} W_{N+1+l}\exp(-j\omega l\Delta)\right]\cdot\exp(j\omega k\Delta)\right\}=0 \qquad \text{for } k=-N,\ldots,0,\ldots,N \tag{11.92}$$

Note that

$$E\{\exp[-j\omega(l-k)\Delta]\}=\frac{\sin[\pi B\Delta(l-k)]}{\pi B\Delta(l-k)} \tag{11.93}$$

it follows that

$$E\left\{\left[\sum_{l=-N}^{N} W_{N+1+l}\exp(-j\omega l)\right]\exp(j\omega l\Delta)\right\}=\sum_{l=-N}^{N} W_{N+1+l}\frac{\sin[\pi B\Delta(l-k)]}{\pi B\Delta(l-k)} \tag{11.94}$$

so that (11.92) can be rewritten in matrix form as

$$\mathbf{v}=\mathbf{C}\mathbf{w} \tag{11.95}$$

where

$$v_k = E\{A_0(\omega)\exp[j(\omega k\Delta - \phi_0(\omega))]\} \tag{11.96}$$

and

$$C_{k,l} = \frac{\sin[\pi B\Delta(l-k)]}{\pi B\Delta(l-k)} \tag{11.97}$$

Consequently the complex weight vector must satisfy the relation

$$\mathbf{w} = \mathbf{C}^{-1}\mathbf{v} \tag{11.98}$$

Using (11.98) to solve for the optimum complex weight vector, we can find the output residue signal power by using

$$R_{ee}(0) = \frac{1}{2\pi B}\int_{-\pi B}^{\pi B} |A(\omega) - F(\omega)|^2 \phi_{JJ}(\omega)\, d\omega \tag{11.99}$$

where $\phi_{JJ}(\omega)$ is the constant interference signal power spectral density. Assume the interference power spectral density is unity across the bandwidth of concern, then the output residue power due only to main channel amplitude variations is given by

$$R_{ee_A} = \frac{1}{2\pi B}\int_{-\pi B}^{\pi B} |A_0(\omega) - F(\omega)|^2\, d\omega \tag{11.100}$$

Since $A(\omega) - \mathrm{F}(\omega)$ is orthogonal to $F(\omega)$, it follows that [7]

$$E\{|A(\omega) - F(\omega)|^2\} = E\{|A(\omega)|^2\} - E\{|F(\omega)|^2\} \tag{11.101}$$

and hence

$$R_{ee_A} = \frac{1}{2\pi B}\int_{-\pi B}^{\pi B} [A_0^2(\omega) - |F(\omega)|^2]\, d\omega \tag{11.102}$$

It likewise follows from (11.99) that the output residue power contributed by main channel phase variations is given by

$$R_{ee_p} = \frac{1}{2\pi B}\int_{-\pi B}^{\pi B} |e^{-j\phi_0(\omega)} - F(\omega)|^2 \phi_{JJ}(\omega)\, d\omega \tag{11.103}$$

where $\phi_0(\omega)$ represents the main channel phase variation. Once again assuming that the input signal spectral density is unity across the signal bandwidth and noting that $[e^{-j\phi_0(\omega)} - F(\omega)]$ must be orthogonal to $F(\omega)$, we see it immediately follows that

$$R_{ee_p} = \frac{1}{2\pi B}\int_{-\pi B}^{\pi B}\left[1 - |F(\omega)|^2\right]d\omega$$

$$= 1 - \sum_{j=-N}^{N}\sum_{k=-N}^{N} w_k w_j^* \frac{\sin\left[\pi B\Delta(k-j)\right]}{\pi B\Delta(k-j)} \tag{11.104}$$

where the complex weight vector elements must satisfy (11.95)–(11.98).

If it is desired to evaluate the effects of both amplitude and phase mismatching simultaneously, then the appropriate expression for the output residue power is given by (11.99), which (because of orthogonality) may be rewritten as

$$R_{ee}(0) = \frac{1}{2\pi B}\int_{-\pi B}^{\pi B}\left\{|A(\omega)|^2 - |F(\omega)|^2\right\}\phi_{JJ}(\omega)\,d\omega \tag{11.105}$$

where the complex weights used to obtain $F(\omega)$ must again satisfy (11.96)–(11.98), which now involve both a magnitude and a phase component and it is assumed that $\phi_{JJ}(\omega)$ is a constant.

### 11.3.1 Example: Effects of Amplitude Mismatching

To evaluate (11.102) it is necessary to adopt a channel amplitude model corresponding to $A(\omega)$. One possible channel amplitude model is given in Figure 11.23 for which

$$A(\omega) = \begin{cases} 1 + R\cos\omega T_0 & \text{for } |\omega| \leqslant \pi B \\ 0 & \text{otherwise} \end{cases} \tag{11.106}$$

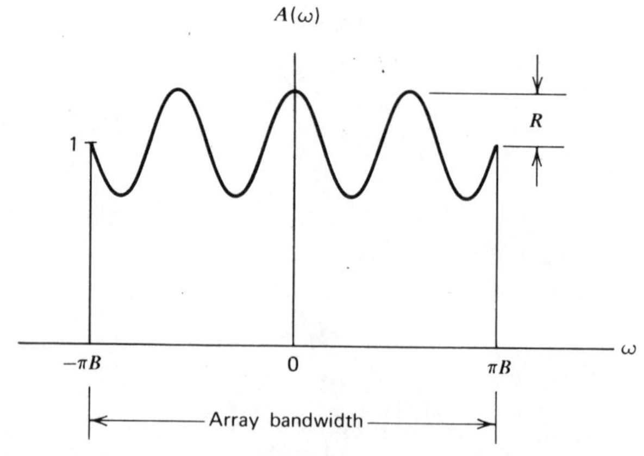


**Figure 11.23** Channel amplitude model having $3\frac{1}{2}$ cycles of ripple for evaluation of amplitude mismatch effects.

where

$$T_0 = \frac{2n+1}{2B} \qquad \text{for } n=0,1,2,\ldots$$

and the integer $n$ corresponds to $(2n+1)/2$ cycles of amplitude mismatching across the bandwidth $B$. Letting the phase error $\phi_0(\omega)=0$, it follows from (11.96) that

$$v_k = \frac{1}{2\pi B}\int_{-\pi B}^{\pi B}\left[1+R\cos\omega T_0\right]e^{j\omega k\Delta}\,d\omega \tag{11.107}$$

or

$$v_k = \frac{\sin(\pi Bk\Delta)}{\pi Bk\Delta} + \frac{R}{2}\left[\frac{\sin(\pi B[T_0+k\Delta])}{\pi B[T_0+k\Delta]} + \frac{\sin(\pi B[T_0-k\Delta])}{B[T_0-k\Delta]}\right]$$

$$\text{for } k=-N,\ldots,0,\ldots,N \tag{11.108}$$

Evaluation of (11.108) permits the complex weight vector to be found, which in turn may be used to determine the residue power by way of (11.102).

Now

$$|F(\omega)|^2 = F(\omega)F^*(\omega) = \mathbf{w}^\dagger\boldsymbol{\beta}\boldsymbol{\beta}^\dagger\mathbf{w} \tag{11.109}$$

where

$$\boldsymbol{\beta} = \begin{bmatrix} e^{j\omega N\Delta} \\ e^{j\omega(N-1)\Delta} \\ \vdots \\ e^{-j\omega N\Delta} \end{bmatrix} \tag{11.110}$$

Carrying out the vector multiplications indicated by (11.109) then yields

$$|F(\omega)|^2 = \sum_{i=1}^{2N+1}\sum_{k=1}^{2N+1} w_i w_k^* e^{j\omega(k-i)\Delta} \tag{11.111}$$

The output residue power is therefore given by [see 11.102]

$$R_{ee_A} = \int_{-\pi B}^{\pi B}\left[1+R\cos\omega T_0\right]^2 d\omega$$

$$-\int_{-\pi B}^{\pi B}\sum_{i=1}^{2N+1}\sum_{k=1}^{2N+1} w_i w_k^* e^{j\omega(k-i)\Delta}\,d\omega \tag{11.112}$$

Equation (11.112) may be evaluated using the following expressions:

$$\frac{1}{2\pi B}\int_{-\pi B}^{\pi B}[1+R\cos\omega T_0]^2 d\omega=\left(1+\frac{R^2}{2}\right)+2R\frac{\sin\pi[(2n+1)/2]}{\pi[(2n+1)/2]}+\frac{R^2}{2}\frac{\sin\pi(2n+1)}{\pi(2n+1)} \tag{11.113}$$

and

$$\frac{1}{2\pi B}\int_{-\pi B}^{\pi B}\sum_{i=1}^{2N+1}\sum_{k=1}^{2N+1} w_i w_k^* e^{j\omega(k-i)\Delta}d\omega=\sum_{i=1}^{2N+1}\sum_{k=1}^{2N+1} w_i w_k^* \frac{\sin\pi(k-i)B\Delta}{\pi(k-i)B\Delta} \tag{11.114}$$

### *11.3.2 Results for Compensation of Selected Amplitude Mismatch Model*

The evaluation of (11.112) requires that the ripple amplitude $R$, the number of cycles of amplitude mismatching across the bandwidth, and the product of

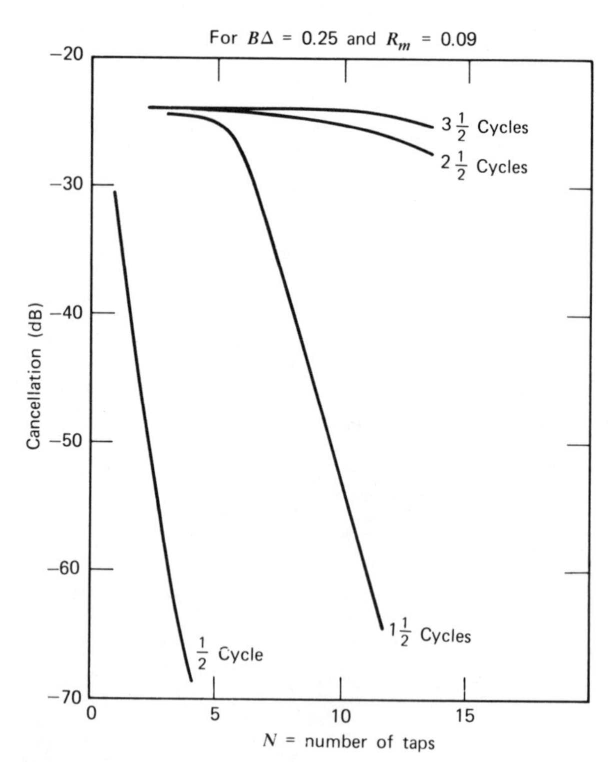


**Figure 11.24** Decibel cancellation versus number of taps for selected amplitude mismatch models with $B\Delta=0.25$.

$B\Delta$ (where $B$ is the cancellation bandwidth and $\Delta$ is the intertap delay spacing) be specified. The results of a computer evaluation of the output residue power are summarized in Figures 11.24–11.27 for $B\Delta=0.25$, 0.5, 0.75, and 1, and $R=0.09$. Each of the figures presents a plot of the decibel cancellation (of the undesired interference signal) achieved as a function of the number of taps in the transversal filter and the number of cycles of ripple present across the cancellation bandwidth. No improvement (over the cancellation that can be achieved with only one tap) is realized until a sufficient number of taps is present in the transversal filter to achieve the resolution required by the amplitude versus frequency variations in the amplitude mismatch model. The sufficient number of taps for the selected amplitude mismatch model was found empirically to be given by

$$N_{\text{sufficient}} \approx \left(\frac{N_r - 1}{2}\right)\left[7 - 4(B\Delta)\right] + 1 \tag{11.115}$$

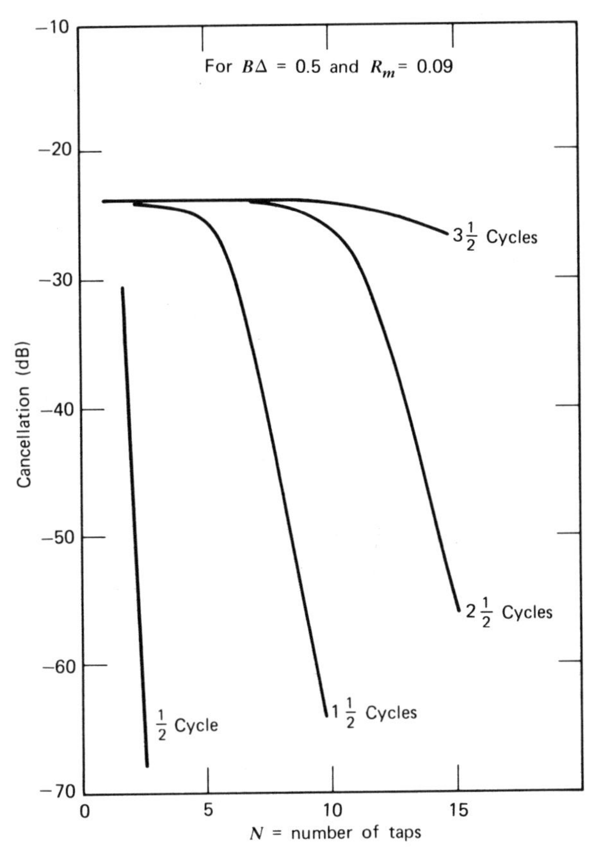


**Figure 11.25** Decibel cancellation versus number of taps for selected amplitude mismatch models with $B\Delta=0.5$.

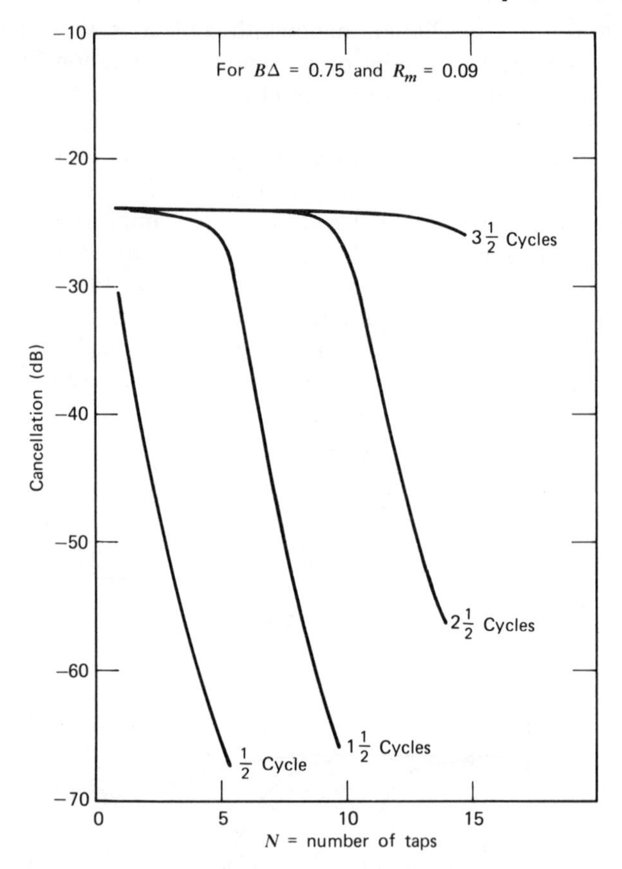


**Figure 11.26** Decibel cancellation versus number of taps for selected amplitude mismatch models with $B\Delta = 0.75$.

where $N_r$ is the number of half-cycles of ripple appearing in the mismatch model.

Once the sufficient number of taps in the transversal filter has been reached, the degree of improvement in cancellation performance realized by adding additional taps depends on how well the resulting transfer function of the transversal filter matches the gain and phase variations of the channel mismatch model. Since the transversal filter transfer function resolution depends in part on the product $B\Delta$, a judicious selection of this parameter can ensure that providing additional taps will provide a more exact match (and hence a significant improvement in cancellation performance), whereas a poor choice of this parameter will result in very poor transfer function matching even with the addition of more taps.

Taking the inverse Fourier transform of (11.106) $\mathfrak{F}^{-1}\{A(\omega)\}$ yields a time function corresponding to an autocorrelation function $f(t)$ that can be ex-

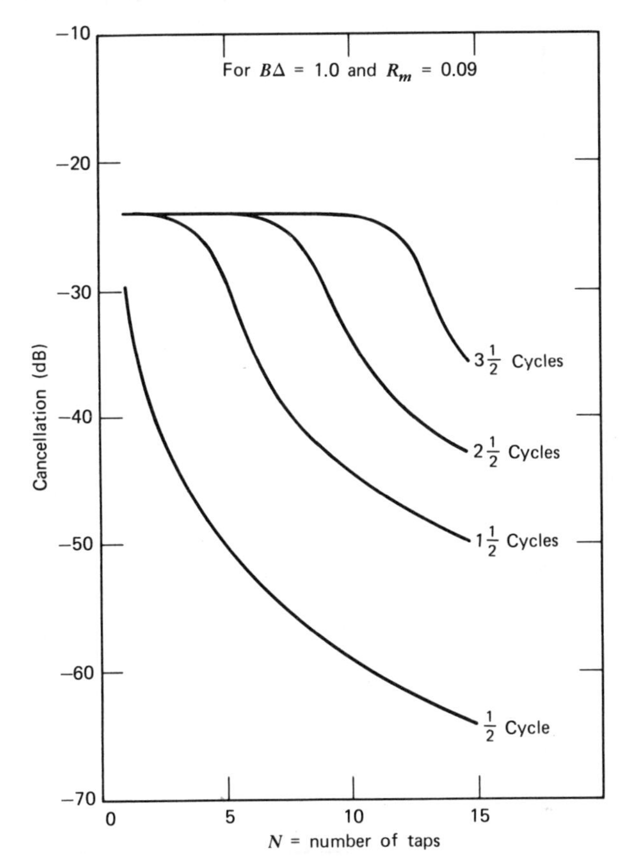


**Figure 11.27** Decibel cancellation versus number of taps for selected amplitude mismatch models with $B\Delta = 1.0$.

pressed as

$$f(t) = s(t) + Ks(t \pm T_0) \tag{11.116}$$

From (11.116) and the results of Section 11.2.3, it is seen that it is desirable to maintain $\Delta = T_0$ (or equivalently, $B\Delta$ = number cycles of ripple mismatch) if the product $B\Delta$ is to "match" the amplitude mismatch model. This result is illustrated in Figure 11.28 where decibel cancellation is plotted versus $B\Delta$ for a one-half-cycle ripple mismatch model. A pronounced minimum occurs at $B\Delta = \frac{1}{2}$ for $N = 3$ and $R_m = 0.9$.

When the number of cycles of mismatch ripple exceeds unity, the foregoing rule of thumb leads to the spurious conclusion that $B\Delta$ should exceed unity. Suppose, for example, there were two cycles of mismatch ripple for which it was desired to compensate. By setting $B\Delta = 2$ (corresponding to $B_f = \frac{1}{2}B$), two complete cycles for the transversal filter transfer function will be found to

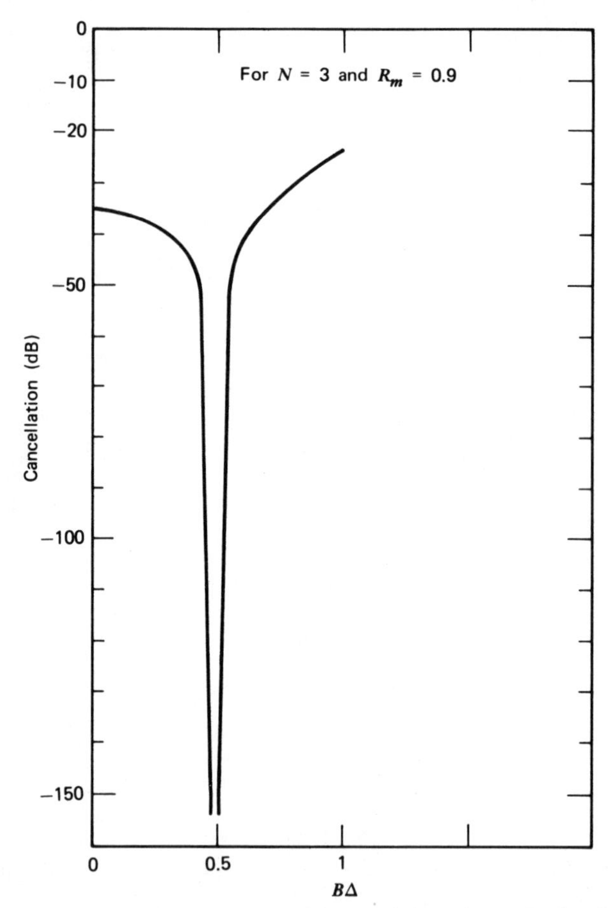


**Figure 11.28** Decibel cancellation versus $B\Delta$ for one-half cycle amplitude mismatch model.

occur across the cancellation bandwidth. By matching only one cycle of the channel mismatch quite good matching of the entire mismatch characteristic will occur, but at the price of sacrificing the ability to independently adjust the complex weights across the entire cancellation bandwidth, thereby reducing the ability to appropriately process broadband signals. Consequently, if the number of cycles of mismatch ripple exceeds unity, it is usually best to set $B\Delta = 1$ and accept whatever improvement in cancellation performance can be obtained with that value, or increase the number of taps.

### 11.3.3 Example: Effects of Phase Mismatching

Let $\phi(\omega)$ corresponding to the phase error be characterized by

$$\phi(\omega) = \begin{cases} A\cos\omega T_0 & \text{for } |\omega| \leqslant \pi B \\ 0 & \text{otherwise} \end{cases} \tag{11.117}$$

where $A$ represents the peak number of degrees associated with the phase error ripples. This model corresponds to the error ripple model of (11.106) (with zero average value present).

Since

$$v_k = \frac{1}{2\pi B}\int_{-\pi B}^{\pi B} \exp(j\{A\cos\omega T_0 + \omega k\Delta\})\,d\omega \qquad \text{for } k = -N,\ldots,0,\ldots,N \tag{11.118}$$

it can easily be shown by defining

$$f(K,\text{sgn}) \triangleq \frac{\sin\pi[K + \text{sgn}\cdot(i-(N+1))B\Delta]}{\pi[K+\text{sgn}\cdot(i-(N+1))B\Delta]}$$

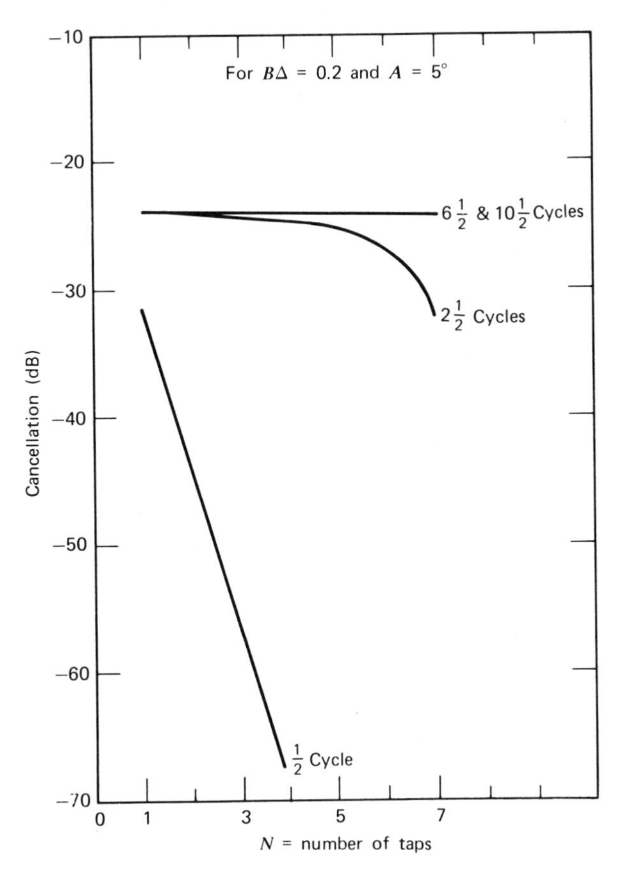


**Figure 11.29** Decibel cancellation versus number of taps for selected phase mismatch models with $B\Delta = 0.2$.

and
$$g(K) \triangleq f(K,+)+f(K,-) \text{ that}$$

$$v_i = J_0(A)f(0,+)+jJ_1(A)g\left[\frac{2n+1}{2}\right]$$
$$+\sum_{k=1}^{\infty}(-1)^k\left\{J_{2k}(A)g\left[k(2n+1)\right]+jJ_{2k+1}(A)g\left[(2k+1)\left(\frac{2n+1}{2}\right)\right]\right\} \tag{11.119}$$

where $J_n(\cdot)$ denotes a Bessel function of the $n$th order for $i=1,2,\ldots,2N+1$.

### *11.3.4 Results for Compensation of Selected Phase Mismatch Model*

The compensation results for the selected phase mismatch model of the previous section can now be presented. The computer evaluation of the output residue power resulted in the performance summarized in Figures

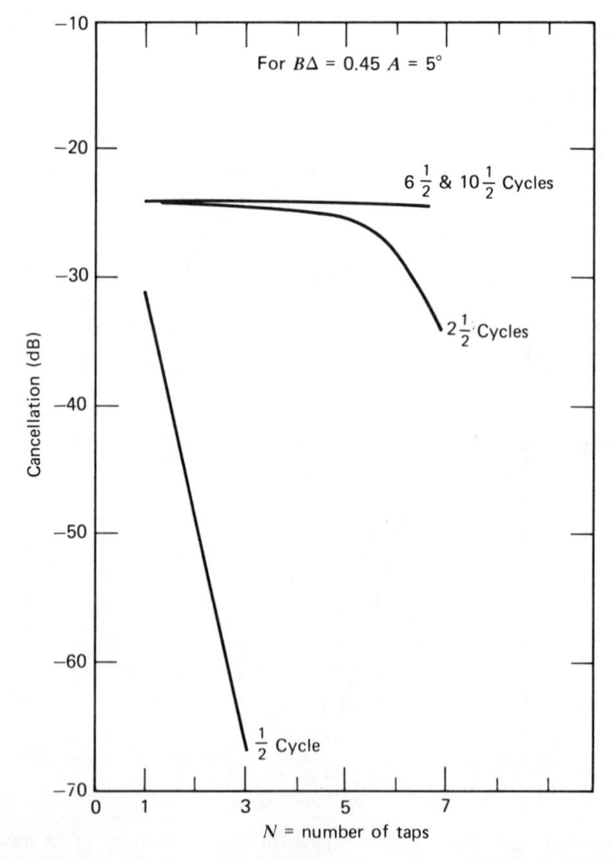


**Figure 11.30** Decibel cancellation versus number of taps for selected phase mismatch models with $B\Delta=0.45$.

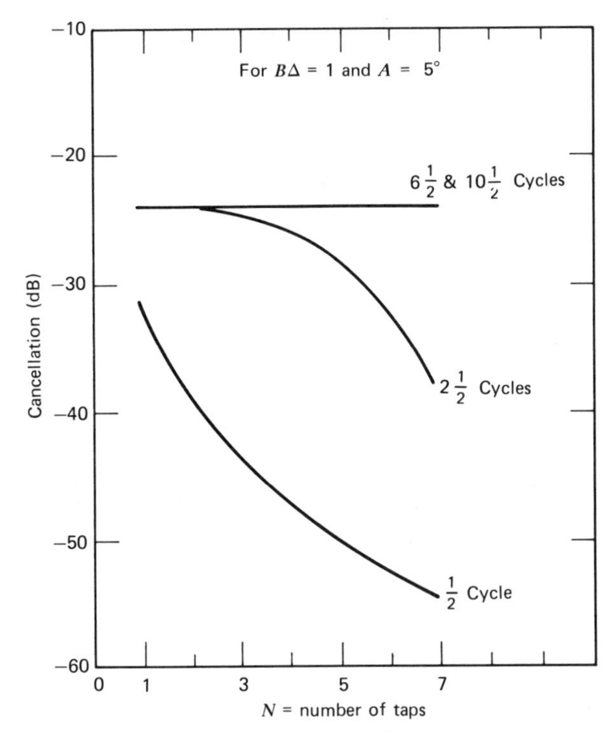


**Figure 11.31** Decibel cancellation versus number of taps for selected phase mismatch models with $B\Delta = 1.0$.

11.29–11.31 for $B\Delta = 0.2$, 0.45, and 1.0 and $A = 5°$. These figures present the decibel cancellation achieved as a function of the number of taps in the transversal filter and the number of cycles of phase ripple present across the cancellation bandwidth. The general nature of the curves appearing in Figures 11.29–11.31 is the same as that of Figures 11.24–11.27 for amplitude mismatching. Furthermore, just as in the amplitude mismatch case, a better channel transfer function fit can be obtained with the transversal filter when the mismatch characteristic has a fewer number of ripples.

## 11.4 SUMMARY AND CONCLUSIONS

The transversal filter consisting of a sequence of weighted taps with intertap delay spacing offers a practical means for achieving the variable amplitude and phase weighting as a function of frequency that is required if an adaptive array system is to perform well against to wideband interference signal sources. The distortionless channel transfer functions for a two-element array were derived. It was found that to ensure distortion-free response to a broadband signal, the channel phase should be a linear function of frequency, while the channel amplitude function should be nearly flat over a 40% bandwidth. Quadrature hybrid processing may provide adequate broadband signal response for signals having as much as 20% bandwidth. If 20% or more bandwidth signals must be accommodated, however, then tapped-delay line processing is a practical necessity. A transversal filter provides an attractive means of compensating the system auxiliary channels for the undesirable effects of the following:

1. Multipath interference.
2. Interchannel mismatch.
3. Propagation delay across the array.

It was found that for multipath interference it is desirable to maintain the value of the intertap delay in the neighborhood of the delay time associated with the multipath ray. If the intertap delay time exceeds the multipath delay time by more than about 30% and the multipath delay time is appreciable, a severe loss of compensation capability is incurred. If the intertap delay is too small, then an excessive number of taps will be required for effective cancellation to occur. Since multipath delay having "small" values of associated time delay do not severely degrade the array performance, it is reasonable to determine the most likely values of multipath delay that will occur for the desired application and base the multipath compensation design on those delay times (assuming $B\Delta \leqslant 1$). For reflection coefficients of 0.5 and $BD_m = 0.45$, the use of five taps will ensure a $-30$ dB cancellation capability.

The results shown in Figures 11.19 and 11.20 indicate that array propagation delay effects are usually much easier to compensate than are multipath effects. This result occurs because multipath in effect introduces two (or more) signals in each channel (that are essentially uncorrelated if $BD_m \gg 1$) which require more degrees-of-freedom to adequately compensate.

The problem presented by interchannel mismatch is to obtain a transfer function with the transversal filter that succeeds in matching the amplitude and phase error characteristics exhibited among the various sensor channels. As might be expected, the more severe the mismatching between channels, the more difficult it is to achieve an acceptable degree of compensation. In particular, it is highly undesirable for more than $2\frac{1}{2}$ cycles of mismatch ripple to occur over the cancellation bandwidth; even this degree of mismatch requires seven taps on the transversal filter before a truly effective degree of compensation can be achieved. It may very well result that the best choice of intertap delay spacing for the interchannel mismatch characteristic of concern is far different from the optimum choice of intertap delay selected for multipath compensation; should this actually occur, it is necessary to adopt a compromise value for the intertap delay spacing. Such a compromise value for the intertap delay spacing hopefully results in an acceptable degree of compensation for both multipath and interchannel mismatch effects.

## PROBLEMS

### *Distortionless Transfer Functions*

**1** From (11.5) and (11.6) it immediately follows that $|H_1(\omega)=|H_2(\omega)|$, thereby yielding the pair of equations

$$f_1\{|H_1|, \alpha_1, \alpha_2, \theta_s\} = \exp(-j\omega T_1)$$

and

$$f_2\{|H_1|, \alpha_1, \alpha_2, \theta_i\} = 0$$

(a) Show from the above pair of equations that $\alpha_1(\omega)$ and $\alpha_2(\omega)$ must satisfy

$$\alpha_2(\omega) - \alpha_1(\omega) = \frac{\pi\omega}{\omega_0}\sin\theta_i \pm n\pi$$

where $n$ is any odd integer.

(b) Since the magnitude of $\exp(-j\omega T_1)$ must be unity, show using $f_1\{\ \} = \exp(-j\omega T_1)$ that (11.7) results.

(c) Show that the angle condition associated with $f_1\{\ \} = \exp(-j\omega T_1)$ yields (11.8).

(d) Show that substituting (11.8) into the results from part (a) above yields (11.9).

**2** For a three-element linear array, the overall transfer function encountered by the desired signal in passing through the array is

$$H_d(\omega) = H_1(\omega) + H_2(\omega)\exp\left(-j\frac{\omega d}{c}\sin\theta_s\right) + H_3(\omega)e\left(-j\frac{\omega 2d}{c}\sin\theta_s\right)$$

and the overall transfer function seen by the interference signal is

$$H_I(\omega) = H_1(\omega) + H_2(\omega)\exp\left(-j\frac{\omega d}{c}\sin\theta_i\right) + H_3(\omega)e\left(-j\frac{\omega 2d}{c}\sin\theta_i\right)$$

What does imposing the requirements (11.3) and (11.4) now imply for the three-channel transfer functions?

**3** ***Hilbert Transform Relations.*** Prove the Hilbert transform relations given by (11.19)–(11.22).

**4** Using (11.53), (11.54), and the results of (11.55)–(11.57), show that $\mathbf{R}_{ee}$ is given by (11.58).

**5** Derive the correlation functions given by (11.70)–(11.72) for the signal environment assumptions (11.67) and (11.68).

**6** Show that as the time-bandwidth product $B\Delta$ approaches zero, then the matrix $\mathbf{R}_{xx}(0)$ [whose elements are given by (11.72)] becomes singular so that matrix inversion cannot be accomplished.

**7** For the phase error $\phi(\omega)$ given by (11.117), show that $v_k$ given by (11.119) follows from the application of (11.118).

*Compensation for Channel Phase Errors*

**8** Let $\phi(\omega)$ correspond to the phase error model be given by

$$\phi(\omega) = \begin{cases} A\left[1 - \cos\dfrac{2\omega}{B}\right] & \text{for } |\omega| \leqslant \pi B \\ 0 & \text{otherwise} \end{cases}$$

Show that $v_k$ of (11.118) is given by

$$v_k = \int_{-\pi B}^{\pi B}\left[\cos\left\{A\left(1 - \cos\omega\frac{2}{B}\right)\right\} + j\sin\left\{A\left(1 - \cos\omega\frac{2}{B}\right)\right\}\right]\exp(j\omega k\,\Delta_{d\omega})$$

Use the trigonometric identities

$$\cos\left[A - A\cos\omega\frac{2}{B}\right] = \cos A\cos\left[A\cos\omega\frac{2}{B}\right] + \sin A\sin\left[A\cos\omega\frac{2}{B}\right]$$

$$\sin\left[A - A\cos\omega\frac{2}{B}\right] = \sin A\cos\left[A\cos\omega\frac{2}{B}\right] - \cos A\sin\left[A\cos\omega\frac{2}{B}\right]$$

and the fact that

$$\cos(A\cos\omega T_0) = J_0(A) + 2\sum_{k=1}^{\infty}(-1)^k \cdot J_{2k}(A)\cos[(2k)\omega T_0]$$

$$\sin(A\cos\omega T_0) = 2\sum_{k=0}^{\infty}(-1)^k J_{2k+1}(A) \cdot \cos[(2k+1)\omega T_0]$$

where $J_n(\cdot)$ denotes a Bessel function of the $n$th order and define

$$f(n, \text{sgn}) \triangleq \frac{\sin\pi[n + \text{sgn}\cdot(i-(N+1))B\Delta]}{\pi[n + \text{sgn}\cdot(i-(N+1))B\Delta]}$$

$$g(n) \triangleq f(n, +) + f(n, -)$$

to show that

$$v_i = J_0(A)\cdot f(0,+)[\cos A + j\sin A] + J_1(A)\cdot g(2)[\sin A - j\cos A] + \sum_{k=1}^{\infty}(-1)^k\{J_{2k}(A)\cdot g(4k)[\cos A + j\sin A] + J_{2k+1}(A)\cdot g[(2k+1)2][\sin A - j\cos A]\}$$

for $i = 1, 2, \ldots, 2N+1$

**9** Let $\phi(\omega)$ corresponding to the phase error model be given by

$$\phi(\omega) = \begin{cases} b\omega^2(\pi B - |\omega|) & \text{for } |\omega| \leqslant \pi B \\ 0 & \text{otherwise} \end{cases}$$

As before, it follows that

$$v_i = \frac{1}{2\pi B}\int_{-\pi B}^{\pi B} \exp\{j[b\omega^2(\pi B - |\omega|) + \omega i\Delta]\}\, d\omega$$

Letting $u = \omega/\pi B$, applying Euler's formula, and ignoring all odd components of the resulting expression, show that

$$v_i = \int_0^1 \exp\left\{j\left[\frac{27A}{4}u^2(1-u)\right]\right\}\cos\pi[u(i-(N+1))B\Delta]\, du$$

where $A = 4b(\pi B/3)^3$ for $i = 1, 2, \ldots, 2N+1$. The foregoing equation for $v_i$ can be evaluated numerically to determine the output residue power contribution due to the above phase error model.

## REFERENCES

[1] W. E. Rodgers and R. T. Compton, Jr., "Tapped Delay-Line Processing in Adaptive Arrays," Report 3576–3, April 1974; prepared by The Ohio State University Electro Science Laboratory, Department of Electrical Engineering under Contract N00019-73-C-0195 for Naval Air Systems Command.

[2] ______,"Adaptive Array Bandwidth with Tapped Delay-Line Processing," Report 3832-3, May 1975; prepared by The Ohio State University Electro Science Laboratory, Department of Electrical Engineering under Contract N00019-74-C-0141 for Naval Air Systems Command.

[3] ______,"Adaptive Array Bandwidth with Tapped Delay-Line Processing," *IEEE Trans. Aerosp. Electron. Syst.*, Vol. AES-15, No. 1, January 1979, pp. 21–28.

[4] T. G. Kincaid, "The Complex Representation of Signals," General Electric Report No. R67EMH5, October 1966, HMED Publications, Box 1122 (Le Moyne Ave.), Syracuse, NY, 13201.

[5] R. A. Monzingo, "Transversal Filter Implementation of Wideband Weight Compensation for CSLC Applications," unpublished Hughes Aircraft Interdepartmental Correspondence Ref. No. 78-1450.10/07, March 1978.

[6] A. M. Vural, "Effects of Perturbations on the Performance of Optimum/Adaptive Arrays," *IEEE Trans. Aerosp. Electron. Syst.*, Vol. AES-15, No. 1, January 1979; pp. 76–87.

[7] A. Papoulis, *Probability, Random Variables, and Stochastic Processes*, McGraw-Hill, New York, 1965, Ch. 7.

# Chapter 12. Current Trends in Adaptive Array Research

In considering new adaptive array concepts to include for discussion in this chapter, it was impossible to describe all significant contributions. The scope of attention has accordingly been restricted so that only a few exceptionally novel and promising avenues of investigation are presented. Two of the topics to be considered here involve adaptation algorithms that have not been described in earlier sections but which have desirable features that make them attractive candidates for certain applications. One algorithm is the maximum entropy method (MEM) for obtaining spectral estimates, while the other algorithm concerns the recursive Bayesian approach to detection and estimation. The third topic to be discussed concerns partially adaptive array concepts. Partial adaptivity is of interest when only a portion of the total number of elements are controlled, thereby reducing the number of processors required to achieve an acceptable level of adaptive array performance.

## 12.1 THE MAXIMUM ENTROPY METHOD (MEM) FOR SPECTRAL ESTIMATION

It was found in Chapter 3 that the broadband optimal array processor may be determined from the cross power spectral density of the signal environment. Consequently spectral estimation can be considered an essential step in the control process and the MEM algorithm is becoming a widely accepted tool for spectral estimation [1]–[4]. MEM spectral estimation is well known in geophysics and in recent years has also received attention in the fields of speech processing, sonar, and radar [5]–[8]. In addition to spectrum analysis, MEM can also be applied to bearing estimation problems for signals received by an array of sensors [9],[10].

MEM provides a procedure for the estimation of higher resolution power spectra than can be obtained with more conventional techniques when the information about the signal process is limited. This limited information may be in the form of the first few lags of the autocorrelation function or simply the available time series data. High resolution is achieved by extrapolating the partially known autocorrelation function beyond the last known lag value in a manner that maximizes the entropy of the corresponding power spectrum at each step of the extrapolation [1],[11]. If the autocorrelation function is unknown, then the power spectrum can be estimated directly from the available time series data using a method devised by Burg [12]. Excellent power spectrum estimates may be obtained from relatively short time series data record lengths, and the approach has been reported to have a rapid rate of convergence [13].

In its most elementary form the maximum entropy principle for estimating the power spectrum of a single-channel, stationary, complex time series can be stated as a problem of finding the spectral density function $\phi_{xx}(f)$ that maximizes

$$\text{entropy} = \int_{-W}^{W} \ln \phi_{xx}(f)\, df \tag{12.1}$$

under the constraint that $\phi_{xx}(f)$ satisfies a set of $N$ linear measurement equations

$$\int_{-W}^{W} \phi_{xx}(f) G_n(f)\, df = g_n, \qquad n = 1, \ldots, N \tag{12.2}$$

where the time series is sampled with the uniform period $\Delta t$ so the Nyquist fold-over frequency is $W = 1/2\Delta t$ and the power spectrum of the time series is band-limited to $\pm W$. The functions $G_n(f)$ in the measurement equations are known test functions, and the $g_n$ are the observed values resulting from the measurements. Two cases can now be considered: the first where the autocorrelation function is partially known and the second where the autocorrelation function is unknown.

### *12.1.1 Partially Known Autocorrelation Function*

Let $x(t)$ represent the time series of a stationary (scalar) random process with an associated autocorrelation function $r(\tau)$ for which $N$ discrete lag values $\{r(0), r(1), \ldots, r(N-1)\}$ are known. It is now desired to estimate the value of $r(N)$, which lies outside the interval of known autocorrelation function values. The basic autocorrelation function theorem states that $r(N)$ must have a value such that the $(N+1)\times(N+1)$ Hermitian Toeplitz autocorrelation matrix given by

$$\mathbf{R}_N \triangleq \begin{bmatrix} r(0) & r(-1) & \cdots & r(-N) \\ r(1) & r(0) & \cdots & r(1-N) \\ \vdots & \cdots & \cdots & \vdots \\ r(N) & r(N-1) & \cdots & r(0) \end{bmatrix} \tag{12.3}$$

must be positive semidefinite (i.e., all subdeterminants of $\mathbf{R}_N$ must be nonnegative). Since $\det(\mathbf{R}_N)$ is a quadratic function of $r(N)$, there are two values of $r(N)$ that make the determinant equal to zero. These two values of $r(N)$ define boundaries within which the predicted value of $r(N)$ must fall. The MEM procedure seeks to select that value of $r(N)$ that maximizes $\det(\mathbf{R}_N)$. For a Gaussian random process this procedure is equivalent to maximizing (12.1) subject to the constraint equations [12]:

$$r(n)=\int_{-W}^{W}\phi_{xx}(f)\exp(j2\pi fn\,\Delta t)\,df, \qquad -N\leqslant n\leqslant N \tag{12.4}$$

The problem of maximizing $\det(\mathbf{R}_N)$ is equivalent to finding coefficients for a prediction error filter, and these coefficients play an important role in finding the MEM spectral estimate. Before finding these coefficients, it is instructive to examine the role played by a prediction error filter in obtaining the MEM spectral estimate for a partially known autocorrelation function. Suppose there are $N$ samples of $x(t)$ denoted by $x_0, x_1, \ldots, x_{N-1}$ where each of the samples are taken $\Delta t$ seconds apart. A linear predicted estimate of $x_N$ based on the previous sampled values of $x(t)$ can be obtained from an $(N+1)$-point prediction filter as follows:

$$\hat{x}_N=-\sum_{i=1}^{N}a(N,i)x_{N-i} \tag{12.5}$$

The error associated with $\hat{x}_N$ is then given by

$$\varepsilon_N=x_N-\hat{x}_N=x_N+\sum_{i=1}^{N}a(N,i)x_{N-i} \tag{12.6}$$

Equation (12.6) can be written in matrix form as

$$\varepsilon_N=\mathbf{a}_N^T\mathbf{x} \tag{12.7}$$

where $\mathbf{x}^T=[x_N, x_{N-1}, \ldots, x_0]$ and $\mathbf{a}_N^T=[1, a(N,1), a(N,2), \ldots, a(N,N)]$ where the coefficient $a(N,N)=C_N$ is called the reflection coefficient of order $N$. The error $\varepsilon_N$ may therefore be regarded as the output of an $N$th order prediction error filter whose coefficients are given by the vector $\mathbf{a}_N$ and whose power output is

$$P_N=E\{\varepsilon_N^2\} \tag{12.8}$$

It is clearly desirable to minimize the MSE of (12.8) by appropriately selecting the prediction filter coefficients contained in $\mathbf{a}_N$. In order to obtain the minimum mean square estimate $\hat{x}_N$, the error $\varepsilon_N$ must be orthogonal to the past data so that

$$E\{x_i\varepsilon_N\}=0 \qquad \text{for } i=0,1,\ldots,N-1 \tag{12.9}$$

Furthermore, the error $\varepsilon_N$ must be uncorrelated with all past estimation errors so that

$$E\{\varepsilon_N \varepsilon_{N-k}\} = 0 \qquad \text{for } k = 1, 2, \ldots, N-1 \tag{12.10}$$

Equations (12.8) and (12.10) are just the conditions required for a random process to have a white power spectrum of total power $P_N$ (or a power density level of $P_N/2W$ where $W = 1/2\Delta t$). The prediction error filter may therefore be regarded as a whitening filter that operates on the input data $\{x_0, x_1, \ldots, x_{N-1}\}$ to produce output data having a white power density spectrum of level $P_N/2W$ as indicated in Figure 12.1. It immediately follows that the estimate $\hat{\phi}_{xx}(f)$ of the input power spectrum $\phi_{xx}(f)$ is given by

$$\hat{\phi}_{xx}(f) = \frac{P_N/2W}{\left|1 + \sum_{n=1}^{N} a(N,n) \exp(-j2\pi f n \Delta t)\right|^2} \tag{12.11}$$

where the denominator of (12.11) is recognized as the power response of the prediction error filter. Equation (12.11) yields the MEM estimate of $\phi_{xx}(f)$ provided that the coefficients $a(N,n)$, $n = 1, \ldots, N$, and the power $P_N$ can be determined.

A relationship between the coefficients $a(N,n)$, $n = 1, \ldots, N$, the power $P_N$, and the autocorrelation function values $r(-N), r(-N+1), \ldots, r(0), \ldots, r(N-1), r(N)$ is provided by the well-known prediction error filter matrix equation [12]:

$$\begin{bmatrix} r(0) & r(-1) & \cdots & r(-N) \\ r(1) & r(0) & \cdots & r(-N+1) \\ \vdots & \vdots & & \vdots \\ r(N) & r(N-1) & \cdots & r(0) \end{bmatrix} \begin{bmatrix} 1 \\ a(N,1) \\ \vdots \\ a(N,N) \end{bmatrix} = \begin{bmatrix} P_N \\ 0 \\ \vdots \\ 0 \end{bmatrix} \tag{12.12}$$

Equation (12.12) can be derived [as done in reference [12] by maximizing the entropy of (12.1)] subject to the constraint equations (12.4). If we know the autocorrelation values $\{r(-N), r(-N+1), \ldots, r(-1), r(0), r(1), \ldots, r(N-1), r(N)\}$, the coefficients $a(N,n)$ and the power $P_N$ may then be found using

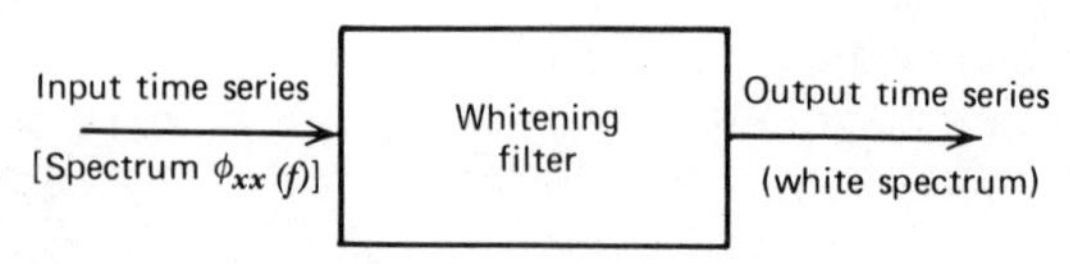


**Figure 12.1** Whitening filter representation of prediction error filter.

(12.12). Equation (12.12) can be written in matrix form as

$$\mathbf{R}_N \mathbf{a}_N = \begin{bmatrix} P_N \\ 0 \\ \vdots \\ 0 \end{bmatrix} \quad \text{or} \quad \mathbf{a}_N = \mathbf{R}_N^{-1} \begin{bmatrix} P_N \\ 0 \\ \vdots \\ 0 \end{bmatrix} \tag{12.13}$$

Let

$$\mathbf{R}_N^{-1} = \begin{bmatrix} z_{11} & z_{12} & \cdots \\ z_{21} & z_{22} & \\ \vdots & & \\ z_{N1} & & \end{bmatrix} \tag{12.14}$$

It then follows that

$$\mathbf{a}_N^T = \left[ 1, \frac{z_{21}}{z_{11}}, \frac{z_{31}}{z_{11}}, \ldots, \frac{z_{N1}}{z_{11}} \right] \tag{12.15}$$

Ulrych and Bishop [2] also give a convenient recursive procedure for determining the coefficients in $\mathbf{a}_N$.

Having determined the prediction error filter coefficients and the corresponding MEM spectral estimate, we must now consider how the autocorrelation function can be extended beyond $r(N)$ to $r(N+1)$ where the autocorrelation values $r(0), r(1), \ldots, r(N)$ are all known. Suppose for example that $r(0)$ and $r(1)$ are known, and it is desired to extrapolate the autocorrelation function to the unknown value $r(2)$. The prediction error filter matrix equation for the known values $r(0)$ and $r(1)$ is given by

$$\begin{bmatrix} r(0) & r(-1) \\ r(1) & r(0) \end{bmatrix} \begin{bmatrix} 1 \\ a(1,1) \end{bmatrix} = \begin{bmatrix} P_1 \\ 0 \end{bmatrix} \tag{12.16}$$

where $r(-1) = r^*(1)$. To determine the estimate $\hat{r}(2)$, append one more equation to the above matrix equation by incorporating $\hat{r}(2)$ into the autocorrelation matrix to yield [2]

$$\begin{bmatrix} r(0) & r(-1) & \hat{r}(-2) \\ r(1) & r(0) & r(-1) \\ \hat{r}(2) & r(1) & r(0) \end{bmatrix} \begin{bmatrix} 1 \\ a(1,1) \\ 0 \end{bmatrix} \begin{bmatrix} P_1 \\ 0 \\ 0 \end{bmatrix} \tag{12.17}$$

Solving (12.17) for $\hat{r}(2)$ then yields $\hat{r}(2) + a(1,1) r(1) = 0$ or

$$\hat{r}(2) = -a(1,1) r(1) = -C_1 r(1) \tag{12.18}$$

Continuing the extrapolation procedure to still more unknown values of the autocorrelation function simply involves the incorporation of these additional $r(n)$ into the autocorrelation matrix along with additional zeros appended to the two vectors to give the appropriate equation set that yields the desired solution. Since the prediction error filter coefficients remain unchanged by this extrapolation procedure, it follows that the spectral estimate given by (12.11) remains unchanged by the extrapolation as well.

### *12.1.2 Unknown Autocorrelation Function*

The discussion of the previous section assumed that the first $N$ lag values of the autocorrelation function were precisely known. Quite often, however, only the basic time series data is available, and any information concerning the autocorrelation function must be estimated directly from the time series data. To obtain estimates of the autocorrelation function, quite often the average given by

$$\hat{r}(\tau)=\frac{1}{N}\sum_{i=1}^{N-\tau} x_i x_{i+\tau}^* \tag{12.19}$$

is used. With finite data sets, however, (12.19) implicitly assumes that any data that may exist outside the finite data interval is zero. The application of Fourier transform techniques to a finite data interval likewise assumes that any data that may exist outside the data interval is periodic with the known data. These unwarranted assumptions about unknown data represent "end effect" problems that may be avoided using the MEM approach, which makes no assumptions about any unmeasured data.

The MEM approach to the problem of spectral estimation when the autocorrelation function is unknown is to estimate the coefficients of a prediction error filter that never runs off the end of a finite data set, thereby making no assumptions about data outside the data interval. The prediction error filter coefficients are then used to estimate the maximum entropy spectrum. This approach exploits the autocorrelation reflection-coefficient theorem, which states that there is a one-to-one correspondence between an autocorrelation function and the set of numbers $\{r(0), C_1, C_2, \ldots, C_N\}$ [12].

To obtain the set $\{r(0), C_1, C_2, \ldots, C_N\}$ first consider the problem of estimating $r(0)$. The estimate of the autocorrelation function $r(0)$ is simply given as the average square value of the data set, that is,

$$\hat{r}(0)=\frac{1}{N}\sum_{i=1}^{N}|x_i|^2 \tag{12.20}$$

Now consider how a two-point prediction error filter coefficient can be estimated from an $N$-point long data sample. The problem is to determine the two-point filter (having a first coefficient of unity) that has the minimum

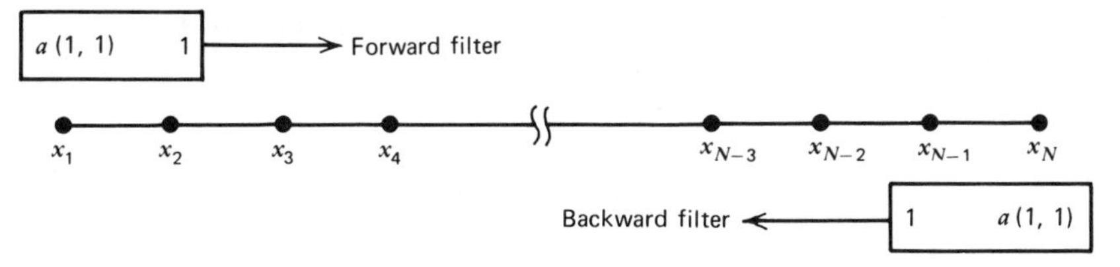


**Figure 12.2** Two-point prediction error filter operating forward and backward over an $N$-point data set.

average power output where the filter is not run off the ends of the data sample. For a two-point filter running forward over the data sample as shown in Figure 12.2, the average power output is given by

$$P_1^f = \frac{1}{N-1} \sum_{i=1}^{N-1} |x_{i+1} + a(1,1)x_i|^2 \tag{12.21}$$

Since a prediction filter can be operated equally well running backward over a data set as well as forward, the average power output for a backward running two-point filter is given by

$$P_1^b = \frac{1}{N-1} \sum_{i=1}^{N-1} |x_i + a^*(1,1)x_{i+1}|^2 \tag{12.22}$$

Since there is no reason to prefer a forward running filter over a backward running filter, and (12.21) and (12.22) represent different estimates of the same quantity, averaging the two estimates should result in a better estimator than either one alone [and also guarantees that the estimate of the reflection coefficient $a(1,1)$ is bounded by unity—a fact whose significance will be seen shortly] so that

$$\begin{aligned} P_1 &= \tfrac{1}{2}(P_1^f + P_1^b) \\ &= \frac{1}{2(N-1)} \left[ \sum_{i=1}^{N-1} |x_{i+1} + a(1,1)x_i|^2 + \sum_{i=1}^{N-1} |x_i + a^*(1,1)x_{i+1}|^2 \right] \end{aligned} \tag{12.23}$$

It is now desired to minimize $P_1$ by selecting the coefficient $a(1,1)$. By setting the derivative of $P_1$ with respect to $a(1,1)$ equal to zero, it is easy to show that the minimizing value of $a(1,1)$ is given by [14]

$$a(1,1) = \frac{-2 \sum_{i=1}^{N-1} x_i^* x_{i+1}}{\sum_{i=1}^{N-1} \left(|x_i|^2 + |x_{i+1}|^2\right)} \tag{12.24}$$

Having $\hat{r}(0)$ and $a(1,1)$, we can find the remaining unknown parts of the prediction error filter matrix equation using (12.12), that is,

$$\begin{bmatrix} \hat{r}(0) & \hat{r}(-1) \\ \hat{r}(1) & \hat{r}(0) \end{bmatrix} \begin{bmatrix} 1 \\ a(1,1) \end{bmatrix} = \begin{bmatrix} P_1 \\ 0 \end{bmatrix} \tag{12.25}$$

so that

$$\hat{r}(1) = -a(1,1)\hat{r}(0) \tag{12.26}$$

and the output power of the two-point filter is given by

$$P_1 = \hat{r}(0)\left[1 - |a(1,1)|^2\right] \tag{12.27}$$

The result expressed by (12.27) implies that $|a(1,1)| \leqslant 1$, which is also the necessary and sufficient condition that the filter defined by $\{1, a(1,1)\}$ be a prediction error filter.

Next consider how to obtain the coefficients for a three-point prediction error filter from the two-point filter just found. The prediction error filter matrix equation now takes the form

$$\begin{bmatrix} \hat{r}(0) & \hat{r}(-1) & \hat{r}(-2) \\ \hat{r}(1) & \hat{r}(0) & \hat{r}(-1) \\ \hat{r}(2) & \hat{r}(1) & \hat{r}(0) \end{bmatrix} \begin{bmatrix} 1 \\ a(2,1) \\ a(2,2) \end{bmatrix} = \begin{bmatrix} P_2 \\ 0 \\ 0 \end{bmatrix} \tag{12.28}$$

From the middle row it follows that

$$\hat{r}(1) + a(2,1)\hat{r}(0) + a(2,2)\hat{r}^*(1) = 0 \tag{12.29}$$

On substituting (12.26) into (12.29) it immediately follows that

$$a(2,1) = a(1,1) + a(2,2)a^*(1,1) \tag{12.30}$$

The corresponding output power $P_2$ is then

$$P_2 = P_1(1 - |a(2,2)|^2) \tag{12.31}$$

Consequently the coefficient vector for the three-point filter takes the form

$$\mathbf{a}_2^T = \left[1, a(1,1) + a(2,2)a^*(1,1),\ a(2,2)\right] \tag{12.32}$$

An equation for the estimated power output of a three-point filter can be written corresponding to (12.23) as $P_2 = \frac{1}{2}(P_2^f + P_2^b)$ where now

$$P_2^f = \frac{1}{N-2} \sum_{i=1}^{N-2} |x_{i+2} + a(2,1)x_{i+1} + a(2,2)x_i|^2 \tag{12.33}$$

and

$$P_2^b = \frac{1}{N-2} \sum_{i=1}^{N-2} |x_i + a^*(2,1)x_{i+1} + a^*(2,2)x_{i+2}|^2 \tag{12.34}$$

Since $a(2,1)$ is given by (12.30) and $a(1,1)$ is already known, it follows that the minimization of $P_2$ can be carried out by varying only $a(2,2) = C_2$. As was the case with the two-point filter, the magnitude of the three-point filter coefficient $a(2,2)$ must not exceed unity. On carrying out the minimization of $P_2$, it turns out that $|a(2,2)| \leqslant 1$, which is also the necessary and sufficient condition that the filter defined by $\{1, a(1,1) + a(2,2)a^*(1,1), a(2,2)\}$ be a prediction error filter. The corresponding Hermitian Toeplitz autocorrelation matrix will then be nonnegative definite as required by the basic autocorrelation function theorem.

The four-point filter can likewise be formed from the three-point filter by carrying out the minimization of $P_3$ with respect to $a(3,3)$ only. The above procedure is continued until all the coefficients $a(N,n)$, $n = 1,\ldots,N$, are found and (12.11) then yields the MEM spectral estimate. The general solution for the reflection coefficients $C_N = a(N,N)$ is given in reference [14]. Equations (12.30) and (12.31) may be expressed in general terms as

$$a(N,k) = a(N-1,k) + a(N,N)a^*(N-1,N-k) \tag{12.35}$$

and

$$P_N = P_{N-1}\left[1 - |a(N,N)|^2\right] \tag{12.36}$$

Since $|a(N,N)| \leqslant 1$, it follows that $0 \leqslant P_N \leqslant P_{N-1}$, so the error decreases with increasing filter order $N$. The choice of $N$ is determined by the desired resolution of the estimated spectrum.

Burg [4] has shown that the inverse of the covariance matrix $\mathbf{R}_{xx}$ can be determined directly from the prediction error filter coefficients $a(m,k)$ and their corresponding error powers $P_k$ for $k = 1,2,\ldots,L$ where $L \leqslant N-1$ is the filter length. The resulting MEM estimate of $\mathbf{R}_{xx}^{-1}$ will differ from the inverse of the sample covariance matrix used in Chapter 6, and there is evidence which suggests that the MEM estimate $\hat{\mathbf{R}}_{xx}^{-1}$ converges to $\mathbf{R}_{xx}^{-1}$ more rapidly than the DMI estimate. This fast convergence feature makes the Burg algorithm particularly attractive in situations where only a small number of independent data samples are available.

### *12.1.3 Extension to Multichannel Spectral Estimation*

Since the processing of array sensor outputs involves multichannel complex signals, it is important to generalize the single-channel Burg process to multiple channels in order to exploit the MEM procedure for adaptive array applications. There are several multichannel generalizations of the scalar

MEM approach that have been proposed [16]–[22], and the generalization provided by Strand [23], [24] will be outlined here.

Suppose there are $N$ observations of the $p$-channel vector $\mathbf{x}(t)$ denoted by $\{\mathbf{x}_0, \mathbf{x}_1, \ldots, \mathbf{x}_{N-1}\}$. A linear prediction of $\mathbf{x}_N$ based on the previous observations of $\mathbf{x}(t)$ can be obtained as

$$\hat{\mathbf{x}}_N = -\sum_{i=1}^{N} \mathbf{A}^{\dagger}(N,i)\mathbf{x}_{N-i} \tag{12.37}$$

where $\mathbf{A}(N,i)$ now denotes the matrix of $N$-long forward prediction filter coefficients. The error associated with $\hat{\mathbf{x}}_N$ is then given by

$$\boldsymbol{\varepsilon}_N = \mathbf{x}_N - \hat{\mathbf{x}}_N = \mathbf{x}_N + \sum_{i=1}^{N} \mathbf{A}^{\dagger}(N,i)\mathbf{x}_{N-i} = \mathbf{A}_N^{\dagger}\mathbf{x}_f \tag{12.38}$$

where

$$\mathbf{A}_N^{\dagger} = \left[\mathbf{I}, \mathbf{A}^{\dagger}(N,1), \mathbf{A}^{\dagger}(N,2), \ldots, \mathbf{A}^{\dagger}(N,N)\right]$$

and

$$\mathbf{x}_f^T = \left[\mathbf{x}_N^T, \mathbf{x}_{N-1}^T, \ldots, \mathbf{x}_0^T\right]$$

A prediction error filter running backward over a data set in general will not have the same set of matrix prediction filter coefficients as the forward running prediction filter, so the error vector associated with a backward running prediction filter is denoted by

$$\mathbf{b}_N = \mathbf{x}_0 - \hat{\mathbf{x}}_0 = \mathbf{x}_0 + \sum_{i=1}^{N} \mathbf{B}^{\dagger}(N,i)\mathbf{x}_i = \mathbf{B}_N^{\dagger}\mathbf{x}_b \tag{12.39}$$

The fact that $\mathbf{A}(N,i) \neq \mathbf{B}^{\dagger}(N,i)$ reflects the fact that the multichannel backward prediction error filter is not just the complex conjugate time reverse of the multichannel forward prediction error filter (as it was in the scalar case).

The matrix generalization of (12.12) is given by

$$\mathbf{R}^f\mathbf{A}_N = \begin{bmatrix} \mathbf{R}(0) & \mathbf{R}(-1) & \cdots & \mathbf{R}(-N) \\ \mathbf{R}(1) & \mathbf{R}(0) & \cdots & \mathbf{R}(-N+1) \\ \vdots & & & \\ \mathbf{R}(N) & \mathbf{R}(N-1) & \cdots & \mathbf{R}(0) \end{bmatrix} \begin{bmatrix} \mathbf{I} \\ \mathbf{A}(N,1) \\ \vdots \\ \mathbf{A}(N,N) \end{bmatrix}$$

$$= \begin{bmatrix} \mathbf{P}_N^f \\ 0 \\ \vdots \\ 0 \end{bmatrix} \tag{12.40}$$

where the $p \times p$ block submatrices $\mathbf{R}(k)$ are defined by

$$\mathbf{R}(k) = E\{\mathbf{x}(t)\mathbf{x}^{\dagger}(t - k\Delta t)\}, \qquad k = 0, 1, \ldots, N \tag{12.41}$$

so that

$$\mathbf{R}(-k) = \mathbf{R}^{\dagger}(k) \tag{12.42}$$

The forward power matrix $\mathbf{P}_N^f$ for the prediction error filter satisfying (12.40) is given by

$$\mathbf{P}_N^f = E\{\boldsymbol{\varepsilon}_N \boldsymbol{\varepsilon}_N^{\dagger}\} = \mathbf{A}_N^{\dagger} \mathbf{R}^f \mathbf{A}_N \tag{12.43}$$

The optimum backward prediction error filter likewise satisfies

$$\mathbf{R}^b \mathbf{B}_N = \begin{bmatrix} \mathbf{R}(0) & \mathbf{R}(1) & \cdots & \mathbf{R}(N) \\ \mathbf{R}(-1) & \mathbf{R}(0) & \cdots & \mathbf{R}(N-1) \\ \vdots & & & \\ \mathbf{R}(-N) & \mathbf{R}(-N+1) & \cdots & \mathbf{R}(0) \end{bmatrix} \begin{bmatrix} \mathbf{I} \\ \mathbf{B}(N,1) \\ \vdots \\ \mathbf{B}(N,N) \end{bmatrix} = \begin{bmatrix} \mathbf{P}_N^b \\ 0 \\ \vdots \\ 0 \end{bmatrix} \tag{12.44}$$

The backward power matrix $\mathbf{P}_N^b$ for the prediction error filter satisfying (12.44) is then

$$\mathbf{P}_N^b = E\{\mathbf{b}_N \mathbf{b}_N^{\dagger}\} = \mathbf{B}_N^{\dagger} \mathbf{R}^b \mathbf{B}_N \tag{12.45}$$

The matrix coefficients $\mathbf{A}(N,N)$ and $\mathbf{B}(N,N)$ are referred to as the forward and backward reflection coefficients, respectively as follows:

$$\mathbf{C}_N^f = \mathbf{A}(N,N) \qquad \text{and} \qquad \mathbf{C}_N^b = \mathbf{B}(N,N) \tag{12.46}$$

The maximum entropy power spectral density matrix can be computed either in terms of the forward filter coefficients using

$$\boldsymbol{\Phi}_{xx}(f) = \Delta t \left[ \mathbf{A}^{-1}\left(\frac{1}{z}\right) \right]^{\dagger} \mathbf{P}_N^f \left[ \mathbf{A}^{-1}\left(\frac{1}{z}\right) \right] \tag{12.47}$$

where

$$\mathbf{A}(z) = \mathbf{I} + \mathbf{A}(N,1)z + \cdots + \mathbf{A}(N,N)z^N \tag{12.48}$$

and $z \triangleq e^{-j2\pi f \Delta t}$, or in terms of the backward filter coefficients using

$$\mathbf{\Phi}_{xx}(f) = \Delta t\left[\mathbf{B}^{-1}(z)\right]^{\dagger}\mathbf{P}_N^b\left[\mathbf{B}^{-1}(z)\right] \tag{12.49}$$

where

$$\mathbf{B}(z) = \mathbf{I} + \mathbf{B}(N,1)z + \cdots + \mathbf{B}(N,N)z^N \tag{12.50}$$

The forward and backward reflection coefficients are furthermore related by

$$\mathbf{C}_N^f = \left(\mathbf{P}_{N-1}^b\right)^{-1}\mathbf{C}_N^{b\dagger}\mathbf{P}_{N-1}^f \qquad \text{or} \qquad \mathbf{C}_N^b = \left(\mathbf{P}_{N-1}^f\right)^{-1}\mathbf{C}_N^{f\dagger}\mathbf{P}_{N-1}^b \tag{12.51}$$

If an observation process has collected $N_d$ consecutive vector samples $\{\mathbf{x}_1, \mathbf{x}_2, \ldots, \mathbf{x}_{N_d}\}$, then a prediction error filter of length $N$ will have $M = N_d - N$ consecutive $(N+1)$-tuples from the data set to operate on. It is convenient to array each $(N+1)$-tuple of data as an extended vector defined by

$$\mathbf{x}_m^N \triangleq \begin{bmatrix} \mathbf{x}_{m+N} \\ \mathbf{x}_{m+N-1} \\ \vdots \\ \mathbf{x}_{m+1} \\ \mathbf{x}_m \end{bmatrix}, \qquad \begin{array}{l} N = 0, 1, \ldots, N_d - 1 \\ m = 1, 2, \ldots, M = N_d - N \end{array} \tag{12.52}$$

The residual outputs of the forward error filter (denoted by $\mathbf{u}_m$) and the backward error filter (denoted by $\mathbf{v}_m$) when these filters are applied to the $m$th $(N+1)$-tuple of data are given by

$$\mathbf{u}_m = \left[\mathbf{A}_{N-1}^{\dagger} \,|\, 0\right]\mathbf{x}_m^N + \mathbf{C}_N^{f\dagger}\left[0 \,|\, \mathbf{B}_{N-1}^{\dagger}\right]\mathbf{x}_m^N = \boldsymbol{\varepsilon}_m^N + \mathbf{C}_N^{f\dagger}\mathbf{b}_m^N \tag{12.53}$$

and

$$\mathbf{v}_m = \mathbf{C}_N^{b\dagger}\left[\mathbf{A}_{N-1}^{\dagger} \,|\, 0\right]\mathbf{x}_m^N + \left[0 \,|\, \mathbf{B}_{N-1}^{\dagger}\right]\mathbf{x}_m^N = \mathbf{C}_N^{b\dagger}\boldsymbol{\varepsilon}_m^N + \mathbf{b}_m^N \tag{12.54}$$

Using (12.51), however, we can rewrite (12.54) as

$$\mathbf{v}_m = \mathbf{P}_{N-1}^b\mathbf{C}_N^f\mathbf{P}_{N-1}^{f\,-1}\boldsymbol{\varepsilon}_m^N + \mathbf{b}_m^N \tag{12.55}$$

Equations (12.53) and (12.55) show that the forward and backward prediction error filter residual outputs depend only on the forward reflection coefficient $\mathbf{C}_N^f$. The coefficient $\mathbf{C}_N^f$ will therefore be chosen to minimize a weighted sum of squares of the forward and backward residual outputs of the filter of length $N$; that is, minimize $SS(\mathbf{C}_N^f)$ where

$$SS(\mathbf{C}_N^f) \triangleq \frac{1}{2}\sum_{m=1}^{M} K_m\left[\mathbf{u}_m^{\dagger}\left(\mathbf{P}_{N-1}^f\right)^{-1}\mathbf{u}_m + \mathbf{v}_m^{\dagger}\left(\mathbf{P}_{N-1}^b\right)^{-1}\mathbf{v}_m\right] \tag{12.56}$$

and $K_m$ is a positive scalar weight $=1/M$. The equation that yields the optimum value of $\mathbf{C}_N^f$ for (12.56) is then

$$\mathbf{B}\mathbf{C}_N^f + \mathbf{P}_{N-1}^b \mathbf{C}_N^f (\mathbf{P}_{N-1}^f)^{-1} \mathbf{E} = -2\mathbf{G} \tag{12.57}$$

where

$$\mathbf{B} \triangleq \sum_{m=1}^{M} K_m \mathbf{b}_m^N \mathbf{b}_m^{N\dagger} \tag{12.58}$$

$$\mathbf{E} \triangleq \sum_{m=1}^{M} K_m \boldsymbol{\varepsilon}_m^N \boldsymbol{\varepsilon}_m^{N\dagger} \tag{12.59}$$

and

$$\mathbf{G} \triangleq \sum_{m=1}^{M} K_m \mathbf{b}_m^N \boldsymbol{\varepsilon}_m^{N\dagger} \tag{12.60}$$

After obtaining $\mathbf{C}_N^f$ from (12.57) (which is a matrix equation of the form $\mathbf{AX}+\mathbf{XB}=\mathbf{C}$) then $\mathbf{C}_N^b$ can be obtained from (12.51), and the desired spectral matrix estimate can be computed from (12.47) or (12.49).

## 12.2 SEQUENTIAL IMPLEMENTATION OF BAYES OPTIMAL ARRAY PROCESSOR

It was noted in Chapter 3 that when the overall goal of good signal detection is the principal concern, then array processors based on the likelihood ratio are optimum in the sense of minimizing the risk associated with an incorrect decision concerning signal presence or absence. Furthermore, when the signal to be detected has one or several uncertain parameters (due to location uncertainty, etc.) then the sufficient test statistic for decision can be obtained from the ratio of marginal probability density functions in the following manner.

Let the observation vector $\mathbf{x}$ consist of the observed random processes appearing at the array elements (or the Fourier coefficients representing the random processes). If any uncertain signal parameters are present, model these parameters as additional random variables and summarize any prior knowledge about them with an *a priori* probability density function $p(\boldsymbol{\theta})$. The likelihood ratio can then be written as the ratio of the following marginal probability density functions:

$$\Lambda(\mathbf{x}) = \frac{\int_{\Theta} p(\mathbf{x}/\boldsymbol{\theta}, \text{signal present})\, p(\boldsymbol{\theta})\, d\boldsymbol{\theta}}{p(\mathbf{x}/\text{signal absent})} \tag{12.61}$$

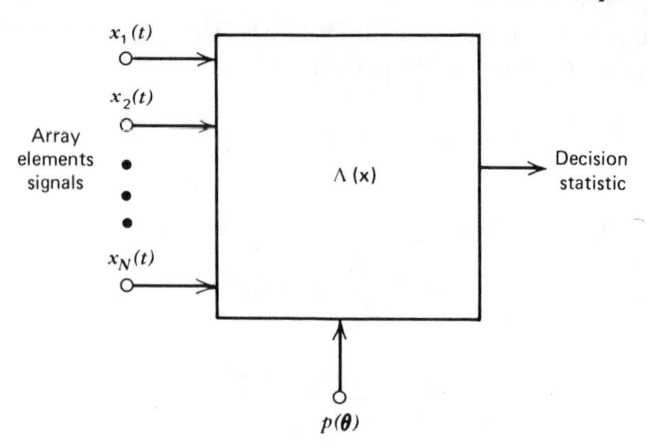


**Figure 12.3** Block diagram for one-shot array processor.

where $\boldsymbol{\theta} \in \Theta$. Equation (12.61) can be implemented by means of a suboptimal "one-shot array processor" [25] for which a block diagram is given in Figure 12.3.

### *12.2.1 Estimate and Plug Array Processor*

An intuitively appealing approach to the array detection problem when uncertain signal parameters exist is to directly estimate these parameters and then plug them into the conditional likelihood ratio as though they were exactly known. A block diagram of such an estimate and plug array processor is given in Figure 12.4. Quite naturally the merit of any estimate and plug structure depends on the accuracy of the estimates generated by the parameter estimation scheme, and the question of the processor's sensitivity to mismatch between the true parameter values and those assumed (or estimated) has received some attention [26], [27].

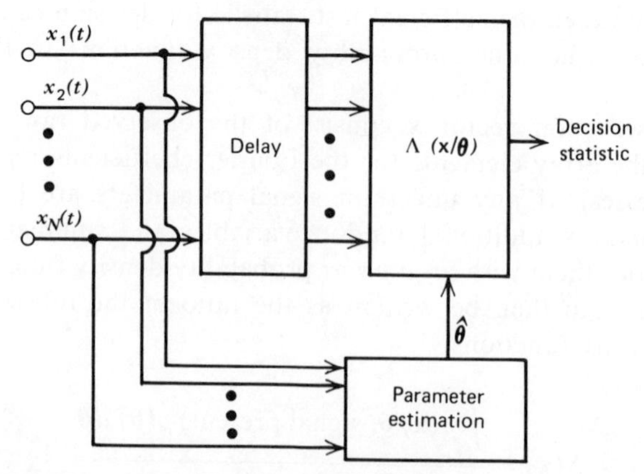


**Figure 12.4** Block diagram of estimate and plug array processor.

Two popular approaches to obtaining estimates of random signal parameters for use in an estimate and plug array processor are the maximum likelihood (ML) approach and the Bayes optimum approach. In the ML approach, the maximum likelihood estimate (MLE) of the unknown signal parameters is formed by solving

$$\left.\frac{\partial p(\mathbf{x}/\boldsymbol{\theta}, \text{signal present})}{\partial \boldsymbol{\theta}}\right|_{\boldsymbol{\theta}=\hat{\boldsymbol{\theta}}_{\text{MLE}}} = 0 \tag{12.62}$$

and then using the resulting estimate in the likelihood ratio test statistic as if it were known exactly. The Bayes approach to the parameter estimation problem incorporates any *a priori* knowledge concerning the signal parameters in the form of a probability density function $p(\boldsymbol{\theta}/\text{signal present})$. To obtain a signal parameter estimate for an estimate and plug array processor, note that

$$\frac{\Lambda(\mathbf{x})}{\Lambda(\mathbf{x}/\hat{\boldsymbol{\theta}})} = \frac{\int_{\Theta} p(\mathbf{x}/\boldsymbol{\theta}, \text{signal present})\, p(\boldsymbol{\theta}/\text{signal present})\, d\boldsymbol{\theta}}{p(\mathbf{x}/\hat{\boldsymbol{\theta}}, \text{signal present})} \tag{12.63}$$

The optimal Bayes processor explicitly incorporates a priori knowledge about the unknown signal parameters $\boldsymbol{\theta}$ into the likelihood ratio $\Lambda(\mathbf{x})$ through the averaging process expressed by the numerator of (12.63). To obtain an estimate $\hat{\boldsymbol{\theta}}$ to use in a suboptimal estimate and plug structure, require

$$\frac{\Lambda(\mathbf{x})}{\Lambda(\mathbf{x}/\hat{\boldsymbol{\theta}})} = 1 \tag{12.64}$$

Having evaluated $\Lambda(\mathbf{x})$ using the averaging process, we find $\hat{\boldsymbol{\theta}}$ as the solution to (12.64), and it is referred to as a "pseudo-estimate" $\hat{\boldsymbol{\theta}}_{\text{PSE}}$ [28]. The performance of the Bayes optimal processor for the case of a signal known except for direction (SKED) was investigated by Gallop and Nolte [29]. A comparison between the two estimate and plug structures obtained with a MLE and a Bayes pseudo-estimate is given in reference [30] for the case of target location unknown. The results indicated that the ML detector performs the same as the Bayes pseudo-estimate detector when the a priori knowledge about the uncertain parameter is uniformly distributed. When the a priori knowledge available is more precise, however, the performance of the Bayes pseudo-estimate detector improves whereas that of the ML detector does not, and this difference between the two processors becomes more pronounced as the array size becomes larger.

### *12.2.2 Sequential Optimal Array Processor*

When implemented in the form of an estimate and plug structure, the Bayes optimal array processor processes all the observed data at the same time. By implementing the same processor sequentially, the resulting array processor will exhibit adaptive (or learning) features naturally.

To see how to implement a Bayes optimal processor sequentially, let $\mathbf{x}^i$ denote the vector of observed outputs (or the Fourier coefficients thereof) from the array elements for the $i$th sample period. The sequence $\{\mathbf{x}^1,\mathbf{x}^2,\ldots,\mathbf{x}^L\}$ then represents $L$ different observation vector samples. The joint probability density function of the observed vector sequence can be written as

$$p(\mathbf{x}^1,\mathbf{x}^2,\ldots,\mathbf{x}^L)=\prod_{i=1}^{L} p(\mathbf{x}^i/\mathbf{x}^{i-1},\ldots,\mathbf{x}^1) \tag{12.65}$$

When unknown signal parameters are present, application of the averaging process described in the previous section to $p(\boldsymbol{\theta})$ yields

$$P(\mathbf{x}^1,\mathbf{x}^2,\ldots,\mathbf{x}^L)=\int_{\Theta}\prod_{i=1}^{L} p(\mathbf{x}^i/\mathbf{x}^{i-1},\ldots,\mathbf{x}^1,\boldsymbol{\theta})p(\boldsymbol{\theta})\,d\boldsymbol{\theta} \tag{12.66}$$

Now assume parameter conditional independence of the various $\mathbf{x}^i$ so that

$$p(\mathbf{x}^i/\mathbf{x}^{i-1},\ldots,\mathbf{x}^1,\boldsymbol{\theta})=p(\mathbf{x}^i/\boldsymbol{\theta}) \tag{12.67}$$

it follows that (12.66) can be rewritten as

$$P(\mathbf{x}^1,\mathbf{x}^2,\ldots,\mathbf{x}^L)=\int_{\Theta}\prod_{i=1}^{L} p(\mathbf{x}^i/\boldsymbol{\theta})p(\boldsymbol{\theta})\,d\boldsymbol{\theta} \tag{12.68}$$

According to Baye's rule it follows that

$$p(\boldsymbol{\theta}/\mathbf{x}^1)=\frac{p(\mathbf{x}^1/\boldsymbol{\theta})p(\boldsymbol{\theta})}{p(\mathbf{x}^1)} \tag{12.69}$$

$$p(\boldsymbol{\theta}/\mathbf{x}^1,\mathbf{x}^2)=\frac{p(\mathbf{x}^2/\boldsymbol{\theta})p(\boldsymbol{\theta}/\mathbf{x}^1)}{p(\mathbf{x}^2/\mathbf{x}^1)} \tag{12.70}$$

$$\vdots$$

and

$$p(\boldsymbol{\theta}/\mathbf{x}^{i-1},\ldots,\mathbf{x}^1)=\frac{p(\mathbf{x}^{i-1}/\boldsymbol{\theta})p(\boldsymbol{\theta}/\mathbf{x}^{i-2},\ldots,\mathbf{x}^1)}{p(\mathbf{x}^{i-1}/\mathbf{x}^{i-2},\ldots,\mathbf{x}^1)} \tag{12.71}$$

so that

$$p(\mathbf{x}^1,\ldots,\mathbf{x}^L)=\prod_{i=1}^{L}\int_{\Theta} p(\mathbf{x}^i/\boldsymbol{\theta})p(\boldsymbol{\theta}/\mathbf{x}^{i-1},\ldots,\mathbf{x}^1)\,d\boldsymbol{\theta} \tag{12.72}$$

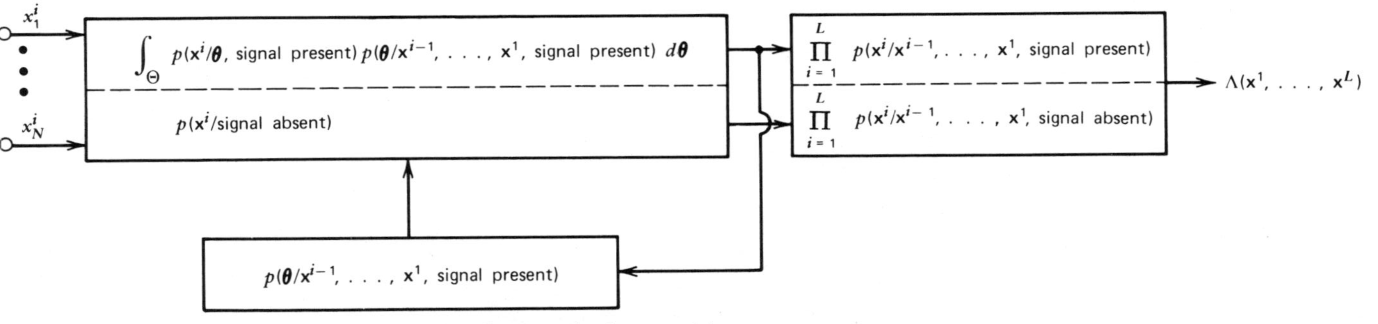


**Figure 12.5** Bayes optimal sequential array processor.

where $p(\boldsymbol{\theta}/\mathbf{x}^{i-1},\ldots,\mathbf{x}^1)$ represents the updated form of the *a priori* information contained in $p(\boldsymbol{\theta})$.

A sequential processor may now be implemented using (12.71) and (12.72) to form the marginal density functions required in the likelihood ratio as follows:

$$\Lambda(\mathbf{x}^1,\ldots,\mathbf{x}^L)=\frac{p(\mathbf{x}^1,\ldots,\mathbf{x}^L/\text{signal present})}{p(\mathbf{x}^1,\ldots,\mathbf{x}^L/\text{signal absent})} \tag{12.73}$$

A block diagram of the resulting sequential array processor based on (12.71)–(12.73) is given in Figure 12.5. The sequential Bayesian updating of $p(\boldsymbol{\theta})$ represented by (12.71) results in an optimal processor having an adaptive capability.

Performance results using an optimal sequential array processor have been reported in reference [25] for a detection problem involving a signal known exactly imbedded in Gaussian noise where the noise has an additive component arising from a noise source with unknown direction. The adaptive processor in this problem must both succeed in detecting the presence or absence of the desired signal and in "learning" the actual direction of the noisy signal source. The results obtained indicated that even though the directional noise to thermal noise ratio was relatively low, the optimal sequential processor could nevertheless determine the directional noise source's location.

## 12.3 PARTIALLY ADAPTIVE ARRAY CONCEPTS

Consider the block diagram of an interference cancelling adaptive array system given in Figure 12.6. For this system a main array output is formed by merely combining the $N$ element outputs. An adaptive processor combines only $M$ of the element signals (where $M \leqslant N$) to form an adaptive output that is then subtracted from the main array output. In the event that $M=1$, a simple sidelobe canceller system results. On the other hand if $M=N$, then the array is said to be fully adaptive. Cases that fall in between where $1<M<N$ are referred to as a partially adaptive array (or as a multiple sidelobe canceller for the given configuration).

A fully adaptive array in which every element is individually adaptively controlled is obviously preferred since it affords the greatest control over the array response. For large arrays containing thousands of elements, individually controlling every element can prove a prohibitive implementation expense. Furthermore, signal processor implementation becomes much more difficult and costly as the dimensionality of the signal vector to be processed increases. Consequently it is highly desirable to reduce the dimensionality of the signal processor while maintaining a high degree of control over the array response by adopting one of the following adaptive control philosophies:

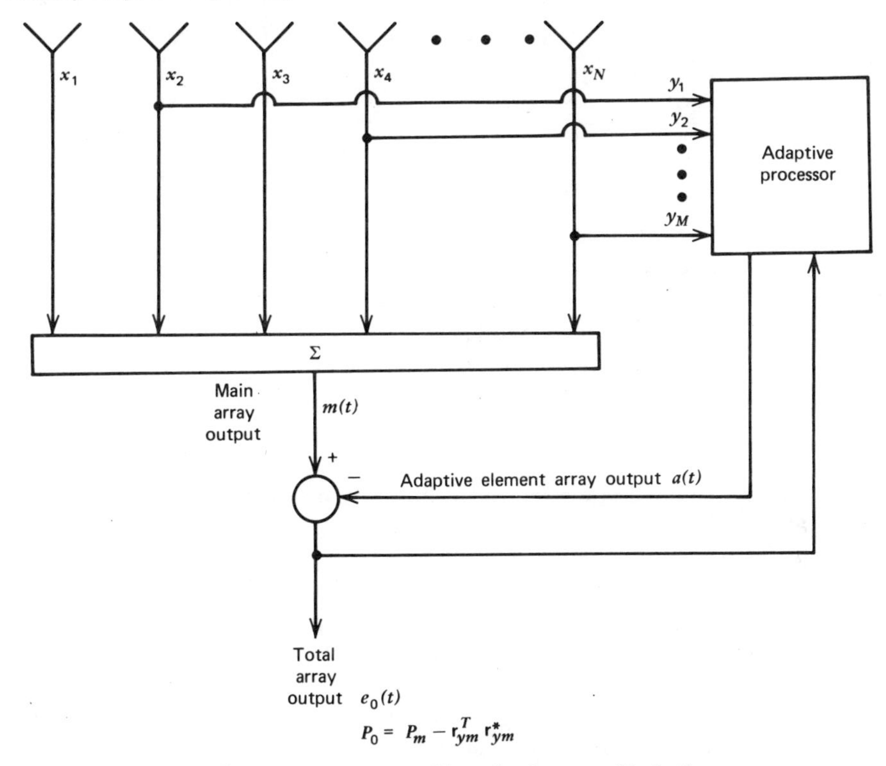


**Figure 12.6** Interference cancelling adaptive array block diagram.

1. Judiciously select only a fraction of the array elements to adaptively control thereby resulting in elemental level adaptivity [31], [32].
2. Combine the $N$ elements in the entire array into a collection of $M$ subarrays by means of a subarray beamformer transformation $\mathbf{Q}$ and adaptively control each of the resulting subarray outputs [33], [37].

Each of these approaches will now be discussed to determine the characteristics that typify these partially adaptive array concepts.

### *12.3.1 Adaptive Subarray Beamforming*

The fundamental concept of adaptive subarray beamforming discussed by Chapman [33] is to reduce the required dimensionality of the signal processor by introduction of an $N \times M$ transformation matrix $\mathbf{Q}$ (called a subarray beamformer) as shown in Figure 12.7. The subarray signals resulting from the elements so combined in subarrays are then all adaptively controlled to produce the total array response. From Figure 12.7 it follows that

$$\mathbf{y} = \mathbf{Q}^T \mathbf{x} \tag{12.74}$$

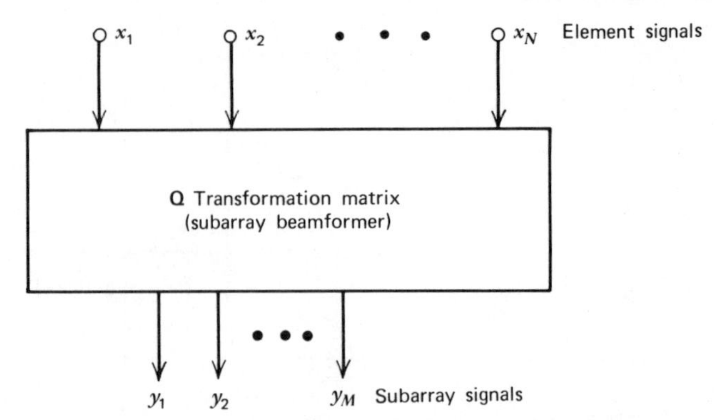


**Figure 12.7** Use of a subarray beamformer to reduce the signal vector dimensionality.

Consequently the covariance matrix of the subarray signal vector is given by

$$\mathbf{R}_{yy} = \mathbf{Q}^{\dagger}\mathbf{R}_{xx}\mathbf{Q} \tag{12.75}$$

If the adaptive processor employs the Howells-Applebaum SNR performance measure algorithm, then the optimum weight vector solution is given by

$$\mathbf{w}_{y_{\text{opt}}} = \alpha\mathbf{R}_{yy}^{-1}\mathbf{v}_y^* \tag{12.76}$$

where the beam-steering vector for the subarray $\mathbf{v}_y$ is related to the beam-steering vector for the total array $\mathbf{v}_x$ by

$$\mathbf{v}_y = \mathbf{Q}^T\mathbf{v}_x \tag{12.77}$$

Consequently the resulting array beam pattern can be computed by using the implied relationship

$$\mathbf{w}_x = \mathbf{Q}\mathbf{w}_{y_{\text{opt}}} \tag{12.78}$$

It should be noted that $\mathbf{w}_x$ of (12.78) is not the same as $\mathbf{w}_{x_{\text{opt}}} = \alpha\mathbf{R}_{xx}^{-1}\mathbf{v}_x^*$ for the fully adaptive array, but is rather a suboptimum solution that is constrained by the subarray configuration.

Now consider the effect of element-level amplitude and phase errors in a fully adaptive array and in a partially adaptive array. Amplitude and phase errors between any two channels can be characterized by a mean offset signal level (over the bandwidth of interest) plus a variation about the mean as a function of frequency. The adjustment of a single complex weight in one channel will remove any mean error that exists between a pair of channels, so a fully adaptive array will only be sensitive to the error variance about the mean.

A subarray configuration, however, does not have as many degrees of freedom as the fully adaptive array and consequently may not succeed in removing all the mean errors that may exist between all the elements of the entire array. Consequently the performance of an array that is partitioned into subarrays is more susceptible to element-level errors than a fully adaptive array.

The effect of element random phase errors on a subarray structure can be determined by appropriately modifying the **Q** subarray beamformer matrix. Define an "errored" **Q** matrix $\mathbf{Q}^{\epsilon}$, which is given by

$$\mathbf{Q}^{\epsilon} = \mathbf{E}^{\rho}\mathbf{Q} \tag{12.79}$$

where $\mathbf{E}^{\rho}$ is an $N \times N$ diagonal matrix with elements given by

$$e_{ii}^{\rho} = \exp\left[ j2\pi(1-\rho_e)z_i \right] \tag{12.80}$$

The parameter $\rho_e (0 \leqslant \rho_e \leqslant 1)$ represents the severity of the error (with no error corresponding to $\rho_e = 1$) while $z_i$ is a uniformly distributed random variable $(-0.5 \leqslant z_i \leqslant 0.5)$.

This errored **Q** matrix model results in a random sidelobe structure (RSL) for the total array whose mean level (with respect to isotropic) is approximately given by [34]

$$|g_{\text{RSL}}(\theta)|^2 \cong \frac{\pi^2(1-\rho_e)^2}{3} |g_e(\theta)|^2 \tag{12.81}$$

where $g_e(\theta)$ denotes the directional voltage gain of an array element.

Choosing subarrays by merely grouping physically contiguous elements is termed a "simple subarray" approach. Any simple subarray configuration has the immediate problem that the phase centers of the subarrays are usually well separated (by several wavelengths) so that grating lobes between the subarrays (or echelon lobes) result that cannot further be altered by the adaptive processor. A more attractive configuration that avoids the echelon lobe problem is provided by the "beam-space" subarray in which the full array aperture is used for each subarray so the resulting structure can be regarded as a multibeam processor. Vural [35] likewise has shown that a beam-based processor realizes superior performance in partially adaptive operations under diverse interference conditions. To introduce constraints into beam space adaptive algorithms, the only difference as compared with element space data algorithms is in the mechanization of the constraint requirements [37].

Subarray groups for planar array designs may be chosen by combining row (or column) subarrays. Such a choice obviously results in adaptivity for only one principal plane of the array beam pattern and may therefore be inadequate. A more realistic alternative to the row-column subarray approach is a configuration called the row-column precision array (RCPAA) [36] in which each element signal is split into two paths: a row path and a column path. All the element signals from a given row or column are summed together, and all the row outputs and column outputs are then adaptively combined. The number of degrees of freedom in the resulting adaptive processor is then equal to the number of rows plus the number of columns in the actual array.

When ideal operating conditions are assumed with perfect element channel matching, simulation studies have shown that the subarray configurations discussed above are all capable of yielding array performance that is nearly the same as that of fully adaptive arrays [33]. When independent random errors are introduced into the array elements, however, the resulting random sidelobe structure severely deteriorates the quality of the array response. This performance deterioration results from the need for precision in the subarray beamforming transformation since this precision is severely affected by the random element errors.

### *12.3.2 Element-Level Adaptivity*

Adaptive subarray beamforming is a highly attractive solution to the partially adaptive array problem, particularly if the application inherently requires multiple beamforming. However, the beamforming matrix transformation introduces an additional expense into the realization that can be avoided if multiple beams are not required by simply directly controlling only a fraction of the array elements in an element-level partial adaptivity scheme. With this approach, the question arises as to which elements of the original array should be adaptively controlled in order to achieve the best possible array performance. To gain some insight into the behavior of the system depicted in Figure 12.6, the approach taken by Morgan [32] will be followed; and an explicit solution for a narrowband two-jammer problem will be obtained.

Using the MMSE performance measure, we find the optimum weight vector for the adaptive processor is given by

$$\mathbf{W}=\mathbf{R}_{yy}^{*-1}\mathbf{r}_{ym} \tag{12.82}$$

where $\mathbf{R}_{yy} \triangleq E\{\mathbf{y}\mathbf{y}^{\dagger}\}$ is the $M \times M$ covariance matrix of the adaptive element signal vector and $\mathbf{r}_{ym} \triangleq E\{m\mathbf{y}^*\}$ is the cross-correlation vector of the main array signal and the adaptive element signal vector. It is worth noting that the solution for $\mathbf{w}$ given by (12.82) results instead of the more familiar expression $\mathbf{w}=\mathbf{R}_{yy}^{-1}\mathbf{r}_{ym}$ due to Morgan's use of $E\{\mathbf{y}\mathbf{y}^{\dagger}\}$ instead of the more common $E\{\mathbf{y}^*\mathbf{y}^T\}$ as the definition of $\mathbf{R}_{yy}$. The minimum total array output power that results when (12.82) is used is then

$$P_0=P_m-\mathbf{r}_{ym}^T\mathbf{R}_{yy}^{-1}\mathbf{r}_{ym}^* \tag{12.83}$$

where $P_m$ is the main array output power. The straightforward evaluation of

(12.83) does not yield much insight into the relationship between adaptive performance and the array configuration and jamming environment. A concise mathematical expression of (12.83) is accordingly more desirable to elucidate the problem.

Consider two narrowband jammers of frequencies $\omega_1$ and $\omega_2$. The composite signal vector $\mathbf{y}$ can then be written as

$$\mathbf{y}(t)=\mathbf{n}(t)+J_1\exp(j\omega_1 t)\mathbf{v}_1+J_2\exp(j\omega_2 t)\mathbf{v}_2 \tag{12.84}$$

where $J_1, J_2$ represent the reference amplitude and phase of the jammers, and $\mathbf{n}$ is the additive noise vector with independent components of equal power $P_n$. The spatial vectors $\mathbf{v}_1, \mathbf{v}_2$ have components given by

$$V_{k,i}=\exp\left\{j\frac{2\pi}{\lambda}(\mathbf{r}_i\cdot\mathbf{u}_k)\right\}, \qquad i\in A,\ k=1,2 \tag{12.85}$$

where $A$ denotes the subset of $M$ elements that have been selected to adaptively control, $\mathbf{r}_i$ is the $i$th element location vector, and $\mathbf{u}_k$ is a unit vector pointed in the direction of arrival of the $k$th jammer. For a linear array having elements aligned along the $x$-axis, (12.85) reduces to

$$V_{k,i}=\exp\left\{j\frac{2\pi}{\lambda}x_i\sin\theta_k\right\} \tag{12.86}$$

where $\theta_k$ is the angle of arrival of the $k$th jammer measured from the array boresight.

The main array output signal can likewise be written as

$$m(t)=\sum_{i=1}^{N} n_i(t)+J_1 e^{j\omega_1 t}h_1+J_2 e^{j\omega_2 t}h_2 \tag{12.87}$$

where

$$h_k=\sum_{i=1}^{N} v_{k,i}, \qquad k=1,2 \tag{12.88}$$

is the main array factor, which is computed by forming the sum of all spatial vector components for all the array elements $i=1,\cdots,N$. It follows that the array output signal can then be written as

$$e_0(t)=m(t)-\mathbf{w}^T\mathbf{y}(t) \tag{12.89}$$

where $\mathbf{w}$ is the vector of adaptive weights.

The minimum total array output power given by (12.83) requires the covariance matrix $\mathbf{R}_{yy}$ and the cross-correlation vector $\mathbf{r}_{ym}$, both of which can be computed from (12.84) and (12.85), to yield

$$\mathbf{R}_{yy}=E\{\mathbf{y}\mathbf{y}^\dagger\}=P_n\mathbf{I}+P_1\mathbf{v}_1\mathbf{v}_1^\dagger+P_2\mathbf{v}_2\mathbf{v}_2^\dagger \tag{12.90}$$

and

$$\mathbf{r}_{ym} = E\{\mathbf{y}^* m\} = P_n \mathbf{1} + P_1 h_1 \mathbf{v}_1^* + P_2 h_2 \mathbf{v}_2^* \tag{12.91}$$

where $P_1$ and $P_2$ denote the jammer power levels and $\mathbf{1}$ is an $M \times 1$ vector of ones.

Since (12.82) requires the inverse of $\mathbf{R}_{yy}$, this inverse can be explicitly obtained by the twofold application of the matrix inversion identity (D.10) of Appendix D to (12.90). Substitution of (12.91) along with $\mathbf{R}_{yy}^{-1}$ into (12.82) then yields

$$\begin{aligned}\mathbf{w}_{\text{opt}} = \mathbf{1} &+ \gamma\left[F_1 - \rho^* F_2 + \frac{P_n}{MP_2} F_1\right]\mathbf{v}_1^* \\ &+ \gamma\left[F_2 - \rho F_1 + \frac{P_n}{MP_1} F_2\right]\mathbf{v}_2^*\end{aligned} \tag{12.92}$$

where

$$F_k = \sum_{i \notin A} v_{k,i} \tag{12.93}$$

denotes the complementary array factor that is computed by summing the spatial vector components for all the elements that are not adaptively controlled, the factor

$$\rho = \frac{1}{M} \sum_{i \subset A} v_{1,i}^* v_{2,i} \tag{12.94}$$

is the complex correlation coefficient of the adaptive element spatial vectors, and

$$\gamma = \left[M(1 - |\rho|^2) + \frac{P_n}{P_1} + \frac{P_n}{P_2} + \frac{P_n^2}{MP_1P_2}\right]^{-1} \tag{12.95}$$

For a fully adaptive array where $N = M$, then (12.92) reduces to $\mathbf{w}_{\text{opt}} = \mathbf{1}$ in the absence of any constraints on the mainlobe. If it is assumed that both jammers are much stronger than the noise so that $P_1, P_2 \gg P_n$, then (12.92) becomes

$$\mathbf{W}_{\text{opt}}\big|_{P_1, P_2 \to \infty} \to \mathbf{1} + \frac{F_1 - \rho^* F_2}{M(1 - |\rho|^2)} \mathbf{v}_1^* + \frac{F_2 - \rho F_1}{M(1 - |\rho|^2)} \mathbf{v}_2^* \tag{12.96}$$

The above expression shows that the adaptive weights will be ill-conditioned whenever $|\rho| \approx 1$. This condition physically corresponds to the situation that occurs when the adaptive array pattern cannot resolve the two jammers. Such a failure to resolve can occur either because the two jammers are too close together or because of their simultaneous appearance on distinct grating lobes.

Recognizing that the output residue power is given by

$$P_0 = P_m - \mathbf{r}_{ym}^T \mathbf{w}_{\text{opt}}^* \tag{12.97}$$

we see that it follows that the normalized output residue power can be expressed as

$$\frac{P_0}{P_n} = N - M + \gamma\left\{ |F_1|^2 + |F_2|^2 - 2\,\text{Re}(\rho F_1 F_2^*) + \frac{P_n}{M}\left(\frac{|F_1|^2}{P_2} + \frac{|F_2|^2}{P_1}\right)\right\} \tag{12.98}$$

It will once again be noted that when $N = M$ (a fully adaptive array) then, in the absence of mainlobe constraints, $P_0 = 0$. Equation (12.98) immediately yields an upper bound for the maximum residue power as

$$P_0 \leqslant P_{0_{\max}} = \frac{P_1 P_2(|F_1| + |F_2|)^2 + \dfrac{P_n}{M}(P_2|F_1|^2 + P_1|F_2|^2)}{P_1 + P_2 + \dfrac{P_n}{M}} \tag{12.99}$$

and $P_0 = P_{0_{\max}}$ when $\rho_{\max} = -\exp[\arg(F_1^* F_2)]$.

In the case of strong jammers for $|\rho| < 1$, then (12.98) reduces to

$$\left.\frac{P_0}{P_n}\right|_{P_1, P_2 \to \infty} \to N - M + \frac{|F_1|^2 + |F_2|^2 - 2\text{Re}(\rho F_1 F_2^*)}{M(1 - |\rho|^2)}, \qquad |\rho| < 1 \tag{12.100}$$

This expression emphasizes the fact that the residue power takes on a large value whenever $|\rho| = 1$. Furthermore, (12.100) surprisingly does not depend on the jammer power levels $P_1$ and $P_2$ but instead depends only on the geometry of the array and the jammer angles. The maximum residue power upper bound (12.99) does depend on $P_1$ and $P_2$ however, since

$$P_{0_{\max}}\Big|_{P_1, P_2 \to \infty} \to (|F_1| + |F_2|)^2 \frac{P_1 P_2}{P_1 + P_2} \tag{12.101}$$

The foregoing expressions for output residue power demonstrate the central role that is played by the correlation coefficient between the adaptive element spatial vectors in characterizing the performance of a partially adaptive array. In simple cases this correlation coefficient $\rho$ can be related to the array geometry and jammer angles of arrival thereby demonstrating that it makes a significant difference which array elements are chosen for adaptive control in a partially adaptive array. In most cases the nature of the relationship between $\rho$ and the array geometry is so mathematically obscured that only computer solutions yield meaningful solutions. For the two-jammer case it appears that the most favorable adaptive element locations for a linear array are edge-clustered positions at both ends of the array [32].

The foregoing results described for a partially adaptive array and a two-jammer environment did not consider the effect of errors in the adaptive weights. It has been clearly established, however, that the effects of errors in the adaptive weight elements are a major consideration in partially adaptive array design [31]. The variation in the performance improvement achieved with a partially adaptive array with errors in the nominal pattern weights is extremely sensitive to the choice of adaptive element location within the entire array. Consequently, for a practical design the best choice of adaptive element location depends principally upon the error level experienced by the nominal array weights and only secondarily on the optimum theoretical performance that can be achieved. It has been found that spacing the elements in a partially adaptive array using elemental level adaptivity so the adaptively controlled elements are clustered toward the center is desirable since then the adapted pattern interference cancellation tends to be independent of errors in the nominal adaptive weight values [31].

## 12.4 SUMMARY AND CONCLUSIONS

The MEM approach to spectral estimation is introduced for obtaining high resolution power spectral density estimates of stationary time series from limited data records. Extension of Burg's scalar MEM algorithm to the multiple channel problem is required for array processing problems, and the extension provided by Strand was outlined for this purpose. When both signal detection and parameter estimation must be accomplished, then the Bayes optimal array processor can be implemented sequentially to provide adaptive capabilities in a natural manner. Reported applications of these algorithms to actual adaptive array problems have been sparse, yet the desirable properties of these algorithms make them highly likely candidates for future attention and development.

Partially adaptive array concepts are highly important because they offer the possibility of realizing near optimal array performance with only a fraction of the control elements (and hence the cost) required for a fully adaptive array. This area is only now starting to be exploited, but the economies that can be realized using this approach are substantial and guarantee that further developments and results can be expected before very long.

## PROBLEMS

**1** The prediction error filter matrix equation for a scalar random process may be developed by assuming that two sampled values of a random process $x_0$ and $x_1$ are known and it is desired to obtain an estimate of the next sampled value $\hat{x}_2$ using a second-order prediction error filter

$$\hat{x}_2 = a(2,2)x_0 + a(2,1)x_1$$

so that $\varepsilon = x_2 - \hat{x}_2 = x_2 - a(2,1)x_1 - a(2,2)x_0$

(a) Using the fact that the optimal linear predictor must provide estimates for which the error is orthogonal to the data (i.e., $\overline{x_0\varepsilon} = 0$ and $\overline{x_1\varepsilon} = 0$), show that

$$\begin{bmatrix} r(1) & r(0) & r(2) \\ r(2) & r(1) & r(0) \end{bmatrix} \begin{bmatrix} 1 \\ -a(2,1) \\ -a(2,2) \end{bmatrix} = \begin{bmatrix} 0 \\ 0 \end{bmatrix}$$

where $r(n) \triangleq \overline{x_i x_j}$, $|i-j| = n$.

(b) If $P_2 \triangleq \overline{\varepsilon^2} = \overline{(x_2 - \hat{x}_2)(x_2 - \hat{x}_2)}$, use the fact that the error in the estimate $\hat{x}_2$ is orthogonal to the estimate itself (i.e., $\overline{(x_2 - \hat{x}_2)\hat{x}_2} = 0$) and the expression given above for $\hat{x}_2$ to show that

$$P_2 = r(0) - a(2,1)r(1) - a(2,2)r(2)$$

The results of part (a) combined with the result from part (b) then yield the prediction error filter matrix equation for this case.

**2** [15] ***The Relationship between MEM Spectral Estimates and ML Spectral Estimates.*** Assume the correlation function of a random process $\mathbf{x}(t)$ is known at uniformly spaced, distinct sample times. Then the ML spectrum (for an equally spaced line array of $N$ sensors) is given by

$$\mathrm{MLM}(k) = \frac{N\Delta x}{\mathbf{v}^{\dagger}(k)\mathbf{R}_{xx}^{-1}\mathbf{v}(k)}$$

where $k$ = wavenumber (reciprocal of wavelength)
$\Delta x$ = spacing between adjacent sensors
$\mathbf{v}(k)$ = beamsteering column vector where $v_n(k) = e^{-j2\pi nk\Delta x}$, $n = 0, 1, \ldots, N-1$
$\mathbf{R}_{xx}$ = $N \times N$ correlation matrix of $\mathbf{x}(t)$

(a) First define the lower triangular matrix $\mathbf{L}$ by

$$\mathbf{L} \triangleq \begin{bmatrix} 1 & 0 & \cdots & 0 \\ c(2,N) & 1 & \cdots & 0 \\ c(3,N) & c(2,N-1) & \cdots & 0 \\ \vdots & & & \\ c(N,N) & c(N-1,N-1) & \cdots & 1 \end{bmatrix}$$

where 1, $c(2,M),\ldots,c(M,M)$ are the weights of the $M$-long prediction error filter whose output power is $P(M)$. Note that

$$\mathbf{R}_{xx}\mathbf{L} = \begin{bmatrix} P(N) & - & - & - & - \\ 0 & P(N-1) & - & - & - \\ 0 & 0 & & - & - \\ \vdots & \vdots & \ddots & & \\ 0 & 0 & & & P(1) \end{bmatrix}$$

The maximum entropy spectrum estimate corresponding to the $M$-long prediction error filter is given by

$$\text{MEM}(k,M) = \frac{P(M)\Delta x}{\left| \sum_{i=1}^{M} c(i,M) \exp(j2\pi k(i-1)\Delta x) \right|^2}$$

where $c(1,M) \equiv 1$. Define the matrix $\mathbf{P}$ according to

$$\mathbf{P} \equiv \mathbf{L}^{\dagger}\mathbf{R}_{xx}\mathbf{L}$$

Show that $\mathbf{P}$ is an $N \times N$ diagonal matrix whose diagonal elements are given by $P(N), P(N-1),\ldots,P(1)$.

(b) Using the fact that $\mathbf{R}_{xx}^{-1} = \mathbf{L}\mathbf{P}^{-1}\mathbf{L}^{\dagger}$, show that

$$\mathbf{v}^{\dagger}\mathbf{R}_{xx}^{-1}\mathbf{v} = (\mathbf{L}^{\dagger}\mathbf{v})^{\dagger}\mathbf{P}^{-1}(\mathbf{L}^{\dagger}\mathbf{v}) = \sum_{n=1}^{N} \frac{\Delta x}{\text{MEM}(k,n)}$$

This result then gives the desired relationship:

$$\frac{1}{\text{MLM}(k)} = \frac{1}{N} \sum_{n=1}^{N} \frac{1}{\text{MEM}(k,n)}$$

Therefore the reciprocal of the ML spectrum is equal to the average of the reciprocals of the maximum entropy spectra obtained from the one-point up to the $N$-point prediction error filter. The lower resolution of the ML method therefore results from the "parallel resistor network averaging" of the lowest to the highest resolution maximum entropy spectra.

3 [11] ***Equivalence of MEM Spectral Analysis to Least-Squares Fitting of a Discrete-Time All-Pole Model to the Available Data.*** Assume that the first $(N+1)$ points $\{r(0),r(1),\ldots,r(N)\}$ of the autocorrelation function of a stationary Gaussian process are known exactly, and it is desired to estimate $r(N+1)$. Consider the Toeplitz covariance matrix

$$\mathbf{R}_{N+1}=\begin{bmatrix} r(0) & r(1) & \cdots & r(N) & r(N+1) \\ r(1) & r(0) & \cdots & r(N-1) & r(N) \\ \cdots & \cdots & \cdots & & \\ r(N+1) & r(N) & & r(1) & r(0) \end{bmatrix}$$

The basic autocorrelation function theorem states that $\mathbf{R}_{N+1}$ must be semi-positive definite if the quantities $r(0),r(1),\ldots,r(N+1)$ are to correspond to an autocorrelation function. Consequently $\det[\mathbf{R}_{N+1}]$ must be nonnegative.

MEM spectral analysis seeks to select that value of $r(N+1)$ that maximizes $\det[\mathbf{R}_{N+1}]$. The entropy of the $(N+2)$ dimensional probability density function with covariance matrix $\mathbf{R}_{N+1}$ is given by

$$\text{entropy}=\ln(2\pi e)^{[N+2/2]}\det\left[\mathbf{R}_{N+1}\right]^{1/2}$$

and the choice for $r(N+1)$ maximizes this quantity. In order to obtain $r(N+2)$, the value of $r(N+1)$ just found is substituted into $\mathbf{R}_{N+2}$ to find $\det[\mathbf{R}_{N+2}]$, and the corresponding entropy is maximized with respect to $r(N+2)$. Likewise substituting the values of $r(N+1)$ and $r(N+2)$ found above into $\det[\mathbf{R}_{N+3}]$ and maximizing yields $r(N+3)$. The estimates for additional values $r(N+4),r(N+5),\cdots$ may then be evaluated by following the same procedure.

(a) Show that maximizing $\det[\mathbf{R}_{N+1}]$ with respect to $r(N+1)$ is equivalent to the relation

$$\det\begin{bmatrix} r(1) & r(0) & \cdots & r(N-1) \\ r(2) & r(1) & \cdots & r(N-2) \\ \vdots & \vdots & & \vdots \\ r(N+1) & r(N) & \cdots & r(1) \end{bmatrix}=0$$

(b) Consider the all-pole data prediction error model given by

$$y(n)+a_1'y(n-1)+\cdots+a_N'y(n-N)=e(n)$$

or

$$\mathbf{y}^T\mathbf{a}'=e(n)$$

where $\mathbf{a}'^T=[1,a'_1,a'_2,\ldots,a'_N]$
$\mathbf{y}^T=[y(n),y(n-1)\cdots y(n-N)]$
$N$ = order of all-pole model
$n$ = number of data samples, $n>N$

and where $e(n)$ is a zero-mean random variable with $E\{e(i)e(j)\}=0$ for $i\neq j$. Assuming that $E\{e(n)y(n-k)\}=0$ for $k>0$, show that multiplying both sides of the above equation for $e(n)$ by $y(n-k)$ and taking expectations yields

$$r'(k)+a'_1r'(k-1)+a'_2r'(k-2)+\cdots+a'_Nr'(k-N)=0, \qquad \text{for } k>0$$

where $r'(k)\triangleq E\{y(n)y(n-k)\}$.

(c) Using the results of part (b) and the fact that $r(\tau)=r(-\tau)$, it follows that

$$\begin{aligned} r'(1)+a'_1r'(0)+\cdots+a'_Nr'(N-1)&=0\\ r'(2)+a'_1r'(1)+\cdots+a'_Nr'(N-2)&=0\\ &\vdots\\ r'(N+1)+a'_1r'(N)+\cdots+a'_Nr'(1)&=0 \end{aligned}$$

or $\mathbf{R}'_{N+1}\mathbf{a}'=\mathbf{0}$ in matrix notation. Use this result to show that

$$\det\begin{bmatrix} r'(1) & r'(0) & \cdots & r'(N-1)\\ r'(2) & r'(1) & \cdots & r'(N-2)\\ \vdots & \vdots & & \vdots\\ r'(N+1) & r'(N) & \cdots & r'(1) \end{bmatrix}=0$$

If the first $N+1$ exact values $\{r(0),r(1),\ldots,r(N)\}$ of any autocorrelation function are available, then substituting these values into the first $N$ of the simultaneous linear equations corresponding to $\mathbf{R}_{N+1}\mathbf{a}=\mathbf{0}$ yields a unique solution for the coefficients $\{a_1,a_2,\ldots,a_N\}$. Consequently the value for $r(N+1)$ for a discrete-time all-pole model having the coefficients $\{a_1,a_2,\ldots,a_N\}$ is uniquely determined by

$$\det\begin{bmatrix} r(1) & r(0) & \cdots & r(N-1)\\ r(2) & r(1) & \cdots & r(N-2)\\ \vdots & \vdots & & \\ r(N+1) & r(N) & \cdots & r(1) \end{bmatrix}=0$$

The above result is identical to the relation obtained in part (a), and hence the same solution would have been obtained from maximum entropy spectral analysis.

**4** ***Angle of Arrival Estimation*** The MEM technique has superior capability for resolving closely spaced spectral peaks that may be exploited for estimating the angular distribution of received signal power.

(a) Using a time/space dualism, reformulate equation (12.11) to give a spatial spectrum $\hat{\phi}_{xx}(\mu)$, where $\mu = \cos\theta$ and $\theta$ is the angle from array endfire for an $N+1$ element linear array. Assume narrowband signals with spacing $d$ between elements.

(b) Reformulate equation (12.12) for the spatial estimation problem of part (a). What correspondence exists between the prediction error filter coefficients and the weights of a coherent sidelobe canceller with $N$ auxiliary antennas?

**5** Recognizing that

$$(\mathbf{A}+\mathbf{b}\mathbf{b}^{\dagger})^{-1} = \mathbf{A}^{-1} - \frac{\mathbf{A}^{-1}\mathbf{b}^{\dagger}\mathbf{b}\mathbf{A}^{-1}}{(1+\mathbf{b}^{\dagger}\mathbf{A}^{-1}\mathbf{b})}$$

let $\mathbf{A} = P_n\mathbf{I} + P_1\mathbf{v}_1\mathbf{v}_1^{\dagger}$ in (12.90) and apply the above matrix identity once. Finally apply the above matrix identity to the resulting expression and show that (12.92) results from the above operations on substitution of (12.90) into (12.82).

**6** [32] ***Partially Adaptive Linear Array Having Two Adaptive Elements for a Two-Jammer Scenario***

(a) Assume that the $N$ elements in a uniformly spaced linear array are spaced at half-wavelengths. Show that the elements of the spatial vectors are given by

$$v_{k,i} = \exp(ji\pi\sin\theta_k)$$

Furthermore show that the complementary array factors are given by

$$F_1 = \exp\left[\frac{j(N+1)\alpha}{2}\right]S_1 - 2\exp\left[\frac{j(p+q)\alpha}{2}\right]\cos\left(\frac{m\alpha}{2}\right)$$

and

$$F_2 = \exp\left[\frac{j(N+1)\beta}{2}\right]S_2 - 2\exp\left[\frac{j(p+q)\beta}{2}\right]\cos\left(\frac{m\beta}{2}\right)$$

where $p, q$ denote the adaptive element numbers, $m = q-p$, and $\alpha, \beta = \pi\sin\theta_{1,2}$ are the phase shifters per element for the two jammers. The two factors

$$S_1 = \frac{\sin(N\alpha/2)}{\sin(\alpha/2)} \qquad \text{and} \qquad S_2 = \frac{\sin(N\beta/2)}{\sin(\beta/2)}$$

denote the array factor moduli for the two jammers.

(b) Using the results obtained in part (a), show that the jammer correlation coefficient becomes

$$\rho = \exp\left[\frac{j(p+q)(\beta-\alpha)}{2}\right]\cos\frac{m(\beta-\alpha)}{2}$$

and

$$\gamma = \frac{P_1 P_2}{P_n(P_1 + P_2 + P_n/2) + 2P_1 P_2 \sin^2 m(\beta-\alpha)/2}$$

(c) Using the results obtained in parts (a) and (b) show that normalized output residue power of the system in Figure 12.6 is given by

$$\frac{P_0}{P_n} = N + \frac{P_1 P_2}{P_n(P_1 + P_2) + 2P_1 P_2 \sin^2 m(\beta-\alpha)/2}$$
$$\cdot\left\{ S_1^2 + S_2^2 - 2S_1 S_2 \cos\frac{n(\beta-\alpha)}{2}\right.$$
$$\cdot\cos\frac{m(\beta-\alpha)}{2} + 4\sin\frac{m(\beta-\alpha)}{2}$$
$$\cdot\left[ S_2\cos\frac{n\beta}{2}\sin\frac{m\alpha}{2} - S_1\cos\frac{n\alpha}{2}\right.$$
$$\left.\left.\cdot\sin\frac{m\beta}{2} + \sin\frac{m(\beta-\alpha)}{2}\right]\right\} + O(P_n/P_{1,2})$$

where $n = N + 1 - p - q$, and the notation $O(P_n/P_{1,2})$ denotes terms of the order at most $P_n/P_1$ or $P_n/P_2$. Note that the term in the brackets of the above expression reduces to zero for $\alpha = \beta$ (i.e., $\cos\theta_1 = \cos\theta_2$), which is to be expected since then the two jammers effectively act as one.

(d) Show that the maximum residue power is attained when

$$|\beta - \alpha| = \frac{k2\pi}{m}, \qquad k = 1, 2, \cdots$$

or when

$$|\cos\theta_2 - \cos\theta_1| = \frac{2k}{m}, \qquad k = 1, 2, \cdots$$

This condition corresponds to jammers spaced between $k$ grating lobes of the adaptive element array pattern.

(e) Show that the maximum residue power that can result is given by

$$P_{0_{\max}} = NP_n + \frac{P_1 P_2}{P_1 + P_2}\left[ S_1^2 + S_2^2 - 2S_1 S_2 \cdot \cos k\pi\left(1 + \frac{n}{m}\right)\right] + O\left(\frac{P_n}{P_{1,2}}\right)$$

It is obviously desirable to minimize the above bracketed quantity. Show that this condition is realized if the two adaptive elements are located so that

$$\frac{n}{m} = \begin{cases} 1 \pmod 2, & S_1 S_2 > 0 \\ 0 \pmod 2, & S_1 S_2 < 0 \end{cases}$$

for all jammer angles.

## REFERENCES


[1] R. T. Lacoss, "Data Adaptive Spectral Analysis Methods," *Geophysics*, Vol. 36, August 1971, pp. 661–675.

[2] T. J. Ulrych and T. N. Bishop, "Maximum Entropy Spectral Analysis and Autoregressive Decompositions," *Rev. Geophys. Space Phys.*, Vol. 13, 1975, pp. 183–200.

[3] J. P. Burg, "Maximum Entropy Spectral Analysis," Paper presented at the 37th Annual Meeting of the Society of Exploration Geophysicists, October 31, 1967, Oklahoma City, OK.

[4] ______, "A New Analysis Technique for Time Series Data," NATO Advanced Study Institute on Signal Processing with Emphasis on Underwater Acoustics, Vol. I, paper no. 15, Enschede, The Netherlands, 1968.

[5] T. J. Ulrych and R. W. Clayton, "Time Series Modelling and Maximum Entropy," *Phys. Earth Planet. Inter.*, Vol. 12, 1976, pp. 188–200.

[6] S. B. Kesler and S. Haykin, "The Maximum Entropy Method Applied to the Spectral Analysis of Radar Clutter," *IEEE Trans. Inf. Theory*, Vol. IT-24, No. 2, March 1978, pp. 269–272.

[7] R. G. Taylor, T. S. Durrani, and C. Goutis, "Block Processing in Pulse Doppler Radar," Radar-77, Proceedings of the 1977 IEE International Radar Conference, October 25–28, London, pp. 373–378.

[8] G. Prado and P. Moroney, "Linear Predictive Spectral Analysis for Sonar Applications," C. S. Draper Report R-1109, C. S. Draper Laboratory, Inc., 555 Technology Square, Cambridge, MA, September 1977.

[9] R. N. McDonough, "Maximum-Entropy Spatial Processing of Array Data," *Geophysics*, Vol. 39, December 1974, pp. 843–851.

[10] D. P. Skinner, S. M. Hedlicka, and A. D. Matthews, "Maximum Entropy Array Processing," *J. Acoust. Soc. Am.*, Vol. 66, No. 2, August 1979, pp. 488–493.

[11] A. Van Den Bos, "Alternative Interpretation of Maximum Entropy Spectral Analysis," *IEEE Trans. Inf. Theory*, Vol. IT-17, No. 4, July 1971, pp. 493–494.

[12] J. P. Burg, "Maximum Entropy Spectral Analysis," Ph.D. dissertation, Stanford University, Department of Geophysics, May 1975.

[13] J. H. Sawyers, "The Maximum Entropy Method Applied to Radar Adaptive Doppler Filtering," Proceedings of the 1979 RADC Spectrum Estimation Workshop, October 3–5, Griffiss AFB, Rome, NY.

[14] S. Haykin and S. Kesler, "The Complex Form of the Maximum Entropy Method for Spectral Estimation," *Proc. IEEE*, Vol. 64, May 1976, pp. 822–823.

[15] J. P. Burg, "The Relationship Between Maximum Entropy Spectra and Maximum Likelihood Spectra," *Geophysics*, Vol. 37, No. 2, April 1972, pp. 375–376.

[16] R. H. Jones, "Multivariate Maximum Entropy Spectral Analysis," presented at the Applied Time Series Analysis Symposium, May 1976, Tulsa, OK.

[17] ______, "Multivariate Autoregression Estimation Using Residuals," presented at Applied Time Series Analysis Symposium, May 1976, Tulsa, OK; also published in *Applied Time-Series Analysis*, edited by D. Findley, Academic Press, New York, 1977.

[18] A. H. Nutall, "Fortran Program for Multivariate Linear Predictive Spectra Analysis Employing Forward and Backward Averaging," Naval Underwater System Center, NUSC Tech. Doc. 5419, New London, CT, May 9, 1976.

[19] ______, "Multivariate Linear Predictive Spectral Analysis Employing Weighted Forward and Backward Averaging: A Generalization of Burg's Algorithm," Naval Underwater System Center, NUSC Tech. Doc. 5501, New London, CT, October 13, 1976.

[20] ______, "Positive Definite Spectral Estimate and Stable Correlation Recursion for Multivariate Linear Predictive Spectral Analysis," Naval Underwater System Center, NUSC Tech. Doc. 5729, New London, CT, November 14, 1976.

[21] M. Morf, A. Vieira, D. T. Lee, and T. Kailath, "Recursive Multichannel Maximum Entropy Spectral Estimation," *IEEE Trans. Geosci. Electron.*, Vol. GE-16, No. 2, April 1978, pp. 85–94.

[22] R. A. Wiggins and E. A. Robinson, "Recursive Solution to the Multichannel Filtering Problem," *J. Geophys. Res.*, Vol. 70, 1965, pp. 1885–1891.

[23] O. N. Strand, "Multichannel Complex Maximum Entropy Spectral Analysis," 1977 IEEE International Conference on Acoustics, Speech, and Signal Processing, pp. 736–741.

[24] ______, "Multichannel Complex Maximum Entropy (Autoregressive) Spectral Analysis," *IEEE Trans. Autom. Control*, Vol. AC-22, No. 4, August 1977, pp. 634–640.

[25] L. W. Nolte and W. S. Hodgkiss, "Directivity or Adaptivity?," EASCON 1975 Record, IEEE Electronics and Aerospace Systems Convention, September 1975, Washington, DC, pp. 35.A–35.H.

[26] H. Cox, "Sensitivity Considerations in Adaptive Beamforming," Proceedings of the NATO Advanced Study Institute on Signal Processing with Emphasis on Underwater Acoustics, August 21–September 1, 1972, Loughborough, U.K.; also appeared in *Signal Processing*, edited by J. W. R. Griffiths, P. L. Stocklin, and C. van Schooneveld, Academic Press, New York, 1973.

[27] ______, "Resolving Power and Sensitivity to Mismatch of Optimum Array Processors," *J. Acoust. Soc. Am.*, Vol. 54, No. 3, September 1973, pp. 771–785.

[28] D. Jaarsma, "The Theory of Signal Detectability: Bayesian Philosophy, Classical Statistics, and the Composite Hypothesis," Cooley Electronics Laboratory, University of Michigan, Ann Arbor, Tech. Report TR-200, 1969.

[29] M. A. Gallop and L. W. Nolte, "Bayesian Detection of Targets of Unknown Location," *IEEE Trans. Aerosp. Electron. Syst.*, Vol. AES-10, No. 4, July 1974, pp. 429–435.

[30] W. S. Hodgkiss and L. W. Nolte, "Bayes Optimum and Maximum-Likelihood Approaches in an Array Processing Problem," *IEEE Trans. Aerosp. Electron. Syst.*, Vol. AES-11, No. 5, September 1975, pp. 913–916.

[31] R. Nitzberg, "OTH Radar Aurora Clutter Rejection when Adapting a Fraction of the Array Elements," EASCON 1976 Record, IEEE Electronics and Aerospace Systems Convention, September 1976, Washington DC, pp. 62.A–62.D.

[32] D. R. Morgan, "Partially Adaptive Array Techniques," *IEEE Trans. Antennas Propag.*, Vol. AP-26, No. 6, November 1978, pp. 823–833.

[33] D. J. Chapman, "Partial Adaptivity for the Large Array," *IEEE Trans. Antennas Propag.*, Vol. AP-24, No. 5, September 1976, pp. 685–696.

[34] D. J. Chapman, "Adaptive Array Techniques Study (U)," ECM/ECCM Studies, Vol. 3, final report prepared by the Syracuse University Research Corporation for the U.S. Army Advanced Ballistic Missile Defense Agency under Contract DAHC60-73-C-0044, SURC TR 73-186, February 1974 (Secret, this volume unclassified).

[35] A. M. Vural, "A Comparative Performance Study of Adaptive Array Processors," presented at the 1977 IEEE International Conference on Acoustics, Speech, and Signal Processing, May 9–11, Hartford, CT., paper 20.6.

[36] P. W. Howells, "High Quality Array Beamforming with a Combination of Precision and Adaptivity," prepared by Syracuse University Research Corporation, SURC TN74-150, June 1974 (Unclassified).

[37] A. M. Vural, "An Overview of Adaptive Array Processing for Sonar Applications," EASCON 1975 Record, IEEE Electronics and Aerospace Systems Convention, Sept. 29–Oct. 1, Washington, D.C., pp. 34.A–34.M.

# Appendix A. Frequency Response Characteristics of Tapped-Delay Lines

The frequency response characteristics of the tapped-delay line filter shown in Figure A.1 can be developed by first considering the impulse response $h(t)$ of the network. For the input signal $x(t)=\delta(t)$ it follows that

$$h(t)=\sum_{i=1}^{2n+1} w_i\delta\left[t-(i-1)\Delta\right] \tag{A.1}$$

where $w_i, i=1,\ldots,2n+1$ denotes the various complex weights located at the taps in the tapped-delay line having intertap delay spacing equal to $\Delta$.

Taking the Laplace transform of (A.1) yields

$$\mathfrak{L}\{h(t)\}=H(s)=\sum_{i=1}^{2n+1} w_i e^{-s(i-1)\Delta} \tag{A.2}$$

Equation (A.1) represents a sequence of weighted impulse signals that are summed to form the output of the tapped-delay line. The adequacy of the tapped-delay line structure to represent frequency dependent amplitude and phase variations by way of (A.2) depends on signal bandwidth considerations.

Signal bandwidth considerations are most easily introduced by discussing the continuous input signal depicted in Figure A.2. With a continuous input signal, the signals appearing at the tapped-delay line taps (after time $t=t_0+2n\Delta$ has elapsed where $t_0$ denotes an arbitrary starting time) are given by the sequence of samples $x(t_0+(i-1)\Delta), i=1,2,\ldots,2n+1$. The sample sequence $x(t_0+(i-1)\Delta), i=1,\ldots,2n+1$, uniquely characterizes the corresponding continuous waveform from which it was generated provided that the signal $x(t)$ is band-limited with its highest frequency component $f_{\max}$ less than or equal to one-half the sample frequency corresponding to the time delay, that is,

$$f_{\max}\leqslant\frac{1}{2\Delta} \tag{A.3}$$

Equation (A.3) expresses the condition that must be satisfied in order for a

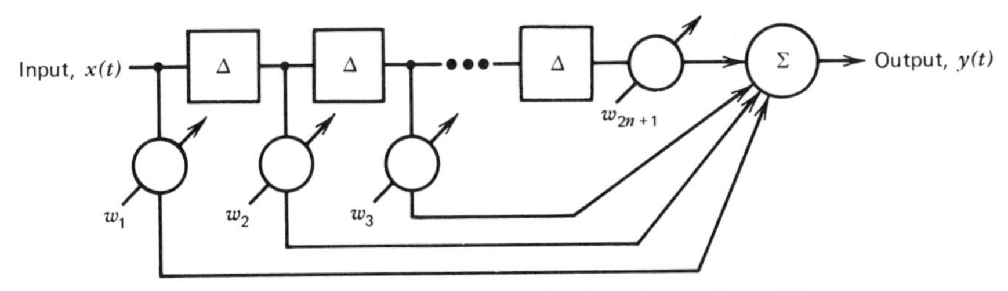


**Figure A.1** Transversal filter having $2n+1$ complex weights.

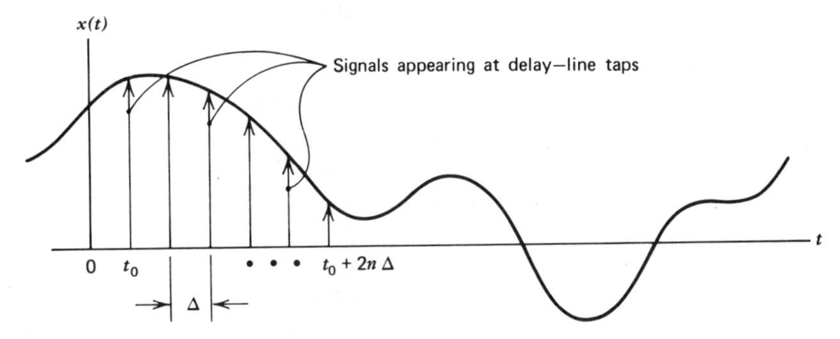


**Figure A.2** Continuous input signal.

continuous signal to be uniquely reconstructed from a sequence of discrete samples spaced $\Delta$ seconds apart, and it is formally recognized as the "sampling theorem" [A.1]. Since the total (two-sided) bandwidth of a band-limited signal $x(t)$ is $BW = 2f_{\max}$, it follows that a tapped-delay line can uniquely characterize any continuous signal having $BW \leqslant 1/\Delta$ (Hz), so $1/\Delta$ can be regarded as the "signal bandwidth" of the tapped-delay line.

Since the impulse response of the transversal filter consists of a sequence of weighted impulse functions, it is convenient to adopt the $z$-transform description for the filter transfer function instead of the (more cumbersome) Laplace transform description of (A.2) by defining

$$z \triangleq e^{s\Delta} \tag{A.4}$$

so that

$$\mathfrak{Z}\{h(t)\} = H(z) = \sum_{i=1}^{2n+1} w_i z^{-(i-1)} \tag{A.5}$$

The frequency response of the transversal filter can then be obtained by setting $s = j\omega$ and considering how $H(j\omega)$ behaves as $\omega$ varies. Letting $s = j\omega$ corresponds to setting $z = e^{j\omega\Delta}$, which is a multiple-valued function of $\omega$ since it is impossible to distinguish $\omega\Delta = +\pi$ from $\omega\Delta = -\pi$, and consequently

$$H(j\omega) = H\left(j\omega \pm k\frac{2\pi}{\Delta}\right) \tag{A.6}$$

Equation (A.6) expresses the fact that the tapped-delay line transfer function is a periodic function of frequency having a period equal to the signal bandwidth capability of the filter. The periodic structure of $H(s)$ is easily seen in the complex $s$-plane, which is divided into an infinite number of periodic strips as shown in Figure A.3 [A.2]. The strip located between $\omega = -\pi/\Delta$ and $\omega = \pi/\Delta$ is called the "primary strip," while all other strips occurring at higher frequencies are designated as "complementary strips." Whatever behavior of $H(j\omega)$ obtains in the primary strip, this behavior is merely repeated in each succeeding complementary strip. It is seen from (A.5) that for $2n+1$ taps in the tapped-delay line, there will be up to $2n$ roots of the resulting polynomial in $z^{-1}$ that describes $H(z)$. It follows that there will be up to $2n$ zeros in the transfer function corresponding to the $2n$ delay elements in the tapped-delay line.

We have seen how the frequency response $H(j\omega)$ is periodic with period determined by the signal bandwidth $1/\Delta$ and that the number of zeros that can occur across the signal bandwidth is equal to the number of delay elements in the tapped-delay line. It remains to show that the resolution associated with each of the zeros of $H(j\omega)$ is approximately $1/N\Delta$, where $N$ = number of taps in the tapped-delay line. Consider the impulse response of

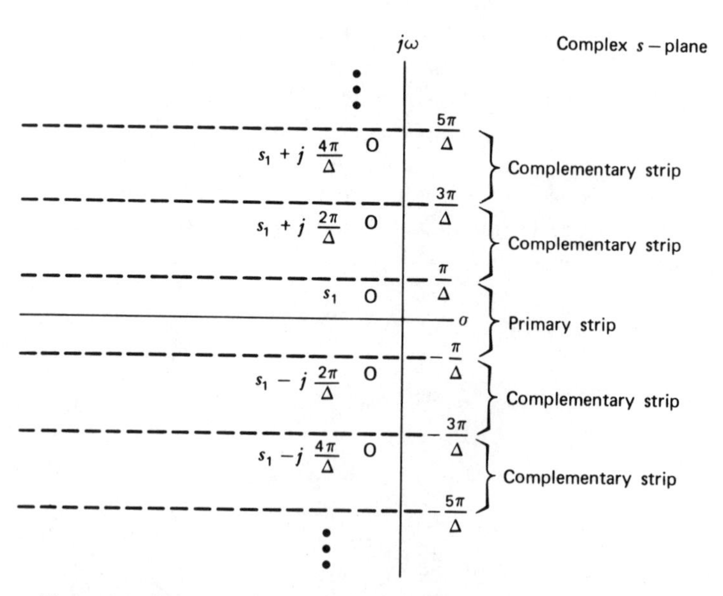


**Figure A.3** Periodic structure of $H(s)$ as seen in the complex $s$-plane.

a transversal filter having $N$ taps and all weights set equal to unity so that

$$H(j\omega)=\sum_{i=0}^{N-1} e^{-j\omega i\Delta}=\frac{1-e^{-j\omega N\Delta}}{1-e^{-j\omega\Delta}} \tag{A.7}$$

Factoring $e^{-j\omega(\Delta/2)N}$ from the numerator and $e^{-j\omega(\Delta/2)}$ from the denominator of (A.7) then yields

$$H(j\omega)=\exp\left[j\omega\frac{\Delta}{2}(1-N)\right]\frac{\sin[\omega(\Delta/2)N]}{\sin[\omega(\Delta/2)]}$$

$$=N\exp\left[j\omega\frac{\Delta}{2}(1-N)\right]\frac{\left\{\dfrac{\sin[\omega(\Delta/2)N]}{[\omega(\Delta/2)N]}\right\}}{\left[\dfrac{\sin[\omega(\Delta/2)]}{[\omega(\Delta/2)]}\right]} \tag{A.8}$$

The denominator of (A.8) has its first zero occurring at $f=1/\Delta$, which is outside the range of periodicity of $H(j\omega)$ for which $\frac{1}{2}BW=1/2\Delta$. The first zero of the numerator of (A.8), however, occurs at $f=1/N\Delta$ so the total frequency range of the principal lobe is just $2/N\Delta$; it follows that the 3 dB frequency range of the principal lobe is very nearly $1/N\Delta$ so the resolution (in frequency) of any root of $H(j\omega)$ may be said to be approximately the inverse of the total delay in the tapped-delay line. In the event that unequal weighting is employed in the tapped-delay line, the width of the principal lobe merely broadens, so the above result gives the best frequency resolution that can be achieved.

The discussion so far has assumed that the frequency range of interest is centered about $f=0$. In most practical systems, the actual signals of interest are transmitted with a carrier frequency component $f_0$ as shown in Figure A.4. By mixing the actual transmitted signal with a reference oscillator having the carrier frequency, the transmitted signal can be reduced to baseband by

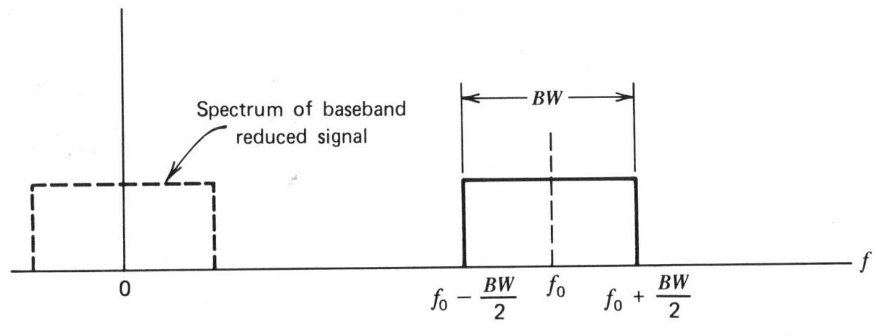


**Figure A.4** Signal bandwidth $BW$ centered about carrier frequency $f_0$.

removal of the carrier frequency component, thereby centering the spectrum of the information carrying signal component about $f=0$. By writing all signals as though they had been reduced to baseband, no loss of generality results since this merely assumes that the signal spectrum is centered about $f=0$; the baseband reduced information-carrying component of any transmitted signal is referred to as the "complex envelope," and complex envelope notation is discussed in Appendix B.

## REFERENCES

[A.1] R. M. Oliver, J. R. Pierce, and C. E. Shannon, "The Philosophy of Pulse Code Modulation," *Proc. IRE*, Vol. 36, No. 11, November 1948, pp. 1324–1331.

[A.2] B. C. Kuo, *Discrete-Data Control System*, Prentice-Hall, Englewood Cliffs, NJ, 1970, Ch. 2.

# Appendix B. Complex Envelope Notation

Complex envelope notation provides a convenient means of characterizing the information-carrying component of modulated carrier signals so the carrier frequency component of the signal (which does not carry information) does not appear in the signal description. The properties of complex signal descriptions outlined in this appendix are discussed in some detail in references [B.1] and [B.2].

We are interested in characterizing signals that propagate toward the array elements as uniform plane waves through a nondispersive transmission medium that only introduces propagation delay to the transmitted signals. Consequently the output of any array element can be modeled by time-advanced and time-delayed versions of the signal arriving at the array phase center with an additive thermal noise component.

For band-limited real signals whose spectrum is located between zero frequency and twice the array center frequency $2f_c$, then the signals present at the array phase center may be represented as amplitude- and phase-modulated carrier signals of the form

$$v_i(t) = \alpha_i(t)\cos(\omega_c t + \phi_i(t) + \Theta_i) \tag{B.1}$$

where the subscript $i$ denotes the signal from the $i$th source. There is an analytic signal associated with each real signal that is defined by

$$z_i(t) \triangleq \left[ v_i(t) + j\check{v}_i(t) \right] \tag{B.2}$$

where $\check{v}_i$ is the Hilbert transform [B.3] of $v_i(t)$, that is,

$$\check{v}_i(t) \triangleq \frac{1}{\pi}\int_{-\infty}^{\infty} \frac{v_i(\tau)}{t-\tau}\, d\tau = v_i(t) * \frac{1}{\pi t} \tag{B.3}$$

where $*$ here denotes the convolution operation. Clearly then, $\check{v}_i(t)$ can be regarded as the output of a quadrature filter having the input signal $v_i(t)$.

As a consequence of (B.3) and (B.1) it follows that

$$\check{v}_i(t)=\alpha_i(t)\sin(\omega_c t+\phi_i(t)+\Theta_i) \tag{B.4}$$

The baseband-reduced complex envelope of $v_i(t)$, denoted by $\tilde{v}_i(t)$, is then defined in terms of the analytic signal by

$$\tilde{v}_i(t)=z_i(t)\exp(-j\omega_c t)=\alpha_i(t)\exp\{j[\phi_i(t)+\Theta_i]\} \tag{B.5}$$

Consequently

$$v_i(t)=\text{Re}\{\tilde{v}_i(t)\exp(j\omega_c t)\} \tag{B.6}$$

gives the relationship for real (in-phase) signal components in terms of their corresponding complex envelopes. Likewise

$$\check{v}_i(t)=\text{Im}\{\tilde{v}_i(t)\exp(j\omega_c t)\} \tag{B.7}$$

gives the relationship for imaginary (quadrature) signal components in terms of their corresponding complex envelopes. Let the output of the $k$th array element in response to several signal sources be denoted by $x_k(t)$ so that

$$x_k(t)=\sum_i \underbrace{v_i(t-\tau_{ik})}_{\substack{\text{delayed}\\ \text{real}\\ \text{signal}\\ \text{component}\\ \text{of } i\text{th signal}}} + \underbrace{n_k(t)}_{\substack{\text{thermal}\\ \text{noise}\\ \text{component}}} \tag{B.8}$$

The complex envelope of $x_k(t)$ can then be written as

$$\tilde{x}_k(t)=\sum_i \tilde{v}_i(t-\tau_{ik})\exp\{-j\omega_c\tau_{ik}\}+\tilde{n}_k(t) \tag{B.9}$$

where $\tau_{ik}$ represents the differential delay of the $i$th signal to the $k$th array element relative to the array phase center, and $\tilde{v}_i(t-\tau_{ik})$ denotes the complex envelope $\tilde{v}_i(t)$, time delayed by $\tau_{ik}$.

The power spectral density relationships that exist among $v(t)$, $z(t)$, and $\tilde{v}(t)$ are illustrated in Figure B.1. These relationships are readily derived from

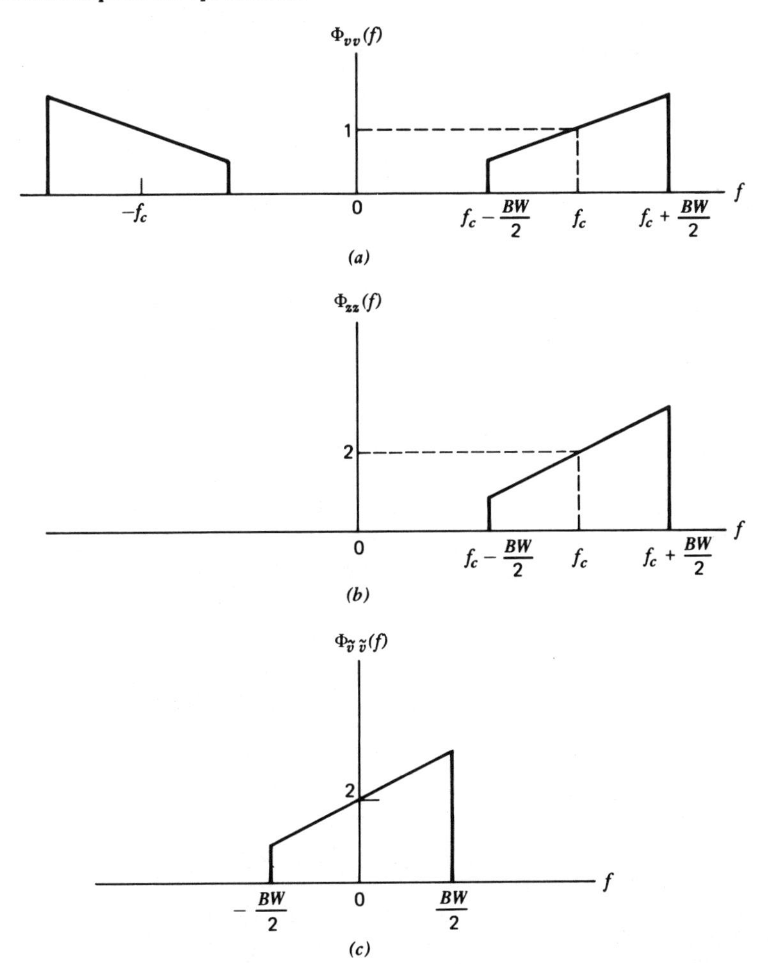


**Figure B.1** Spectrum relationships among the real signal representation, the analytic signal representation, and the complex envelope representation of the signal $v(t)$. (*a*) Spectrum of the real signal $v(t)$. (*b*) Spectrum of $z(t)$, the analytic signal associated with $v(t)$. (*c*) Spectrum of $\tilde{v}(t)$, the complex envelope representation of $v(t)$.

the power spectral density relationships between input and output signals for linear filters [B.3].

It is also instructive to consider two signals $v_1(t)$ and $v_2(t)$ and determine the cross product relationships that exist between them for the implications this has for covariance matrix entries. Using the notation of (B.1), we see that it follows

$$v_1(t)v_2(t) = \frac{\alpha_1(t)\alpha_2(t)}{2}\{\cos[\phi_1(t)-\phi_2(t)+\Theta_1-\Theta_2] + \cos[2\omega_c t+\phi_1(t)+\phi_2(t)+\Theta_1+\Theta_2]\} \tag{B.10}$$

$$v_1(t)\check{v}_2(t)=\frac{\alpha_1(t)\alpha_2(t)}{2}\{\sin[\phi_1(t)-\phi_2(t)+\Theta_1-\Theta_2]$$
$$+\sin[2\omega_c t+\phi_1(t)+\phi_2(t)+\Theta_1+\Theta_2]\} \tag{B.11}$$

$$\check{v}_1(t)\check{v}_2(t)=\frac{\alpha_1(t)\alpha_2(t)}{2}\{\cos[\phi_1(t)-\phi_2(t)+\Theta_1-\Theta_2]$$
$$-\cos[2\omega_c t+\phi_1(t)+\phi_2(t)+\Theta_1+\Theta_2]\} \tag{B.12}$$

$$\tilde{v}_1(t)\tilde{v}_2^*(t)=v_1(t)v_2(t)+\check{v}_1(t)\check{v}_2(t)+j[\check{v}_1(t)v_2(t)$$
$$-v_1(t)\check{v}_2(t)]$$
$$=\alpha_1(t)\alpha_2(t)\{\cos[\phi_1(t)-\phi_2(t)+\Theta_1-\Theta_2]$$
$$-j\sin[\phi_1(t)-\phi_2(t)+\Theta_1-\Theta_2]\} \tag{B.13}$$

If the products of the real valued waveforms are now low-pass filtered to remove the double frequency terms, then

$$2\,\overline{v_1(t)v_2(t)}=\mathrm{Re}\{\tilde{v}_1(t)\tilde{v}_2^*(t)\}=\mathrm{Re}\{z_1(t)z_2^*(t)\} \tag{B.14}$$

and

$$2\,\overline{v_1(t)\check{v}_2(t)}=-\mathrm{Im}\{\tilde{v}_1(t)\tilde{v}_2^*(t)\}=-\mathrm{Im}\{z_1(t)z_2^*(t)\} \tag{B.15}$$

## REFERENCES

[B.1] K. L. Reinhard, "Adaptive Antenna Arrays for Coded Communication Systems," M. S. Thesis, The Ohio State University, 1973.

[B.2] T. G. Kincaid, "The Complex Representation of Signals," General Electric Company Report No. R67EMH5, October 1966, and HMED Technical Publications, Box 1122, Le Moyne Ave., Syracuse, NY, 13201.

[B.3] A. Papoulis, *Probability, Random Variables, and Stochastic Processes*, McGraw-Hill, New York, 1965, Ch. 10.

# Appendix C. Convenient Formulas for Gradient Operations

The scalar function trace $[\mathbf{AB}^T]$ has all the properties of an inner product, so the formulas for differentiation of the trace of various matrix products that appear in reference [C.1] are of interest for computing the gradients of certain inner products that appear in adaptive array optimization problems. For many problems, the extension of vector and matrix derivative operations to matrix functions of vectors and matrices is important; this generalization is described in reference [C.2].

$$\frac{\partial}{\partial \mathbf{X}} \operatorname{tr}[\mathbf{X}] = \mathbf{I} \tag{C.1}$$

$$\frac{\partial}{\partial \mathbf{X}} \operatorname{tr}[\mathbf{AX}] = \mathbf{A}^T \tag{C.2}$$

$$\frac{\partial}{\partial \mathbf{X}} \operatorname{tr}[\mathbf{AX}^T] = \mathbf{A} \tag{C.3}$$

$$\frac{\partial}{\partial \mathbf{X}} \operatorname{tr}[\mathbf{AXB}] = \mathbf{A}^T\mathbf{B}^T \tag{C.4}$$

$$\frac{\partial}{\partial \mathbf{X}} \operatorname{tr}[\mathbf{AX}^T\mathbf{B}] = \mathbf{BA} \tag{C.5}$$

$$\frac{\partial}{\partial \mathbf{X}^T} \operatorname{tr}[\mathbf{AX}] = \mathbf{A} \tag{C.6}$$

$$\frac{\partial}{\partial \mathbf{X}^T} \operatorname{tr}[\mathbf{AX}^T] = \mathbf{A}^T \tag{C.7}$$

$$\frac{\partial}{\partial \mathbf{X}^T} \operatorname{tr}[\mathbf{AXB}] = \mathbf{BA} \tag{C.8}$$

$$\frac{\partial}{\partial \mathbf{X}^T} \operatorname{tr}[\mathbf{AX}^T\mathbf{B}] = \mathbf{A}^T\mathbf{B}^T \tag{C.9}$$

$$\frac{\partial}{\partial \mathbf{X}} \operatorname{tr}[\mathbf{X}^n] = n(\mathbf{X}^{n-1})^T \tag{C.10}$$

$$\frac{\partial}{\partial \mathbf{X}} \mathrm{tr}[\mathbf{AXBX}] = \mathbf{A}^T\mathbf{X}^T\mathbf{B}^T + \mathbf{B}^T\mathbf{X}^T\mathbf{A}^T \qquad \text{(C.11)}$$

$$\frac{\partial}{\partial \mathbf{X}} \mathrm{tr}[\mathbf{AXBX}^T] = \mathbf{A}^T\mathbf{XB}^T + \mathbf{AXB} \qquad \text{(C.12)}$$

$$\frac{\partial}{\partial \mathbf{X}} \mathrm{tr}[\mathbf{X}^T\mathbf{AX}] = \mathbf{AX} + \mathbf{A}^T\mathbf{X} = 2\mathbf{AX} \qquad \text{if } \mathbf{A} \text{ is symmetric} \qquad \text{(C.13)}$$

## REFERENCES

[C.1] M. Athans and F. C. Schweppe, "Gradient Matrices and Matrix Calculations," M.I.T. Lincoln Laboratory, Lexington, MA, Technical Note 1965-53, November 1965.

[C.2] W. J. Vetter, "Derivative Operations on Matrices," *IEEE Trans. Autom. Control*, Vol. AC-15, No. 2, April 1970, pp. 241–244.

# Appendix D. Useful Matrix Relations and the Schwartz Inequality

This appendix summarizes some matrix properties that are especially useful for solving optimization problems arising in adaptive array processing. Derivations for these relations as well as a treatment of the basic properties of matrixes may be found in the references [D.1]–[D.3] for this appendix.

The trace of a square matrix is defined to be the sum of its diagonal elements so that

$$\text{trace}(\mathbf{A}) \triangleq \sum_i a_{ii} \tag{D.1}$$

From the definition (D.1) it is easily verified that

$$\text{trace}(\mathbf{AB}) = \text{trace}(\mathbf{BA}) \tag{D.2}$$

and

$$\text{trace}(\mathbf{A}+\mathbf{B}) = \text{trace}(\mathbf{A}) + \text{trace}(\mathbf{B}) \tag{D.3}$$

The length or norm of a vector $\mathbf{x}$ is

$$\|\mathbf{x}\|^{1/2} = \sqrt{\mathbf{x}^\dagger \mathbf{x}} = \sqrt{\text{trace}(\mathbf{x}\mathbf{x}^\dagger)} = \sqrt{\sum_i |x_i|^2} \tag{D.4}$$

The norm of a matrix is in turn defined by

$$\|\mathbf{A}\|^{1/2} = \sqrt{\text{trace}(\mathbf{A}\mathbf{A}^\dagger)} = \sqrt{\sum_{i,j} |a_{ij}|^2} \tag{D.5}$$

A square matrix is called Hermitian if

$$\mathbf{A}^\dagger = \mathbf{A} \tag{D.6}$$

The Hermitian matrix **A** is said to be positive definite if the Hermitian form satisfies

$$\mathbf{x}^{\dagger}\mathbf{A}\mathbf{x}=\sum_{i,j} a_{ij}x_i^*x_j>0 \qquad \text{for all } \mathbf{x}\neq\mathbf{0} \tag{D.7}$$

If, however, $\mathbf{x}^{\dagger}\mathbf{A}\mathbf{x}\geqslant 0$ for all $\mathbf{x}\neq\mathbf{0}$, then the Hermitian form is merely positive semidefinite.

For any arbitrary matrix **D** the product $\mathbf{D}^{\dagger}\mathbf{D}$ is Hermitian and positive semidefinite. If in addition **D** is square and nonsingular (so that $\mathbf{D}^{-1}$ exists) then $\mathbf{D}^{\dagger}\mathbf{D}$ is positive definite. Any positive definite Hermitian matrix **A** may be factored into the form

$$\mathbf{A}=\mathbf{D}\mathbf{D}^{\dagger} \tag{D.8}$$

In certain optimization problems involving the trace of a matrix, the following formula for completing the square is very useful:

$$\mathbf{x}^{\dagger}\mathbf{A}\mathbf{x}-\mathbf{x}^{\dagger}\mathbf{y}-\mathbf{y}^{\dagger}\mathbf{x}=-\mathbf{y}^{\dagger}\mathbf{A}^{-1}\mathbf{y}+[\mathbf{x}-\mathbf{A}^{-1}\mathbf{y}]^{\dagger}\mathbf{A}[\mathbf{x}-\mathbf{A}^{-1}\mathbf{y}] \tag{D.9}$$

The following matrix inversion identities are very useful in many of the derivations. First,

$$[\mathbf{P}^{-1}+\mathbf{M}^{\dagger}\mathbf{Q}^{-1}\mathbf{M}]^{-1}\equiv\mathbf{P}-\mathbf{P}\mathbf{M}^{\dagger}[\mathbf{M}\mathbf{P}\mathbf{M}^{\dagger}+\mathbf{Q}]^{-1}\mathbf{M}\mathbf{P} \tag{D.10}$$

is easily shown by direct multiplication. The second identity is

$$[\mathbf{P}^{-1}+\mathbf{M}^{\dagger}\mathbf{Q}^{-1}\mathbf{M}]^{-1}\mathbf{M}^{\dagger}\mathbf{Q}^{-1}\equiv\mathbf{P}\mathbf{M}^{\dagger}[\mathbf{M}\mathbf{P}\mathbf{M}^{\dagger}+\mathbf{Q}]^{-1} \tag{D.11}$$

The third identity is

$$\mathbf{Q}^{-1}-[\mathbf{M}\mathbf{P}\mathbf{M}^{\dagger}+\mathbf{Q}]^{-1}\equiv\mathbf{Q}^{-1}\mathbf{M}[\mathbf{P}^{-1}+\mathbf{M}^{\dagger}\mathbf{Q}^{-1}\mathbf{M}]^{-1}\mathbf{M}^{\dagger}\mathbf{Q}^{-1} \tag{D.12}$$

The Schwartz inequality for two scalar functions $f(t)$ and $g(t)$ is as follows:

$$\left|\int f^*(t)g(t)\,dt\right|\leqslant\int|f(t)|^2\,dt\int|g(t)|^2\,dt \tag{D.13}$$

Equality in (D.13) obtains if and only if $f(t)$ is a scalar multiple of $g(t)$.

For two vector functions $\mathbf{f}(t)$ and $\mathbf{g}(t)$, the inequality may be written in the form

$$\left|\int \mathbf{f}^{\dagger}(t)\mathbf{g}(t)\,dt\right|^2\leqslant\int\mathbf{f}^{\dagger}(t)\mathbf{f}(t)\,dt\int\mathbf{g}^{\dagger}(t)\mathbf{g}(t)\,dt \tag{D.14}$$

Equality in (D.14) obtains if and only if $\mathbf{f}(t)$ is a scalar multiple of $\mathbf{g}(t)$. Note that (D.14) is analogous to

$$|\mathbf{x}^\dagger \mathbf{y}|^2 \leqslant (\mathbf{x}^\dagger \mathbf{x})(\mathbf{y}^\dagger \mathbf{y}) \tag{D.15}$$

where $\mathbf{x}$ and $\mathbf{y}$ represent two vectors.

Equation (D.14) may also be generalized to matrix functions by introducing the trace of a matrix [defined in (D.1)]. Let the complex matrixes $\mathbf{F}(t)$ and $\mathbf{G}(t)$ be $m\times n$ matrixes so that $\mathbf{F}^\dagger(t)\mathbf{G}(t)$ is square and $n\times n$. Then the Schwartz inequality can be written as

$$\left|\int \operatorname{trace}\left[\mathbf{F}^\dagger(t)\mathbf{G}(t)\right] dt\right|^2 \leqslant \int \operatorname{trace}\left[\mathbf{F}^\dagger(t)\mathbf{F}(t)\right] dt \cdot \int \operatorname{trace}\left[\mathbf{G}^\dagger(t)\mathbf{G}(t)\right] dt \tag{D.16}$$

Equality in (D.16) obtains if and only if $\mathbf{F}(t)$ is a scalar multiple of $\mathbf{G}(t)$. Note that (D.16) is analogous to

$$|\operatorname{trace}(\mathbf{A}^\dagger \mathbf{B})|^2 \leqslant \operatorname{trace}(\mathbf{A}^\dagger \mathbf{A})\operatorname{trace}(\mathbf{B}^\dagger \mathbf{B}) \tag{D.17}$$

Yet another useful generalization of the Schwartz inequality for matrixes is

$$\operatorname{trace}(\mathbf{B}^\dagger \mathbf{A}^\dagger \mathbf{A}\mathbf{B}) \leqslant \operatorname{trace}(\mathbf{A}^\dagger \mathbf{A})\operatorname{trace}(\mathbf{B}^\dagger \mathbf{B}) \tag{D.18}$$

## REFERENCES


[D.1] R. Bellman, *Introduction to Matrix Analysis*, McGraw-Hill, New York, 1960.

[D.2] E. Bodewig, *Matrix Calculus*, North-Holland Publishing Co., Amsterdam, 1959.

[D.3] K. Hoffman and R. Kunze, *Linear Algebra*, Prentice-Hall, Englewood Cliffs, NJ, 1961.

# Appendix E. Multivariate Gaussian Distributions

Useful properties that apply to real Gaussian random vectors are briefly discussed and extensions to complex Gaussian random vectors are presented in this appendix. Detailed discussions of these properties will be found in the references [E.1]–[E.4] for this appendix.

## E.1 REAL GAUSSIAN RANDOM VECTORS

Let $\mathbf{x}$ denote a real random vector of dimension $n\times 1$ whose components are all scalar random variables and having a mean value

$$E\{\mathbf{x}\}=\mathbf{u} \tag{E.1}$$

and an associated covariance matrix

$$\operatorname{cov}(\mathbf{x}) \triangleq E\{(\mathbf{x}-\mathbf{u})(\mathbf{x}-\mathbf{u})^T\}=\mathbf{V} \tag{E.2}$$

The covariance matrix $\mathbf{V}$ is then symmetric and positive definite (or at least positive semidefinite). A random vector $\mathbf{x}$ is said to be Gaussian if it has a characteristic function given by

$$C_x(j\boldsymbol{\Theta}) \triangleq E\{\exp(j\boldsymbol{\Theta}^T\mathbf{x})\}=\exp\{j\boldsymbol{\Theta}^T\mathbf{u}-\tfrac{1}{2}\boldsymbol{\Theta}^T\mathbf{V}\boldsymbol{\Theta}\} \tag{E.3}$$

It follows from (E.3) that the probability density function for $\mathbf{x}$ has the form

$$p(\mathbf{x})=[2\pi]^{-n/2}(\det\mathbf{V})^{-1/2}\exp\{-\tfrac{1}{2}(\mathbf{x}-\mathbf{u})^T\mathbf{V}^{-1}(\mathbf{x}-\mathbf{u})\} \tag{E.4}$$

Consider a linear transformation of the random vector $\mathbf{x}$ as follows:

$$\mathbf{y}=\mathbf{A}\mathbf{x}+\mathbf{b} \tag{E.5}$$

where $\mathbf{A}$ and $\mathbf{b}$ are nonrandom. Then the vector $\mathbf{y}$ is also Gaussian and

$$E\{\mathbf{y}\}=\mathbf{A}\mathbf{u}+\mathbf{b} \tag{E.6}$$

$$\operatorname{cov}\{\mathbf{y}\}=\mathbf{A}\mathbf{V}\mathbf{A}^T \tag{E.7}$$

A vector $\mathbf{w}$ is referred to as a "normalized Gaussian random vector" if it is Gaussian with zero mean and its components are all independent having unit variance. In other words,

$$E\{\mathbf{w}\}=\mathbf{0} \tag{E.8}$$

and

$$E\{\mathbf{w}\mathbf{w}^T\}=\mathbf{I} \tag{E.9}$$

Since any positive semidefinite matrix $\mathbf{V}$ can be factored into the form

$$\mathbf{V}=\mathbf{Z}\mathbf{Z}^T \tag{E.10}$$

it follows that a Gaussian vector $\mathbf{x}$ having mean $\mathbf{u}$ and covariance matrix $\mathbf{V}$ can be obtained from a normalized Gaussian vector $\mathbf{w}$ by means of the following linear transformation:

$$\mathbf{x}=\mathbf{Z}\mathbf{w}+\mathbf{u} \tag{E.11}$$

Let

$$\begin{bmatrix} \mathbf{x}_1 \\ \mathbf{x}_2 \end{bmatrix}$$

be a random vector $\mathbf{x}$ comprised of two real jointly Gaussian random vectors with mean

$$E\left\{\begin{matrix} \mathbf{x}_1 \\ \mathbf{x}_2 \end{matrix}\right\}=\begin{bmatrix} \mathbf{u}_1 \\ \mathbf{u}_2 \end{bmatrix} \tag{E.12}$$

and

$$\operatorname{cov}\left\{\begin{matrix} \mathbf{x}_1 \\ \mathbf{x}_2 \end{matrix}\right\}=\begin{bmatrix} \mathbf{V}_{11} & \mathbf{V}_{12} \\ \mathbf{V}_{21} & \mathbf{V}_{22} \end{bmatrix} \tag{E.13}$$

Then the conditional distribution for $\mathbf{x}_1$ given $\mathbf{x}_2$ is also Gaussian with

$$E\{\mathbf{x}_1/\mathbf{x}_2\}=\mathbf{u}_1+\mathbf{V}_{12}\mathbf{V}_{22}^{-1}(\mathbf{x}_2-\mathbf{u}_2) \tag{E.14}$$

and

$$\operatorname{cov}\{\mathbf{x}_1/\mathbf{x}_2\}=\mathbf{V}_{11}-\mathbf{V}_{12}\mathbf{V}_{22}^{-1}\mathbf{V}_{21} \tag{E.15}$$

If $x_1, x_2, x_3, x_4$ are all jointly Gaussian scalar random variables having zero mean, then

$$\begin{aligned} E\{x_1x_2x_3x_4\}=&\,E\{x_1x_2\}E\{x_3x_4\}+E\{x_1x_3\}E\{x_2x_4\} \\ &+E\{x_1x_4\}E\{x_2x_3\} \end{aligned} \tag{E.16}$$

The result expressed by (E.16) can be generalized to jointly Gaussian random vectors (all having zero means). Let the covariance matrix $E\{\mathbf{wx}^T\}$ be denoted by

$$E\{\mathbf{wx}^T\} = \mathbf{V}_{wx} \tag{E.17}$$

and denoted similarly for the other covariances. The desired generalization can then be expressed as

$$E\{\mathbf{w}^T\mathbf{xy}^T\mathbf{z}\} = (\text{trace } \mathbf{V}_{wx})(\text{trace } \mathbf{V}_{yz}) + \text{trace}(\mathbf{V}_{wy}\mathbf{V}_{zx}) + \text{trace}(\mathbf{V}_{wz}\mathbf{V}_{yx}) \tag{E.18}$$

Quadratic forms are important in detection problems. Let $\mathbf{x}$ be a real Gaussian random vector with zero mean and covariance matrix $\mathbf{V}$. The quadratic form defined by

$$y = \mathbf{x}^T\mathbf{KK}^T\mathbf{x} \tag{E.19}$$

then has the associated characteristic function

$$C_y(j\Theta) \triangleq E\{e^{j\Theta y}\} = E\{\exp[j\Theta\mathbf{x}^T\mathbf{KK}^T\mathbf{x}]\} \tag{E.20}$$

The expectation in (E.20) can be expressed as the integral

$$C_y(j\Theta) = \int (2\pi)^{-n/2}(\det \mathbf{V})^{-1/2}\exp\{-\tfrac{1}{2}\mathbf{x}^T\mathbf{V}^{-1} \cdot [\mathbf{I} - 2j\Theta\mathbf{VKK}^T]\mathbf{x}\}\, d\mathbf{x} \tag{E.21}$$

Noting that

$$(\det \mathbf{V})^{1/2} = \int (2\pi)^{-n/2}\exp\{-\tfrac{1}{2}\mathbf{x}^T\mathbf{V}^{-1}\mathbf{x}\}\, d\mathbf{x} \tag{E.22}$$

it immediately follows from (E.21) that

$$C_y(j\Theta) = [\det(\mathbf{I} - 2j\Theta\mathbf{VKK}^T)]^{-1/2} \tag{E.23}$$

The corresponding distribution of $y$ is referred to as "generalized chi-squared." The first two moments of $y$ are then given by

$$E\{y\} + E\{\mathbf{x}^T\mathbf{KK}^T\mathbf{x}\} = \text{trace}\{E[\mathbf{K}^T\mathbf{xx}^T\mathbf{K}]\} = \text{trace}(\mathbf{K}^T\mathbf{VK}) = \text{trace}(\mathbf{KK}^T\mathbf{V}) \tag{E.24}$$

and

$$E\{y^2\} = E\{\mathbf{x}^T\mathbf{KK}^T\mathbf{xx}^T\mathbf{KK}^T\mathbf{x}\} \tag{E.25}$$

Applying (E.18) to the above expression yields

$$E\{y^2\} = \left[\text{trace}(\mathbf{KK}^T\mathbf{V})\right]^2 + 2\text{trace}\left[(\mathbf{KK}^T\mathbf{V})^2\right] \tag{E.26}$$

Combining the results of (E.24) and (E.26) then yields

$$\text{var}(y) \triangleq E\{y^2\} - E^2\{y\} = 2\text{trace}\left[(\mathbf{KK}^T\mathbf{V})^2\right] \tag{E.27}$$

## E.2 COMPLEX GAUSSIAN RANDOM VECTORS

A complex Gaussian multivariate distribution can be defined in such a way that many of the useful properties of real Gaussian random vectors also hold in the complex case. To develop some of the properties of complex Gaussian random vectors, first consider two real random $n$-dimensional vectors $\mathbf{x}$ and $\mathbf{y}$ having zero mean and, in addition, having the special properties

$$\text{cov}(\mathbf{x}) = \text{cov}(\mathbf{y}) = \frac{\mathbf{V}}{2} \tag{E.28}$$

and

$$\text{cov}(x_i y_j) = -\text{cov}(y_i x_j) = -\frac{w_{ij}}{2} \tag{E.29}$$

As a consequence of (E.28) and (E.29), the joint covariance matrix has the special form

$$E\left\{\begin{bmatrix}\mathbf{x}\\ \mathbf{y}\end{bmatrix}\begin{bmatrix}\mathbf{x}^T\mathbf{y}^T\end{bmatrix}\right\} = \frac{1}{2}\begin{bmatrix}\mathbf{V} & -\mathbf{W}\\ \mathbf{W} & \mathbf{V}\end{bmatrix} \tag{E.30}$$

where $\mathbf{W}^T = -\mathbf{W}$. A one-to-one relationship (known as an isomorphism) exists between multiplication of real matrixes of the form

$$\begin{bmatrix}\mathbf{V} & -\mathbf{W}\\ \mathbf{W} & \mathbf{V}\end{bmatrix} \tag{E.31}$$

and multiplication of complex matrixes of the form

$$\mathbf{C} = \mathbf{V} + j\mathbf{W} \tag{E.32}$$

For example, if one defines the complex vector

$$\mathbf{z} = \mathbf{x} + j\mathbf{y} \tag{E.33}$$

then it is easy to verify that

$$\begin{bmatrix}\mathbf{x}^T\mathbf{y}^T\end{bmatrix}\begin{bmatrix}\mathbf{V} & -\mathbf{W}\\ \mathbf{W} & \mathbf{V}\end{bmatrix}\begin{bmatrix}\mathbf{x}\\ \mathbf{y}\end{bmatrix} = \mathbf{z}^\dagger\mathbf{C}\mathbf{z} \tag{E.34}$$

and

$$[\mathbf{x}^T\mathbf{y}^T]\begin{bmatrix}\mathbf{V} & -\mathbf{W}\\ \mathbf{W} & \mathbf{V}\end{bmatrix}^{-1}\begin{bmatrix}\mathbf{x}\\ \mathbf{y}\end{bmatrix}=\mathbf{z}^{\dagger}\mathbf{C}^{-1}\mathbf{z} \tag{E.35}$$

A complex random vector $\mathbf{z}$ as defined in (E.33) will be defined to be a complex Gaussian random vector if its real part $\mathbf{x}$ and imaginary part $\mathbf{y}$ are jointly Gaussian and the joint covariance matrix has the special form given by (E.30).

When $\mathbf{z}$ has zero mean, its covariance matrix is given by

$$E\{\mathbf{z}\mathbf{z}^{\dagger}\}=E\{(\mathbf{x}+j\mathbf{y})(\mathbf{x}^T-j\mathbf{y}^T)\}=\mathbf{V}+j\mathbf{W}=\mathbf{C} \tag{E.36}$$

If $z_1, z_2, z_3, z_4$ are all complex jointly Gaussian scalar random variables having zero mean, then

$$E\{z_1 z_2^* z_3^* z_4\}=E\{z_1 z_2^*\}E\{z_3^*\ z_4\}+E\{z_1 z_3^*\}E\{z_2^*\ z_4\} \tag{E.37}$$

The probability density function for a complex Gaussian random vector $\mathbf{z}$ having mean $\mathbf{u}$ and covariance matrix $\mathbf{C}$ can be written as

$$p(\mathbf{z})=\pi^{-n}(\det\mathbf{C})^{-1}\exp\{-(\mathbf{z}-\mathbf{u})^{\dagger}\mathbf{C}^{-1}(\mathbf{z}-\mathbf{u})\} \tag{E.38}$$

Any linear transformation of a complex Gaussian random vector also yields a complex Gaussian random vector. For example, if

$$\mathbf{s}=\mathbf{A}\mathbf{x}+\mathbf{b} \tag{E.39}$$

then $\mathbf{s}$ is also complex Gaussian with mean

$$E\{\mathbf{s}\}=\mathbf{A}E\{\mathbf{x}\}+\mathbf{b} \tag{E.40}$$

and covariance

$$\operatorname{cov}(\mathbf{s})=\mathbf{A}\mathbf{C}\mathbf{A}^{\dagger} \tag{E.41}$$

If $\mathbf{z}_1$ and $\mathbf{z}_2$ are two jointly Gaussian complex random vectors with mean values $\mathbf{u}_1$ and $\mathbf{u}_2$ and joint covariance matrix

$$\operatorname{cov}\left\{\begin{bmatrix}\mathbf{z}_1\\ \mathbf{z}_2\end{bmatrix}\right\}=\begin{bmatrix}\mathbf{C}_{11} & \mathbf{C}_{12}\\ \mathbf{C}_{21} & \mathbf{C}_{22}\end{bmatrix} \tag{E.42}$$

then the conditional distribution for $\mathbf{z}_1$ given $\mathbf{z}_2$ is also complex Gaussian with

$$E\{\mathbf{z}_1/\mathbf{z}_2\}=\mathbf{u}_1+\mathbf{C}_{12}\mathbf{C}_{22}^{-1}[\mathbf{z}_2-\mathbf{u}_2] \tag{E.43}$$

and

$$\text{cov}(\mathbf{z}_1/\mathbf{z}_2) = \mathbf{C}_{11} - \mathbf{C}_{12}\mathbf{C}_{22}^{-1}\mathbf{C}_{21} \tag{E.44}$$

Hermitian forms are the complex analog of quadratic forms for real vectors and play an important role in many optimization problems. Suppose that $\mathbf{z}$ is a complex Gaussian random vector with zero mean and covariance matrix $\mathbf{C}$. Then the Hermitian form

$$y = \mathbf{z}^\dagger \mathbf{K}\mathbf{K}^\dagger \mathbf{z} \tag{E.45}$$

has the following characteristic function:

$$C_y(j\Theta) \triangleq E\{e^{j\Theta y}\} = \left[\det(\mathbf{I} - j\Theta\mathbf{C}\mathbf{K}\mathbf{K}^\dagger)\right]^{-1} \tag{E.46}$$

The mean value of $y$ is given by

$$\begin{aligned} E\{y\} &= E\{\mathbf{z}^\dagger\mathbf{K}\mathbf{K}^\dagger\mathbf{z}\} = \text{trace}\left[E\{\mathbf{K}^\dagger\mathbf{z}\mathbf{z}^\dagger\mathbf{K}\}\right] = \text{trace}(\mathbf{K}^\dagger\mathbf{C}\mathbf{K}) \\ &= \text{trace}(\mathbf{K}\mathbf{K}^\dagger\mathbf{C}) \end{aligned} \tag{E.47}$$

The second moment of $y$ is given by

$$E\{y^2\} = E\{\mathbf{z}^\dagger\mathbf{K}\mathbf{K}^\dagger\mathbf{z}\mathbf{z}^\dagger\mathbf{K}\mathbf{K}^\dagger\mathbf{z}\} = \text{trace}\left[(\mathbf{K}\mathbf{K}^\dagger\mathbf{C})^2\right] + \left[\text{trace}(\mathbf{K}\mathbf{K}^\dagger\mathbf{C})\right]^2 \tag{E.48}$$

From (E.47) and (E.48) it immediately follows that

$$\text{var}(y) = \text{trace}\left[(\mathbf{K}\mathbf{K}^\dagger\mathbf{C})^2\right] \tag{E.49}$$

It is convenient to introduce a parameter $\alpha$ such that $\alpha = 1$ for real random vectors and $\alpha = 2$ for complex random vectors. Common expressions that are valid for both real and complex Gaussian random vectors can then be written using the parameter $\alpha$. For example, the probability density function of $\mathbf{x}$, the characteristic function of the quadratic form $y$, and the variance of the quadratic form can be written respectively as

$$p(\mathbf{x}) = \left(\frac{2\pi}{\alpha}\right)^{-n\alpha/2} (\det \mathbf{V})^{-\alpha/2} \exp\left\{-\tfrac{1}{2}\alpha(\mathbf{x}-\mathbf{u})^\dagger\mathbf{V}^{-1}(\mathbf{x}-\mathbf{u})\right\} \tag{E.50}$$

$$C_y(j\Theta) = \left[\det\left(\mathbf{I} - j\frac{\Theta\mathbf{V}\mathbf{K}\mathbf{K}^\dagger 2}{\alpha}\right)\right]^{-\alpha/2} \tag{E.51}$$

and

$$\text{var}(y) = \left(\frac{2}{\alpha}\right)\text{trace}\left\{(\mathbf{K}\mathbf{K}^\dagger\mathbf{V})^2\right\} \tag{E.52}$$

One result that applies to certain transformations of non-Gaussian complex random variables is also sometimes useful. Let $\mathbf{z}=\mathbf{x}+j\mathbf{y}$ be a complex random vector that is not necessarily Gaussian but has the special covariance properties (E.28), (E.29). Now consider the real scalar,

$$r=\mathrm{Re}\{\mathbf{k}^{\dagger}\mathbf{z}\} \tag{E.53}$$

where Re$\{\cdot\}$ denotes "real part of" and

$$\mathbf{k}^{\dagger}=\mathbf{a}^{T}-j\mathbf{b}^{T} \tag{E.54}$$

It follows that the scalar $r$ can be written as

$$r=\mathbf{a}^{T}\mathbf{x}+\mathbf{b}^{T}\mathbf{y} \tag{E.55}$$

and

$$E\{r\}=\mathbf{a}^{T}E\{\mathbf{x}\}+\mathbf{b}^{T}E\{\mathbf{y}\}=\mathrm{Re}\left[\mathbf{k}^{\dagger}E\{\mathbf{z}\}\right] \tag{E.56}$$

The variance of $r$ can also be found as

$$\begin{aligned}\mathrm{var}(r) &\triangleq E\left\{\left[r-E(r)\right]^{2}\right\}\\ &=\tfrac{1}{2}\left[\mathbf{a}^{T}\mathbf{V}\mathbf{a}+\mathbf{b}^{T}\mathbf{V}\mathbf{b}-\mathbf{a}^{T}\mathbf{W}\mathbf{b}+\mathbf{b}^{T}\mathbf{W}\mathbf{a}\right]\\ &=\tfrac{1}{2}\mathbf{k}^{\dagger}\mathbf{C}\mathbf{k}\end{aligned} \tag{E.57}$$

## REFERENCES

[E.1] T. W. Anderson, *An Introduction to Multivariate Statistical Analysis*, Wiley, New York, 1959.

[E.2] D. Middleton, *An Introduction to Statistical Communication Theory*, McGraw-Hill, New York, 1960.

[E.3] N. R. Goodman, "Statistical Analysis Based on a Certain Multivariate Complex Distribution (An Introduction)," *Ann. Math. Stat.*, Vol. 34, March 1963, pp. 152–157.

[E.4] I. S. Reed, "On a Moment Theorem for Complex Gaussian Processes," *IEEE Trans. Info. Theory*, Vol. IT-8, April 1962.

# Appendix F. Geometric Aspects of Complex Vector Relationships

For purposes of interpreting array gain expressions, which initially appear to be quite complicated, it is very convenient to introduce certain geometric aspects of various complex vector relationships and thereby obtain quite simple explanations of the results. The discussion in this appendix is based on the development given by Cox [F.1].

The most useful concept to be explored is that of a generalized angle between two vectors. Let the matrix $\mathbf{C}$ be positive definite Hermitian, and let $\mathbf{a}$ and $\mathbf{b}$ represent two $N$-component complex vectors. A generalized inner product between $\mathbf{a}$ and $\mathbf{b}$ may be defined in the complex vector space $H(\mathbf{C})$ as

$$\text{inner product } (\mathbf{a}, \mathbf{b}) \triangleq \mathbf{a}^{\dagger}\mathbf{C}\mathbf{b} \tag{F.1}$$

In this complex vector space the cosine-squared function of the generalized angle can be defined as

$$\cos^2(\mathbf{a}, \mathbf{b}; \mathbf{C}) \triangleq \frac{|\mathbf{a}^{\dagger}\mathbf{C}\mathbf{b}|^2}{\{(\mathbf{a}^{\dagger}\mathbf{C}\mathbf{a})(\mathbf{b}^{\dagger}\mathbf{C}\mathbf{b})\}} \tag{F.2}$$

Applying the Schwartz inequality to (F.2) yields the bounds

$$0 \leqslant \cos^2(\mathbf{a}, \mathbf{b}; \mathbf{C}) \leqslant 1 \tag{F.3}$$

In the complex vector space $H(\mathbf{C})$, the length of the vector $\mathbf{b}$ is $(\mathbf{b}^{\dagger}\mathbf{C}\mathbf{b})^{1/2}$. The vectors $\mathbf{a}$ and $\mathbf{b}$ are orthogonal in $H(\mathbf{C})$ when $\cos^2(\mathbf{a}, \mathbf{b}; \mathbf{C}) = 0$. Likewise the vectors $\mathbf{a}$ and $\mathbf{b}$ are in exact alignment (in the sense that one is a scalar multiple of the other) when $\cos^2(\mathbf{a}, \mathbf{b}; \mathbf{C}) = 1$. Furthermore the sine-squared function can be naturally defined using the trigonometric identity

$$\sin^2(\cdot) = 1 - \cos^2(\cdot) \tag{F.4}$$

Cases that are of special interest for adaptive array applications are $\mathbf{C} = \mathbf{I}$, the identity matrix, and $\mathbf{C} = \mathbf{Q}^{-1}$ (where the matrix $\mathbf{Q}$ represents the normalized noise cross-spectral matrix). When $\mathbf{C} = \mathbf{I}$ it is convenient to simply write $\cos^2(\mathbf{a}, \mathbf{b})$ in place of $\cos^2(\mathbf{a}, \mathbf{b}; I)$.

To obtain a clearer understanding of the effect of the matrix $\mathbf{Q}^{-1}$ on the generalized angle between $\mathbf{a}$ and $\mathbf{b}$, it is helpful to further compare $\cos^2(\mathbf{a},\mathbf{b})$ and $\cos^2(\mathbf{a},\mathbf{b};\mathbf{Q}^{-1})$. To obtain such a comparison involves a consideration of the eigenvalues and corresponding orthonormal eigenvectors of $\mathbf{Q}$ denoted by $\{\lambda_1,\ldots,\lambda_N\}$ and $\{\mathbf{e}_1,\ldots,\mathbf{e}_N\}$, respectively. Since $\mathbf{Q}$ is normalized (and its trace is consequently equal to $N$), $\Sigma_i\lambda_i = N$. The two vectors $\mathbf{a}$ and $\mathbf{b}$ can be represented in terms of their projections on the eigenvectors of $\mathbf{Q}$ as:

$$\mathbf{a}=\sum_{i=1}^{N} h_i\mathbf{e}_i \qquad \text{where } h_i=\mathbf{e}_i^\dagger\mathbf{a} \tag{F.5}$$

and

$$\mathbf{b}=\sum_{i=1}^{N} g_i\mathbf{e}_i \qquad \text{where } g_i=\mathbf{e}_i^\dagger\mathbf{b} \tag{F.6}$$

Substituting (F.5) and (F.6) into (F.2) and using the orthonormal properties of the $\mathbf{e}_i$ vectors then yields

$$\cos^2(\mathbf{a},\mathbf{b})=\frac{\left|\sum_{i=1}^{N} h_i^* g_i\right|^2}{\left(\sum_{i=1}^{N}|h_i|^2\right)\left(\sum_{i=1}^{N}|g_i|^2\right)} \tag{F.7}$$

$$\cos^2(\mathbf{a},\mathbf{b};\mathbf{Q}^{-1})=\frac{\left|\sum_{i=1}^{N} h_i^* g_i/\lambda_i\right|^2}{\left(\sum_{i=1}^{N}|h_i|^2/\lambda_i\right)\left(\sum_{i=1}^{N}|g_i|^2/\lambda_i\right)} \tag{F.8}$$

It will be noted from equations (F.7) and (F.8) that the only effect of the matrix $\mathbf{Q}^{-1}$ is to scale each of the factors $h_i$ and $g_i$ by the quantity $(1/\lambda_i)^{1/2}$. This scaling weights more heavily those components of $\mathbf{a}$ and $\mathbf{b}$ corresponding to small eigenvalues of $\mathbf{Q}$ and weights more lightly those components corresponding to large eigenvalues of $\mathbf{Q}$. Since the small eigenvalues of $\mathbf{Q}$ correspond to components having less noise, it is natural that these components would be emphasized in an optimization problem.

## REFERENCE

[F.1] H. Cox, "Resolving Power and Sensitivity to Mismatch of Optimum Array Processors," *J. Acoust. Soc. Am.*, Vol. 54, No. 3, September 1973, pp. 771–785.

# Index